MICROBIOLOGY RESEARCH ADVANCES

MICROBIAL CATALYSTS

VOLUME 1

MICROBIOLOGY RESEARCH ADVANCES

Additional books and e-books in this series can be found on Nova's website under the Series tab.

MICROBIOLOGY RESEARCH ADVANCES

MICROBIAL CATALYSTS

VOLUME 1

SHADIA M. ABDEL-AZIZ
NEELAM GARG
ABHINAV AERON
CHIAITANYA KUMAR JHA
S. CHANDRA NAYAK
AND
VIVEK KUMAR BAJPAI
EDITORS

nova
science publishers
New York

Library of Congress Cataloging-in-Publication Data

ISBN: 978-1-53614-679-0

Published by Nova Science Publishers, Inc. † New York

CONTENTS

PREFACE

Microbial catalysts are enzymes that contribute to many reactions and widely used in food and industrial products. Enzymes were discovered in the second half of the nineteenth century, since then have been extensively used in several industrial processes. Enzymes from microorganisms are extremely efficient and highly specific biocatalysts. An enzyme is a protein that catalyzes a specific reaction. The action of enzymes depends on their ability to bind the substrate at a domain of enzyme molecule called "active site". Enzymes are more specific macromolecular biological catalysts that accelerate or catalyze chemical reactions. Some enzymes are not, however, reactive on their own, and often need cofactors. Chemically, enzymes like any catalyst, are not consumed in chemical reactions, nor do alter the equilibrium of a reaction. Enzyme activity can be affected by other molecules such as inhibitors and activators. Inhibitors are molecules that decrease enzyme activity, while activators are molecules that increase the enzyme activity. Many drugs and poisons are enzyme inhibitors. Activity of enzymes decrease markedly outside the optimal temperature and pH. Microbial enzymes (*biocatalysts*) are known to play a crucial role as metabolic catalysts, leading to their use in various applications and industrial fields. With the advancement in biotechnology over last three decades, especially in the area of genetics and protein engineering, enzymes have found their way into many new industrial processes.

Over 500 industrial products are being made using enzymes. Microorganisms have served and continue to serve as one of the largest and useful sources of many enzymes. In biocatalytic processes, natural catalysts such as protein enzymes perform chemical transformations on organic compounds. Microbial catalysts are environmentally friendly, greatly cost-effective, and consume lower energy. There is an increasing demand to replace some traditional chemical processes with biotechnological processes involving microorganisms or their enzymes such as pectinases, xylanases, cellulases, mannanase, α-galactosidase, laccases and ligninases which not only provide an economically viable alternative but are also more environmentally friendly. Uses of microbial enzymes in foods, pharmaceuticals, textile, paper, leather, and other industries are numerous and increased rapidly. Supplementary enzymes are added to the dough to ensure high bread quality in the form of a uniform crumb structure and better volume. Special enzymes can also increase the shelf life of bread by preserving its freshness longer. Enzymes have also contributed greatly to the development and improvement of modern household and industrial detergents, the largest application area for enzymes today. Moreover, one of the prime roles of enzymes is to improve the quality of leather.

Over the past century, there has been a tremendous increase in awareness of the effects of pollution. Enzymes can minimize the impact of pollution and waste on the environment. Replacement of some traditional chemical processes with biotechnological processes involving microorganisms is an advantage. All types of living organisms, where metabolic reactions occur, produce enzymes. A wide range of sources is used for production of commercial enzymes. Out of the total enzymes being used industrially, over half are extracted from fungi and yeast, one third are obtained from bacterial systems, and the remaining from animal (8%) and plant (4%) sources. Microorganisms are generally preferred than plant and animal as sources of industrial enzymes due to their low production costs, the enzyme content is more predictable and controllable, and more so because of the availability of raw materials with constant composition for their cultivation.

The current book consists of five different sections on *Microbial Catalyst* with topics include: Applied Microbiology, Industrial Microbiology, Agricultural Microbiology, Food Microbiology, and Structural Microbiology. All the sections discuss microbial catalysts with respect to the enzymes, their function, and benefits in general and to biotechnology as a subject of which ENZYMOLOGY is an integral part. Through this book we wish to bridge the gap between various literature that exists on enzymes, which is either too high on the caliber of written technical language or too low to be of scientific interest. Therefore, experts across the globe were invited to contribute in this comprehensive collection making it unique to be understandable, lucid, easy to increase our understanding of ENZYMOLOGY.

Collection of articles included in this book can be a useful material for Research Students, Post Graduate Students, Lecturers, Professors, and Academicians related to Microbiology, Biotechnology, Biochemistry, Environmental Sciences, Fermentation Technology, Health Professionals, Agricultural Sciences, Applied Microbiologists and many more alike.

Shadia M. Abdel-Aziz
Neelam Garg
Abhinav Aeron
Chiaitanya Kumar Jha
S. Chandra Nayak
Vivek Kumar Bajpai

Editors

5 June 2018

SECTION A. APPLIED MICROBIOLOGY

In: Microbial Catalysts. Volume 1
Editors: S. M. Abdel-Aziz, N. Garg, A. Aeron et al.
ISBN: 978-1-53614-679-0

Chapter 1

INDUSTRIAL APPLICATION OF MICROBIAL AMYLASES

Hasan Ümit Öztürk*[1,*]*, Kamil Serkan Uzyol*[2]*,
***Aziz Akın Denizci*[1] *and Dilek Kazan* [3,†]**

[1]The Scientific and Technological Research Council of Turkey (TUBITAK),
Marmara Research Center (MAM),
Genetic Engineering and Biotechnology Institute, Gebze-Kocaeli, Turkey
[2]Marmara University, Institute of Pure and Applied Science, Istanbul, Turkey
[3]Marmara University, Faculty of Engineering,
Department of Bioengineering, Istanbul, Turkey

ABSTRACT

Amylases, glycoside hydrolyses, catalyze the hydrolysis of starch by randomly cleaving the 1,4-α-D-glycosidic bonds to produce α-anomeric products. Approximately 25-30% of the total enzyme market is composed by amylases, and they are widely used in sugar, textile, paper, brewing, baking, distilling industries, production of cakes, fruit juices, starch syrups, and pharmaceuticals. Although various sources as animal, plant and microbial sources are used for amylase production, fungal and bacterial sources are mostly preferred for production of amylases used in industrial applications. Catalytic and functional properties of amylases are important parameters for amylase applications. Growing interest for production of fuels from plant materials effect the requirements for amylolytic enzymes in the production of the fermentable sugar from starchy crops. In this article, microbial sources for amylase production, substantial properties required for their industrial applications and examples include industrial application of amylases were reported.

Keywords: α-amylase, industrial application, production

* Corresponding Author's Email: umit.ozturk@tubitak.gov.tr.
† Corresponding Author's Email: dkazan@marmara.edu.tr.

1. INTRODUCTION

Amylases (α-amylases, β-amylases and glucoamylases), a group of hydrolases, are one of the most important and oldest industrial enzymes which holds approximately %25-30 of market share of enzyme sales. They are widely used in sugar, brewing, baking, distilling industries, textile, paper, fruit juices, starch syrups, detergent and pharmaceuticals [1-5]. Distribution of the amylase usage in different industries is shown in Figure 1.1.

α-amylases (EC 3.2.1.1, endo-1,4-α-D-glucan glucanohydrolase), endo-acting enzymes, in glycosyl hydrolases family 13, catalyze the hydrolysis of starch by randomly cleaving the 1,4-α-D-glycosidic linkages between adjacent glucose units in the linear amylose chain to produce α-anomeric products [6, 7]. β-amylase (EC 3.2.1.2, α-1,4-glucan maltohydrolase) hydrolyses the non-reducing ends of starch and glycogen to liberate β-anomeric maltose [8]. Glucoamylase (EC 3.2.1.3, α-1,4-glucan glucanohydrolases), exoenzymes, catalyse the hydrolysis of starchy materials and other oligosaccharides by cleaving 1,4-α-glucosidic bonds from the non-reducing ends of starch and 1,6-α-glucosidic bond in polysaccharides yielding glucose [9,10].

The first starch-degrading enzyme was found by Kirchoff in 1811. This finding was followed by very different and detailed researchs related to amylases. Nowadays, amylases used in very different industrial processes. Although amylases are produced by plants, animals and microorganisms, microbial sources especially fungal and bacterial sources are the most preferred one for large scale production because of short growth period, biochemical diversity and the economical bulk production capacity [7].

Several different amylase preparations are available for specific use in varied industries [9]. The major obstacle to use amylases are the yield, stability and cost of enzyme production [7]. In the light of present day biotechnology, amylases and related amylolytic enzymes are still the most important group of enzymes. In this review, we emphasize the various aspects of amylases used in different applications.

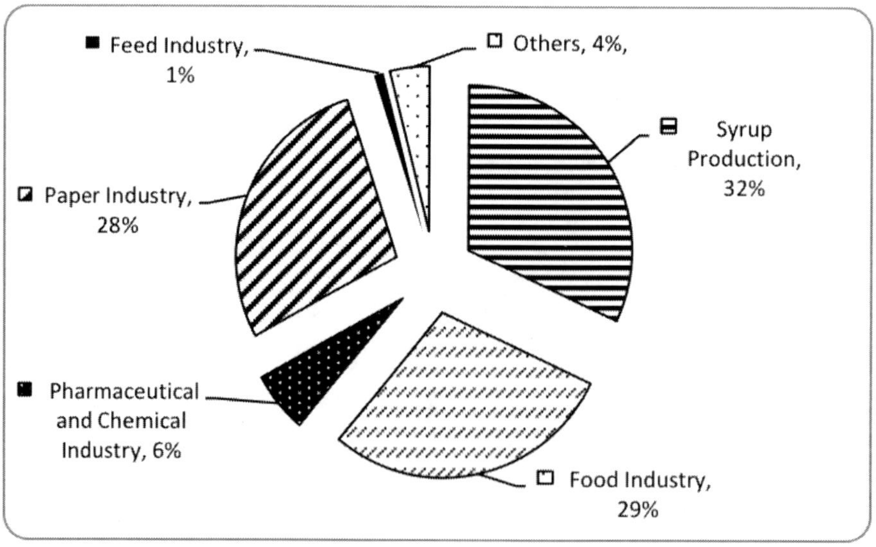

Figure 1.1. Distribution of the amylase usage in different industries.

2. Structure and Properties of Starch

Starch is a major reserve polymer, consisting of a large number of α-D-glucose units linked by glycosidic bonds in the C1 conformation [11]. Starch is synthesized by most green plants in consequence of photosynthesis, which is the conversion of light energy from the sun into chemical energy, in plastids and especially in amyloplasts (found in tubers, seeds, and roots) through the polymerization of glucose [12].

Starch is an important component of the human diet and due to its unique physicochemical behavior; starch and its derivatives become important functional components in food and non-food industries [13-16]. Although, most plants are able to synthesize starch, only few economically important plant species such as maize, rice, wheat, potato and tapioca contain high starch content and can be used in industrial starch processing [17]. Sources of starch production show differences according to the geographical locations in the world for example, in USA starch produced mainly from maize (98%), from cassava in Asia and from different sources such as maize, wheat, potatoes, peas etc. in Europe [18] (Figure 2.1). In Europe, ten million tons of starch were produced from maize (47.4%), wheat (38.6%) and starch potato (14%) in 2011 (http://www.aaf-eu.org/european-starch-industry/).

Starch mainly consists of two high molecular weight glucose polymers, amylose and amylopectin, which represent approximately 98-99% of the dry weight [19]. The ratio of amylose and amylopectin in starch show variation depending on the botanical origin. Amylopectin, is the major contributor of the starch, ranging from 75-85% [20]. These two polymers have different structural and physical properties which have been discussed and reviewed by many authors, [13, 21-25].

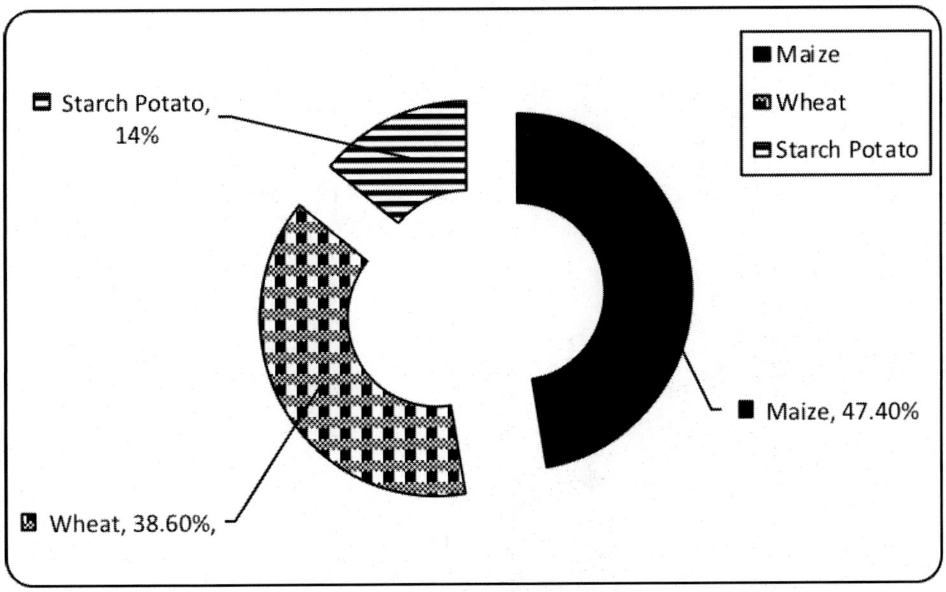

Figure 2.1. Amount of starch produced from different source [18].

Amylose is essentially a linear polymer consisting about 6000 D-glucopyranosyl units linked with α,1-4 glycosidic bonds and slightly branched by α,1-6 linkages [24, 26, 27]. Amylose has a molecular weight of approximately 1×10^5–1×10^6 [28, 29].

Due to the bond angles in the α,1-4 glycosidic bonds, amylose tends to twist gradually that resulting in a left handed single helical conformation or single helices come together to form stiffer parallel left handed double helicoidal junction zones [30, 31]. The inner surface of the single helix amylose having hydrophobic nature holds water molecules so that lipids, alcohols, aroma molecules and iodine like complexing agents can easily be replaced with water and form amylose-agent complex. For iodine color formation, amylose chain must contain at least 18 glucose units. The intensity and the corresponding λ max increase linearly with chain length to about 70 glucose units and becomes blue at DP 45 [32].

Amylopectin is one of the largest polymers in nature with its average 2 million glucose units [33]. Amylopectin, highly branched fraction of the starch, is composed of short linear chains of α-(1,4) linked D-glucopyranosyl units that are joined together by α-(1→ 6) linked site chains [26]. The average number of branching points in amylopectin is 5-6%, but varies with the botanical origin [24].

The cluster base structure is widely accepted model for amylopectin [24, 26]. In this model, linear A-chains, containing 12-16 glucose units, are linked to a linear B-chain only through their reducing ends and thus are unbranched. The B-chains linked to the A-chains are also linked to other B chains or the linear C-chain through the C6 of one or more glucosyl units and so that they are branched. Moreover, the B-chains can be further divided into B1 (DP 20-24), B2 (DP 42-48), B3 (DP 69-75) and B4-chains depending on the number of clusters they are involved [13, 20, 34].

Starch granules are roughly spherical shapes with a diameter ranging from 2 to 100 μm [29]. Starch granules are synthesized as semi-crystalline matrices in plants. The crystallinity of native starch granules varies between 15 and 45% depending on the botanical origin [35, 11] or 8-51% [29]. Since amylose found in amorphous form, amylopectin fraction of starch is responsible from the crystallinity of starch granules [16, 36]. Structural studies with X-ray diffraction have shown that native starch mainly has 3 different patterns. A-type patterns of monoclinic symmetry is for cereal starch, B-type patterns of hexagonal symmetry is for tuber and root starches and C-type which is the mixture of A- and B- type is for leguminosae starches [37].

3. Microorganisms Used for Amylase Production

Microorganisms had a significant role in the production of various type of enzymes. Amylases, can be produced from microbial sources and in order to obtain enzymes having desired characteristics, a wide variety of fungi and bacteria have been used. Among other bacterial strains, genus *Bacillus. B. subtilis, B. stearothermophilus, B. licheniformis* and *B. amyloliquefaciens* are good producers of commercial production of α-amylase [2]. A large number of literatures are available on various microbial sources from different environments especially extreme environments for the production of amylases [1, 2, 7, 38-47].

In this part, some of the most recent literature related to microbial sources for amylase production was summarized. Since *Bacillus* are the most important group for amylase production, different *Bacillus* strains have been reported. Sharma and Satyanarayana [47] isolated acidophilic *Bacillus acidicola*, produced acid stable-calcium independent amylase. Moradi et al. [48] reported *Bacillus cereus, Bacillus licheniformis, Bacillus amyloliquefaciens* and *Penibacillus lautus* as amylolytic bacterial strains. Another *Bacillus* sp. strain PM1 which is an alkalo-thermostable amylase producer was noted by Sharma et al. [4]. *Bacillus* species such as *Bacillus megaterium* strain KAN1 [49], *Bacillus licheniformis* RT7PE1 [50], *Geobacillus stearothermophilus* [51] and *Bacillus subtilis* MTCC 121 were also mentioned as an amylase producer [52].

In addition to thermophilic bacteria, the most thermoactive α-amylases from hyperthermophilic archaea have attracted increasing attention and have been characterized as *Pyrococcus woesei, P. furiosus, Thermococcus profundus* and *T. hydrothermalis* [53]. Li et al. [53] reported *Flavobacteriaceae* α-amylases from a novel *Flavobacteriaceae* species. The structure of the α-amylase produced from *Flavobacteriaceae* were closest to plant amylases whereas fungi and bacteria were located at further distances.

Fungi are the important component of the soil microbiota and there is significant interest in their biotechnological utilization for amylase production. Besides *Bacillus* strains, fungi especially belonging to genus *Aspergillus* have been used for amylase production [54-56]. Recently, Jebor et al. [57] reported *Aspergillus niger* for glucoamylase production. Alpha-amylase produced by *Aspergillus oryzae* has been reported by Shah et al. [5]. Singh et al. [58] isolated fungal strains and they noted *Aspergillus fumigates* NTCC1222 as the best amylase producer. Welday et al. [59] isolated thermophilic fungi belonging to *Aspergillum* to produce α-amylase for starch industry. *Aspergillus flavus, A. niger* and *A. fumigatus* as amylase producer has been isolated by Oseni et al. [60].

Besides *Aspergillus* strains, *Rhizopus* strains were also reported as amylase producers [61]. Recently, Freitas et al. [62] produced amylases from *Rhizopus oryzae* and *Rhizopus microspores* var. *oligosporus*.

Since almost 70% of the earth is covered by ocean, different types of microorganisms having capability of produce amylase have been also isolated from ocean and glacier. These microorganisms are mostly psychrophiles and psychrotrophs which belong to bacteria, archaea, yeast and fungi. Kuddus et al. [63], rewieved the psychrophilic and psychrotrophic microorganisms for production of cold-active microbial amylases. Kuddus et al. [63] reported isolation of 30 cold α-amylase producing bacteria and they identified two novel microorganisms as *Microbacterium foliorum* GA2 and *Bacillus cereus* GA6 from glacier. Another marine bacterium *Zunongwangia profunda* has been announced by Quin et al. [64] as cold-active and salt-tolerant alpha amylase producer.

In addition to thermophiles and psychrophiles, halophiles are important group of extremophiles for production of amylases and other enzymes [65]. Amylases from *Halobacterium, Halobacillus* and *Halothermothrix* have been reported by Adrio et al. [65].

Although different type of microorganisms has been isolated and identified, less than 1% were cultivated for enzyme production. Therefore, metagenome screening and genome mining are new tools to find novel amylase producers from different habitants [65].

4. Characterization of Microbial Amylases

4.1. Catalytic Mechanism of Amylases

Enzymatic hydrolysis of the glycosidic bond takes place through general acid catalysis that requires two critical amino acid residues: glutamic acid as a proton donor and an aspartate as a nucleophile/base [66, 67]. This hydrolysis occurs via two major mechanisms end up with either an overall retention, or an inversion of the anomeric configuration [12, 68]. The position of the proton donor is within hydrogen bonding distance of the glycosidic oxygen, and this distance is identical in both retaining and the inverting mechanisms. In retaining enzymes, the nucleophilic catalytic base is located nearby of the anomeric carbon of sugar, whereas it is more distant in inverting enzymes. The average distance between two catalytic residues ~9.5 å in the inverting enzymes, which is just right for the simultaneous binding of water and substrate, while in retaining enzymes shorter distance (~5.5 å) is enough for the direct attack of the nucleophile [69, 70]. α-amylases use retaining mode-of-action, while both β-amylases and glucoamylases are belong to an inverting glycosidase [71].

The inverting enzymes proceed via single displacement mechanism in which protonation of the glycosidic oxygen from the leaving group with the help of acid residue is accompanied by a concomitant attack of a water molecule that is deprotonating by general base residue. The reaction proceeds through an oxocarbenium ion-like transition state [68]. This single nucleophilic displacement yields a product with opposite stereochemistry to the substrate.

On the other hand, retaining enzymes like α-amylases catalyze the hydrolysis of α-1,4-glycosidic bonds via a generally accepted double displacement mechanism as proposed by Koshland [66]. With the binding of substrate to the active site first, glutamic acid in acid form donates a proton to the oxygen glycosidic bond and the nucleophilic aspartate attacks the anomeric centre (C1 of glucose at subsite -1). This results the formation of an oxocarbonium ion-like transition followed by the formation of a covalent intermediate. Secondly, a water molecule attacks the covalent bond between the anomeric centre and the nucleophilic amino acid and an oxocarbenium ion-like transition state formed again. Then the base catalyst amino acid accepts hydrogen from the incoming water molecule and the oxygen from the water molecule replaces the covalent bond. This second nucleophilic substitution at the anomeric carbon generates a product with the same stereochemistry as the substrate [12, 68].

Although, three catalytic amino acid residues (Glu, Asp, Asp) conserved in the active site only two of them plays a role directly in the hydrolysis. The third residue, a second aspartate, plays an important role in the positioning of the substrate in the correct orientation into the active site through hydrogen bonding to the OH-2 and OH-3 groups of the substrate [12, 72].

Based on the amino acid sequence similarities between the α-amylases, the two acidic residues Glu283 and Asp252 are predicted to be involved in the catalytic function of GTHase as a general acid/base and nucleophile, respectively. To investigate the catalytic and substrate recognition mechanisms, Okazaki et al. [73], produced two single and one double mutant KM1-GTHase proteins via mutated Glu283 and Asp252 located at the active site. Glu283 was belived to have a general acid-base function and responsible for the GTHase activity. Glu 283 was mutated to the corresponding amide (Gln) while Asp252 was mutated to Ser to change its catalytic function from anomer retaining mechanism to inverting. As a result, D252S mutant showed very little enzymatic activity 3.4 but D252E and E283Q mutants had almost no enzymatic activity when they compared with wild type GTHase. Takase et al. [74], were mutated the three amino acid residues (D176, E208 and D269) in the active site of the *Bacillus subtilis* N7 α-amylase to their amide forms in order to evaluate the possible roles of these residues in activity. Mutations results in a 15.000-fold decrease in the specific activity of the enzyme but substrate binding ability of the mutant did not change. In another study, D331E mutation in the active site of the thermostable α-amylase from *Bacillus stearothermophilus* by site-directed mutagenesis showed almost complete inactivation [47, 75].

4.2. Properties of Microbial Amylases

Amylases play a significant role in starch, detergent, textile and beverage industries. They have an ability to use in brewing, baking, sugar production, paper industry, desizing and in detergent manufacturing. The properties of amylases such as resistant to denaturing agent, optimum temperature and pH, amino acid sequence effect the usage of them in different industrial processes. Although amylases are grouped depending on structural similarities, they have different biochemical properties [76].

Among the exoamylases (β-amylase, glucoamylase), endoamylase especially α-amylase are the most widely used amylolytic enzyme. Based on thermal and pH stability of α-amylase, they are classified as thermophilic, psychrophilic, alkaliphilic, acidophilic and mesophilic α-amylases. Most of the α-amylases repoted in the literature are active at neutral pH [1, 41, 77]. Some of them also reported in the literature are tabulated in Table 4.2.1. Neutral and slightly alkaline characters of α-amylase from moderately halophiles were also noted [78-80].

However, there are still increasing attention to develop novel amylases having unique and desirable properties such as thermal and pH stability, specificity, pH, temperature and metal ion dependence [81]. Especially, amylase having catalytic activity at pH ≥ 8.0 and temperature ≥75°C are preferred [45]. Tabassum et al. [50] reported thermophilic and alkaliphilic α-amylase from *Bacillus megaterium* strain KAN1 having optimum temperature as 60°C and pH as 11.0. This enzyme kept 84% of its original activity at 70°C. The microbial α-amylase for industrial purposes are mainly produced from *Bacillus subtilis, B. amyloliquefaciens and B. licheniformis* [50]. Another thermostable α–amylase produced from *Bacillus* sp. isolate DM-15 [39] has optimum pH and temperature as 60°C and pH as 5.5-6.0. The DM-15 enzyme was stable for 15 min at 60°C while 24% of the original activity was lost at 100°C. Saxena et al. [82] reported *Bacillus* sp. PN5 α-amylase retained 80% of its activity at pH 10.0 for 1 h pre-incubation while Behal et al. [83] reported *Bacillus* sp. AB 04 α-amylase stable between pH 7.0 and 10.0. Liu and Xu [84] also noted *Bacillus* sp. YX-1 α-amylase stable between pH 4.5 and 11.0. Corderio et al. [85] reported α-amylase from thermophilic *Bacillus* sp. having stability at 50°C, over 2 h. Additionally, Asoodeh et al. [86] studied α-amylase from *Bacillus ferdowsicous* having 75% of its activity at 75°C after 45 min incubation. Behal et al. [83] determined that α-amylase enzyme they obtained from *Bacillus* sp. AB 04 was stable between 50 and 80°C. Alkaline α-amylase from *Bacillus licheniformis* RT7PE1 display their highest activity at 55°C and pH 9.8 [50]. The half-life of RT7PE1 was 17 h at 95°C. Uzyol et al. [80] also reported thermostable α-amylase from moderately halophilic *Halomonas* sp. AAD21 that it retained 70% of its original activity after 2 hours of incubation at 90°C. Recently, Welday isolated fungi belonging to genus *Aspergillus* which produces thermostable α-amylase. Some of the properties of amylases are tabulated in Table 4.2.1.

Besides alkali stable α-amylases, several researches have been carried out for acid-stable amylases in recent years [47]. Sharma and Satyanarayana [47] reviewed acid stable amylases from different microbial sources as bacteria, fungi and yeast. pH and thermostability of acid-stable amylases represent considerable potential for large scale industrial applications [61].

Although α-amylase from genus *Bacillus* have been extensively studied, enzyme from certain strain *Geobacillus thermodenitrificans* HRO10, *Geobacillus* sp. LH8 [87-89] and *G. stearothermophilus* (Fincan and Enez 2014). Optimum temperature and pH of enzymes were 5.5 and 80°C, 5.0–7.0 and 80°C, 7.0 and 70°C respectively. Optimum temperature of *G. stearothermophilus* are in the range of 50-70°C [51]. *G. stearothermophilus* α-amylase was also stable between 50°C and 60°C for 2 h incubation and that it lost 50% of its activity at 70°C [51].

MW of most of the α-amylase enzymes between 45 and 70 kDa [90,51] have been reported: 63 kDa for *Bacillus* sp. IMD 435, 52 kDa for *Geobacillus* sp. LH8, 58 kDa for *Bacillus licheniformis* NH1, 58 kDa for *B. licheniformis*, 58 kDa for *Geobacillus thermodenitrificans* HRO10, 63 kDa for *G. stearothermophilus* and 67 kDa for *Bacillus megaterium* strain KAN1 [91-93,6,87,55,49]. Some α-amylase having molecular weight higher than 70 kDa such as 94.5 kDa for *Alicyclobacillus acidocaldarius* [38] 112 kDa for *Bacillus licheniformis* RT7PE1 [50] and 126 kDa *for Bacillus* sp. DM-15 [39] were studied.

Substrate specificity is also very important parameter for enzymes. α-amylases have their highest specificity towards starch compared to the amylose, amylopectine and maltotriose [47]. Km and Vm values of some α-amylases are tabulated in Table 4.2.2.

Ca^{2+} ion is one of the most important cations for α-amylases since it alters the catalytic activity and enhanced the thermal stability of α-amylases [38, 39, 41, 47, 50]. Although activity of α-amylases requires Ca^{2+}ions, dependent, several reports have been published of Ca^{2+}-independent nature of amylases [86, 90, 94, 95]. Optimum temperature, pH, stability and molecular weight of some α-amylases from different microbial sources.

Additionally, effect of different additives such as metal ions, surface active agents are very important for usage of amylases in industrial applications. Depending on the functional differences, effectiveness of additives varies among α-amylases. 5 mM Na_2SO_3 and KCL cause to increase enzyme activity *from of Bacillus* sp. isolate DM-15 [39], however 5 mM EDTA, $CuSO_4$, $FeSO_4$, SDS (81%) urea (8M) and Co^{+2} inhibited the activity of DM-5 enzyme. Satheesh et al. [38] reported that thermophilic *Alicyclobacillus acidocaldarius* retained its activity in the presence of SDS, Triton X-100, Tween-20, Tween-80 and it was inhibited by EDTA, urea, Hg^{2+}, Zn^{2+} and Co^{2+} inhibits the enzyme. Hg^{2+}, Pb^{2+}, Zn^{2+}, Cu^{2+}, EDTA and SDS also caused to decrease activity of *Bacillus licheniformis* RT7PE1 amylase. Fe^{2+}, Zn^{2+}, Cu^{2+} and Hg^{2+} were also inhibited amylase from *Geobacillus stearothermophilus* [51] from *Bacillus* sp. TS-23 [114], from thermophilic *Bacillus* [115] and from *Nesterenkonia* sp strain [116].

In the literature cold active amylases were also reported [63, 64]. Cold active α- amylases can be used in energy saving processes due to their activity and stability at low temperatures. Most of the cold adapted amylases have optimal temperature at 30-35°C [63] and they have wide range of pH activity and stability changed from 5.0 to 10.0. More recently, Qin et al. [64] reported cold active-salt tolerant α-amylase which showed maximum activity at 35°C and pH 7.0. Having distinct characteristics make cold active amylases good candidate for basic and applied biology research. However, more efforts have to be made to increase the usage of cold active amylases in industry.

Table 4.2.1. Optimum temperature, pH, stability and molecular weight of some α-amylases from different microbial sources

Microbial Source	Molecular weight (kDa)	Optimum pH	Optimum Temperature (°C)	References
Bacillus amyloliquefaciens	-	7.0	33	[96]
Bacillus licheniformis	22.5	9.0	76	[97]
Bacillus licheniformis	28.0	9.0	90	[98]
Bacillus licheniformis NCIB 6346	62-65	7.0	70-90	[99]
Bacillus sp. 1-3	-	7.0	70	[100]
Bacillus sp. PS-7	71	6.5	60	[101]
Bacillus subtilis JS-2004	-	7.0	50	[102]
Bacillus ferdowsicous	53	4.5	70	[86]
Bacillus sp. KR-8104	59	4.0-6.0	70-75	[94]
Bacillus dipsosauri DD1	80	6.1	60	[103]
Lactobacillus manihotivorans LMG 18010T	135	5.5	55	[104]
Bacillus stearothermophilus ATCC 12980	59	5.0-6.0	70	[105]
Bacillus subtilis 65	68	6.0	60	[106]
Esherichia coli	48	6.5	50	[107]
Streptococcus bovis JB1	77	5.0-6.0	50	[108]
Micromonospora melanosporea	45	7.0	55	[109]
Thermomonospora profundus DT5432	42	5.5-6.0	80	[110]
Aspergillus awamori ATCC 22342	54	4.8-5.0	50	[111]
Aspergillus niger van Tieghem CFTRI 1105	56.2	5.0	60	[112]
Aspergillus niger UO-1	-	4.95	50	[113]
Cryptococcos flavus	75	5.5	50	[114]

Table 4.2.2. Km and Vm values of some α-amylases

Microbial Source	Km	Vm	References
Bacillus megaterium strain KAN1	0.65 mg/ml	1.568 mg/min	[45]
Alicyclobacillus acidocaldarius	2.9 mg/ml	7936 U/ml	[38]
Bacillus licheniformis RT7PE1.	3.4 mg/ml	19.5 IU/mg	[50]
Bacillus subtilis KCX 006	0.291 mg/ml	23.6 U/ml	[41]
Geobacillus stearothermophilus	0.051 mM	1.424 mmol/min	[51]
Bacillus subtilis KIGBE HAS	2.6 mg/ml	1.77 U/ml	[117]
Bacillus subtilis DM-03	1.2 mg/ml	1.96 (μmol· min^{-1}· mg^{-1})	[118]

Based on the end products of starch hydrolysis, α-amylases are classified as maltose, maltotetrose, maltopentose and malto-hexose forming amylases [47]. Many bacterial α-amylases have been reported to produce maltooligosaccharides, as maltohexose-forming α-amylase from *B. stearothermophilus* US100 [119] and *B. halodrans* LBK [120] maltopentaose-forming amylase from *Pseudomonas sp.* KO-8940 [121] and *B. circulans* [99]. Only a few strains as *Bacillus* sp. GM8901 [122] and *B. subtilis* KCC103 [123] and *B. subtilis* SDP1 [90] were reported to produce small maltooligosaccharides as maltotriose and maltotetraose.

Besides α-amylase, glucoamylases are the main enzyme used for the production of glucose syrup, high fructose corn syrup and alcohol [54, 56, 62, 124]. Although many studies carried out related to glucoamylase we here reported the more recent publications for glucoamylase. The production of glucose syrups is the primary application of glucoamylase and this prosess requires high activity and thermostability. Therefore, several researches carried out to find novel microbial sources that produce amylase resistant to thermal inactivation [54]. The commercial glucoamylase are mostly produced by *Aspergillus sp.* [56] or *Rhizopus* sp [62]. Bhatti et al. [54] reported glucoamylase from *Fusarium solani*. The optimum temperature and pH of the enzyme was found as 40-45°C and pH 4.5. The half-life of enzyme was 26 min. *F. solani* glucoamylase activity was increased in the presence of Cu^{2+} and Mg^{2+} ions, while Hg^{2+}, Pb^{2+}, Zn^{2+}, Ni^{2+} and Fe^{3+} ions cause to decrease in enzyme activity. It was also noted that the Km values of *F. solani* glucoamylase were 1.9 mg/ml which is 2 and 5 times lower than the glucoamylase produced by *Aspergillus niger* NCIM [56, 125, 126] and *Acremonium sp.* [127]. Koç and Metin [56] reported *Aspergillus flavus* HBF34 glucoamylase which has optimum pH and temperature as 6.0 and 60°C. The enzyme retained 56% of its original activity during 72 hr incubation. They noted that Mn^{2+}, Ca^{2+}, Co^{2+} and Ba^{2+} activates the HBF34 glucoamylase activity. However, Hg^{2+}, Fe^{3+}, Al^{3+}, Zn^{2+} and Cu^{2+} inhibit the enzyme activity. The glucoamylase from *Aspergillus niger* was also reported by Deepika and Rangabhashiyam [125]. Optimum pH and temperature of enzyme were 5.2 and 45°C, which is lower than HBF34 glucoamylase. More recently, Freitas et al. [62] presented glucoamylase from *Rhizopus oryzae*. They obtained highest activity at pH 5.0 and temperature 50°C. The enzyme has great range of operation for enzymatic processes, since the enzyme retained 80% of its original activity at 55° and 60°C. Glucoamylase from *R. oryzae* has an ability to produce glucose from starch in 3.5 hours of reaction.

Third major classes of amylase enzyme are β-amylases. Although β-amylase are mostly produced from plant materials, few β-amylases produced from microbial sources were also reported in the literature. Although, *Actinomicetes, Pseudomonas, Clostridium* and *Bacillus* are β-amylase producers, β-amylase from *Bacillus* strains have high saccharifying activity to convert starch to maltose [43, 128-131]. Recently, one of the β-amylase having optimal temperature and pH as 50°C and 6.0 produced by *Bacillus subtilis* was reported by Femi-Ola et al. [43]. Molecular weight of many β-amylases ranged from 69 kDa to 209 kDa. However, molecular weight of β-amylase from *B. subtilis* was reported as 27.6 kDa, which is much lower than the literature data. Km and Vm values of *B. subtilis* β-amylase were reported as 4.6 mg/ml and 47.62 U/mg. Hg^{2+}, Ca^{2+} and Fe^{3+} inhibit the enzyme activity like other amylases [43]. Recently, Oseni and Ekperigin [60] reported β-amylase from fungi strains. They found that *Aspergillus niger, A. flavus, A. fumigatus* and *A. flavus* produced β-amylase. β-amylase produced by these fungi are stable up to 80°C. but β-amylases lost their original activities between 90° and 100°C. In other words, β-amylase from fungal strains represents higher thermal stability.

According to very diverse properties of α-, β- and gluco amylases, they are used for different industrial sectors such as production of high fructose corn syrup, de-starching detergents, production of oligosaccharides etc. The following part of this work includes the some of the application of amylases in different industries.

5. APPLICATION OF AMYLASES IN INDUSTRY

In all over the World, enzymes are required in increasing quantities for different applications such as bakery industry, detergent industry, textile industry, fruit juices and wine. The global market for industrial enzymes has been estimated to reach USD 7.1 billion by 2018 [132]. It is also reported that the growth of enzyme production is increased % 10 per year. Amylases constitute 30% of the world enzyme and they have diverse applications in a different type of industries as summarized below [133].

5.1. Industrial Production of Glucose and Fructose Syrup

The major use of processed starch is for the production of glucose, which is used for the production of crystalline dextrose, dextrose syrups or high fructose syrup. However, the first sweet-tasting syrup from starch-water suspension was achieved by acid hydrolysis treatment using dilute HCl by German scientist Kirchhoff in 1811, development of large-scale starch hydrolyzing industry took several decades. With the introducing the industrial scale jet-cookers in the 1950s, instead of boiling starch is mixing with the steam under pressure at 100-175°C and stuch slurry is rapidly heated within a few seconds and pumped directly into the hydrolysis treatment [12].

Due to the development in the industrial microbiology and biochemical engineering after the Second World War [134]. In 1946, Dale and Langois patented the use of commercially available enzymes and hydrolyzed starch in the manufacture of commercial corn syrups of unique properties [135, 136]. In the following years, acid hydrolysis for the production of dextrose or high fructose syrups (HFS) has been replaced by enzymatic treatment. Today, the commercial starch processing become totally enzyme based and depends on the availability of three major enzyme such as α-amylase, glucoamylase and glucose isomerase and if needed pullunase [137, 138].

Complete conversion of high fructose syrup happens in three steps. The first step, liquefaction, is the conversion of concentrated starch suspension into soluble, short chain length dextrins. The pH of the starch slurry that contains 30-35% dry solids increased to pH 6, mixed with α-amylase and then passed through a jet cooker that instantly raises the temperature of the slurry to the gelatinization temperature (95-105°C) for 90 min. In order to ensure the removal of lipid-amylose complexes the gelatinization temperature must be raised over 100°C [136]. In the beginning, *Bacillus amyloliquefaciens* α-amylase was used but it has been replaced with more thermostable enzymes from *B. licheniformis* and *B. stearothermophilus* [9, 12]. The amount of enzyme addition and incubation time differs the DE value of starch-hydrolysate syrup. DE value 8-10 is needed for glucose production.

One of the shortcomings of currently used α-amylases is they are not performed well below pH 5.9 at high temperatures. To overcome this problem the starch slurry is adjusted from its natural pH of 4.5 to pH ~6.0 by adding NaOH. Ca^{+2} must be added because these enzymes are calcium-dependent for their activity and stability. However, the cost of the process increases with these modifications and require additional ion-exchange desalting of the final product [139].

There were several approaches of protein engineering and screening of novel α-amylases having thermal stability at low pH, and Ca-independent nature. For example, Mitchinson et al. [140] engineered *Bacillus licheniformis* and succeeded to operate the enzyme at lower pH, lower calcium level and higher temperatures. Jorgensen et al. [141], cloned and expressed a novel extracellular α-amylase from *Pyrococcus furiosus*, hyperthermophilic archaeon, in *E. coli*. This novel α-amylase having promising characteristics like high thermostability in the absence of metal ions and having unique product pattern and substrate specificity. In another study, site directed mutagenesis on mesophilic α-amylase from *Bacillus megaterium* WHO resulted in mutant H58I possessing high thermostability and calcium independency [142].

The next step is saccharification in which the single glucose residues sequentially removing from the soluble starch-hydrolyzate syrup by the action of exo-acting glucoamylase (E.C.3.1.2.3) that hydrolyzes α,1-4 glycosidic bonds from the non-reducing end of the chain. Glucoamylases are produced industrially from *Aspergillus niger* or related species, *A. awamori* and *A. niger* GA and also *Rhizopus* and *Endomyces* species having pH optimum of 4.2 and stable at 60°C [135].

For an efficient saccharification, the pH of the liquefact first adjusted down to 4.2-4.5 with hydrochloric acid in order to maintain an optimum pH environment for glucoamylase and inactivate the further action of α-amylase. Depending on the specification of the final product, this process takes 12-96 hours at 60°C.

There are also two practical problems in saccharification process due to the nature of the glucoamylase. Glucoamylase is predominantly hydrolysing α,1-4 glycosidic linkages but slows down when reaches to the α,1-6 branch point which will result in the accumulation of isomaltose. To solve this problem, addition of debranching enzyme pullulanase, having the same pH and temperature optima like glucoamylase is required. The second problem comes from the process condition which requires high dry-solid content (% 32) to make economically feasible high glucose syrup containing >95% glucose. Under these high glucose concentration, glucoamylase tends to form reversion products like maltose and isomaltose so balancing the dosage of enzyme, the temperature of the reaction and incubation time is the only solution [12, 135, 136].

The third step of industrial starch processing is isomerization, which is the conversion of high glucose syrup into two times sweet isomer high fructose syrup. The enzyme used in this step is called D-xylose-ketol isomerase (E.C 5.3.1.5) or better-known glucoisomerase commonly isolated from bacterial species like *Bacillus, Microbacterium, Streptomyces, or Actinoplanes* [135, 143]. While α-amylase and glucoamylase are inexpensive so they are added directly to the processing slurry and used once, pricey glucose isomerase is immobilized by package into columns and reused until it loses most of its activity. Before passing the dextrose mixture over immobilized column, it is first refined, carbon filtered, concentrated to >40% dry solids and adjusted to pH 7-8. The major product of this reaction is HFCS-90, which means that after isomerization of glucose the mixture composed of 90% fructose and 10% glucose [144]. After clarification and removal of the impurities HFCS-90 is blended with glucose syrup to obtain syrup containing 42% and 55% fructose accepted as commercial product [145].

Both HFCS-55 and HFCS-42 has unique properties that make them attractive to specific food manufacturers. HFCS-55 becomes a liquid alternative for sucrose with its high fructose content and used as a major sweetener in soft and fruit juices, and carbonated drinks. On the other hand, HFCS-42 is preferred extensively in canned fruits, baked goods, soups, sauces, and many other processed foods, since it does not mask the natural flavors of food [144].

Today, genetically modified α-amylases and glucoamylases are used in HFCS processing in order to improve heat stability. On the other hand, raw starch digesting enzymes, which are capable of digesting raw starch granules, would be a value to reduce cost of gelatinization which is the high-energy-consuming process to simplify the entire process of starch conversion.

5.2. Detergent Industry

Primary consumers of enzymes are detergent industries. The feasibility of the modern detergents depends on the use of specific enzymes. The main advantage of enzyme application in detergents is due to much milder conditions than with enzyme free detergents [9]. Proteases, amylases and lipases, are the major enzymes used in detergent formulations. Amylases are one of the most important enzymes used in detergents, the huge part of liquid detergents (90%) contains amylases [40, 42, 146]. They are used in order to remove the strains of starchy foods such as potatoes, gravies, pasta, chocolate, etc. by converting to dextrins and other smaller oligosaccharides [40, 147]. Starch contained surfaces act as a good attractant for particular soils, therefore, removing of starch from surfaces is also necessary to provide a white appearance [40].

Despite the washing environment is very oxidative, α-amylases that are used in detergent formulations are sensitive to oxidants. Additionally, the stability of amylases at low temperatures, at high pH, under detergent conditions are the main criteria to use them in detergent formulations. Enzyme stability also depends on calcium level and low calcium concentrations causes low stability [9]. Chelators such as EDTA are used in detergent formulations, but they also remove calcium, which is essential for amylase. Therefore, there is a need for calcium independent amylases [148].

5.3. Textile Industry

For fabrics, made from cotton and blends, adhesive substances known as size are used to coat the warp threads [149]. The layer of sizing agent protects the yarn, for this purpose a removable protective layer is applied to the threads. The sizing agents must be removed afterward from the raw fabric [150]. Although different type of materials have been used for sizing, starch is generally preferred as sizing agent because it is cheap, available in everywhere, besides it can be easily removed from the fabric [9, 58, 148].

Effective desizing of starch sized fabrics can be done by using of α-amylase, which selectively hydrolyze the size to soluble dextrins and do not attack the fibers [9, 58, 149-151].

Controlling of pH, temperature and water hardness, is required for good enzymatic desizing. Amylases produced from Bacillus strains were used in the textile industry for a long time [42]. Singh et al. [58] studied the fungal amylases to use in textile applications. They reported the amylase from *Aspergillus fumigatus* NTCC1222 that is used under harsh conditions of textile wet processing [54].

5.4. Pulp and Paper Industry

Starch is used for coating treatment to increase the strength of paper against mechanical damage during processing [9]. Additionally, coating makes the surface of the paper smooth and strong [40]. While starch provides a good coating for paper, starch sizing improves the quality and erasability of paper [40]. Amylases used for the starch modification process in pulp and paper industry. The viscosity of starch is not suitable for sizing of paper. Paper sizing requires lower viscosity, therefore natural starch's viscosity has to be adjusted by partially degrading the starch with α-amylases. The conditions of degradation depend on the source of starch and the α-amylase [9, 42] reported that cold active α- amylases can be used in paper and pulp industry.

5.5. Fuel Industry

In recent years, there has been great interest an efficient utilization of agro-industrial by-products to obtain valuable bio-products as biofuels, biochemicals and biomaterials. Minimization of environmental problems and reducing dependence on petroleum resources are main results of bioprocessing of agro-industrial residues when they are used to produce biofuels. The development of novel and efficient bioprocesses for biomass utilization is the forefront of biotechnological research and industrial application [152-154]. In recent years, various starchy materials such as corn, wheat, cassava root, and starch have received much attention for biofuel especially bioethanol production [53, 155, 156]. The conventional bioethanol production process from starch-based materials includes two enzymatic steps: liquefaction using thermal-stable α-amylase and saccharification by addition of amyloglucosidase [157, 158].

The starch-processing industry used Commercial enzyme preparations such as Thermamyl and BAN 480L (Novozymes, Denmark). Thermamyl is heat-stable α-amylase produced from genetically modified *Bacillus* strains. BAN 480L (Novozymes, Denmark) effective only towards gelatinized starch, is also bacterial α-amylase produced by *Bacillus amyloliquefaciens* [159, 160]. These enzymes were suitable for applications in processes carried out at elevated temperatures to improve the reaction kinetics. STARGEN 001 developed by Genencor International (Palo Alto, CA), has been recently introduced to the market. STARGEN 001 has the ability to use it in mediums of high density. Additionally, there is no need for an activator, such as calcium ions for most amylases.

A lot of factors as the processing method, hydrolyzing enzymes, amylose/amylopectin ratio, protein and lipid content, granular size and surface area, and starch granular structure affects the hydrolysis of starch [29, 161]. The amounts of enzymes adsorbed to the granule surface is proportional to the surface area of the starch granule and the hydrolysis rate of starch is impacted by the granular surface area [162]. Additionally, small starch granules show less crystallinity and greater water affinity than that of large granules [161].

Hydrolysis of dissolved starch is significantly faster than that of granular starch. The efficiency of enzymatic hydrolysis of granular starch depends on its origin because different starches vary considerably in terms of their supramolecular structure [163]. A comparison of the four cheapest commercial native starches in terms of their susceptibility to amylolysis results in diminishing order as corn > wheat > cassava > potato [159].

Since bioethanol producing microorganisms such as *Saccharomyces cerevisiae* cannot utilize starch and starchy materials, amylolytic enzymes, glucoamylase (EC3.2.1.3) and ⟨-amylase (EC 3.2.1.1) are required to hydrolyze the starchy materials to glucose [164]. In order to obtain glucose from starch economically, it is desired to have α- amylase active at high temperatures [165]. To obtain amylase produced by a wide variety of microorganisms, including thermophiles and mesophiles suitable for starch hydrolysis at high temperatures, many comparative analyses and directed evolutions have been performed [166-168,53].

Since a large amount of enzyme is necessary to hydrolyze biomass to sugars, a lot of works have been carried out to solve this problem by using recombinant glucoamylase-expressing yeasts having the ability to convert starch to ethanol directly [164]. Additionally, recombinant yeasts which co-produce glucoamylase and α-amylase have been developed to improve the efficiency of starch fermentation [169-172]. Shigechi et al. [164] studied to express ⟨-amylase from the lactic acid bacterium *Streptococcus bovis* 148 together with *Rhizopus oryzae* glucoamylase. They investigated direct and efficient production of ethanol from raw corn starch using the yeast *Saccharomyces cerevisiae* co-displaying *Rhizopus oryzae* glucoamylase and *Streptococcus bovis* ⟨-amylase. Shigechi et al. [164] reported that 61.8 g of ethanol/liter, with 86.5% of theoretical yield from raw corn starch was produced after 72 fermentation.

Although different research works have been carried out to find microorganisms having ability to ferment starchy materials to glucose directly, improvement the efficiency of bioprocesses for amylase enzymes are still continuing using different biomass. Besides other food crops, corn has been the most widely studied starchy biomaterials [173]. Ethanol from corn has been found to reduce carbon emissions over gasoline by as much as 60%, while others may increase greenhouse gas (GHG) emissions over gasoline by as much as 20% [174].

Besides conventional corn, Dale and Kim [173] reported a new variety of corn that expresses α-amylase directly in the seed endosperm. This enzyme was named corn amylase (CA) and it represents another approach to improving upon the efficiency, cost, and environmental footprint of biofuels.

More recently cassava, a starch-containing root crop is used extensively as a raw material for bioethanol production [154]. Cassava (*Manihot esculenta* Cranz) is the world's sixth most important crop and is grown in many African, Asian, and Latin American countries. In addition to ethanol production, cassava is also used as a raw material for animal feed industry and starch industry. Cassava pulp composed of 50%–70% starch, cellulose and non-starch polysaccharides [152]. Comparing to the other starch materials, cassava starch is easily swollen, highly viscous, highly pure and neutral flavor [175]. In order to hydrolyse cassava to glucose, multi-enzyme preparations consisting of cellulase, D-xylanase, β-D-glucosidase, α-amylase, amyloglucosidase, and pectinase have been used. It was reported that amyloglucosidase in combination with pectinase was used to digest raw starch from cassava roots [176, 177]. Paolucci-Jeanjean et al. [178] used only amylase for the hydrolysis of raw cassava starch and found that the amount of reducing sugar was about 25 g/L after 1-hour incubation. Combination of raw starch degrading amylolytic activity with multi nonstarch polysaccharide hydrolyzing enzymes were used by Rattanachomsri et al. [152] to evaluated the direct hydrolysis and saccharification of cassava pulp. They obtained a maximum efficiency of hydrolysis of about 90% after 30 h of reaction. Sriroth et al. [179] obtained a maximum efficiency of 40% by using physical disruption through sonication and enzymatic treatment using cellulase, pectinase, and α-amylase to hydrolyze starch from cassava. Collares et al. [154] studied the hydrolysis of cassava starch using the pectinase, α-amylase, and amyloglucosidase. They noted that the maximum productivity was 22.9 g/(L·h) was obtained by using 0.52 g/g starch of amyloglucosidase, 2.35 μl/g dry matter of pectinase and amyloglucosidase,

Although, sugarcane (Brazil) and corn grain (USA) are main feedstocks for bioethanol production, alternative and conventional raw materials are under research [180]. Because of higher starch yield, sweet potato has been considered a promising substrate for alcohol production [180, 182]. Lareo et al. [180] studied the two-step enzymatic hydrolysis of the sweet potato starch using amylase and glucoamylase. They found that, after the addition of glucoamylase, 100% starch hydrolysis was performed for 90 min of the α-amylase action.

Another alternative starchy material hydrolysed enzymatically to produce glucose for biofuel production is Sago Hampas [183]. Sago hampas is a starchy lignocellulosic by-product generated from pith of *Metroxylon sagu,* an increasingly important socio-economic crop in Southeast Asia (sago palm) [184, 185]. Bandaru et al. [186] studied simultaneous saccharification and fermentation of ethanol from sago starch with co-immobilized amyloglucosidase and *Zymomonas mobilis* MTCC 92 and they obtained 55.3 g/L ethanol from a starch concentration of 150 g/L. Awg-Adeni et al. [183] studied the hydrolysis of waste starch of Sago hampas using three-cycle enzymatic hydrolysis process which involved reusing the hydrolysate for subsequent enzymatic hydrolysis process. They used commercial enzyme dextrozyme (5.56 U/mL), a mixture of glucoamylase from *Aspergillus niger* and pullulanase from *Bacillus acidopullulyticus*. Awg-Adeni [183] reported that, after the completion of three cycles of hydrolysis, greater improvement of glucose concentration (138.45 g/L) and better conversion yield (52.72%) were obtained.

Microalgae are used as an ideal third generation biofuel feedstock. Since they can be produced on non-arable land, microalgae do not compete with food or feed crops, and [187]. Microalgal biomass composed of three main components: proteins, carbohydrates, and lipids (oil) [188, 189]. Since the industrially used *Saccharomyces cerevisiae* is non-cellulolytic and non amylolytic, the enzymes were employed to hydrolyse and produce sugars. Singh and Triversi [187] studied *Spirogyra* biomass for bioethanol production. They used α-amylase to convert algal starch to fermentable sugar and they reported that the presence of 0.09 grams of α-amylase enzyme ethanol production is recorded maximum by using *Zymomonasmobilis* than *Saccharomyces cerevisiae*.

5.6. Bread and Bakery Industry

The bread making process is very important event in mankind and originated in ancient times. It is believed that Egyptians knew how to make fermented bread till 2,000 BC [190]. Baking industry evolve very quickly with the introduction of enzymes in the baking process. Today, bakery enzymes occupy the relevant segment of the enzyme market. According to the Novozymes S/A report 2011 [191] food and beverage enzymes represented 29% of enzyme business and biobusiness sales by the industry [191].

Malt and microbial α-amylases have been widely used in the baking industry to produce various high-quality products for decades [134, 192]. Higher volume, better colour and softer crumb is achieved by using these enzymes in breads and rolls. The success of the malt preparation has opened the way for various enzymes to be used in baking industry [9]. Different enzyme preparations such as proteases, xylanases, cellullases, lipases, pullulanases, pentosanases, glucose oxidases, etc. are being used for different purposes in bread industry today [193, 194]. However, none of them had been able to replace α-amylase.

Although barley malt is rich in α-amylases it is better to utilize commercial enzyme because fungal α-amylases supplied with standardized activity whereas the enzymatic activity in the malt preparations vary widely. The use of fungal α-amylases as bread additives have been permitted in US (1955) and in UK (1963) after confirmation of their GRAS status and today, they are used all over the world [9].

α-amylases degrade the damaged starch in wheat flour into small dextrins via cleavage of α,1-4 glycosidic bonds. The increased level of fermentable and reducing sugars allows yeast to ferment these sugars into alcohols and carbon dioxide which causes rising of the dough. The addition of malt or fungal α-amylases results in improved loaf volume and crump texture [12]. Moreover, hydrolysis generates additional glucose and maltose that enhance the Maillard reactions, which is responsible for the browning of the crust and the development of an attractive baked flavor [195, 196].

Certain amylases are able to reduce the firmness of bread crump, acting as anti-stalling agent that is increase the shelf-like of the baked products [197]. This anti-stalling ability of amylases is very important because bread and cake staling is responsible for significant financial loss. For instance, annually only in the USA bread worth more than US$ 1 billion was discarded due to staling. [134]. Staling is the loss of freshness in terms of increased crumb firmness, decreased in the flavour, aroma, loss of moisture and crump elasticity. Typically, amylases having intermediate thermostability display anti-staling function. *Bacillus stearothermophilus*'s maltogenic α-amylase is one of the most effective anti-staling amylases [196]. Novozymes developed a bacterial maltogenic α-amylase called Novamyl® has been found to have a substantial anti-staling effect. Novamyl® acts on starch during baking at the temperature when most of the starch starts to gelatinize. During storage these modified granules remain more flexible. Martin and Hoseney [198] reported that bacterial amylases producing mainly dextrins of DP (3-9) were stopped retrogradation of bread. Moreover, small MOS like maltopentaose and maltotriose are acting as an antifirming agent and able to decrease retrogradation [199]. On the other hand, due to its small size and ability to diffuse easily, maltose is not an effective agent in the prevention of bread staling. In bakery, the recent trend is to use intermediate temperature stable enzymes instead of thermostable enzymes since thermostable enzymes can survive at oven temperature, cause over hydrolyzation of starch, and result in gumminess of the final product. Although many different microbial α-amylases have been reported, very few amylases are known to produce small maltooligosaccharides (DP 3-4) and ITS property [90, 122, 123, 200].

CONCLUDING REMARKS AND FUTURE PERSPECTIVES

Amylases mainly α-amylase, β-amylase and glucoamylase are an important group of enzymes because of their great application in different industries. The properties of each amylases are very important to carry out reactions and/or synthesize products. With the advent of new frontiers in enzyme technology, the spectrum of amylase application has expanded into many other fields. Developing modified amylases open the door for development of economical hydrolysis process and they would play an important role in various industrial and biotechnological applications.

REFERENCES

[1] Pandey, A., Nigam, P., Soccol, C., et al. (2000). Advances in microbial amylases. *Biotechnol Appl Biochem., 31*: 135–52.

[2] Sivaramakrishnan, S., Gangadharan, D., Nampoothirl, K., et al. (2006). α-amylase from microbial sources- An overview on recent developments. *Food Technol Biotechnol., 44*: 173-184.

[3] Raul, D., Biswas, T., Mukhopadhyay, S., et al. (2014). Production and partial purification of α amylase from *Bacillus subtilis* (MTCC 121) using solid state fermentation. *Biochem Res Intr., 2014*: 1-5.

[4] Sharma, K., Bhutty, S., Paul, K. S., et al. (2014). Isolation, identification and optimization of culture conditions of *Bacillus sp.* strain PM1 for alkalo-thermostable amylase production. *Brit Microb Res J., 4*: 369-380.

[5] Shah, I. J., Gami, P. N., Shukla, R. M., Acharya, D. K. 2014. "Optimization of α-amylase production by *Aspergillus oryzae* using submerged fermentation technology." *Basic Res J., 1*: 1-10.

[6] Hmidet, N., Bayoudh, A., Berrin, J. G., et al. (2008). Purification and biochemical characterization of a novel α-amylase from *Bacillus licheniformis* NH1 Cloning, nucleotide sequence and expression of amyN gene in *Escherichia coli. Process Biochem., 43*: 499–510.

[7] Naidu, M. A. and Saranraj, P. (2013). Bacterial Amylase: A Review. *Intr J Pharm Biol Arch., 4*: 274-287.

[8] Hirata, A., Adachi, M., Sekine, A., et al. (2004). Structural and enzymatic analysis of soybean β-amylase mutants with increased pH optimum. *J Biol Chem., 279*: 7287–7295.

[9] Gupta, R., Gigras, P., Mohapatra, H., et al. (2003). Microbial α-amylases: a biotechnological perspective. *Process Biochem., 38*: 1599-1616.

[10] Jebor, M. A., Ali, Z. M. and Hassan, B. A. (2014). Purification and characterization of the glucoamylase from *Aspergillus niger*." *Int J Cur Microbiol Appl Sci., 3*: 63-75.

[11] El-Fallal, A., Dobara, M., El-Sayed, A., et al. (2012). Starch and microbial α-amylases: from concepts to biotechnological application. Carbohydrates – Comprehensive Studies on Glycobiology and Glycotechnology. *InTech*, pp. 459-488. Doi.org/10.5772/51571.

[12] Van der Maarel, M. J., Van der Veen, B., Uitdehaag, J. C., et al. (2002). Properties and applications of starch-converting enzymes of the α-amylase family. *J Biotechnol., 94*: 137-155.

[13] Fredriksson, H., Silverio, J., Andersson, R., et al. (1998). The influence of amylose and amylopectin characteristics on gelatinization and retrogradation properties of different starches. *CarbohydrPolym., 35*: 119-134.

[14] Vandeputte, G., Vermeylen, R., Geeroms, J., et al. (2003*a*). Rice starches. I. Structural aspects provide insight into crystallinity characteristics and gelatinisation behaviour of granular starch. *J Cereal Sci., 38*: 43-52.

[15] Alavi, S. (2003). Starch research over the years. *Food Res Int., 36*: 307-308.

[16] Bijttebier, A., Goesaert, H. and Delcour, J. A. (2007). Temperature impacts the multiple attack action of amylases. *Biomacromolecules*, *8*: 765–772.

[17] Singh, N., Singh, J., Kaur, L., et al. (2003). Morphological, thermal and rheological properties of starches from different botanical sources. *Food Chem.*, *81*: 219-223.

[18] *Agrosynergie Report- European Commision* http://ec.europa.eu/agriculture/eval/reports/decoupling/exec_sum_en.pdf.

[19] Alberton, C., Colman, T., De Souza, J., et al. (2014). Thermal analysis, rheology, x-ray diffractometry and atomic force microscopy in the evaluation of binary mixtures of starch-hydrocolloids. *J Microbiol Biotech Food*, *3*:305-309.

[20] Manners, D. J. (1989). Recent developments in our understanding of amylopectin structure. *Carbohyd Polym.*, *11*: 87-112.

[21] Morrison, W. R. and Karkalas, J. (1990). Starch. In: *Methods in Plant Biochemistry*. Dey PM (ed.) vol 2 Academic Press, New York, pp 323-352.

[22] Kent, N. L. and Evers, A. D. (1994). *Kent's Technology of Cereals*, fourth ed., Pergamon Press, Oxford England.

[23] Zobel, H. F. and Stephen, A. M. (1996). Starch: structure, analysis and application. In: *Food Polysaccharides and their application*. Stephen AM (ed.). Marcel Dekker, New York, pp 19-66.

[24] Buléon, A., Colonna, P., Planchot, V., et al. (1998). Starch granules: structure and biosynthesis. *Int J Bio Macromol., 23*: 85-112.

[25] Tester, R. F., Karkalas, J. 2002. "Starch". In: Steinbüchel A (series ed) Vandamme E. J., De Baets S., Steinbüchel A. (vol eds) *Biopolymers, vol 6. Polysaccharides. II. Polysaccharides from Eukaryotes*, Wiley-VCH, Weinheim, pp 381-438.

[26] Myers, A., Morell, M., James, M., et al. (2000). Recent progress towards understanding biosynthesis of the amylopectin crystal. *Plant Physiol., 122*: 989-997.

[27] Lacerda, L., Carvalho-Filho, M., Demiate, I., et al. (20080. Thermal behavior of corn starch granules under action of fungal α-amylase. *J Therm Anal Calorim.*, *93*: 445-449.

[28] Mua, J. P. and Jackson, D. S. (1997). Fine structure of corn amylose and amylopectin fractions with various molecular weights. *J Agr Food Chem., 45*: 3840-3847.

[29] Tester, R. F., Karkalas, J. and Qi, X. (2004). Starch-composition, fine structure and architecture. *J Cereal Chem., 39*: 151-165.

[30] Zobel, H. F. (1988). Molecules to granules: A comprehensive review. *Starch/Stärke*, *40*: 1-7.

[31] Imberty, A., Imberty, H. and Chanzy, P. (1988). The double-helical nature of the crystalline part of A-starch. *J Mol Biol., 201*: 365-378.

[32] Hong, M., Soini, H. and Novotny, M. (2000). Affinity capillary electrophoretic studies of complexation between dextrin oligomers and polyiodides. *Electrophoresis*, *21*: 1513-1520.

[33] Singh, A. V. (2011). Biopolymers in Drug Delivery: A Review. *Pharmacologyonline*, *1*: 666-674.

[34] Hizukuri, S. (1986). Polymodal distribution of the chain lengths of amylopectin and its significance. *Carbohydr Res., 147*: 342-347.

[35] Zobel, H. F. (1992). Starch granule structure. In: *Developments in Carbohydrate chemistry*. Alexander RJ, Zobel HF (eds.). American association of cereal chemistry, Minnesota, USA. pp 1-36.

[36] Marc, J., Van der Maarel, C., Joost, B., et al. (2002). Properties and applications of starch-converting enzymes of the α-amylase family. *J Biotechnol.*, *94*: 137-155.

[37] Gerard, C., Planchot, V., Colonna, P., et al. (2000). Relationship between branching density and crystalline structure of A- and B-type maize mutant starches. *Carbohydr Res., 326*: 130-144.

[38] Satheesh, K., Subhosh, C., Mallaiah, K., et al. (2010). Purification and characterization of highly thermostable α amylase from thermophilic *Alicyclobacillus acidocaldarius. Biotechnol Bioproc E.*, *15*: 435-440.

[39] Ozcan, B., Baylan, M., Ozcan, N., et al. (2010). Characterization of thermostable α-amylase from thermophilic and alkaliphilic *Bacillus sp.* isolate DM-15. *Res J Biol Sci., 5*: 118-124.

[40] De Souza, P. M. and Magalhaes, P. D. E. (2010). Application of microbial α-amylase in industry- a review. *Braz J Microbiol.*, *41*: 850-861.

[41] Amutha, K. and Priya, K. J. (2011). Effect of pH, temperature and metal ions on amylase activity from *Bacillus subtilis* KCX 006. *Int J Pharm Bio Sci.*, *2*: 407-413.

[42] Mobini-Dehkordi, M. and Javan, F. A. (2012). Application of α-amylase in biotechnology. *J Biol Today World, 1* :39-50.

[43] Femi-Ola, T., Oshokoya, A. and Bamidele, O. (2013). Kinetic properties of beta-amylase from *Bacillus subtilis. IOSR J Env Sci Toxicol Food Technol.*, *2*: 19-23.

[44] Deb, P., Talukdar, S., Mohsina, K., et al. (2013). *Production and partial characterization of extracellular amylase enzyme from Bacillus amyloliquefaciens* P-001. Springer Plus, *2*: 154-165.

[45] Roy, J. and Mukherjee, A. (2013). Application of a high maltose forming, thermo-stable alpha amylase from an extremely alkalophilic *Bacillus licheniformis* strain AS08E in food and laundry detergent industries. *Biochem Eng J.*, *77*: 220-230.

[46] Alariya, S., Sethi, S., Gupta, S., et al. (2013). Amylase activity of a starch degrading bacteria isolated from soil. *Arch Appl Sci Res.*, *5*: 15-24.

[47] Sharma, A. and Satyanarayana, T. (2013). Microbial acid-stable α-amylases: Characteristics, genetic engineering and applications. *Process Biochem.*, *48*: 201-211.

[48] Moradi, M., Shariati, P., Tabandeh, F., et al. (2014). Screening and isolation of powerful amylolytic bacterial strains. *Int J Curr Microbiol Appl Sci.*, *3*: 758-768.

[49] Roy, A., Khanra, K., Mishra, A., et al. (2014). Partial purification and characterization of amylase from a newly isolated *Bacillus megaterium* strain KAN1 from fermented rice Handia. *Amr J Curr Microbiol., 2*: 2-7.

[50] Tabassum, R., Khaliq, S., Rajoka, M., et al. (2014). Solid state fermentation of a raw starch digesting alkaline α-amylase from *Bacillus licheniformis* RT7PE1 and its characteristics. *Biotech Res Intr.,* doi.org/10.1155/2014/495384.

[51] Fincan, S. A. and Enez, F. (2004). Production, purification and characterization of thermostable α-amylase from thermophilic *Geobacillus stearothermophilus. Starch, 66*: 182-189

[52] Raul, D., Biswas, T., Mukhopadhyay, S., et al. (2014). Production and partial purification of α amylase from *Bacillus subtilis* (MTCC 121) using solid state fermentation. *Biochem Res Intr.*, doi.org/10.1155/2014/568141.

[53] Li, C., Du, M., Cheng, B., et al. (2014). Close relationship of a novel Flavobacteriaceae α-amylase with archaeal α-amylases and good potentials for industrial applications. *Biotechnol Biofuels*, *7*: 18-29.

[54] Bhatti, H., Rashid, M., Nawaz, R., et al. (2007). Purification and characterization of a novel glucoamylase from *Fusarium solani*. *Food Chem.*, *103*: 338-343.

[55] Sarojini, R., Deepika, K. and Rangabhashiyam, S. (2012). Production, characterization, kinetic studies of glucoamylase through solid-state fermentation by *Aspergillus niger* using agricultural residues as substrate. *Intr J Cur Res.*, *4*: 198-201.

[56] Koc O. and Metin, K. (2012). Purification and characterization of a thermostable glucoamylase produced by *Aspergillus flavus* HBF34. *Afr J Biotechnol.*, *9*: 3414-3424.

[57] Jebor, M., Ali, Z. and Hassan, B. (2014). Purification and characterization of the glucoamylase from *Aspergillus niger*. *Intr J Curr Microbiol Appl Sci.*, *3*: 63-75.

[58] Singh, S., Singh, S., Bali, V., et al. (2014). Production of fungal amylases using cheap, readily available agriresidues for potential application in textile industry. *BioMed Res Intr.*, 2014: 1-9. Doi.org/10.1155/2014/215748.

[59] Welday, T., Haki, G., Abera, S., et al. (2014). Isolation and characterization of thermostable fungal α-amylase from geothermal sites of Afar, Ethiopia. *Intr J Adv Pharm Biol Chem.*, *3*: 120-127.

[60] Oseni, O. and Ekperigin, M. (2014). Activity of β-amylase in some fungi strains isolated from forest soil in south-western Nigeria. *Brit Biotechnol J.*, 4: 1-9.

[61] Shen, H., Mo, X., Chen, X., et al. (2012). Purification and enzymatic identification of an acid stable and thermostable α-amylase from *Rhizopus microspores*. *J Inst Brew.*, *118*: 309-314.

[62] Freitas, T., Pereira, C. and Pereira, L. (2014). Syzygium sp. (Myrtaceae) Extracts: Inhibition of Alpha Amylase. *Eur J Med Plants*, *4*: 116-125.

[63] Kuddus, M., Jamal, M. A., Ramteke, W (2011). An overview of cold-active microbial α-amylase: Adaptation strategies and biotechnological potentials. *Biotechnology*, *10*: 246-258.

[64] Quin, Y., Huand, Z. and Liu, Z. (2014). A novel cold-active and salt-tolerant α-amylase from marine bacterium Zunongwangia profunda: molecular cloning, heterologous expression and biochemical characterization. *Extremophiles*, *18*: 271-281.

[65] Adrio, J. L. and Demain, A. L. (2014). Microbial enzymes: Tools for biotechnological Processes. *Biomolecules*, 4:117-139.

[66] Koshland, D.E. (1953). Stereochemistry and the mechanism of enzymatic reactions. *Biol Rev Camb Philos Soc.*, *28*: 416-436.

[67] Sinnott, M. L. (1990). Catalytic mechanisms of enzymic glycosyl transfer. *Chem Rev.*, *90*: 1171-1202.

[68] Davies, G. and Henrissat, B. (1995). Structures and mechanisms of glycosyl hydrolases. *Structure*, *3*: 853-859.

[69] Horváthová, V., Janecek, S. and Sturdík, E. (2000). Amylolytic enzymes: their specicities, origins and properties. *Biologia Bratislava*, *55*: 605-615.

[70] Janecek, S. (1997). α-amylase family: molecular biology and evolution." *Prog Biophys Mol Biol.*, 67: 67-97.

[71] McCarter, J. D. and Withers, S. G. (1994). Mechanisms of enzymatic glycoside hydrolysis. *Curr Opin Struct Biol.*, *4*: 885-892.

[72] Uitdehaag, J., Mosi, R., Kalk, K., (1999). X-ray structures along the reaction pathway of cyclodextrin glycosyltransferase elucidate catalysis in the α-amylase family. *Nature Struct Biol.*, 6: 432–436.

[73] Okazaki, N., Tamada, T., Feese, M., et al. (2012). Substrate recognition mechanism of a glycosyltrehalose trehalohydrolase from *Sulfolobus solfataricus* KM1. *Protein Sci., 21*: 539-552.

[74] Takase, K., Matsumoto, T., Mizuno, H., et al. (1992). Site-directed mutagenesis of active site residues in *Bacillus subtilis* α-amylase. *Biochem Biophys Acta, 1120*: 281-288.

[75] Vihinen, M. and Mantsala, P. (1990). Characterization of a thermostable *Bacillus stearothermophilus* α-amylase. *Biotechnol Appl Biochem., 12*: 427-435.

[76] Cosa de Freitas, A., Escaramboni, B., Carvalho, A., et al. (2014). Production and application of amylases of *Rhizopus oryzae* and *Rhizopus microsporus var. Oligosporus* from industrial waste in acquisition of glucose. *Chem Pap., 68*: 442-450.

[77] Michelin, M., Silva, T., Benassi, V., et al. (2010). Purification and characterization of a thermostable α-amylase produced by the fungus *Paecilomyces variotii. Carbohydr Res., 345*: 2348-2353.

[78] Gomes, J. and Steiner, W. (2004). The biocatalytic potential of extremophiles and extremozymes. *Food Technol Biotechnol, 42*: 223-235.

[79] Coronado, M., Vargas, C., Hofemeister, J., et al. (2000). Production and biochemical characterization of an α-amylase from the moderate Halophile *halomonas meridiana. FEMS Microbiol Lett., 183*: 67-71.

[80] Uzyol, K., Sarıyar-Akbulut, B., Denizci, A., et al. (2012). Thermostable α amylase from moderately halophilic *Halomonas* sp. AAD21. *Turk J Biol., 36*: 327-338

[81] Burhan, A., Nisa, U., Gokhan, C., et al. (2003). Enzymatic properties of a novel thermostable, thermophilic, alkaline and chelator resistant amylase from an alkaliplilic *Bacillus sp.* isolate ANT-6. *Process Biochem., 38*: 1397-1403.

[82] Saxena, R., Dutt, K., Agarwal, L., et al. (2007). Highly thermostable and alkaline amylase from a *Bacillus sp.* PN5. *Biores Technol., 98*: 260-265.

[83] Behal, A., Singh, J., Sharma, M., et al. (2006). Characterization of alkaline α-amylase from *Bacillus sp.* AB 04. *Int J Agric Biol., 8*: 80-83.

[84] Liu, X.D. and Xu, Y. (2008). A novel raw starch digesting α-amylase from a newly isolated *Bacillus sp.* YX-1: Purification and characterization. *Biores Technol., 99*: 4315–4320.

[85] Corderio, C., Martins, M. and Luciano, A. (2002). Production and properties of α-amylase from thermophilic *Bacillus sp. Braz J Microbiol., 33*: 57-61.

[86] Asoodeh, A., Chamani, J. and Lagzian, M. (2010). A novel thermostable, vacidophilic α-amylase from a new thermophilic, *Bacillus sp. ferdowsicous* isolated from Ferdows hot mineral spring in Iran: Purification and biochemical characterization. *Int J Biol Macromol., 46*: 289-297.

[87] Ezeji, T. C. and Bahl, H. (2006). Purification, characterization and synergistic action of phytate resistant α-amylase and α-glucosidase from *Geobacillus thermodenitrificans* HRO10. *J Biotechnol., 125*: 27–38.

[88] Kolcuoğlu, Y., Colak, A., Faiz, O., et al. (2010). Cloning, expression and characterization of highly thermo- and pH stable maltogenic amylase from a thermophilic bacterium *Geobacillus caldoxylosilyticus* TK4. *Process Biochem., 45*: 821-828.

[89] Mollania, N., Khajeh, K., Hosseinkhani, S., et al. (2010). Purification and characterization of a thermostable phytate resistant α-amylase from *Geobacillus sp.* LH8. *Int J Biol Macromol., 46*: 27-36.

[90] Ozturk, H. U. (2013). *Investigating of industrial enzymes from Bacillus sp isolated from Acacia cyanophylla Lindley*. Ph.D. Thesis.

[91] Ivanova, V., Dopreva, E. and Emanuilova, E. (1993). Purification and characterization of a thermostable α-amylase from *Bacillus licheniformis*. *J Biotechnol., 28*: 277-289.

[92] Hamilton, L., Kelly, C. and Fogarty, W. (1999). Production and properties of the raw starch-digesting α-amylase of *Bacillus sp.* IMD 435. *Process Biochem., 35*: 27-31.

[93] Carvalho, R., Côrrea, T., Silva, J., et al. (2008). Properties of an amylase from thermophilic *Bacillus sp*. *Braz J Microbiol.,* 39: 102-107.

[94] Sajedi, R., Naderi-Manesh, H., Khajeh, K., et al. (2005). A Ca-independent α--amylase that is active and stable at low pH from the *Bacillus sp* KR-8104. *Enz Microbial Technol., 36*: 666-671.

[95] Mehta, M. and Satyanarayana, T. (2013). Biochemical and molecular characterization of recombinant acidic and thermostable raw-starch hydrolyzing α-amylase from an extreme thermophile *Geobacillus thermoleovorans*. *J Mol Catal B: Enz., 85-86*: 229-238.

[96] Tanyildizi, M., Ozer, D. and Elibol, M. (2007). Production of bacterial α-amylase by *B. amyloliquefaciens* under solid substrate fermentation. *Biochem Eng J., 37*: 294-297.

[97] Saito, N. (1973). Thermophilic extracellular α-amylase from *Bacillus licheniformis*. *Arch Biochem Biophys., 155*: 290-298.

[98] Krishnan, T. and Chandra, A. (1983). Correlation between α-amylase production and sporulation in *Bacillus licheniformis* Cumc-305 with respect to the effect of some carbohydrates and phenylmethylsulfonyl fluoride treatment. *Zentralb Mikrobiol, 138*: 475-485.

[99] Morgan, F. J. and Priest, F. G. (1981). Characterization of a thermostable α-amylase from *Bacillus licheniformis* NCIB-6346. *J Appl Bacteriol., 50*: 107-114.

[100] Goyal, N., Gupta, J. and Soni, S. (2005). A novel raw starch digesting thermostable α-amylase from *Bacillus sp.* 1-3 and its use in the direct hydrolysis of raw potato starch. *Enz Microbial Technol., 37*: 723-734.

[101] Sodhi, H., Sharma, K., Gupta, J., et al. (2005). Production of a thermostable α-amylase from *Bacillus sp.* PS-7 by solid state fermentation and its synergistic use in the hydrolysis of malt starch for alcohol production. *Process Biochem., 40*: 525-534.

[102] Asgher, M., Asad, M., Bhatti, H., et al. (2007). Hyperactivation and thermostabilization of *Phanerochaete chrysosporium* lignin peroxidase by immobilization in xerogels. *World J Microbiol Biotechnol., 23*: 525-531.

[103] Deutch, C. E. (2002). Characterization of a salt tolerant extracellular α-amylase from *Bacillus dipsosauri*. *Lett Appl Microbiol., 35*: 78-84.

[104] Aguilar, G., Morlon-Guyot, J., Trejo-Aguilar, B., et al. (2000). Purification and characterization of an extracellular α--amylase produced by *Lactobacillus manihotivorans* LMG 18010(T), an amylolytic lactic acid bacterium. *Enz Microbial Technol., 27*: 406-413.

[105] Vihinen, M., Ollikka, P., Niskanen, J., et al. (1990). Site-directed mutagenesis of a thermostable α-amylase from *Bacillus stearothermophilus:* putative role of three conserved residues. *J Biochem., 107*: 267-72.

[106] Hayashida, S., Teramoto, Y. and Inoue, T. (1988). Production and characteristics of raw-potato-starch-digesting α-amylase from *Bacillus subtilis* 65. *Appl Env Microbiol., 54*: 1516-1522.

[107] Marco, J., Bataus, L., Valencia, F., et al. (1996). Purification and characterization of a truncated *Bacillus subtilis* α-amylase produced by *Escherichia coli. Appl Microbiol Biotechnol., 44*: 746-752.

[108] Freer, S. N. (1993). Purification and characterization of the extracellular α-amylase from *Streptococcus bovis* jb1. *Appl Env Microbiol., 59*: 1398-1402.

[109] Kelly, C., Collins, B., Fogarty, W., et al. (1993). Mechanisms of action of the α-amylase of *Micromonospora melanosporea. Appl Microbiol Biotechnol.*, 39: 599-603.

[110] Chung, Y., Kobayashi, T., Kanai, H., et al. (1995). Purification and properties of extracellular amylase from the hyperthermophilic archaeon *Thermococcus profundus* Dt5432. *Appl Env Microbiol, 61*: 1502-1506.

[111] Bhella, R. S. and Altosaar, I. (1985). Purification and some properties of the extracellular α-amylase from *Aspergillus awamori. Can J Microbiol., 31*: 149-153.

[112] Ramachandran, N., Sreekantiah, K. and Murthy, V. (1979). Influence of media composition on the production of α-amylase and amyloglucosidase by a strain of *Aspergillus niger. Starke, 31*: 134-138.

[113] Hernandez, M., Rodriguez, M., Guerra, N., et al. (2006). Amylase production by *Aspergillus niger* in submerged cultivation on two wastes from food industries. *J Food Eng., 73*: 93-100.

[114] Lin, L., Chyau, C. and Hsu, W. 1998. Production and properties of a raw-strach degrading amylase from thermophilic and alkaliphilic *Bacillus sp.* TS-23. *Biotechnol Appl Biochem., 28*: 61-68.

[115] Mamo, G. and Gessesse, A. (1999). Purification and characterization of two rawstarch-digesting thermostable α-amylases from a thermophilic *Bacillus. Enz Microbial Technol., 25*: 433-438.

[116] Shafiei, M., Ziaee, A. and Amoozegar, M. (2010). Purification and biochemical characterization of a novel SDS and surfactant stable, raw starch digesting and halophilic α-amylase from a moderately halophilic bacterium *Nesterenkonia sp* strain F. *Process Biochem., 45*: 694-699.

[117] Qader, S., Aman, A., Syed, M., et al. (2011). Purification and Characterization of Novel α-amylase from *Bacillus subtilis* KIBGE HAS AAPS. *Pharm Sci Technol., 12*: 255-261.

[118] Das, K., Doley, R. and Mukherjee, A. (2004). Purification and biochemical characterization of a thermostable, alkaliphilic, extracellular α-amylase from *Bacillus subtilis* DM-03, a strain isolated from the traditional fermented food of India. *Biotech Appl Biochem., 40*: 291-8.

[119] Ben-Ali, M., Mhiri, S., Mezghani, M., et al. (2001). Purification and sequence analysis of the a typical maltohexaose-forming α-amylase of the *B. stearothermophilus* US100. *Enz Microbial Technol, 28*: 537-542.

[120] Hashim, S., Delgado, O., Martinez, M., et al. (2005). Alkaline active maltohexaose-forming α-amylase from *Bacillus halodurans* LBK 34. *Enz Microbial Technol., 36*: 139-146.

[121] Shida, O., Takano, T., Takagi, H., et al. (1992). Cloning and nucleotide sequence of the maltopentaose-forming amylase gene from *Pseudomonas sp* KO-8940. *Biosci Biotech Biochnol., 56*: 76-80.

[122] Kim, T., Gu, B., Jeong, J., et al. (1995). Purification and characterization of a maltotetraose-forming alkaline α-amylase from an alkalophilic *Bacillus strain*, Gm8901. *Appl Env Microbiol., 61*: 3105-3112.

[123] Nagarajan, D., Rajagopalan, G. and Krishnan, C. (2006). Purification and characterization of a maltooligosaccharide-forming α-amylase from a new *Bacillus subtilis* KCC103. *Appl Microbiol Biotechnol., 73*: 591-597.

[124] Sarojini, R., Deepika, K. and Rangabhashiyam, S. (2012). Production, characterization, kinetic studies of glucoamylase through solid-state fermentation by *Aspergillus niger* using agricultural residues as substrate. *Intr J Cur Res, 4*: 198-201.

[125] Selvakumar, P., Ashakumary, L., Helen, A., et al. (1996). Purification and characterization of glucoamylase produced by *Aspergillus niger* in solid state fermentation. *Lett Appl Microbiol, 23*: 403-406.

[126] Deepika, A. and Rangabhashiyam, S. (2012). Production, characterization, kinetic studies of glucoamylase through solid-state fermentation by *Aspergillus niger* using agricultural residues as substrate. *Intr J Curr Res, 4*: 198-201.

[127] Marlida, Y., Hasan, S., Radu, S., et al. (2000). Purification and characterization of sago starch degrading of glucoamylase from *Acremonium sp. endophtic fungus. Food Chem., 71*: 221-227.

[128] Sanni, G. (2003). Thermostable amylolytic enzymes from a new *Clostridium* isolate. *Appl Env Microbiol., 53*: 1661-1667.

[129] Ray, R.R. (2000). Purification and characterization of extracellular β - amylase of *Bacillus megaterium B. Acta Microbiol Immunol Hung., 47*: 29-40.

[130] Beckord, J. (2005). Selection of microorganism which produce raw starch degrading amylase. *Appl Microbiol Biotechnol., 27*: 443-446.

[131] Boldon, M. and Effront, P. (2000). Microbial amylolytic enzymes. *Crit Rev Biochem Mol., 24*: 329-418.

[132] Rigoldi, F., Donini, S., Redaelli, A., et al. (2018). Engineering of thermostable enzymes for industrial applications. *APL Bioeng., 2*: 011501.

[133] http://www.rnrmarketresearch.com/world-enzymes-to-2017-market-report.html.

[134] Miguel, A., Martins-Meyer, T., Da Costa Figueiredo, E., et al. (2013). Enzymes in Bakery: Current and Future Trends, Food Industry. Muzzalupo I (ed.) *InTech*. DOI: 10.5772/53168.

[135] Nigam, P. and Singh, D. (1995). Enzyme and microbial systems involved in starch processing. *Enz Microbial Technol., 17*: 770-778.

[136] Crabb, W. and Mitchinson, C. (1997). Enzymes involved in the processing of starch to sugars. *Trends Biotechnol., 15*: 349-352.

[137] Crabb, W. D. and Shetty, J. K. (1999). Commodity scale production of sugars from starches. *Curr Opin Microbiol, 2*: 252-256.

[138] Leveque, E., Janecek, S., Haye, B., et al. (2000). Thermophilic archaeal amylolytic enzymes. *Enz Microbial Technol., 26*: 3-14.

[139] Tonkova, A. (2006). Microbial starch converting enzymes of the α-Amylase family. In: *Microbial Biotechnology in Horticulture*. Ray RC (ed.). Science Publishers, Enfield, New Hampshire, USA, pp 421-472.

[140] Mitchinson, C., Requadt, C., Ropp, T., et al. (1997). Bacillus α-amylase mutant recombinant production, improved low pH starch liquefaction, thermal stability and activity and use as laundry detergent or dishwashing detergants EU Patent WO 9639528 *Chem Abstr., 126*: 105762.

[141] Jorgensen, S., Vorgias, C. and Antranikian, G. (1997). Cloning, sequencing, characterization, and expression of an extracellular α-amylase from the

hyperthermophilic archaeon *Pyrococcus furiosus* in *Escherichia coli* and *Bacillus subtilis*. *J Biol Chem.*, *272*: 16335-16342.
[142] Ghollasi, M., Khajeh, K., Naderi-Manesh, H., et al. (2010). Engineering of a Bacillus α-amylase with improved thermostability and calcium independency. *Appl Biochem Biotechnol.*, *162*: 444-459.
[143] Bhosale, S. H., Rao, M. B. and Deshpande, V. V. (1996). Molecular and industrial aspects of glucose isomerase. *Microbiol Rev.*, *60*: 280-300.
[144] Parker, K., Salas, M. and Nwosu, C. V. (2010). High fructose corn syrup: Production, uses and public health concerns. *Biotechnol Mol Biol Rev.*, *5*: 71-78.
[145] Tharanathan, R. N. (2005). Starch-value addition by modification. *Crit Rev Food Sci Nutr.*, *45*: 371-384
[146] Mitidieri, S., Martinelli, S., Schrank, A., et al. (2006). Enzymatic detergent formulation containing amylase from *Aspergillus niger*: a comparative study with commercial detergent formulations. *Bioresour Technol.*, *97*: 1217-1224.
[147] Olsen, H. and Falholt, P. (1998). The role of enzymes in modern detergency. *J Surfactants Deterg.*, *1*: 555-567.
[148] Nonaka, T., Fujihashi, M., Kita, A., et al. (2003). Crystal structure of calcium free α-amylase from *Bacillus sp.* strain KSM-K38 (amyK38) and its sodium ion binding sites. *J Biol Chem.*, *278*: 24828-24824.
[149] Araujo, R., Casal, M. and Cavaco-Paulo, A. (2008). Application of enzymes for textile fibres Processing. *Biocatal Biotransfor.*, *26*: 332-349.
[150] Feitkenhauer, H., Fischer, D. and Faeh, D. (2003). Microbial desizing using starch as model compound: enzyme properties and desizing efficiency. *Biotechnol Prog.*, *19*: 874-879.
[151] Ahlawat, S., Dhiman, S., Battan, B., et al. (2009). Pectinase production by *Bacillus subtilis* and its potential application in biopreparation of cotton and micropoly fabric. *Process Biochem.*, *44*: 521-526.
[152] Rattanachomsri, U., Tanapongpipat, S., Eurwilaichitr. L., et al. (2009). Simultaneous non-thermal saccharification of cassava pulp by multi-enzyme activity and ethanol fermentation by *Candida tropicalis*. *J Biosci Bioeng.*, *107*: 488-493
[153] Pan, S. K., Wu, S. J. and Kim, J. M. (2011). Preparation of glucosamine by hydrolysis of chitosan with commercial α-amylase and glucoamylase. *J. Zhejiang Univ-Sci., B, Biomed Biotechnol.*, *12*: 931-934
[154] Collares, R., Miklasevicius, L., Bassaco, M., et al. (2012). Optimization of enzymatic hydrolysis of cassava to obtain fermentable sugars. *J Zhejiang Univ-Sci., B, Biomed Biotechnol.*, *13*: 579-586.
[155] Shetty, J., Chotani, G., Gang, D., et al. (2007). Cassava as an alternative-feed stocks in the production of renewable transportation fuel. *Intr Sugar J.*, *109*: 3-11
[156] Sánchez, O. J. and Cardona, C. A. (2008). Trends in biotechnological production of fuel ethanol from different feedstocks. *Biores Technol.*, *99*: 5270-5295
[157] Gray, K. and Zhao, L., Emptage, M. (2006). Bioethanol. *Curr Opin Chem Biol.*, *10*: 141-146.
[158] Lareo, C., Ferrari, M., Guigou, M., et al. (2013). *Evaluation of sweet potato for fuel bioethanol production: hydrolysis and fermentation*. SpringerPlus, doi:10.1186/2193-1801-2-493.

[159] Sarikaya, E., Higasa, T., Motoyasu, A., et al. (2000). Comparison of degradation abilities of α- and β-amylases on raw starch granules. *Process Biochem., 35*: 711-715.
[160] Szymanowska, P., Lewandowicz, G., Błaszczak, W., et al. (2012). Structural changes of corn starch during fuel ethanol production from corn flour. *J Biot Computational Biol Bionanotech., 93*: 333-341.
[161] Svihus, B., Uhlen, A. and Harstad, O. (2005). Effect of starch granule structure, associated components and processing on nutritive value of cereal starch: A review. *Anim Feed Sci Tech., 122*: 303-320.
[162] Kong, B., Kim, J., Kim, M., et al. (2003). Porcine pancreatic α-amylase hydrolysis of native starch granules as a function of granule surface area. *Biotechnol Prog., 19*: 1162-1166.
[163] Dona, A., Pages, G., Gilbert, R., et al. (2010). Digestion of starch: In vivo and in vitro kinetic models used to characterize oligosaccharide or glucose release. *Carbohydr Polym., 80*: 599-617.
[164] Shigechi, H., Koh, J., Fujita, Y., et al. (2004). Direct production of ethanol from raw corn starch via fermentation by use of a novel surface-engineered yeast strain codisplaying glucoamylase and α-amylase. *Appl Env Microbiol., 70*: 5037-5040.
[165] Prakash, O. and Jaiswal, N. (2010). α-Amylase: an ideal representative of thermostable enzymes." *Appl Biochem Biotechnol., 160*: 2401-2414.
[166] Shiau, R., Hung, H. and Jeang, C. (2003). Improving the thermostability of raw-starch-digesting amylase from a *Cytophaga sp.* by site-directed mutagenesis. *Appl Env Microbiol., 69*: 2383-2385.
[167] Linden, A. and Wilmanns, M. (2004). Adaptation of class-13 α-amylases to diverse living conditions. *Chembiochem., 5*: 231-239.
[168] Alikhajeh, J., Khajeh, K., Ranjbar, B., et al. (2010). Structure of *Bacillus amyloliquefaciens* α-amylase at high resolution: implications for thermal stability. *Acta Crystallogr F., 66*: 121-129.
[169] Birol, G., Onsan, Z., Kirdar, B., et al. (1998). Ethanol production and fermentation characteristics of recombinant *Saccharomyces cerevisiae* strains grown on starch. *Enz Microb Technol., 22*: 672-677
[170] Eksteen, J., Van Rensburg, P., Otero, R., et al. (2003). Starch fermentation by recombinant *Saccharomyces cerevisiae* strains expressing the α-amylase and glucoamylase genes from *Lipomyces kononenkoae* and *Saccharomycopsis fibuligera. Biotechnol Bioeng., 84*: 639-646.
[171] Shigechi, H., Fujita, Y., Koh, J., et al. (2000). Energy-saving direct ethanol production from low-temperature-cooked corn starch using a cell-surface engineered yeast strain co-displaying glucoamylase and α-amylase. *Biochem Eng J., 18*: 149-153.
[172] Shigechi, H., Uyama, F., Matsumoto, T., et al. (2002). Efficient ethanol production from starch through development of novel flocculent yeast strains displaying glucoamylase and co-displaying or secreting α-amylase. *J Mol Cat B, 17*: 179-187.
[173] Dale, B. and Kim, S. (2009). Corn amylase: ımproving the efficiency and environmental footprint of corn to ethanol through plant biotechnology. *AgBioForum, 12*: 149-154.
[174] Wang, M. (2005). *Energy and greenhouse gas emissions impacts of fuel ethanol.* http://www.transportation.anl.gov/pdfs/TA/349.pdf.
[175] Zamora, L., Calderón, J., Vázquez, E., et al. (2010). Optimization of ethanol production process from cassava starch by surface response. *J Mex Chem Soc., 54*: 198-203.

[176] Chitradon, L., Poonpairoj, P., Mahakhan, P., et al. (1996*a*). Pectinases from *Rhizopus* sp. efficient in enhancing the hydrolysis of raw cassava starch: Purification and characterization. In: *Pectins and Pectinases*. Visser J, Voragen AGJ (eds.). *Proceedings of an International Symposium*. Elsevier, Amsterdam, pp. 715-722.

[177] Chitradon, L., Mahakhan, P., Poonpairoj, P., et al. (1996*b*). Production of pectinases from *Rhizopus* sp. in solid substrates. In: *Pectins and Pectinases*. Visser J, Voragen AGJ (eds.) *Proceedings of an International Symposium*. Elsevier, Amsterdam, pp. 853-860.

[178] Paolucci-Jeanjean, D., Belleville, M., Zakhia, N., et al. (2000). Kinetics of cassava starch hydrolysis with Termamyl enzyme. *Biotechnol Bioeng., 68*: 71-77.

[179] Sriroth, K., Chollakup, R., Chotineeranat, S., et al. (2000). Processing of cassava waste for improved biomass utilization. *Bioresour Technol., 71*: 63-69.

[180] Lareo, C., Ferrari, M., Guigou, M., et al. (2013). *Evaluation of sweet potato for fuel bioethanol production: hydrolysis and fermentation.* SpringerPlus, doi:10.1186/2193-1801-2-493.

[181] Duvernay, W., Chinn, M. and Yencho, G. (2013). Hydrolysis and fermentation of sweet potatoes for production of fermentable sugars and ethanol. *Ind Crop Prod., 42*: 527-537.

[182] Lee, W., Chen, I., Chang, C., et al. (2012). Bioethanol production from sweet potato by co-immobilization of saccharolytic molds and *Saccharomyces cerevisiae. Renew Energy, 39*: 216-222.

[183] Awg-Adeni, D., Bujang, K., Hassan, M., et al. (2013). Recovery of glucose from residual starch of Sago hampas for bioethanol production. *BioMed Res Intr., 2013*: 1-8.

[184] Awg-Adeni, D., Abd-Aziz, S., Bujang, K., et al. (2010). Bioconversion of sago residue into value added products. *Afr J Biotechnol., 9*: 2016-2021.

[185] Karim, A., Tie, A., Manan, D., et al. (2008). Starch from the sago (Metroxylon sagu) palm tree—properties, prospects, and challenges as a new industrial source for food and other uses. *Compr Rev Food Sci F, 7*: 215-228.

[186] Bandaru, V., Somalanka, S., Mendu, D., et al. (2006). Optimization of fermentation conditions for the production of ethanol from sago starch by coimmobilized amyloglucosidase and cells of Zymomonas mobilis using response surface methodology. *Enz Microbial Technol., 38*: 209-214.

[187] Singh, D. and Trivedi, R. (2013). Production of biofuel from algae: An economic and eco-friendly resource. *Intr J Sci Res., 2*: 2319-7064.

[188] Um, B., Karim, M. and Henk, L. (2003). Effect of sulfuric and phosphoric acid pretreatments on enzymatic hydrolysis of corn stover. *Appl Biochem Biotechnol., 105*: 115-125.

[189] Khan, S., Rashmi, A., Hussain, M., et al. (2009). Prospects of biodiesel production from microalgae in India. *Renew Sustainable Energy Rev., 13*: 2361-2372.

[190] http://www.bakeinfo.co.nz/Facts/Bread-making/Bread-production-in-New-Zealand.

[191] Novozymes A/S: *The Novozymes Report 2011*. http://report2011.novozymes.com/Service/Download+report/The-Novozymes-Report-2011.pdf.

[192] Hamer, R. J. (1995). Enzymes in the baking industry. In: *Enzymes in food processing*. Tucker GA, Woods LFJ, (eds.). Blackie Academic and Professional, Galsgow, pp. 190-222.

[193] Kulp, K. (1993). Enzymes as dough improvers. In: *Advances in baking technology*. Kamel BS, Stauffer CE (eds.). Blackie Academic and Professional, VCH Publishers, New York pp. 152-178.

[194] Prieto, J., Bort, B., Martinez, J., et al. (1995). Purification and characterization of a new α-amylase of intermediate thermal stability from the yeast *Lipomyces kononen-koae*. *Biochem Cell Biol., 73*: 41-49.

[195] Poutanen, K. (1997). Enzymes: An important tool in the improvement of the quality of cereal foods. *Trends Food Sci Technol., 8*: 300-306.

[196] Goesaert, H., Slade, L., Levine, H., et al. (2009). Amylases and bread firming-an integrated view. *J Cereal Sci., 50*: 345-352.

[197] Cole, M. S. (1982). *Antistaling baking composition*. US Patent, US4320151.

[198] Martin, M. L. and Hoseney, R. C. (1991). A mechanism of bread firming II role of starch hydrolyzing enzymes. *Cereal Chem., 68*: 503-507.

[199] Duran, E., Leon, A., Barber, B., et al. (2001). Effect of low molecular weight dextrins on gelatinization and retrogradation of starch. *Eur Food Res Technol., 212*: 203-207.

[200] Takasaki, Y. (1985). An amylase producing maltotriose from *Bacillus subtilis*. *Agric Biol Chem., 49*: 1091-1097.

In: Microbial Catalysts. Volume 1
Editors: S. M. Abdel-Aziz, N. Garg, A. Aeron et al.
ISBN: 978-1-53614-679-0

Chapter 2

FUNGAL LIPASES PRODUCTION BY SOLID-STATE FERMENTATION: AN OPPORTUNITY FOR INCREASING PROCESS PROFITABILITY

Luis Beltrán Ramos-Sanchez[1,*], Maria Caridad Julián-Ricardo[1], Jesus Córdova[2] and Patrick Fickers[3]

[1]Departamento de Ingeniería Química, Universidad de Camagüey "Ignacio Agramonte Loynaz", Camagüey, Cuba

[2]Departamento de Química, Universidad de Guadalajara, Guadalajara, Jalisco, México

[3]Unité de Biotecnologies et Bioprocédés, Université libre de Bruxelles, Bruxelles, Belgium

ABSTRACT

Lipase is one of the most promising enzymes in the chemical and biopharmaceutical industry. A great number of applications have already been reported including fine chemicals, detergents, biodiesel synthesis and many others. These enzymes are commonly produced by a wide variety of microorganisms such as bacteria, yeast and filamentous fungi. Their production by submerged fermentation (SmF) is at present well established and high titers of these enzymes are reported in some cases. SmF nevertheless has some drawbacks that limit process productivity and the possibility of lowering total production cost down to more competitive levels. These include: managing high amounts of water that increase cost of purification, limitations for increasing cell density in the broth because of oxygen transfer rate, increased deleterious action of proteinase. Solid-state fermentation (SSF) offers the possibility of overcoming most of these problems. Oxygen transfer limitations are solved due to a much-enlarged surface for mass transfer. SSF is characterized by the use of very limited water in the growth medium and hence contributes

* Corresponding Author's Email: luis.ramos@reduc.edu.cu.

solving the associated problems of this factor. Additionally, it requires less care for asepsis control. In addition to these advantages, it is recognized that some microorganisms such as filamentous fungi and yeasts behave more efficiently during SSF and show greater productivities when compared to SmF. Although filamentous fungi adequately growth in SSF, there are some limitations for cultivating them in SSF. These limitations arise during the process of heat evacuation. One of the best ways of evacuating heat released by metabolic activity is by the agitation of the solid medium. This action however is harmful to hyphae and means a great compromise for process viability. This problem is not encountered during the cultivation of yeasts in SSF, so offering a good opportunity for producing lipase with enhanced possibilities of competing in lipase market. In this chapter, a review is made about lipase production by fungi in SSF. This includes the analysis of most used strains, characterization of growth media, and ambient factors such as moisture, pH and temperature. SSF bioreactor designs are also reviewed and some results are summarized. The limited existence of industrial applications of enzyme production by SSF are discussed in regards to difficulties encountered in the scale-up of this process and general approaches for developing SSF technologies are discussed.

Keywords: enzyme, lipase, solid-state fermentation, biocatalysis, bioreactor

1. General Considerations on Lipases

Lipases are serine hydrolases defined as triacylglycerol acylhydrolases (E.C. 3.1.1.3). They catalyze in nature the hydrolysis of the ester bonds of tri-, di- and mono- glycerides into fatty acids and glycerol. They were first isolated in 1901 from bacteria such as *Bacillus prodigiosus*, *B. pyocyanus* and *B. fluorescent* (nowadays denominated *Serratias marcescens*, *Pseudomonas aeruginosa* and *Pseudomonas fluorescent*, respectively) [1]. Since then, lipases were reported in many microorganisms including bacteria, yeasts and molds but also in plants and higher eukaryotes. The biochemical characterization of lipases has been initiated in the late 1950s by the group of Desnuelle in Marseille (France) with their works on the pig pancreatic lipase. Lipases, *sensus stricto*, differ from the closely related esterases (E. C. 3.1.1.1) due to their ability to hydrolyze triglycerides at a lipid-water interface [2]. This activation phenomenon, known as "interfacial activation", was first proposed to distinguish lipases from esterases [3,4]. However, this criterion was not sufficient in many cases to characterize the behavior of several lipases [3]. Therefore, lipases were redefined as long chain fatty acid triglyceride hydrolases in apposition of esterases that are able to hydrolyze only short chain (i.e., less than 10 carbon atoms) fatty acid triglycerides.

The determination of the 3-D structure of lipases highlighted that they belong to the structural super family of α/β-hydrolases whose activities rely mainly on a catalytic triad usually formed by Ser, His and Asp residues [5]. The serine residue usually appears embedded in a conserved pentapeptide Gly-X-Ser-X-Gly that forms a characteristic β–turn-α motif named the nucleophilic elbow. In addition, many lipases show a "lid" controlling the access to the active site. This lid is composed by a α-helix connected to the core of the enzyme by a flexible structure. This mobile element allows, by conformational changes, the substrate to get access to the active site in the presence of a lipid-water interface. This mechanism, responsible of the "interfacial activation" phenomenon, allows the enzyme to fold into an active (open) form [6]. Substrate hydrolysis is accomplished through the formation of two tetrahedral intermediates, which are stabilized by interactions with amino acids composing the so-called oxyanion hole [7]. Beside their hydrolytic activities, lipases are able to catalyze in favorable thermodynamic conditions (for instance low thermodynamic water activity) reactions of synthesis such as esterification or amidation [8]. Lipases also exhibit chemo-selective, regioselective and stereoselective properties [9]. The catalytic properties of lipases have made them increasingly popular in agri-food, chemical and pharmaceutical industries [10].

2. Regulation of Lipase Production in Regard to the Medium Composition

Many microorganisms are known as potent producers of extracellular lipases including bacteria, yeasts and filamentous fungi [11]. Fungi are usually cultured in solid-state fermentation (SSF) while bacteria and yeasts are preferably cultured in submerged fermentation (SmF) [12]. Lipase synthesis yields vary according the strain, the composition of the growth medium (solid support), culture medium (carbon and nitrogen sources) and cultivation conditions (pH, temperature, moisture). A comparison of lipase production yield between SmF and SSF is difficult due to the methods used for the quantification of the lipase activity. However, such comparisons have been reported in the literature. For example, lipase activities of 17 U g^{-1} and 12 U ml^{-1} were obtained for *Penicillium restrictum* in SSF and SmF, respectively [13], whereas lipase production obtained with *Rhizopus homothallicus* in SSF and SmF were reported respectively equal to 1500 U g^{-1} and 50 U ml^{-1} [14]. A similar comparison was also reported for the extracellular lipase lip2pproduced by the non-conventional yeast *Yarrowia lipolytica* by SSF based on barley bran-triturated nut (23 U g^{-1}) [15] and in SmF after optimization of the culture production (3000 U ml^{-1}) [16].

Some of the main important lipases from filamentous fungi produced in SSF belong to the genera *Aspergillus*, *Rhizopus* and *Penicillium* while those from yeast origin belong to *Candida* and *Yarrowia* (see Table 1). Most of the lipases produced by yeast and filamentous fungi are extracellular and their production is modulated by the culture conditions. Expression of lipase encoding genes is regulated by the nutrients from the culture medium and/or by the solid substrate used in SSF. Lipidic carbon sources are essential for high lipase production yields. Oleic acid (cis-9-Octadecenoic acid) was reported as the most suitable inducer for the production of the main extracellular Lip2p lipase in *Yarrowia lipolytica* [17,18]. However, pure oleic acid is not convenient for the development of industrial processes owing to its excessive cost. With the aim to develop a cost-friendly process, different oleic acid esters and triglycerides have been tested for their ability to support lipase synthesis.

Olive oil, mainly constituted by triolein, the main triglyceride, and methyl-oleate is produce a lipase activity between 1130 and 1190 Uml^{-1} [61]. These results are in the range of that obtained with pure oleic acid (i.e., 1045 Uml^{-1}). By contrast, ethyl-oleate led to a significant decrease of lipase production (195 U ml^{-1}) [61]. Expression of *ylLIP2* lipase-encoding gene, and thus Lip2p synthesis, appeared also to be markedly influenced by oleic acid (i.e., oleate) concentration [62]. A significant increase of *ylLIP2* expression was observed for 1% of oleic acid compared to 0.5%. However, no further significant increase was observed in the presence

Table 1. Strain used in solid state fermentation for lipase production. Lipase activity, unless stated otherwise, is expressed in unit per gram of dry solid support

Strain	Solid support/substrate	Lipase activity	Reference
Yeasts			
Candida rugosa	Coconut oil cake	88 U g^{-1}	[19]
	Sesame oil cake	32 U ml^{-1}	[20]
	Groundnut oil cake	57 U ml^{-1}	[20]
	Rice bran	26 U g^{-1}	[21]
	Rice Bran	37 U g^{-1}	[22]
Candida utilis	Olive oil cake	25 U g^{-1}	[23]
Yarrowia lipolitica	Barley bran	23 U ml^{-1}	[24]
	Sugarcane bagasse/wheat bran	9 U g^{-1}	[25]
	Palm Kernal cake	19 U g^{-1}	[26]
	Mahua cake	9 U ml^{-1}	[27]
Colletotrichum gloeosporioides	Pongamia oil cake	983 U g^{-1}	[28]
Endomicopsis burtonii			
Molds			
Aspergillus flavus	Rice husks/rubber wood dust	7 U g^{-1}	[29]
	Wheat bran/castor oil cake	103 U g^{-1}	[30]
A. niger	Wheat bran	650 U g^{-1}	[31]
	Wheat bran/gingely oil cake	384 U g^{-1}	[32]
	Wheat bran	9 U g^{-1}	[33]
	Wheat bran/oleic acid	49 U g^{-1}	[34]
A. versicolor	Jatropha seed cake	1288 U g^{-1}	[35]
A. fumigates	Defatted rice bran	9 U ml^{-1}	[36]
A. terreus	Mustard oil cake	1363 U g^{-1}	[37]
A. tamari	Wheat bran/gingelly oil cake	998 U g^{-1}	[38]
A. carneu	Sunflower	13 U g^{-1}	[39]
Rhizopus homothalicus	Sugarcane bagasse/olive cake	50 U g^{-1}	[40]
	Sugarcane bagasse/olive cake	43 U ml^{-1}	[41]
R. oryzae	Sugarcane bagasse	215 U g^{-1}	[42]
R. rhizopodiformis	Sugarcane bagasse/olive cake	15 U g^{-1}	[40]
R. oligosporus	Almond meal	48 U g^{-1}	[43]
	Sugarcane bagasse	826 U g^{-1}	[44]
	Almond meal	62 U g^{-1}	[45]
	Pisum sativum seed	4 U g^{-1}	]46]
R. chicness	Wheat flour/wheat bran	24 U g^{-1}	[47]
R. delemar	Polymeric resin/dextrin	96 U g^{-1}	[48]
	Wheat bran	2 U g^{-1}	[49]
R. pusillus	Sugarcane bagasse/olive cake	21 U ml^{-1}	[50]
Penicillium chrysogenum	Wheat bran/castor oil cake	20 U g^{-1}	[51]
	Wheat bran	14 U g^{-1}	[49]
Penicillium candidum	Wheat bran	480 U ml^{-1}	[52]
P. brevicompactum	Babassu oil cake	49 U g^{-1}	[53]
	Castor meal	88 U g^{-1}	[53]
P. simplicissimum	Babassu oil cake	26 U g^{-1}	[54]
P. verrucosum	Soybean bran	40 U g^{-1}	[55]
P. restrictum	Babassu oil cake	30 U g^{-1}	[56]
	Babassu oil cake	28 U g^{-1}	[57]
P. camenberti	Wheat bran	8 U ml^{-1}	[58]
P. roqueforti	Wheat bran	8 U g^{-1}	[49]
Pennicillium sp	Soybean meal	140 U g^{-1}	[59]
Rhizomucor pusillus	Sugarcane bagasse	5 U g^{-1}	[50]
	Sugarcane bagasse/olive cake	19 U g^{-1}	[49]

Strain	Solid support/substrate	Lipase activity	Reference
Rhizomucor rhizopodiformis	Sugarcane bagasse	3 U g^{-1}	[50]
Mucor sp	Wheat bran	25 U g^{-1}	[49]
Geotrichum candidum	Wheat bran	1 U g^{-1}	[49]
Trichoderma harzianum	Wheat bran/olive oil	60 U g^{-1}	[51]
Trichoderma sp	Rice bran	0.1 U g^{-1}	[60]

of 3% of oleic acid [62]. By contrast, carbon sources such as glucose and glycerol (one of the end product of triglyceride hydrolysis) were found to repress lipase production [63]. Nitrogen sources used to supplement the culture medium also greatly influence the lipase production. Various mineral and organic nitrogen sources were tested for their capacity to support cell growth and lipase production [62]. Mineral nitrogen sources such as ammonium chloride or ammonium sulfate had no influence on either cell growth or lipase production. By contrast, a significant increase in lipase productivity was observed upon addition of organic nitrogen sources such as urea or casein hydrolysates. For instance, tryptone allowed a 166-fold increased lipase production compared to the non-supplemented medium. However, cultures in the presence of casamino acid, which result from a complete hydrolysis of casein, led to a low lipolytic productivity. These results suggest that specific peptides constituting tryptone could regulate lipase production [62]. This hypothesis was confirmed by Turki et al. [64] by testing separately different tryptone constituting peptides [64].

Lipase production in *Candida rugosa* is also modulated in yield (enzymatic activity) and quality (i.e., isozyme synthesis) by the culture conditions and medium composition. Glucose and glycerol were found to repress the lipase synthesis [65,66]. By contrast, oleic acid, ester of oleic acid and olive oil were found to enhance lipase synthesis [67]. A low concentration, oleic acid seemed to favor the synthesis and secretion of Lip2 and Lip3 lipase isozymes while Lip1 is only produced at high oleic acid concentration [68]. Lipase production was also reported for carbon sources such as sterols [69], hexadecane [69], carboxylic acids [67], vegetal oils [70] and dodecanol. Tween and Span were also reported to favor lipase production [71,72]. By contrast, no lipase production could be detected in the presence of alkane, despite this carbon source could support cell growth [66]. Corn steep liquor [73] was reported has a suitable nitrogen source for lipase production in *C. rugosa* together with urea [19] and peptone [74].

There are several other reports dealing with extracellular lipase production by fungi such as *Endomicopsis burtonii, Saccharomyces cerevisiae, Schwanniomyces castelli* [75,76]. Also, lipases were produced by a microbial consortium derived from a mixture of wastewater sludge in a medium containing solid industrial wastes rich in fats [77].

Considering lipase produced by *Aspergillus niger,* Tween 80, soybean meal and ammonium sulfate were found to be the most suited inductors and carbon and nitrogen sources for lipase synthesis [78]. Urea was also reported as a convenient nitrogen source for lipase production in SSF using wheat bran as solid substrate [79]. Lipase production is also modulated in SSF by the solid support/medium used. Kamini et al. [80] tested twelve solid substrates regarding lipase production in *Aspergillus niger*. The highest lipase production was obtained with wheat bran (198 Ug^{-1}) and gingelly oil cake (169 Ug^{-1}). They also observed that addition of various nitrogen sources, carbohydrates and inducers to the substrate did not yielded to any significant increase of lipase synthesis. This demonstrates that in some cases, the solid substrate used is sufficient to support lipase production at high yield. Beside this, these authors tried to substitute olive oil by corn oil refining byproducts (namely soap stock, stearin and fatty acid) in order to obtain a cost-friendly process. Among them, soap stock lead to a lipase production of 63 U g^{-1} whereas fatty acids presented an inhibitor effect (4 U g^{-1}) [80]. Lipase production in *Aspergillus carneus* was optimized by a response surface approach using a one-at-a-time method [81]. The highest lipase productions were obtained in a medium composed of sunflower oil (1%), glucose (0.8%) and peptone (0.8%). Similarly, Yadav et al. [82] optimized the production of lipase by *A. terrus*. The highest lipases production was obtained in a medium containing corn oil (2%) and casein hydrolysate (0.1%). They also reported that the presence of calcium and magnesium ions enhanced lipase secretion [82]. Beside this, Ohnishi et al. [83] obtained the highest lipase production in *A. oryzae* in a medium containing sunflower oil (3%), glucose (2%), yeast extract (1%) and polypeptone (2%) [83].

Similarly to *Aspergillus* lipases, biosynthesis of lipases from *Rhizopus arrhizus* is inhibited by glucose whereas a significant increase (2.5 fold) was obtained in the presence of corn oil used as inducer [84]. Khayati and Kiyani [85] reported a statistical approach for optimization of lipase production by *Rhizopus oryzae* using rice straw as solid substrate. A Plackett-Burman design was used to evaluate the effect of olive oil, rapeseed oil, palm oil and peanut oil as inducers together with soybean meal, corn steep liquor, urea and ammonium sulfate as nitrogen sources. In a second step, optimal concentration of these compounds was determined by central composite designs and response surface analysis. They found out that olive oil (1.38%) and soybean meal (1.27%) were the optimal inducer and nitrogen source, respectively. In these conditions, the lipase production reached 59 U g^{-1} [85].

Lipase production by *Penicillium restrictum* in SSF on solid waste from the babassu oil industry supplemented with peptone, olive oil or starch at different C:N ratios was investigated by Gombert et al. [56]. The highest lipase production was obtained with1% of peptone enrichment (C/N ratio of 11.7, 28 Ug^{-1}) and with 2% olive oil enrichment (C/N ratio of 14.1, 30 Ug^{-1}). These observations were in agreement with the work of Ohnishi et al. [83] reporting that a high nitrogen content enhance production yield [83]. Lipase production by *Penicillium brevicompactum* in SSF using babassu cake and castor meal was also investigated [53]. For medium optimization, these authors used a four fractional factorials design to test the effect of molasses, urea and soybean oil concentrations as well as moisture content. For babassu cake, the highest lipase activity was obtained for a moisture content of 70% and 2% soybean oil. This result was in accordance with those obtained for *Burkholderia* sp. using olive oil [86]. Similar results were obtained when castor meal was used as solid substrate. Beside this, lipase production by *Penicillium citrinum* was investigated on medium containing 1% ground-nut oil refinery residues as carbon source and various nitrogen contents. Ammonium chloride (0.75%) was found to be the best nitrogen source compared to ammonium sulfate and urea. This latter yielded to a decrease of lipase synthesis by 85% [87]. Compared to oleic acid, a two-fold increased lipase activity was obtained in the presence of this oil refinery residue due to its high carbohydrates content (35%) allowing a rapid growth of the fungus. Production of lipase by *Penicillium simplicissium* in SSF using babassu cake as the basal medium was investigated for supplementation with of olive oil, sugarcane molasses, corn steep liquor and yeast extract. Among these, olive oil and corn steep liquor were found the most suited carbon and nitrogen sources for lipase synthesis [88]. Rigo et al. [59] reported that the medium supplementations with urea and soybean oil significantly increased lipase production by *Penicillium* sp grown in SSF with soybean meal as the solid substrate [59].

All these data highlight clearly that production of fungal lipase depends of several factors such as the solid substrate used and the nature of the carbon and nitrogen sources used as supplement. Additionally, the moisture content, pH and temperature should also be taken into account for process optimization.

3. Catalytic Activity of Lipase

3.1. Common Reactions Catalyzed by Yeast Lipase

As stated in section 1, lipases belong to a class of hydrolases that catalyze the hydrolysis of triglycerides to glycerol and free fatty acids at lipid-water interface. Beside this, they are also able to catalyze the "opposite" reactions of esterification and transesterification (Figure 1) in an enantioselective fashion [89]. Esterification is the reaction where a fatty acid is linked to an alcohol by a covalent bond, releasing a water molecule. Thioesterification and amidation are similar reactions but with a thiol or an amine as substrates. Transesterification reactions include alcoholysis, acidolysis, aminolysis and interesterification [9]. The ability and versatility of lipases to perform very specific chemical transformations has make possible their applications in the food, detergent, cosmetic, organic synthesis and pharmaceutical industries [89].

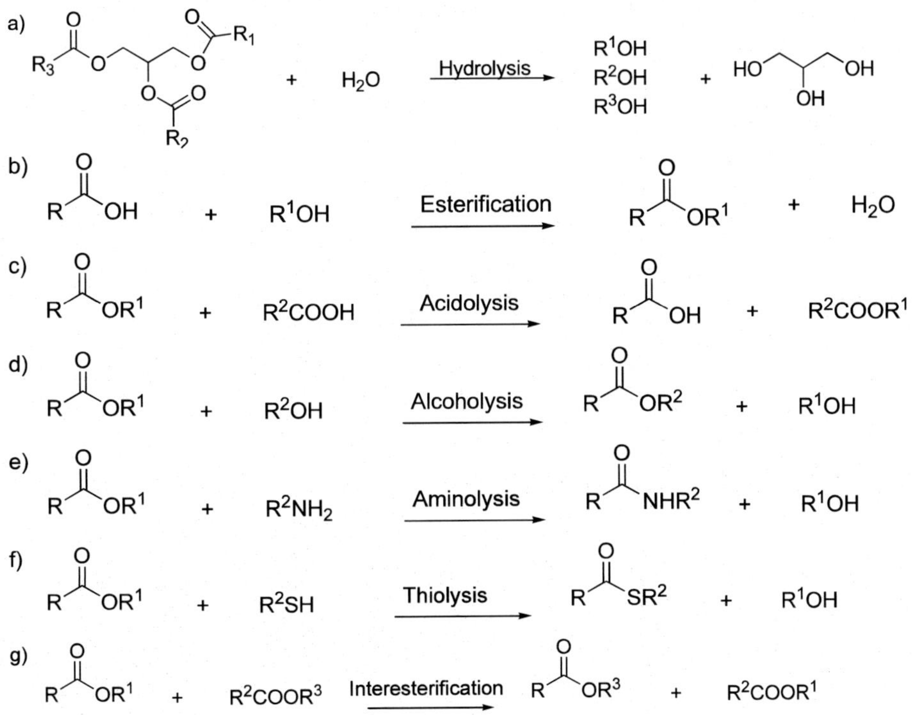

Figure 1. Reactions catalyzed by lipases. a) hydrolysis, b) esterification, c) acidolysis, d) alcoholysis, e) aminolysis, f) thiolysis, g) interesterification.

One of the most used organisms to produce lipases is *Candida antarctica*, which produce two lipases A and B (CAL A and CAL B) with different physicochemical properties. CAL B belongs to the α/β-hydrolase fold superfamily, which are a group of enzymes that have evolved from a common ancestor, catalyzing various types of reactions such as hydrolysis of esters, thioesters, peptides, epoxides and alkyl halides or cleavage of carbon bonds in hydroxynitriles [90]. Successful synthesis of nutraceuticals has been reported by employing CALB immobilized lipase. Lipases are also used in the synthesis of sucralose by regioselective hydrolysis of octaacetylsucrose [91]. The lipases L1 and L3 from *C. rugosa* are adequate for the hydrolysis of methyl esters and steryl esters, respectively [92]. *C. rugosa* lipase is also used for the hydrolysis of ibuprofen esters in aqueous media, where pH has as important role, modifying the substrate selectivity of CRL [93].

C. cylindracea lipase immobilized on Amberlite XAD7 was used to hydrolyze racemic naproxen ethoxyethyl ester in a packed-bed continuous bioreactor, obtaining an enantiomeric excess higher than 95% and overall conversion of 25% at steady-state operation. The bioreactor was operated continuously for 1200 h at 35°C and pH 6.5 with a 20% loss of enzyme activity. On the other hand, *C. rugosa* lipase immobilized on YWG-C6H5 was used to hydrolyze racemic naproxen methyl ester in a trapped aqueous-organic solvent biphasic system, obtaining an enantiomeric excess (*S*-naproxen) higher than 90% and overall conversion of 30% at steady-state operation. The bioreactor was operated continuously for 1440 h at 30°C [94]. *C. rugosa* lipase catalyzed the enantioselective esterification of *S*-naproxen hydroxyalkyl ester in isooctane and cyclohexane, and a high enantiomeric ratio of near 100% was obtained with 1,4-butanediol as the best acyl acceptor [95]. Efficient kinetic resolution processes, using *C. rugosa* lipase [96] and CAL B [97], are performed for the synthesis of Niknomycin-B, non- steroid anti-inflammatory drugs, naproxen, ibuprofen, suprofen and ketoprofen, the antiviral agent lamividine, the antitumor agents alkaloids, antibiotics and vitamins.

3.1.1. Esterification Reactions

C. antarctica lipases were immobilized on Celite 545 and used for the synthesis of monoacylglycerols, obtaining 100% esterification of oleic acid with isopropylidene glycerol [98]. CAL A performed the asymmetric alkoxycarbonylation of different vinyl carbonates with (*R*,*S*)-1-phenyl-ethyl-amine with high yields and enantioselectivities [66]. The solvent-free thioesterification of fatty acids with long-chain thiols and the trans-thioesterification of fatty acid methyl esters with alkane thiols were catalyzed by CAL-B [99,100].

3.1.2. Transesterification Reactions

Lipases from *C. rugosa*, *C. antarctica* and *C. cylindracea* are suitable for transesterification reactions, displaying wide substrates specificity and regiospecificity [101]. Novozym 435 (commercial CAL-B immobilized on acrylic resin) catalyzed the biodiesel synthesis with a yield higher than 90%, using vegetable oil or waste cooking oil as feedstock. Other commercial lipase immobilized on textile membrane (*Candida* sp. 99-125 lipase), catalyzed the biodiesel synthesis with a yield higher than 87%, using lard, waste oil and vegetable oils [102]. CAL-B (immobilized on activated carbon) and *C. rugosa* lipase (immobilized in acurel) were reported for the transesterification of palm oil with isobutanol (100% of conversion) and with ethanol (89% of conversion) [103,104].

Several capric acid rich structured lipids were synthesized using rice bran, groundnut and mustard oils. Acidolysis reaction between oil and capric acid was carried out in 1:3 molar ratios at 45°C, using CAL-B. The incorporation of capric acid was influenced by the nature of the fatty acids present in the original oil. Capric acid mainly replaced fatty acids in Sn-1 and Sn-3 positions of the triglycerides [105].

C. parapsilosis lipase catalyzed the fatty hydroxamic acid biosynthesis in a biphasic liquid/aqueous medium. The substrates of the reaction were acyl donors (fatty acid or fatty acid methyl ester) and a hydroxylamine [106].

Novozym 435 catalyzed the alcoholysis of cottonseed oil with primary and secondary alcohols in a solvent-free medium. Optimal conditions for methanolysis (yield of 91.5%) were: 30% enzyme based on oil weight; oil/alcohol molar ratio 1:4; 50°C and 7 h of reaction time. At the same conditions, cotton seed oil was converted with short-chain primary and secondary alcohols to its corresponding esters with conversions between 72% and 94% [107]. CAL-B catalyzed hydrolysis and alcoholysis of vitamins A and E ester bonds. These reactions were used to determine their level in different food formulae [91] CAL-B catalyzed the alcoholysis, using ionic liquids (1-butyl-3-methylimidazolium tetrafluoroborate or hexafluorophosphate) as reaction media. Reaction rates were comparable or better than those observed in organic media [108].

The use of enantiopure intermediates for drug synthesis is a trend in the pharmaceutical industry, since different physiological effects are associated with the enantiomers of chiral molecules. With lipase catalysis, enantiopure secondary alcohols, used as intermediates in the synthesis of Dorzolamide (an antiglaucoma drug) were obtained. Enantiopure β-hydroxy nitriles as potential intermediates for the synthesis of antidepressant drugs with 1-aryl-3-methylaminopropan-1-ol structures were also obtained with lipases. Kinetic resolution of racemates was the main biocatalytic approach applied. CAL-B catalyzed the acylation of alcohols and the alcoholysis of their esters in organic solvents, such as in diisopropyl ether and tert-butyl methyl ether. Also, CAL-B was used for the acylation of ethyl 3-hydroxybutanoate under solvent free conditions [109].

The application of lipases to interesterification processes for the production of lower value oils and fats, such as margarine and cooking oils, has been investigated. When *C. rugosa* lipase and CAL-B were used for interesterification of oil blends, the triacylglycerides produced were similar to those obtained by chemical interesterification. Therefore, replacement of chemical interesterification by enzymatic process is technically feasible, although it has not been adopted on a commercial scale to date because of the comparative costs [110]. A study was carried out to compare the composition and thermal properties of lard (LD) and engkabang fat (EF) - canola oil (CaO) blend interesterified by CAL-B. A fat blend of 40% EF in CaO was interesterified at 60°C and 200 rpm for different reaction times (6, 12 and 24 h). The slip melting points of the fat blend interesterified for 6 hours were similar to that of LD. The solid fat content (SFC) values of fat blends interesterified for 12 and 24 h were equal to those of LD at the temperature range from 0 to 20°C. Similar to LD, all three interesterified blends had their SFC values approaching zero at the temperature range from 30 to 40°C [111].

3.2. Catalytic Promiscuity of Lipases

In the last decade, there have been vigorous efforts to leverage biotechnological toolbox to develop green chemical approaches with biocatalysis slowly replacing chemistry processes [9,112]. Biocatalysis is an efficient and green tool for modern organic synthesis, due to its high selectivity, mild reaction conditions and potential use of inexpensive renewable resources [113].

Nowadays, one of the most outstanding concepts in biocatalysis is an emerging area called enzymatic promiscuity [89], which refers to enzymes catalyzing unexpected reactions, including those working under unusual conditions (such as in organic solvents) and catalyzing more than one type of chemical reaction (the natural as well as one or more alternative reactions), using the same or different, or induced active site [93,114]. Due to the new promiscuous catalytic activities of enzymes with non-natural substrates is providing useful tools in organic synthesis, biocatalyst applications are expanding rapidly [89,115]. In fact, enzyme promiscuity has attracted much attention from chemists and biochemists in recent years [116].

Enzyme promiscuity has been classed in three major classes: substrate promiscuity (enzymes with relaxed or broad substrate specificity), condition promiscuity (enzymatic activity in conditions different from their natural ones, such as anhydrous media, extreme temperature or pH) and catalytic promiscuity (based on the ability of a single enzyme active site to catalyze several chemical transformations) [93]. Because enzymatic promiscuity concept involves the possibility that one active site of an enzyme can catalyze several different chemical transformations, a rational understanding of the mechanistic reasons for this catalytic performance could develop new practical applications by using "well-known" enzymes [89].

It is interesting to remark that promiscuity of enzymes is considered as the starting point for enzyme divergent evolution; thus, low promiscuous activity towards an unnatural substrate might, by a single point mutation, turn the enzyme into a much more proficient catalyst, affording a survival benefit to the organism [117]. The origin of such promiscuity can be found in the evolution of the enzymes from a common ancestor, which could have performed simultaneously different side-reactions [118,124]. This reveals an impressive potential of natural chemistry to adapt existing proteins to new biochemical pathways [118].

Lipases are the most widely used class of enzymes in biotechnology and organic synthesis [120]. This privileged place has been gained due the availability of large number of commercial preparations, their broad specificity and relatively better stability in non-conventional media (as compared to other enzymes) [121]. They are promiscuous with regard to substrates, accepting wide range of acids and esters, and catalyzing esterification, transesterification and kinetic resolution of racemic compounds [91].

Lipases are also the most extensively studied enzymes with respect to their catalytic promiscuity [10,122], which has only recently been exploited for synthetic applications [94] and large number of promiscuous reactions have been reported [121,122]. These promiscuous reactions include carbon–carbon, carbon–heteroatom, heteroatom–heteroatom bond formations and oxidative processes [10]. Examples of such reactions are: Aldol reactions, Michael additions, Mannich reactions, Markovnikov additions and Henry reactions [116]. With the discovery and development of new of promiscuous reactions, a second renaissance in lipases applications is attested, increasing their already impressive catalytic versatility [89,121]. These new reactions of lipases might provide novel industrial processes [116].

Remarkably, the most used biocatalyst for promiscuous reactions is *Candida antarctica* lipase B (CAL-B). For detailed information in relation to these reactions, three excellent reviews can be consulted [10,89,121].

3.3. Main Industrial Applications of Yeast Lipases

Lipases are essential components in the modern industrial process since they are able to catalyze hydrolysis reaction as well as a large variety of synthesis reactions such as esterification and transesterification, depending on the thermodynamic conditions (mainly water activity) [9]. Remarkably, in the last decade, yeast lipases were also applied in the manufacture of pharmaceuticals, pesticides, biosensors and in waste management [91].

Many species of yeasts have been found to produce lipases, such as: *Candida rugosa, C. tropicalis, C. antarctica, C. cylindracea, C. parapsilopsis, C. deformans, C. curvata, C. valida, Yarrowia lipolytica, Rhodotorula glutinis, R. pilimornae, Pichia bispora, P. mexicana, P. sivicola, P. xylosa, P. burtonii, Saccharomycopsis crataegenesis, Torulaspora globosa and Trichosporon asteroids.* Among the yeasts, Candida species (i.e., *C. antarctica, C. cylindracea, C. lipolytica and C. rugosa*) are clearly the most important lipase producers, used for industrial purposes [89,112,113]. The genes encoding lipases in *Candida sp., Geotrichum sp., Trichosporon sp. and Y. lipolytica* have been cloned and over-expressed [89,123].

C. rugosa lipase is one of the enzymes most frequently used in biotransformation, and has the great advantage of being considered as safe for food applications. The crude enzyme extract contains at least five iso-enzymes, denominated LIP1 to LIP5, that share high sequence homology, but have different catalytic characteristics, substrate specificity and thermal stability [93]. The second most studied yeast lipase is the one produced by *C. antarctica*, particularly, the lipase B (CAL B) [114]. CAL B has several applications, such as synthesis of flavors and fragrance esters, surfactants, biodiesel, waxes, acylated flavonoids and modified glycerides as well as kinetic resolution of racemic esters and amines. Its stability has also been improved using ionic liquids or non-aqueous solvents for enzymatic synthesis [93]. *Y. lipolytica* lipase produces high value compounds, being used for biotransformation of steroids, synthesis of pharmaceutical intermediates and production of fine chemicals. The availability of the genome sequence should lead to advances in genetic manipulation of this yeast [115]. Lipases from yeast are used in many different industrial processes as illustrated in the following sections.

3.3.1. Lipases for Fat and Oleochemical Industry

C. antarctica lipases were used for esterification of free fatty acids in the absence of organic solvent or transesterification of fatty acid methyl esters in hexane with isopropylidene glycerols [116]. Immobilized lipases from *C. antartica, C. cylindracea* and *G. candidum* are used for the esterification of functionalized phenols for synthesis of lipophilic antioxidants [117]. *C. rugosa* lipase is used by Nippon Oils & Fats, a medium-sized company in Tokyo, which seems to be the only firm that operates with an enzyme-based process for the preparation of highly pure unsaturated fatty acids (oleic, linoleic, linolenic, etc.) [118].

3.3.2. Lipases for Detergent Industry

Large amounts, nearly 1000 tons per year, of lipases are used in the detergent industry for removal of oil-based stains [112]. For this purpose, yeast lipases (mainly from *C. cylindracea, Y. lipolytica* and *C. antarctica*) are preferred because they can work at low temperatures and alkaline pH, compared with bacterial and other fungal lipases [120,124]. Miyoshi Yushi, a company in Nagoya, Japan, produces sizable amounts of soap through lipase-based hydrolysis of oils and fats with *C. rugosa* lipase [118]. Enzymatic synthesis of surfactants has been carried out at moderate temperatures (60-80ºC) with excellent regioselectivity. Recently, Unichem International has manufactured isopropyl myristate, isopropyl palmitate and 2-ethyl palmitate (as emollient in personal care products) using *C. cylindracea* lipase [121].

3.3.3. Lipases for Food Processing

Lipases are used for flavor development and improving quality of foods. *C. parapsilosis* and *C. antarctica* lipases were used to synthesize hydroxamic acids (food additive) and short chain flavor thioester, respectively. Additionally, *C. rugosa* lipase was used in the cheese manufacture [91]. The flavor development in concentrated milk and creams by microbial lipases was investigated, finding that each lipase developed a characteristic flavor and *C. rugosa* lipase was rated as the most suitable lipase in this case [10]. Several structural lipids are synthesized by yeast lipases and included in infant formula and nutraceuticals [112,113]. *C. antarctica* lipase is employed as a robust biocatalyst for esterification reactions, obtaining high conversions of flavoring esters of short-chain [122]. *C. cylindracea* lipase is used in bakery products [125].

3.3.4. Synthesis of Compounds and Resolution of Racemic Mixtures

Biocatalytic processes were used to prepare chiral intermediates for pharmaceuticals: (*S*) [1-(acetoxyl)-4-(3-phenyl) butyl] phosphonic acid diethyl ester a key chiral intermediate required for chemical synthesis of BMS-188494 (an anti-cholesterol drug) was prepared by stereoselective acetylation of racemic [1- (hydroxy)-4-(3-phenyl) butyl] phosphonic acid diethyl ester using *G. candidum* lipase [119]. Enzymatic resolution of racemic 2-pentanol and 2-heptanol was performed by CAL B (*S* 2-pentanol is a key chiral intermediate required for synthesis of anti-Alzheimer's drugs) [119]. *C. cylindracea* lipase was applied to the resolution of 2-bromopropionic acids and 2-cloropropionic acids which are starting materials for the synthesis of phenoxy propionic herbicides [126]. *C. cylindracea* and *C. antartica* lipases have been used to resolve the enantiomers of flurbiprofen, naproxen, ibuprofen, suprofen and baclofen [126]. Both enzymes were successfully used to synthesize lobucavir, hepatitis B antiviral and ribavarin antiviral [115]. These lipases are also suitable for transesterification reactions, displaying both wide substrate specificity and regiospecificity [103].

3.3.5. Cosmetics

C. cylindracea lipase is used to synthesize wax esters (esters of fatty acids and fatty alcohols) as additives for personal care products by Croda Universal Ltd. According to the manufacturer, the overall production cost is slightly higher than that of the conventional method, but the cost is justified by the improved quality of the final product [119].

3.3.6. Diagnostic Tool and Medical Applications

C. rugosa lipase was selected to hydrolyze triacylglycerols for clinical lipid analysis and developed as a biosensor [121,126]. *C. rugosa* lipase was also used to synthesize lovastatin, a drug lowing serum cholesterol level. The asymmetric hydrolysis of 3-phenylglycidic acid ester, which is a key intermediate in the synthesis of diltiazem hydrochloride, a coronary vasodilator, was carried out with *S. marcescens* lipase [127].

3.3.7. Use of Lipases in the Production of Biodiesel

Lipases from different sources are able to catalyze biodiesel synthesis, but those from *C. antarctica* and *C. rugosa* are the most used [103]. Simple alkyl ester derivatives of restaurant grease were prepared using immobilized lipases from *T. lanuginosus* and *C. antarctica* [128].

3.3.8. Pulp and Paper Industry

C. rugosa lipase was used by Nippon Paper Industry to hydrolyze 90% of wood triglycerides as a pitch control method [126,129].

3.3.9. Waste, Effluent and Sewage Treatment

Lipases are utilized in activated sludge and other aerobic waste processes, where thin layers of fats must be continuously removed to allow oxygenation. Effective breakdown of solids and the clearing and prevention of fat blockage or filming in waste systems are important in many industrial operations. Examples include: (i) degradation of organic debris (a commercial mixture of lipase, cellulase, protease and amylase is employed for this purpose); (ii) sewage treatment, cleaning of holding tanks, septic tanks, grease traps, etc. *C. rugosa* lipase is a good option to hydrolyze residual fats [124]. The Novozym 435 (Cal-B) lipase was able to convert with high yield into biodiesel, the used olive oil which contained high levels of free fatty acids and water content [103].

4. Lipase Detection and Essays

As with all reactions catalyzed by enzymes, lipase activity measurements are based on different physico-chemical methods by monitoring the disappearance of the substrate or the formation of the hydrolysis products. The main characteristic trait of lipases is the physico-chemical character of the reaction they catalyze. Most lipases are water-soluble enzymes acting on water-insoluble substrates. This implies that the catalysis takes place at a water-lipid interface. The heterogeneous character of this type of catalysis makes it difficult to accurately quantify the lipolytic activity. The quality of the emulsification of the water-insoluble substrates, which is required for the hydrolysis to take place, can drastically influence the lipase activity measurements. Numerous methods have been reported in the literature for the measurement of hydrolytic activity and for the detection of lipases. These could be classified as follows: titrimetry, spectroscopy, chromatography, radioactivity, interfacial tensiometry, turbidimetry, conductrimetry, immunochemistry and microscopy. These techniques have been reviewed by different authors [130,131,132,133]. Among these methods, titrimetric assay using olive oil or triolein as substrate is one of the most commonly used. The pH-stat method is generally recognized as a reference lipase assay [134]. This method is convenient for characterizing lipase activity, specificity and the interfacial activation phenomenon [130]. In the pH-stat method, lipase activity is measured on a mechanically stirred emulsion of natural or synthetic triglycerides by neutralizing the fatty acids released with time by adding a solution of sodium hydroxide of know concentration in order to maintain the pH at a constant fixed value. The amount of sodium hydroxide used is directly related to the lipolytic activity of the sample tested. This method is sensitive at the level of one µmole of released fatty acid per minute. Several colorimetric methods have been also developed. One of them is based on the utilization of *para*-nitrophenyl esters with variable acid chain length [133]. The released of *para*-nitrophenol upon substrate hydrolysis, that can be monitored at 405 nm, is directly related to the lipase activity. Despite this method is rapid and could be used high-throughput, it does not take into account the interfacial activation phenomenon since the substrate is not emulsified *stricto-sensus*. Lipase activity could also be observed directly on agar plate containing an emulsified substrate, usually tributyrin or triolein. Lipase activity is indicated by the formation of clear halos around the growing microbial colonies upon substrate hydrolysis [17]. The sensitivity of this method could be enhanced by using specific dyes such as Victoria blue B, Spirit blue, Nile blue sulfate and night blue [135]. The modification in the color of the dye used traduces the drop in pH due to the release of fatty acid upon substrate hydrolysis, and thus lipase activity [136]. Fluorescent dye Rhodamine B could also be used in plate assay containing emulsified olive oil. In that case, the substrate hydrolysis led to the formation of orange fluorescent halos upon UV illumination [137].

5. Solid-State Fermentation

5.1. Solid-State Fermentation as a Proper System for Fungal Culture

SSF has been known since ancient times in Asian countries [138]. However, this culture process was almost ignored in western countries until the 1940's mainly because SmF was a well-established technology for the production of various compounds, such as antibiotics by molds [139]. SSF offers a quite different environment than SmF for the growth and metabolism of microorganisms. Indeed, SSF could be defined as a fermentation process in which microorganisms grow on solid materials, natural or synthetic, in absence of a free liquid phase [139,140]. Therefore, this type of cultivation is well adapted to yeast and molds since they can grow in low moisture environment by contrast to bacteria [141].

Fungal enzyme production, including lipases, by SmF is well established and high titers of these enzymes are reported in the literature [142]. SmF nevertheless, has some drawbacks that limit the process productivity and the possibility to reduce the total production cost to more competitive levels. These include: managing high amounts of culture broth that increase cost of purification, negative environmental impacts of culture broth wastes, limited cell growth due to the weak oxygen transfer rate capacity, increased deleterious action by protease due to better contact through the broth agitation. SSF offers the possibility to overcome most of these problems [138,139]. Oxygen transfer limitations are solved due to a much more enlarged surface (external and/or internal) of the solid support for mass transfer. In addition, cells and especially aerial hyphae are exposed directly to the gas phase that can, in some cases, make a significant contribution to overall oxygen transfer. SSF is also characterized by the use of very low water amount in the culture medium and hence contributes to solve the associated problems of this factor [139,140]. Indeed, it is known that a low water activity of the solid medium limits the growth of many microorganisms, especially bacteria [141]. The control of the optimal water activity for the specific fungal strain and the utilization of adequate cell inoculum, allow avoiding fungal contamination [143]. This highlights that SSF requires less care for asepsis control than SmF [144].

Beside these advantages, it is recognized that many filamentous fungi and yeasts behave more efficiently in SSF and show greater productivities compared to SmF. Many authors have compared SmF and SSF process for lipase production and all of them have obtained higher enzyme yields in SSF. For example, total capital investment needed for a production scale of 100 $m^3$of lipase concentrate per year is78% higher for SmF compared to SSF [138].

Although filamentous fungi adequately growth in SSF, there are however some limitations in the process. The main one is the generation of heat by the cellular metabolic activity. One way to dissipate this heat consists of mechanical agitation/mixing of the solid medium [143]. However, this action could be harmful for molds and thus, it requires a great compromise between cell survival/heat dissipation for process viability. This problem is not encountered for yeasts, thus offering a good opportunity for producing enzymes such as lipases with enhanced possibilities of competing in enzyme market [141].

5.2. Characterization of Solid Growth Media

One of the main characteristic traits of SSF is the utilization of solid substrates as solid supports but also as sources of nutrients. In some applications, these solid supports are the only sources of carbon and nitrogen available to sustain microbial growth. In the recent years, considerable researches have been carried out using agricultural wastes, which are renewable and abundantly available, to produce value-added products by SSF. Utilization of agroindustrial wastes provides alternative substrates and may help to solve pollution problems, which otherwise might be caused by their disposal. The nature of the substrate is the most important factor affecting fermentative processes. The choice of the substrate depends on several factors, mainly related to cost and availability but also on the nature of the compound to produce (enzyme, metabolite, biomass) [89]. For example rice bran [22], coconut oil cake [19], olive cake and sugar cane bagasse [41], gingelly oil cake [80], babassu oil cake [56], wheat bran [145], *Jatropha curcas* seed cake [146], Niger seed oil cake [147], palm Kernal cake [148], Groundnut oil cake [149] and mustard oil cake [150] have been used successfully for lipase production. In that regard, the most important process parameters that have been investigated and optimized are incubation time, inoculum age, inoculum level, initial moisture content, type and concentration of carbon and nitrogen sources and C/N ratio of the medium [148,149].

5.3. Ambient Factors: pH, Temperature and Moisture

The moisture content of the solid substrate used in SSF can vary by liquid evaporation due to environmental factors, through microbial metabolism and by the water-binding characteristics of the substrate [70]. This moisture content has a significant impact on microbial growth and metabolic activities, including lipase production [25]. In general, the moisture content used in SSF processes vary between 30 and 85% [151]. The influence of initial moisture content on lipase production by *Y. lipolytica* from Niger seed oil cake was studied in the range from50 to 100% (v/w) [147]. An initial moisture content of 60% (v/w) lead to the maximal value of lipase production (18.5 U g^{-1}). At higher moisture content, the lower lipase production observed could be due to a decrease of the substrate porosity and hence in a decrease of the gaseous exchange that lead to suboptimal conditions for cell growth and enzyme production. By contrast, low moisture content could lead to a reduction of the nutrients solubility and availability, a reduced degree of swelling and a higher water tension. For the production of lipase in SSF by *Y. lipolytica* from mustard oil cake, the maximum lipase activity was obtained at 50% v/w initial moisture content [152]. The effect of initial moisture content from 40 to 90% (v/w) on lipase production of *C. rugosa* using Groundnut oilcake was investigated. A moisture content of 60% v/w was reported as the optimal value [149].

Another important factor to consider in SSF is the pH variation during the process, since pH is difficult to control. During cell growth, pH variation is mainly caused by the secretion of organic acids or by the assimilation of nitrogen sources such as ammonium, nitrate or urea. Each microorganism possesses an optimal pH range for growth and metabolic activity, including lipase production/secretion. This value ranges between 4 and 5 for yeasts. However, they are able to grow in a pH range from 2.5 to 8.5. This versatility of fungi in regards to pH can be beneficially exploited to prevent or minimize bacterial contaminations [151]. The lipase production has been investigated for different pH values ranging from 3 to 9. The maximum lipase production has been observed betweenpH6 and7 [19,149].

Another important parameter to consider in SSF is the difficulty of removing the excess of the heat generated by the cellular metabolism, mainly due to the low thermal conductivity of the solid medium [153]. In some cases, this heat is accumulated within the solid medium, ending microbial growth and even, producing cell death. The temperature at the center of the fermenting solid medium (bed) could be higher up to twenty degrees than the incubation temperature [141]. Aerations often used to dissipate metabolic heat from the solid medium and also to supply the oxygen requirements for cultures [141]. Remarkably, optimal temperatures for lipase production have been reported in the range of 28 to 45°C [70,152].

5.4. Most Common Bioreactors and Their Mode of Operation

Several types of bioreactors have been used in SSF processes for enzyme production, including lipase [139,154]. These bioreactors could be operated in batch mode (discontinuous) or in continuous mode (CSSF). In this later case, there is a flow of the solid phase within the reactor [155,156]. There are specific challenges to operate CSSF bioreactors that are not faced in classical continuous SmF processes. Consequently, CSSF bioreactors are currently scarce in industry [156]. However, improved design procedures and sensors development promise a better future for CSSF at large (industrial) scale. In this context, this section will focus only on discontinuous solid-phase bioreactors.

For discontinuous solid phase operations, Ali and Zulkali [157] have proposed a classification of four types of SSF bioreactors: (i) Bioreactors in which the fermenting solid medium (bed) is static or mixed only very infrequently and air is circulated around the bed, but not blown forcefully through it; (ii) Bioreactors in which the bed is static or mixed only very infrequently and air is blown forcefully though the bed, (iii) Bioreactors in which the bed is continuously mixed or mixed intermittently with a frequency of minutes to hours, and air is circulated around the bed, but not blown forcefully through it and (iv) Bioreactors in which the bed is agitated and air is blown forcefully through the bed [157]. Example of first and second types of bioreactors are the so called "tray bioreactors" and "packed-bed bioreactors", respectively. Two bioreactor designs from the third type are "stirred drum bioreactors" and "rotating drum bioreactors". Various bioreactor designs from the fourth type are known as "gas-solid fluidized beds", "rocking drum" and "stirred-aerated bioreactors" [157].

5.4.1. Tray Bioreactors

Tray bioreactors (TB) are the simplest design for SSF [154,157]. They can be considered as a modification of a column bioreactor in which the height of the bed is short enough to prevent temperature profiles to form in the axial direction. Tray bioreactors consist of a chamber where controlled air (flow, temperature and relative humidity) is circulated around a number of trays (Figure 2).

Each tray contains a layer of solid substrate, typically between 5 and 15 cm deep. Trays usually have an open top and perforated bottom to favor exchange with the gaseous phase. Intermittent mixing of the solid substrate by hand could be carried out but in general this will occur only once per day. The main drawback of TB is their low volumetric efficiency compared to column bioreactor [158]. Indeed, the thickness of the bed must remain low while the space between the trays must remain large enough to ensure an optimal gas and heat transfer. There have been no significant advances in tray design over the last decade.

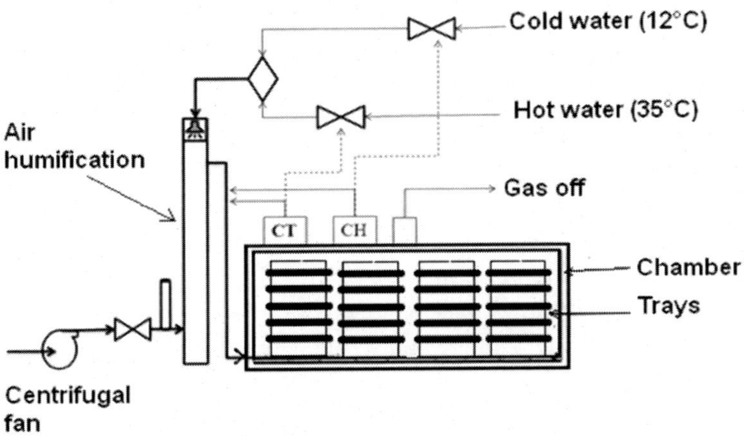

Figure 2. Schematic representation of a tray chamber with air temperature and moisture control.

5.4.2. Packed-Bed Bioreactors

A packed-bed bioreactor (PBB) typically involves a static bed on top of a perforated plate through which conditioned air is blown (Figure 3). In some PBB design, air is blown through a perforated rod inserted into the center of the bed. Over the last 25 years, PBB has received much experimental and modeling attention. The main attractive characteristic is that it has no mechanical (moving) parts, thus reducing the cost of construction, operation and maintenance.

In PBB process, axial dynamical temperature profiles could be observed [159]. Since the water carrying capacity of the air increases with temperature, these axial temperature gradients promote evaporation within the bed even if water-saturated air is used to aerate the column. Despite, this evaporation phenomenon can remove 65% of the heat generated by the cellular metabolism, it lowers the moisture of the solid substrate and thus limit the cellular growth [160]. In PBB, moisture regulation is almost impossible since the bed is unmixed. Therefore, it is desirable to increase heat removal by other means such as water jacket as shown in Figure 3.

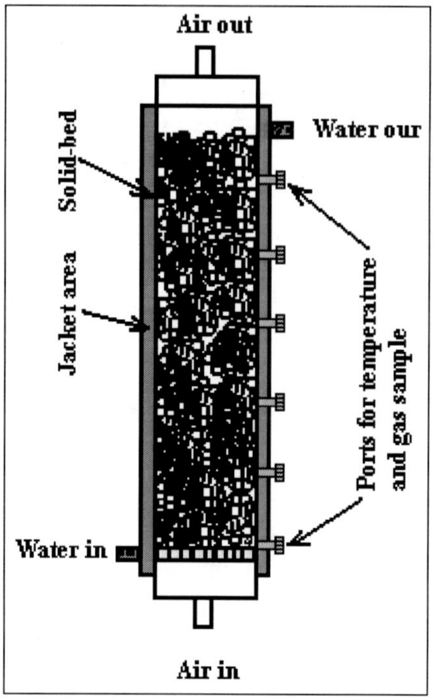

(a)

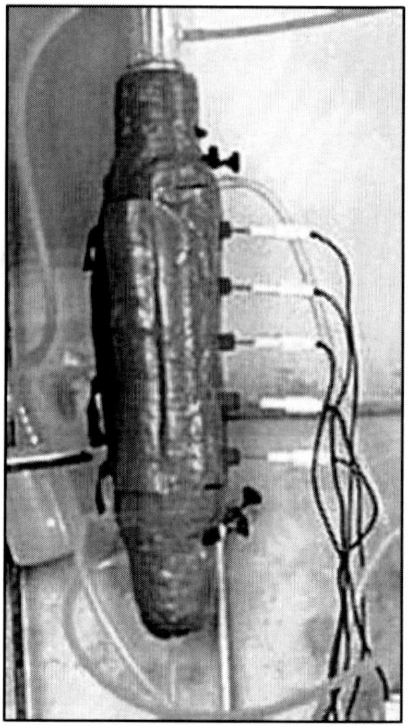

(b)

Figure 3. Packed-bed bioreactors used at different stages of process development: (a) schematic representation; (b) Real lab-scale bioreactor [49].

5.4.3. Stirred Tank Bioreactors

Two main designs of stirred tank bioreactors (STB) are available for solid-state fermentation. In the first one, the vessel remains static while stirrer rotates. In the second design, static stirrers are used while the vessel rotates as in the case of rotating drum bioreactors. In rotating drum bioreactors (Figure 4), the solid substrate is held within a horizontal or near horizontal drum that is continuously rotated at low speed. The drum could be outfit with baffles to increase mixing efficiency. The STB design allows an optimal homogeneity of the system and a suitable aeration of the fermentation medium [158].

Rotating drum bioreactors provide reasonable heat transfer with relatively gentle mixing, since air is blown through a headspace above a tumbling bed of substrate particles [161]. Remarkably, temperature profiles are broken through the agitation of this bioreactor. STB could be operated isothermally at a temperature close to the optimal one. However, this mixing could be harmful, mainly for the fungal growth. Indeed, the sharing forces caused by this mechanical agitation could affect the structure of hyphae and thus negatively impact cell growth and enzyme production [138,162]. By contrast, yeasts are less sensitive to sharing forces and this device could be used for yeast cultures and enzyme production. The complexity of STB electrical and mechanical designs, negatively impact the cost of construction and maintenance. Surprisingly, STB are not frequently applied to lipase production [89], but undoubtedly it has the potential for large scale productions.

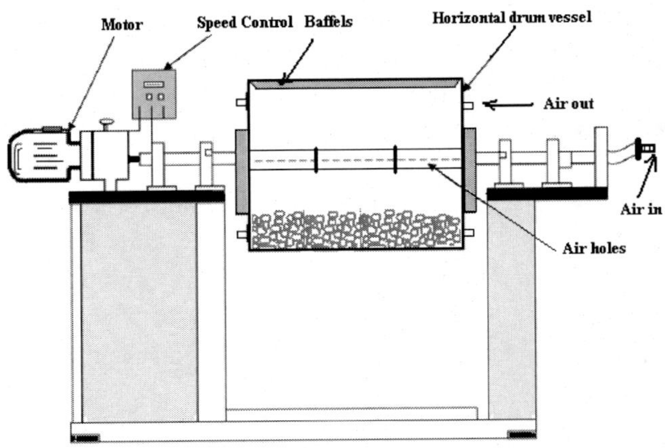

Figure 4. Schematic representation of a rotating drum bioreactor.

5.5. Lipase Production Using Different SSF Bioreactors

Lipase production in SSF has been mainly performed in TB and PBB (Table 2). Literature reports on large-scale production are very scarce and most of them belong to few grams scale of solid support [140].

Table 2. Lipase activity achieved in tray (TB) and packed-bed bioreactors (PBB)

Strain	Bioreactor type	Scale[1] (g)	Lipase activity (Ug^{-1})	Time (h)	Reference
Yeasts					
Yarrowia lipolytica	TB	25	26	96	[147]
Candida rugosa	TB	12	11	120	[149]
	TB	10	88	96	[19]
Molds					
Penicillium Sp.	TB	10	140	48	[59]
Penicillium simplicissimum	TB	10	90	72	[163]
P. chrysogenum	TB	17	65	72	[164]
Aspergillus niger	TB	33	1934	96	[165]
	TB	20	63	48	[34]
	TB	10	623	72	[166]
	TB	100	598	72	[166]
	TB	1000	523	72	[166]
	TB	10	142	96	[167]
Rhizopus chicness	TB	30	350	36	[47]
R. homothallicus	PBB	20	826	12	[44]
	PBB	50	1500	13	[14]
R. microsporus var. tuberosus	PBB	20	74	20	[168]
R. microsporus var. chicness	PBB	20	72	20	[168]
R. oligosporous	TB	10	48	48	[43]
	TB	8	77	120	[169]
R. oryzae	TB	5	236	72	[42]

At small scale, comparison of the yields of lipase productions obtained by using different bioreactors design is not significant. Indeed, the process at that scale is not influenced by axial temperature profiles that form within the solid substrate. Therefore, the differences in lipase productions, presented in Table 2, seem to be more dependent on strains and growth conditions than other factors.

The data shown in Table 2 indicate that 71% are under a lipase production of 500 Ug^{-1}, while less than 10% are over 1500 Ug^{-1}. This indicates that very high productions and productivities are possible to reach, under controlled process operations on SSF. When enzyme production turns to a higher scale, a decrease in process productivity should be expected. For instance, a slight decrease on lipase production was observed when scaling up from 10 g to 1000 g [166].

5.6. Scale-Up of Solid-State Fermentation

5.6.1. Difficulties of Scaling-Up of SSF

Enzyme productions by SSF under different process conditions, microorganisms and substrates have frequently been reported [170]. However, most studies consist of lab-scale experiments using a few grams of solid substrate, constant temperature conditions, sterilized residues and specific microorganisms as the inoculum. Unfortunately, all of these conditions are far from real behavior of SSF processes at higher scale. Interestingly, many workers have reported the exploitation of agricultural by-products in SSF for enzyme production, including lipase production. One of the main drawbacks in SSF scale-up is the absence of commercial scale designs of solid-state bioreactors, due to the difficulty in the standardization of the bioprocess (temperature, pH and humidity gradients are difficult to control) and to the limited reproducibility of the experimental results [158].

Scale-up of SSF process is a huge challenge due to inherent difficulties of handling large quantities of heterogeneous solid substrate and avoiding microbial contamination, and due to the problems related to mass and heat transfe [44,170]. Additionally, absence of appropriate sensors for measuring local variables associated to rates of mass and heat transfers, negatively impact on the development of reliable procedures for the scale-up of SSF processes [171].

5.6.2. Scale-Up Approaches to SSF

Scale-up of SSF bioreactors has received some attention in the past. In general, scaling up of chemical and biotechnological process has been made through three main procedures: empirically, by means of the theory of similarity and by phenomenological mathematical modeling [172].

Normally, empirical approach starts process optimization at low scale and continues at a greater scale using the experience and knowledge acquired in the previous scale. Some examples of this approach in SSF enzyme production could be found in the literature [173,174]. In these works, only small modifications in bioreactor size were made, from few grams to few kilograms. If the size modification is not so large, the loss in process efficiency can be minimized. The main drawback to scale-up SSF process is the difficulty to implement special measurement devices (specific probes) that could help to better understand the different phenomena involved in the process, including biochemical reactions, and mass and heat transfers [171]. Moreover, scaling-up from lab scale to full production may cause limitations in heat and mass transfer that could affect the whole process, such as oxygen transfer limitations [175]. Due to the complexity of SSF processes, the approach based on the theory of similarity is not as common as the mathematical modeling approach. Mathematical modeling of kinetics, heat and mass transfer have received great attention and its importance is frequently recognized for process development [171,173].

ACKNOWLEDGMENTS

This work was supported by WBI (Wallonie-Bruxelles International) grant SUB/2012/86836 to P. Fickers and L. B. Ramos-Sanchez.

REFERENCES

[1] Eijkmann, C. (1901). Uber enzyme bei bakterien und schimmelpilzen [About enzymes in bacteria and mold]. *Zentralbl bakt parasitenk infektionsk*, *29*: 841-848.

[2] Sarda, L. and Desnuelle, P. (1958). Action de la lipase pancréatique sur les ester en emulsion [Action of pancreatic lipase on emulsion esters]. *Biochim Biophys Acta*, *30*: 513-521.

[3] Verger, R. (1997). Interfacial activation of lipases: Facts and artifacts. *Trends Biotechnol.*, *15*: 32-38.

[4] Verger, R. and de Haas, G. H. (1973). Enzyme reactions in a membrane model. 1. A new technique to study enzyme reactions in monolayers. *Chem Phys Lipids*, *10*: 127-136.

[5] Ollis, D., Cheah, E., Dijkstra, B., et al. (1992). The alpha/beta hydrolase fold. *Protein Eng.*, *5*: 197-211.

[6] Brzozowski, A. and Savage, H. (2000). Structural origin of the interfacial activation in Thermomyces (Humicola) lanugosa lipase. *Biochemistry*, *39*: 15071-15082.

[7] Pleiss, J., Fischer, M. and Peiker, M. (2000). Lipase engineering database- understanding and exploiting sequence-structure-function relationships. *J Mol Catal B: Enz.*, *10*: 491-508.

[8] Gupta, N., Shai, V. and Gupta, R. (2007). Alkaline lipase from a novel strain *Burkholderia multivorans*: Statistical medium optimization and production in a bioreactor. *Process Biochem.*, *42*: 518-526.

[9] Casas-Godoy, L., Duquesne, S., Bordes, F., et al. (2012). Lipases: an overview. In: *Lipases and phospholipases: methods and protocols*. Georgina Sandoval.

[10] Pandey, A., Benjamin, S., Soccol, C., et al. (1999). The realm of microbial lipases in biotechnology. *Biotechnol Appl Biochem.*, *29*: 119-131.

[11] Abada, E. (2008). Production and characterization of a mesophile lipase isolated from bacillus stearothermophilus AB-1. *Pakistan J Biol Sci.*, *11*: 1100-1106.

[12] Durta, J., Terzi, S., Bevilaqua, J., et al. (2008). Lipase production in solid state fermentation monitoring biomass growth of *Aspergillus niger* using digital image processing. *Appl Biochem Biotechnol.*, *147*: 63-75.

[13] Azeredo, L., Gomes, P., Sant'Anna, G., et al. (2007). Production and regulation of lipase activity from *Penicillium restrictum* in submerged and solid-state fermentation. *Curr Microbiol., 54*: 361-365.

[14] Diaz, J. C., Terzi, S., Bevilaqua, J., et al. (2006). Lipase from thermotolerant fungus *Rhizopus homothalicus* is more stable when produced using solidstate fermentation than in liquid fermentation procedures. *Enz Microb Technol., 39*: 1042-1050.

[15] Dominguez, A., Costas, M., Longo, M., et al. (2003). A novel application of solid state culture: production of lipases by *Yarrowia lipolytica. Biotechnol Lett., 25*: 1225-1229.

[16] Fickers, P., Destain, J. and Thonart, P. (2008). Improvement of *Yarrowia lipolytica* lipase production by fed-batch fermentation. *J Basic Microbiol., 48*: 1-4.

[17] Fickers, P., Fudalej, F., Nicaud, J., et al. (2005). Selection of new over-producing derivatives for the improvement of extracellular lipase production by the non-conventional yeast *Yarrowia lipolytica. J Biotechnol., 115*: 379-386.

[18] Fickers, P., Nicaud, J., Gaillardin, C., et al. (2004). Carbon and nitrogen sources modulate lipase production in the yeast *Yarrowia lipolytica. J Appl Microbiol., 96*: 742-749.

[19] Benjamin, S. and Pandey, A. (1997). Coconut cake – a potent substrate for the production of lipase by *Candida rugosa* in solid-state fermentation. *Acta Biotechnologica, 17*: 241-251.

[20] Rekha, K., Chandana, M., Devi, C., et al. (2012). Production and optimization of lipase from *Candida rugosa* using groundnut oilcake under solid state fermentation. *Intr J Res Eng Technol., 4*: 571-577.

[21] Venkata Rao, P., Jayaraman, K. and Lakshmi, K. (1993). Production of lipase by *Candida rugosa* in solid state fermentation. 2: Medium optimization and effect of aeration. *Proc Biochem., 28*: 391-395.

[22] Rao, P, Jayaraman, K. and Lakshmanan, C. (1993). Production of lipase by *Candida rugosa* in solid-state fermentation. I: determination of significant process variables. *Pro. Biochem., 28*: 385-389.

[23] Moftah, O., Grbavcic, S., Milena, Z., et al. (2012). Adding value to the oil cake as a waste from oil processing industry: Production of lipase and protease by *Candida utilis* in Solid State Fermentation. *Appl Biochem Biotechnol., 166*: 348-364.

[24] Dominguez, A., Costas, M., Longo, M., et al. (2003). A novel application of solid state culture: production of lipases by *Yarrowia lipolytica. Biotechnol Lett., 25*: 1225-1229.

[25] Babu, I. and Rao, G. (2007). Lipase Production by Yarrowia lipofyäca NCIM 3589 in Solid State Fermentation Using Mixed Substrate. *Res J Microbiol., 2*: 469-474.

[26] Imandi, S., Karanam, S. and Garapati, H. (2010). Optimization of media constituents for the production of lipase in solid state fermentation by *Yarrowia lipolytica* from palm Kernal cake (*Elaeis guineensis*). *Adv Biosci Biotechnol., 1*: 115-121.

[27] Kishan, G., Gopalakannan, P., Muthukumaran, C., et al. (2013). Statistical optimization of critical medium components for lipase production from *Yarrowia lipolytica* (MTCC 35). *J Genet Eng Biotechnol., 11*: 111-116.

[28] Balaji, V. and Ebenezer, P. (2008). Optimization of extracellular lipase production in *Colletotrichum gloeosporioides* by solid state fermentation. *Indian J Sc. Technol., 1*: 1-8.

[29] Pau, H. and Ibrahim, O. (2004). Selection and optimization of lipase production from *Aspergillus flavus* USM A10 via Solid State Fermentation (SSF) on rice husks and wood dusts as substrate. *Paks J Biol Sci*, *7*: 1249-1256.

[30] Toscano, L., Montero, G., Stoytcheva, M., et al. (2013). Lipase production through solid-state fermentation using agro-industrial residues as substrates and newly isolated fungal strains. *Biotechnol Biotechnol Eq.*, *27*: 4074-4077.

[31] Durta, J., Terzi, S., Bevilaqua, J., et al. (2008). Lipase production in solid state fermentation monitoring biomass growth of *Aspergillus niger* using digital image processing. *Appl Biochem Biotechnol.*, *147*: 63-75.

[32] Mala, J., Edwinoliver, N., Kamini, N., et al. (2007). Mixed substrate solid state fermentation for production and extraction of lipase from *Aspergillus niger* MTCC 2594. *J Gen Appl Microbiol.*, *53*: 247-253.

[33] Falony, G., Janny, C., Julio, C., et al. (2006). Production of extracellular lipase from *Aspergillus niger* by solid-state fermentation. *Food Technol Biotechnol.*, *44*: 235-240.

[34] Damaso, M., Passianoto, M., de Freitas, S., et al. (2008). Utilisation of agroindustrial residues for lipase production by solid-state fermentation. *Braz J Microbiol.*, *39*: 676-681.

[35] Veerabhadrappa, M., Shivakumar, S. and Devappa, S. (2013).Solid-state fermentation of Jatropha seed cake for optimization of lipase, protease and detoxification of anti-nutrients in Jatropha seed cake using *Aspergillus versicolor* CJS-98. *J Biosci Bioeng.*, *117*: 208-214.

[36] Rajan, A. and Nair, A. J. (2011). A comparative study on alkaline lipase production by a newly isolated *Aspergillus fumigatus* MTCC 9657 in submerged and solid-state fermentation using economically and industrially feasible substrate. *Turk J Biol.*, *35*: 569-574.

[37] Sethi, B., Rout, J., Das, R., et al. (2013). Lipase production by *Aspergillus terreus* using mustard seed oil cake as a carbon source. *Anal Biochem.*, *63*: 241-252.

[38] Dayanandan, A., Hilda, S., Shanmugavel, M., et al. (2012). Solid state bioprocessing for scale up of *Aspergillus tamari* MTCC5152 lipase and its degreasing effect on cow Hide. *Indian J Sci Technol.*, *5*: 2978-2983.

[39] Kaushik, R., Saran, S., Isar, J., et al. (2006). Statistical optimization of medium components and growth conditions by response surface methodology to enhance lipase production by *Aspergillus carneus*. *J Mol Catal B: Enz.*, *40*: 121-126.

[40] Kademi, A., Morin, A., Ismaili-Alaoui, M., et al. (2003). Growth kinetics and esterase/lipase production by thermophilic fungi grown in solid state fermentation on olive cake. *New Horiz Biotechnol.*, In: *New Horizons in Biotechnology*, Roussos S., Soccol C.R., Pandey A., Augur C. (eds). Springer, Dordrecht, pp. 43-55.

[41] Cordova, J., Nemmaoui, M., Ismaıli-Alaoui, M., et al. (1998). Lipase production by solid state fermentation of olive cake and sugar cane bagasse. *J Mol Catal B: Enz.*, *5*: 75-78.

[42] Vaseghi, Z., Najafpour, G., Mohseni, S., et al. (2013). Production of active lipase by *Rhizopus oryzae* from sugarcane bagasse: solid state fermentation in a tray bioreactor. *Intr J Food Sci Technol.*, *48*: 283-289.

[43] Ul-Haq, I., Idrees, S. and Rajoka, M. (2002). Production of lipase by *Rhizopus oligosporous* by solid state fermentation. *Process Biochem.*, *37*: 637-641.

[44] Rodriguez, J., Mateos, J., Nungaray, J., et al. (2006). Improving lipase production by nutrient source modification using *Rhizopus homothallicus* cultured in solid state fermentation. *Process Biochem.*, *41*: 2264-2269.

[45] Iftikhar, T. and Hussian, A. (2002). Effects of nutrients of the extracellular lipase production by mutant strains of *Rhizopus oligosporous* TVV-31. *Biotechnology*, *1*: 15-20.

[46] Miszkiewicz, H., Rozwandowicz, R. and Bielecki, B. (2004). Physiological and enzymatic activities of *Rhizopus oligosporus* in swing solid-state reactor. *Chem Pap.*, *58*: 424-428.

[47] Sun, S. Y. and Xu, Y. (2008). Solid-state fermentation for 'whole-cell synthetic lipase production from *Rhizopus chicness* and identification of the functional enzyme. *Process Biochem.*, *43*: 219-224.

[48] Christen, P., Angeles, N., Corzo, G., et al. (1995). Microbial lipase production on a polymeric resin. *Biotechnol Tech.*, *9*: 597-600.

[49] Rivera-Munoz, G., Tinoco-Valencia, J., Sanchez, S., et al. (1991). Production of microbial lipase in a solid-state fermentation system. *Biotechnol lett.*, *13*: 277-280.

[50] Cordova, J., Nemmaoui, M., Ismaıli-Alaoui, M., et al. (1998). Lipase production by solid state fermentation of olive cake and sugar cane bagasse. *J Mol Catal B: Enz.*, *5*: 75-78.

[51] Toscano, L., Montero, G., Stoytcheva, M., et al. (2013). Lipase production through solid-state fermentation using agro-industrial residues as substrates and newly isolated fungal strains. *Biotechnol Biotechnol Eq.*, *27*: 4074-4077.

[52] Ortiz-Vazquez, E., Granados-Baeza, M. and Riveramuñoz, G. (1993). Effect of culture conditions on lipolytic enzyme production by *Penicillium candidum* in a solid-state fermentation. *Biotechnol Adv.*, *11*: 409-416.

[53] Silva, M., Freire, D., de Castro, A., et al. (2011). Production of multifunctional lipases by *Penicillium verrucosum* and *Penicillium brevicompactum* under solid state fermentation of babassu cake and castor meal. *Bioproc. Biosyst. Eng.*, *34*: 145-152.

[54] Cavalcanti, E., Gutarra, M., Freire, D., et al. (2005). Lipase production by solid-state fermentation in fixed bed-bioreactor. *Braz Arch Microbiol.*, *48*: 79-84.

[55] Kempka, A., Lipke, N., Pinheiro, T., et al. (2008). Response surface method to optimize the production and characterization of lipase from *Penicillium verrucosum* in solid-state fermentation. *Biop Biosyst Eng.*, *31*: 119-125.

[56] Gombert, A., Pinto, A., Castilho, L., et al. (1999). Lipase production by *Penicillium restrictum* in solid-state fermentation using babassu oil cake as substrate. *Proc Bioch.*, *35*: 85-90.

[57] Palma, M., Annette, L., Gombert, A., et al. (2000). Lipase production by *Penicillium restrictum* using solid waste of industrial babassu oil production as substrate. *Twenty-First Symposium on Biotechnology for Fuels and Chemicals*, pp. 1137-1145.

[58] Malilas, W., Kang, S., Yoo, H., et al. (2013). Lipase from *Penicillium camembertii* KCCM 11268: Optimization of solid state fermentation and application to biodiesel production. *Korean J Chem Eng.*, *30*: 405-412.

[59] Rigo, E., Ninow, J., Di Luccio, M., et al. (2010). Lipase production by solid fermentation of soybean meal with different supplements. *LWT- Food Sci Technol.*, *43*: 1132-1137.

[60] Mohanasrinivasan, V., Dhrisya, P., Dipinsha, K., et al. (2009). A comparative study of the lipase yield by solid state and submerged fermentations using fungal species from biopharmaceutical oil waste. *Afr J Biotechnol.*, *8*: 073-076.

[61] Destain, J., Fickers, P., Weekers, et al. (2005). Utilisation of methyloleate in production of microbial lipase. *Appl Biochem Biotechnol.*, *121-124*: 269-278.

[62] Fickers, P., Nicaud, J., Gaillardin, C., et al. (2004). Carbon and nitrogen sources modulate lipase production in the yeast *Yarrowia lipolytica*. *J Appl Microbiol.*, *96*: 742-749.

[63] Fickers, P., Nicaud, J. M., Destain, J., et al. (2003). Overproduction of lipase by *Yarrowia lipolytica* mutants. *Appl Microbiol Biotechnol.*, *63*: 136-142.

[64] Turki, S., Ines, B., Frederic, W., et al. (2009). Isolation of bioactive peptides from tryptone that modulate lipase production in *Yarrowia lipolytica*. *Biores Technol.*, *100*: 2724-2731.

[65] de Rio, J., Serra, P., Valero, F., et al. (1990). Reaction scheme of lipase production by *Candida rugosa* growing on olive oil. *Biotechnol lett.*, *12*: 835-838.

[66] Domingez de Maria, P., Sanchez-Montero, J., Alcantare, A., et al. (2005). Rational strategy for the production of new crude lipase from *Candida rugosa*. *Biotechnol lett.*, *27*: 499-503.

[67] Wei, D., Zhang, L. and Song, Q. (2004). Studies on a novel carbon source and co-solvent for lipase production by *Candida rugosa*. *J Ind Microbiol Biotechnol.*, *31*: 133-136.

[68] Ferrer, P., Montesinos, J. L., Valero, F., (2001). Production of native and recombinant lipase by *Candida rugosa*. *Appl Biochem Biotechnol.*, *95*: 221-255.

[69] Ota, Y., Miyairi, S. and Yamada, K. (1968). Sterol requirements for the lipase production by *Candida cylindracea*. *Agric Biol Chem.*, *32*: 1476-1478.

[70] Lakshmi, B., Kangueane, B., Pennathur, A., (1999). Effect of vegetable oils in the secretion of lipase from *Candida rugosa* (DSM2031). *Lett Appl Microbiol.*, *29*: 66-70.

[71] Lotti, M., Brocca, S. and Porro, D. (2001). High lipase production by *Candida rugosa* is associated with G1 cells. *Biotechnol lett.*, *23*: 1803-1808.

[72] Akoh, C., Lee, G. and Shaw, J. (2004). Protein engineering and applications of *Candida rugosa* lipase isoform. *Lipids*, *39*: 513-526.

[73] Korbekandi, H., Abedi, D., Pourhossein, M., et al. (2008). Optimisation of *Candida rugosa* lipase esterase activity. *Biotechnology*,*7*: 112-117.

[74] Rajendran, A., Palanisamy, A. and Thangavelu, V. (2008). Evaluation of medium components by Plackett-Burman statistical design for lipase production by *Candida rugosa* and kinetic modeling. *Chinese J Biotechnol.*, *24*: 436-444.

[75] Coradi, G., Visitacao, V., de Lima, E., et al. (2013). Comparing submerged and solid-state fermentation of agro-industrial residues for the production and characterization of lipase by *Trichoderma harzianum*. *Ann. Microbiol.*, *63*: 533-540.

[76] Raimbault, M. (1998). General and microbiological aspects of solid substrate fermentation. *Elect J Biotechnol.*, *1*: 1-15.

[77] Navarro, A., Gea, T. and Sanchez, A. (2011). Production of lipases by solid state fermentation using vegetable oil-refining wastes. *Biores Technol.*, *102*: 10080-10084.

[78] Arputha S. and Kumaresapilai, N. (2013). Isolation, purification and medium optimisation of lipase enzyme producing strains of *Aspergillus niger* isolated from natural sources. *Intr J Pharm Pharma Sci.*, *5*: 321-324.

[79] Sarkar, D. and Laha, S. (2013). Optimization of extracellular lipase enzyme production from *Aspergillus niger* by submerged and solid-state fermentation process. *Intr J Pharm Biol Sci.*, *4*: 978-985.
[80] Kamini, N., GMalla, J. and Puvanakrishnan, P. (1998). Lipase production from *Aspergillus niger* by solid-state fermentation using gingelly oil cake. *Process Biochem.*, *33*: 505-511.
[81] Kaushik, R., Saran, S., Isar, J., et al. (2006). Statistical optimization of medium components and growth conditions by response surface methodology to enhance lipase production by *Aspergillus carneus*. *J Mol Catal B: Enz.*, *40*: 121-126.
[82] Yadav, R., Saxena, R., Gupta, R., et al. (1998). Lipase production by *Aspergillus* and *Penicillium* species. *Folia Microbiol.*, *56*: 479-482.
[83] Ohnishi, K., Yoshida, Y. and Sekiguchi, J. (1994). Lipase production of *Aspergillus oryzae*. *J Ferment Bioing.*, *77*: 490-497.
[84] Elibol, M. and Ozer, D. (2000). Influence of oxygen transfer on lipase production by *Rhizopus arrhizus*. *Process Biochem.*, *36*: 325-329.
[85] Khayati, G. and Kiyani, F. (2012). A statistical approach for optimization of lipase production by using rice straw: analysis of different inducers and nitrogen sources effect. *Minerva Biotechnologica*, *24*: 83-89.
[86] Liu, C., LU, W. and Chang, J. (2006). Optimizing lipase production of *Burkholderia* sp. by response surface methodology. *Process Biochem.*, *41*: 14889-14897.
[87] Miranda, O., Salgueiro, A., Pimentel, M., et al. (1999). Lipase production by a Brazilian strain of *Penicillium citrinum* using an industrial residue. *Biores Technol.*, *69*: 145-147.
[88] Gutara, M., Cavalcanti, E., Castilho, L., et al. (2005). Lipase production by solid-state fermetation. Cultivation condition and operation of tray and packed-bed bioreactors. *Appl Biochem Biotechnol.*, *212-124*: 105-116.
[89] Thakur, S. (2012). Lipases, its sources, Properties and Applications: A Review. *Intr J Sci Energ Res.*, *3*: 1-29.
[90] Joseph, B., Ramteke, P. and Thomas, G. (2008). Cold active microbial lipases: some hot issues and recent developments. *Biotechnol Adv.*, *26*: 457-70.
[91] Aravindan, R., Anbumathi, P. and Viruthagiri, T. (2007). Lipase applications in food industry. *Indian J Biotechnol.*, *6*: 141-158.
[92] Dominguez de Maria, P., Sánchez-Montero, J., Sinisterra, J., et al. (2006). Understanding *Candida rugosa* lipases: an overview. *Biotechnol Adv.*, *24*: 180-96.
[93] Lock, L., Corbellini, V. and Valente, P. (2007). Lipases produced by yeasts: powerful biocatalysts for industrial purposes. *Tecno-Logica*, *11*: 18-25.
[94] Takac, S. and Bakkal, M. (2007). Impressive effect of immobilization conditions on the catalytic activity and enantioselectivity of *Candida rugosa* lipase toward S-Naproxen production. *Process Biochem.*, *42*: 1021-1027.
[95] Chang, C., Hsu, C. and Shang, C. (2003). Lipase-catalyzed enantioselective esterification of (S)-naproxen hydroxyalkyl ester in organic media. *Biotechnol lett.*, *25*: 413-416.
[96] Armenariz Ruiz, M. A. (2012). *Subproductos de carica papaya: fuente de nuevos biocatalizadores con aplicaciones biotecnológicas* (*Byproducts of Carica papaya: Source of new biocatalysts with biotechnological applications*). http://ciatej. repositorionstitucional.mx/jspui/handle/1023/372.

[97] Kim, Y., Cheong, C., Lee, S., et al. (2001). Enzymatic kinetic resolution of ketorolac. *Tetrahedron: Asymmetry*, *12*: 1865-1869.

[98] Adamczak, M. (2003). Synthesis, properties, and application of lipase from *candida antarctica* for high yield monoacylglycerol biosynthesis. *Polish J Food Nut Sci.*, *12*: 3-8.

[99] Sharma, R., Chisti, Y., Banerjee, U., et al. (2001). Production, purification, characterization, and applications of lipases. *Biotechnol Adv.*, *19*: 627-662.

[100] Weber, N., Klein, E. and Mukherjee, K. (1999). Long-chain acyl thioesters prepared by solvent-free thioesterification and transthioesterification catalysed by microbial lipases. *Appl Microbiol Biotechnol.*, *51*:401-404.

[101] Szczesna A., Kubiak, A., Antczak, T., et al. (2009). Enzymatic biodiesel synthesis - Key factors affecting efficiency of the process. *Renew Energ.*, *34*: 1185-1194.

[102] Tan, T., Lu, J., Nie, K., et al. (2010). Biodiesel production with immobilized lipase: A review. *Biotechnol Adv.*, *28*: 628-634.

[103] Yucel, S., Terzioglu, P. and Ozcimen, D. (2012). Lipase Applications in Biodiesel Production. In: *Biodiesel-feedstocks, production and application*. Rijeka: Zhen Fang. InTech., pp. 209-250.

[104] Jang, M., Kim, D., Park, S., et al. (2012). Biodiesel production from crude canola oil by two-step enzymatic processes. *Renew Energ.*, *42*: 99-104.

[105] SilRoy, S. and Ghosh, M. (2011). Enzymatic synthesis of capric acid-rich structured lipids (MUM type) using *Candida antarctica* lipase. *J Oleo Sci.*, *60*: 275-80.

[106] Vaysse, L., Dubreucq, E., Pirat, J., et al. (1997). Fatty hydroxamic acid biosynthesis in aqueous medium in the presence of the lipase-acyltransferase from *Candida parapsilosis*. *J Biotechnol.*, *53*: 41-46.

[107] Kose, O, Melek, T. and Ayse, A. H. (2002). Immobilized *Candida antarctica* lipase-catalyzed alcoholysis of cotton seed oil in a solvent-free medium. *Biores Technol.*, *83*: 125-129.

[108] Madeira L., Van Rantwijk, F., Seddon, K., et al. (2000). Lipase-Catalyzed Reactions in Ionic Liquids. *Org Lett.*, *2*: 4189-4191.

[109] Turcu, M. C. (2010). *Lipase-catalyzed Approaches towards Secondary Alcohols: Intermediates for Enantiopure Drugs*. Univ. Turku., 1-73.

[110] Gordon, M. (1998). Lipid technologies and applications. *Food Chem.*, *62*: 253, doi: 10.1016/S0308-8146(97)00213-6.

[111] Mohamed R., Nazrim M., Key, L., et al. (2014). Interesterification of Engkabang (Shorea macrophylla) Fat - Canola Oil Blend with Lipase from *Candida antarctica* to Simulate the Properties of Lard. *J Oleo Sci.*, *63*: 39-46.

[112] Erik, B., Steinborn, G., Florsch, K., et al. (2009). Yeast Biotechnology: Diversity and Applications. Tulasi, S, Gutthard, K. (eds.). *Yeast Biotechnol Div Appl.*, 323-355. doi: 10.1007/978-1-4020-8292-4.

[113] Ray, A. (2012). Application of Lipase in Industry. *Asian J Pharm Techol.*, *2*: 33-37.

[114] Barbosa, T., Aoki, T., Chung, W., et al. (2009). Improvement of a *Candida antarctica* lipase B-displaying yeast whole-cell biocatalyst and its application to the polyester synthesis reaction. *Appl Microbiol Biotechnol.*, *82*: 59-66.

[115] Johnson, E. (2013). Biotechnology of non-Saccharomyces yeasts--the ascomycetes. *Appl Microbiol Biotechnol.*, *97*: 503-517.

[116] Akoh, C. (1993). Lipase-catalyzed synthesis of partial glyceride. *Biotechnol lett.*, *15*: 949-954.

[117] Bakirci, H., Li, M. and He, B. (1995). Immobilization of *Candida cylindracea* lipase on methyl acrylate-divinyl benzene copolymer and its derivatives. *Enz Microb Technol.*, *17*: 194-199.

[118] Schmid, R. D. and Verger, R. (1998). Lipases: Interfacial enzymes with attractive applications. *Angew Chem Intr Edit.*, *37*: 1609-1633.

[119] Hasan, F., Shah, A. and Hameed, A. (2006). Industrial applications of microbial lipases. *Enz Microb Technol.*, *39*: 235-251.

[120] Pierce, G., Wick, C. and Palmer, D. (1990). Unique Microbial lipases with activity at temperatures and pHs suitable for use in detergents. *Eur Patent.* 0385 401A1.

[121] Vakhlu, J. and Kour, A. (2006). Yeast lipases: enzyme purification, biochemical properties and gene cloning. *Elect J Biotechnol.*, *9*: 69-85.

[122] Larios, A., García, H., Oliart, R., et al. (2004). Synthesis of flavor and fragrance esters using *Candida antarctica* lipase. *Appl Microbiol Biotechnol.*, *65*: 373-376.

[123] Fickers, P., Alain, M. and Jean M. (2011). The lipases from *Yarrowia lipolytica*, Genetics, production, regulation, biochemical characterization and biotechnological applications. *Biotechnol Adv.*, *29*: 632-644.

[124] Hasan, F., Shah, A., Javed, S., et al. (2010). Enzymes used in detergents: Lipases. *J Biotechnol.*, *9*, 4836-4844.

[125] Shay, L. and Fisher, T. (1990). Enhancing the flavor of protein products derived from microorganisms. *EP Patent, 5,071,762*, 1-9.

[126] Sharma, D., Sharma, B. and Shukla, A. K. (2011). Biotechnological Approach of Microbial Lipase: A Review. *Biotechnology*, *10*: 23-40.

[127] Matsumae, H., Furui, M. and Shibatani, T. (1993). Lipase-catalyzed asymmetric hydrolysis of 3-phenylglycidic acid ester, the key intermediate in the synthesis of diltiazem hydrochloride. *J Ferm Bioeng.*, *75*: 93-98.

[128] Hsu, A. F., Jones, K., Foglia, T., et al. (2002). Immobilized lipase-catalysed production of alkyl esters of restaurant grease as biodiesel. *Biotechnol Appl Biochem.*, *36*: 181-186.

[129] Verma, N., Thakur, S. and Bhatt, A. K. (2012). Microbial lipases: Industrial applications and properties: A Review. *Intr Res J Biol Sci.*, *1*: 88-92.

[130] Beisson, F., Tiss, A., Riviere, C., et al. (2000). Methods for lipase detection and assay: a critical review. *Eur J Lipid Sci. Technol.*, *202*: 133-153.

[131] Gupta, R., Rathi, P., Gupta, N., et al. (2003). Lipase assays for conventional and molecular screening: an overview. *Biotechnol Appl Biochem.*, *37*: 63-71.

[132] Schmidt, M. and Bornscheuer, U. (2005). High-throughput assays for lapases and esterases. *Biomol Eng.*, *22*: 51-56.

[133] Hasan, F., Shah, A. and Hameed, A. (2009). Methods for detection and characterization of lipases: A comprehensive review. *Biotechnol Adv.*, *27*: 782-798.

[134] Brockman, H. L. (1981). Triglyceride lipase from porcine pancreas. *Methods Enzymol.*, *71*: 619-627.

[135] Shelley, A., Deeth, H. and MacRae, I. (1987). A numerical taxonomic study of psychrotrophic bacteria associated with lipolytic spoilage of raw milk. *J Appl Bacteriol.*, *62*: 197-207.

[136] Scholze, H., Stutz, H., Paltauf, F., et al. (1999). Fluorescent inhibitors for the qualitative and quantitative analysis of lipolytic enzymes. *Anal Biochem.*, *276*: 72-80.

[137] Kouker, G. and Jaeger, K. E. (1987). Specific and sensitive plate assay for bacterial lipase. *Appl Env Microbiol.*, *53*: 211-213.

[138] Couto, S. R. and Sanroman, M. (2006). Application of solid-state fermentation to food industry—A review. *J Food Eng.*, *76*: 291-302.

[139] Kumar, A. and Kanwar, S. (2012). Lipase production in solid-sate fermentation (SSF): recent developments and biotechnological applications. *Dynam. Biochem. Proc Biotecnol Mol Biol.*, *6*: 13-27.

[140] Bhargav, S., Panda, B., Ali, M., et al. (2008). Solid-state Fermentation: An Overview. *Chem Biochem Eng Q.*, *22*: 49-70.

[141] Pandey, A. (2003). Solid-state fermentation. *Biochem Eng. J.*, *13*: 81-84.

[142] Singh, A. and Mukhopadhyay, R. (2012). Overview of fungal lipase: a review. *Appl Biochem Biotechnol.*, *166*: 486-520.

[143] Mitchell, D., Berovi, M. and Krieger, N. (2006). Introduction to Solid-State Fermentation Bioreactors. In: *Solid-State Fermentation Bioreactors- Fundamentals of Design and Operation*. Mitchell, D., Krieger, N., Berovic, M. (eds), pp. 33-44. Germany.

[144] Holker, U., Hofer, M. and Lenz, J. (2004). Biotechnological advantages of laboratory-scale solid-state fermentation with fungi. *J Appl Microbiol Biotechnol.*, *64*: 175-186.

[145] Mahadik, N., Puntambekar, U., Bastawde, K., et al. (2002). Production of acidic lipase by *Aspergillus niger* in solid state fermentation. *Proc Biochem.*, *38*: 715-721.

[146] Mahanta, N., Gupta, A. and Khare, S. (2008). Production of protease and lipase by solvent tolerant *Pseudomonas aeruginosa* PseA in solid-state fermentation using Jatropha curcas seed cake as substrate. *Biores Technol.*, *99*: 1729-1735.

[147] Imandi, S., Karanam, S. and Garapati, H. (2010). Optimization of Process Parameters for the Production of Lipase in Solid State Fermentation by *Yarrowia Lipolytica* from Niger Seed Oil Cake (Guizotia Abyssinica). *J Microb Biochem Technol.*, *2*: 28-33.

[148] Imandi, S., Karanam, S. and Garapati, H. (2010). Optimization of media constituents for the production of lipase in solid state fermentation by *Yarrowia lipolytica* from palm Kernal cake (Elaeis guineensis). *Adv Biosci Biotechnol.*, *1*: 115-121.

[149] Rekha, K., Chandana, M., Devi, V., et al. (2012). Production and optimization of lipase from *Candida rugosa* using groundnut oilcake under solid state fermentation. *Intr J Res Eng Technol.*, *1*: 571-577.

[150] Imandi, S., Karanam, S. and Garapati, H. (2014). Use of Plackett-Burman design for rapid screening of nitrogen and carbon sources for the production of lipase in solid state fermentation by *Yarrowia lipolytica* from mustard oil cake (*Brassica napus*). *Braz J Microbiol.*, *44*: 915-921.

[151] Krishna, C. (2005). Solid-State Fermentation Systems-An Overview. *Crit Rev Biotechnol.*, *25*: 1-30.

[152] Imandi, S., Karanam, S. and Garapati, H. (2014). Use of Plackett-Burman design for rapid screening of nitrogen and carbon sources for the production of lipase in solid state fermentation by *Yarrowia lipolytica* from mustard oil cake (*Brassica napus*). *Braz J Microbiol.*, *44*: 915-921.

[153] Mitchell, D., Berovi, M., Nopharatana, M., et al. (2006). The Bioreactor Step of SSF: A Complex Interaction of Phenomena. In: *Solid-State Fermentation Bioreactors-*

Fundamentals of Design and Operation. Mictchell, D., Krieger, N., Berovic, M. (eds), pp. 13-32. Germany.

[154] Vaseghi, Z. and Najafpour, G. (2014). An investigation on lipase roduction from soybean meal and sugarcane bagasse in solid state fermentation using *Rhizopus oryzae*. *IJE Trans B: Applications*, *27*: 171-176.

[155] Khanahmadi, M., Mitchell, D., Beheshti, M., et al. (2006). Continuous solid-state fermentation as affected by substrate flowpattern. *Chem Eng Sci.*, *61*: 2675-2687.

[156] Ramos-Sanchez, L. B., Khanahmadi, M. and Mitchell, D. (2006). Continuous Solid-State Fermentation Bioreactors. In *Solid-State Fermentation Bioreactors- Fundamentals of Design and Operation*. Mitchell, D. A., Krieger, N., Berovic, M. (eds), pp. 142-158. Germany.

[157] Ali, H. and Zulkali, M. (2011). Utilization of Agro-Residual Ligno-Cellulosic Substances by Using Solid State Fermentation: A Review. *Croatian J Food Technol Biotechnol Nut.*, *6*: 5-12.

[158] Miron, J., Vazquez, J., Gonzalez, P., et al. (2010). Enhancement glucose oxidase production by solid-state fermentation of *Aspergillus niger* on polyurethane foams using mussel processing wastewaters. *Enz Microb Technol.*, *46*: 21-27.

[159] Ashley, V., Mitchell, D. and Howes, T. (1999). Evaluating strategies for overcoming overheating problems during solid state fermentation in packed bed bioreactors. *Biochem Eng J.*, *3*: 141-150.

[160] Gutierrez, M., Amar, S., Auria, R., et al. (1996). Heat transfer in citric acid production by solid state fermentation. *Proc Biochem.*, *31*: 363-9.

[161] Mitchell, D., Tongta, A., Stuart, D., et al. (2002). The potential for establishment of axial temperature profiles during SSF in rotating drum bioreactors. *Biotechnol Bioeng.*, *80*: 115-122.

[162] Prabhakar, A., Krishnaiah, K., Janaun, J., et al. (2005). An overview of Engineering Aspects of Solid State Fermentation. *Malaysian J Microbiol.*, *1*: 10-16.

[163] Gutarra, M., Godoy, M., Maugeri, F., et al. (2009). Production of an acidic and thermostable lipase of the mesophilic fungus *Penicillium simplicissimum* by solid-state fermentation. *Biores Technol.*, *100*: 5249-5254.

[164] Iftikhar, T., Niaz, M., Nisa, Z., et al. (2011). Optimization of cultural conditions for the biosynthesis of lipases by *Penicillium chrysogenum* (MBL 22) through solid state fermentation. *Paks J Bot.*, *43*: 2201-2206.

[165] Adinarayana, K., Raju, K., Zargar, M., et al. (2004). Optimization of process parameters for production of lipase in solid-state fermentation by newly isolated *Aspergillus species*. *Ind J Biotechnol.*, *3*: 65-69.

[166] Edwinoliver, N., Thirunavukarasu, K., Naidu, R., et al. (2010). Scale up of a novel tri-substrate fermentation for enhanced production of *Aspergillus niger* lipase for tallow hydrolysis. *Biores Technol.*, *101*: 6791-6796.

[167] Mohseni, S., Najafpour, G., Vaseghi, Z., et al. (2012). Solid State Fermentation of Agricultural Residues for Lipase Production in a Tray-Bioreactor. *World Appl Sci J.*, *16*: 1034-1039.

[168] Hernandez, B., Cordova, J., Barzana, E., et al. (2009). Effects of organic solvents on activity and stability of lipases produced by thermotolerant fungi in solid-state fermentation. *J Mol Catal B, Enz.*, *61*: 136-142.

[169] Mahapatra, P., Kumari, A., Garlapati, V., et al. (2010). Optimization of process variables for lipase biosynthesis from *Rhizopus oligosporous* NRRL5905 using evolutionary operation factorial design technique. *Indian J Microbiol.*, *50*: 396-403.

[170] Abraham, J., Gea, T. and Antoni, S. (2013). Potential of the solid-state fermentation of soy fibre residues by native microbial populations for bench-scale alkaline protease production. *Biochem Eng J.*, *74*: 15-19.

[171] Farinas, C., Vitcosque, G., Fonseca, R., et al. (2011). Modeling the effects of solid state fermentation operating conditions on endoglucanase production using an instrumented bioreactor. *Ind Crops Prod.*, *34*: 1186-1192.

[172] Julian-Ricardo, M. C. (2008). *Diseño tecnológico de una planta para el enriquecimiento proteico del bagazo de caña de azúcar* (*Technical information of a plant for the protection of the bagazo Cana azucar*), Universidad de Camaguey, Ignacio Agramonte Loynaz, Camaguey, Cuba.

[173] Alama, M., Mamuna, I., Muyibia, S., et al. (2009). Solid state bioconversion of oil palm empty fruit bunches for cellulase enzyme production using a rotary drum bioreactor. *Biuoch Eng J.*, *46*: 61-64.

[174] Bhavsar, K., Kumar, V. and Khire, J. (2011). High level phytase production by *Aspergillus niger* NCIM 563 in solid state culture: response surface optimization, up-scaling, and its partial characterization. *J Ind Microbiol Biotechnol.*, *38*: 1407-1417.

[175] Botella, C., Diaz, A., Wanga, R., et al. (2009). Particulate bioprocessing: A novel process strategy for biorefineries. *Proc Biochem.*, *44*: 546-555.

In: Microbial Catalysts. Volume 1
Editors: S. M. Abdel-Aziz, N. Garg, A. Aeron et al.
ISBN: 978-1-53614-679-0

Chapter 3

APPLICATION OF LIPASE IN NON-CONVENTIONAL MEDIA BASED CATALYSIS

Maryam Farhana Kamel Ariffin and Ani Idris*
Institute of Bioproduct Development, Department of Bioprocess and Polymer Engineering, School of Chemical and Energy Engineering, Faculty of Engineering, Universiti Teknologi Malaysia, Johor, Malaysia

ABSTRACT

Application of enzyme lipase in the world of synthesis became an alternative in overcoming hazard issues in synthetic catalysis. Robustness in its compatibility towards different working environment as well as wide range of temperature application compared to conventional bio-catalyst made the enzyme of interest to be utilized in chemical processes. Lipase has been known for capability in changing its function depending on the working medium. This further enhances the application of biocatalysis in synthesizing biomedical as well as industrial valued products. This chapter describes the enzyme native structure and changes occurring when applied in non-conventional processes. Presence of minimal amount of water in the working environment changes the function of lipase enzyme drastically. This chapter also explains effect of different working medium towards the lipase efficiency in catalysis followed with methods to enhance its activity and strength when applied in syntheses.

Keywords: lipase, non-conventional media, biocatalysis, green chemistry

1. GENERAL INTRODUCTION OF ENZYME LIPASE

In principle, a catalyst speeds up reaction without being structurally deformed post-catalysis. Utilization of the catalysts in the world of synthesis is so vast due to its numerous benefits. Up till current, chemical based catalysts are preferred as they are easier to handle and

* Corresponding Author's Email: ani@cheme.utm.my.

maintain. However, as imbalance in our ecosystem is now evidenced, researches are shifting toward greener synthesis to reduce the disaster [1]. Enzyme catalyzed reaction emerged an alternative in green chemistry to produce biodiesels, environmental friendly detergents, polyesters and copolymerization of lactones with medicinal properties [2]–[5].

One of the examples of biocatalysts that have been widely explored is enzyme lipases. The enzyme also known as triacylglycerol acylhydrolases (EC 3.1.1.3) was first discovered by Claude Bernard (1856) in pancreatic juices where it hydrolyses insoluble triacylglycerols into simpler products. Lipases are natively found in microbes, plant, animals and their behaviour are determined by the environment properties. This is so as substrate positional specificity, thermostability as well as optimum pH for the enzyme have great influences toward the behavior [6], [7]. A number of microbial producing lipases have been reported with lipases of different types being produced by taxonomically close strains [6]. A unique characteristic of enzyme lipases is that their function can be altered according to the environment; of which becoming one of the favorite biocatalysts manipulated nowadays. The enzyme lipase unique characteristics will be explained in sub- chapters to come.

Microbial lipases would be the focus in this Chapter as they have the most diverse applications to compare with the rest. They have been utilized numerously in driving biocatalyses due to their broad substrate specificity, high enantioselectivity, as well as high stability reported in non-aqueous environment [8].

Structure of lipase

Lipases of different origins have different characteristics that eventually determine their properties. However, they have a shared characteristic in which all of them have a common fold known as alpha / beta (α/β) hydrolase fold [9], [10] with a conserved catalytic triad consisting of Ser, His, Asp/ Glu [8], [10]–[12]. Generally, the enzyme consists of a central hydrophobic β- sheets that is sandwiched between amphiphilic α-helices [13]. An example for general structure of lipase is shown in Figure 1 [9]:

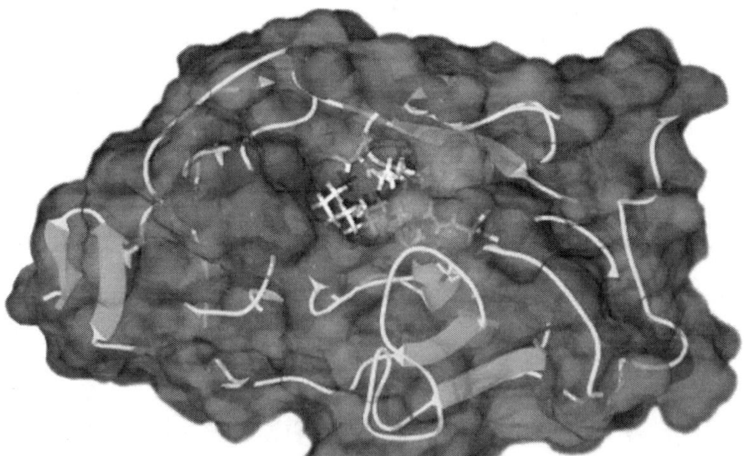

Figure 1. Structure of *Candida antarctica* lipase B that belongs to the α/β - hydrolase- fold super family. The α-helices are shown in red, the β-sheets in pale green, and the enzyme surface is in dark green.

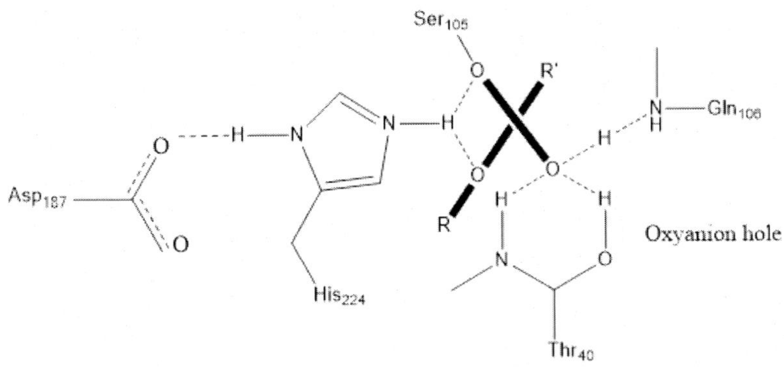

Figure 2. Structure configuration for CAL B active sites that compose of catalytic triad (Asp/Glu, Ser, His) with the bold lines indicating the transition part of substrate (Carlqvist et al., 2003).

It is shown in Figure 1 that the enzyme belongs to α/β-hydrolase fold superfamily of *Candida antarctica* lipase B (CAL B)[11]. The superfamily contains enzymes that evolved from a common ancestor and they are capable of catalyzing various hydrolysis reactions of esters, thioesters, peptides, and dissociate the carbon bonds in hydroxynitriles [14]. A total of 317 amino acid residues with molecular mass of 33 K are involved [12] to make the enzyme CAL B.

Deeper view into the structure of lipase (*eg:* CAL B) exposes the orientation of catalytic triad of amino acids that construct the active sites of the enzyme. It is vital to know the behaviour of site activations as the mechanism of lipase reaction in non-conventional media will be discussed later on. The diagram for active sites of CAL B is shown in Figure 2 [15].

As shown in Figure 2; Ser, His and Asp make up the catalytic triad whereas Gln and Thr together form the oxyanion hole. Thr is important to the stability of the enzyme as it contributes two (2) hydrogen bonds to the oxyanion hole. This involves the NH group of the backbone and OH group of the side chain [15]. Structure of oxyanion hole is made up of spatial arrangement of three (3) hydrogen bond donors; one (1) from Thr and two (2) are originated from the backbone amides of Thr and Gln [9]. Presence of oxyanion hole in attachment to the catalytic triad and negative charge on the carbonyl oxygen through the hydrogen bond [12] stabilizes the transition state.

Presence of stereospecificity pocket in lipase promotes high enantioselectivity of the enzyme. Triglycerides have strong affinity toward four substrate binding pockets positioning at *sn*-1, *sn*-2, *sn*–3, and in the oxyanion hole [16]. The difference in sizes and degree of lipophilicity of these pockets influence the enantio- preference of the enzyme. Thus, understanding the structure and techniques in manipulating lipase in molecular scale will enhance the bio-catalyst utilization in future [17].

It is known that catalytic triad of the active site is positioned in the conserved environment. Generally, the site is covered by a helical surface loop known as lid (or also known as flap). The catalytic site is inaccessible whenever the lid is in a 'closed' form as it is facing towards the catalytic conserved site [18]. Interfacial activation of the enzyme is required in order to shift the lid into the opening state. The lid remains close in a hydrophilic rich environment and will only shift when in contact with the hydrophobic environment (*eg*: oil / water interface) [19], [20].

Opening of the lid resulted in the exposure of hydrophobic core of catalytic site and the enzyme is ready for 'substrate docking' through formation of an oxyanion hole [21], [22]. An example for the mechanism of lid opening and activation of catalytic site of lipase were explained by Tejo and his colleagues [13]. They used lipase from *Candida rugosa* as the model. The lid has two (2) mobile amphiphilic α-helices of 62-69 amino residues and rotated almost 90° in the opening conformation. The helices extended away from the protein surface resulted in accessibility of Ser, His and Glu residues to the substrate. Lipases from *Pseudomonas aeruginosa* also have two (2) lids covering the conserved catalytic triad. This was explained by Cherukuvada and his colleagues in their research [8]. Their α/β hydrolase folding however differ from others as the first two (2) β-strands and one (1) α–helix are missing.

CAL B is the most distinctively different from the rest of lipases due to absence of lid to cover the catalytic site. Having said that, it does not require enzyme interfacial activation to react [23]. Furthermore, the structure arrangement is also different than most of other lipases as hydrophilic and hydrophobic regions are equally distributed over the surface [10].

2. Lipases in Industries

Utilization of bio-catalysts in industries are divided into three main areas where they are applied as technical enzymes, food enzymes and animal-feed enzymes. Summary of global market for industrial enzymes is shown in Figure 3.

As lipase assisted reaction requires milder operating conditions over chemical catalysed synthesis, as well as high affinity of enzyme lipases toward accepting triacylglycerides as substrates and their chemo-, enantio- and regio- specificities, lipases became one of the bio-catalysts of high interest utilized in industries [6]. Lipases have emerged to be the third largest group exploited after proteases and carbohydrases based on the total sales volume worldwide [17].

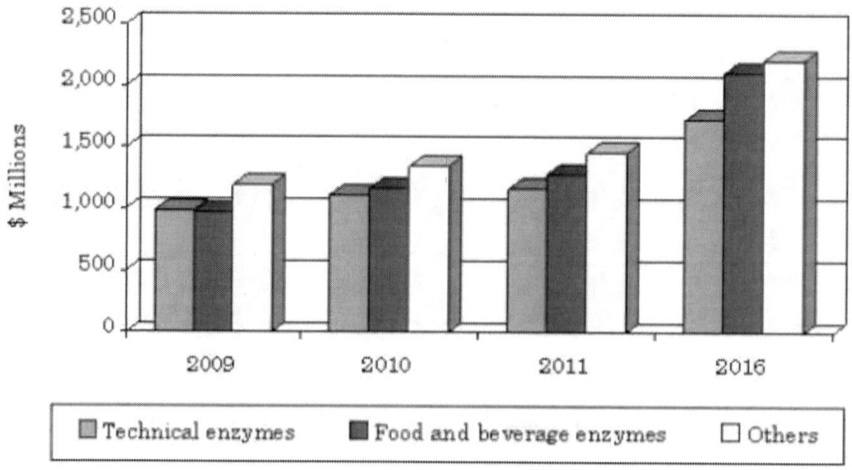

Figure 3. Global market for industrial enzymes ($ Millions) with projection until year 2016.

Table 1. Some of the applications of lipases in industries

Industry	Action	Product / Application
Detergents	Hydrolysis of fats	Fabric cleaning
Foods	Hydrolysis of milk fat, Cheese ripening, butter fat, flavor, aroma enhancers	Improvement of the quality and shelf-life of foods
Chemicals	Synthesis with high enantioselectivity	As a chiral building blocks
Pharmaceuticals	Transesterification	Production of digestive aids and specialty lipids
Cosmetics	Synthesis	Moisturizers, cleanser
Cleaning	Hydrolysis	Removal of fats

Table 2. Some of the commercially available lipases

Type	Source	Application	Company
	C. rugosa	Organic synthesis	Fluka, Sigma
Fungal	*C. antarctica*	Organic synthesis	Novo Nordisk
	R. miehei	Detergent	Novo Nordisk
	T. lanuginosus	Food processing	Novo Nordisk
Bacterial	*Bukholderia cepacia*	Organic synthesis	Fluka
	P. alcaligenes	Detergent	Genencor
	P. mendocina	Organic synthesis	Genencor

A unique characteristic of enzyme lipase *viz.* high lipase stability in non-conventional environments (organic solvents, micro-aqueous systems) allows production of detergents, food emulsifiers, cosmetics, stabilizers, biodiesels, pharmaceutical products that require minimal amount of water during the process. The various major applications of lipase in industries are shown in Table 1 [24] and lipases that are commercialized are shown in Table 2 [17].

Miyoshi Oil & Fat Co., Japan stated that the overall costs for production of soaps using *Candida cylindracea* lipase as catalyst is cheaper than conventional route [25]. One of the breakthroughs in using lipase enzyme as bio-catalyst is production of biodiesel. In emerging towards greener technology and to sustain the world continuous demand for energy source, researches on oleochemicals are shifting to biodiesel production. Lipase used as the driving force for transesterification of selected renewable source of oil to produce biodiesel [26], [27] was reported by Teo and his colleagues where they synthesized biodiesel from microalgae oil using lipase (EC 3.1.1.3) from *C. rugosa* [4].

3. Reaction of Lipase in Non-Conventional Media

Non-conventional media refers to the system with very low water content or uses media other than an aqueous as the reaction solvent. Presence or absence of water greatly influences the behavior of biocatalyst involves in the reaction. For instance, lipases undergone lypolitic, hydrolytic activities in interfacial aqueous based reaction as shown in Figure 4.

However, the role is reversed when lipases are in micro-aqueous environment. Instead of behaving as hydrolases, they shifted their behaviour and became esterase [28]. It has been known that the non- aqueous reaction media increases the enantioselectivity and stability of the enzymes thus increase their versatility in such environment [29], [30]. Apart from that, the micro-aqueous environment in non-conventional media catalysis prevents microbial contamination into the system. The choice of solvent is therefore crucial in determining the behaviour of enzyme as they can cause modifications in the conformation of enzyme thus affecting the catalytic efficiency or enzyme specificity [31]. This is shown in Figure 5.

Mechanism of lipase bio-catalyzed esterification involves two (2) tetrahedral intermediates. Initially binding of substrate occurred due to nucleophilic attack (e.g., acid) towards serine (ser) residue on the active catalytic triad prior to water losing reaction by the intermediate towards the enzyme-substrate complex. The other intermediate is later formed by nucleophilic attack by an alcohol which causes addition of hydroxyl group to the carboxyl group thus causing rearrangement to release the product formed followed with regeneration of lipase enzyme for further usage. The schematic diagram for the mechanism is shown in Figure 6. [16]

Efficiency of lipases bio-catalysed reaction in micro-aqueous environment is greatly influenced by water activity. Water is said to be the most crucial parameter that need to be controlled in order to prevent the reversal behaviour of lipases *viz.* from esterase to hydrolases. Further explanation on effect of water activity and water content toward lipases catalysed reaction will be further explained later on.

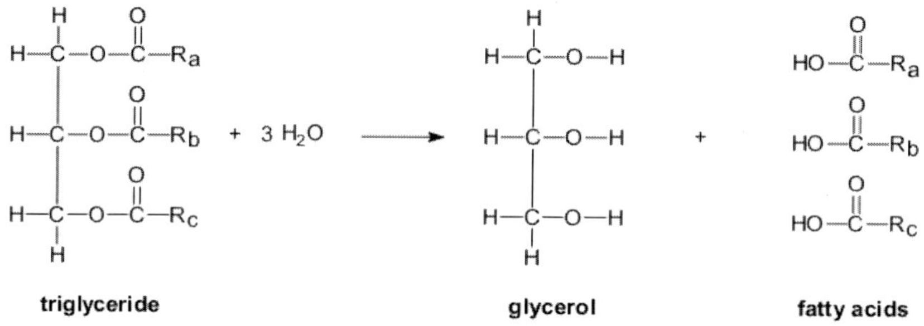

Figure 4. Hydrolysis reaction produces glycerol and fatty acids, which are carboxylic acids derived from fats and oils. In the fatty acids, R_a, R_b, and R_c, represent groups of carbon and hydrogen atoms in which the carbon atoms are attached to each other in an unbranched chain.

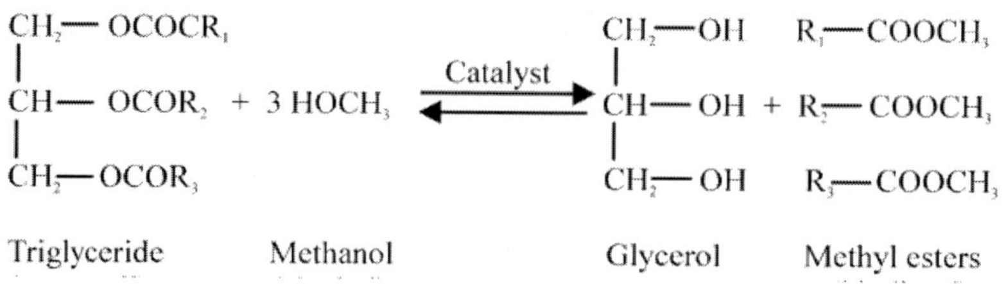

Figure 5. Chemical route showing the transesterification method utilizing triglyceride as acyl group donor and methanol as acyl group acceptor.

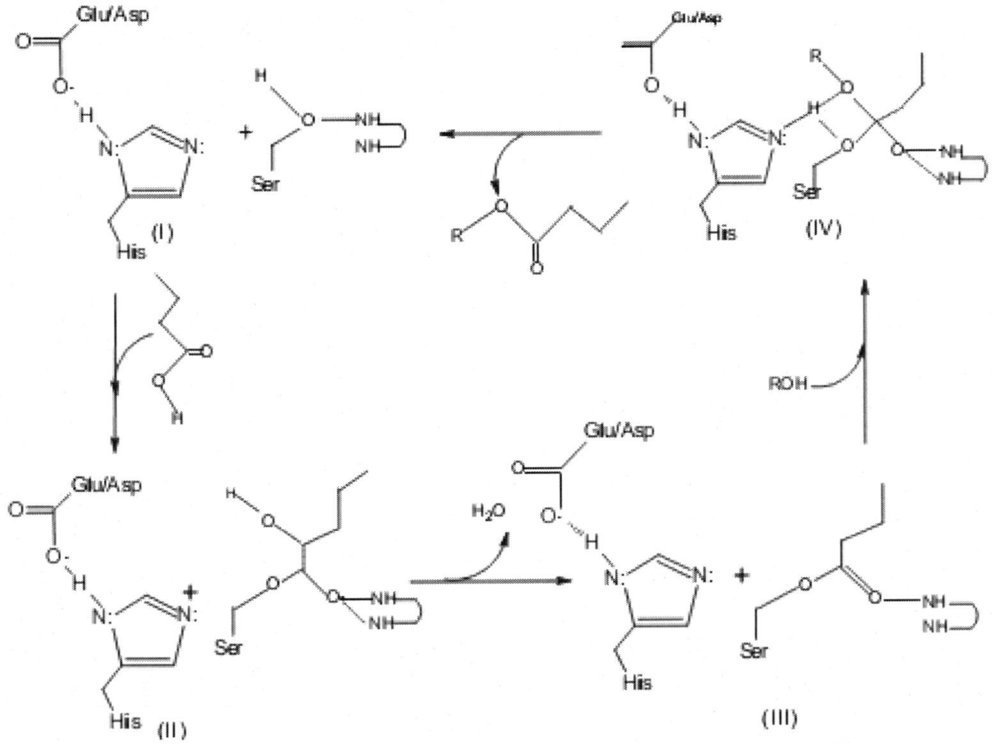

Figure 6. Mechanism of lipase-catalyzed esterification. (1) Binding of substrate, activation of nucleophilic serine residue by neighboring histidine and nucleophilic attack of the substrate's carbonyl carbon atom by Ser. (2) Transient tetrahedral intermediate that loses a water molecule to give an acyl enzyme complex. (3). An alcohol molecule attacks the acyl enzyme complex (nucleophilic attack) to give (4) another tetrahedral intermediate that releases an ester molecule to convert the lipase to its native form.

3.1. Reaction of Lipase in Organic Solvents

Organic solvents are the most well documented media used as substituents to water medium in the field of lipase bio-catalyzed reactions in non-aqueous media. Increase in carbon chain numbers elevates the hydrophobicity of the organic solvents thus making them insoluble in water. For example, hydrophobicity of alcohols was found higher in higher number of carbon chain length. Starting from butanol (4-C saturated chain length alcohol) and onwards, the solvents were observed to form two (2) phase solutions with water.

Organic solvents provide advantages as media for lipase catalyzed biosynthesis of products as they provide better solubility for reactants and products, forcing the forward reaction by reducing water dependent side reaction (Le Chatelier's principle), ease in lipase recovery in post reaction, higher lipase stability in system as water was found to inactivate enzymes at higher temperature as well as it may prevent microbial contaminations during the process (due to a very restricted water present) [32]–[34].

As mentioned earlier in sub-chapter 1.3, the choice of solvents is crucial in determining the behavior of enzyme. Solvents of different *n*-octanol-water partition coefficient (Log *P*) determine the lipase catalytic efficiency as well as the specificity of enzyme [35], [36]. At *Log P* values < 2, the solvent is hydrophilic; at 2 < *log P* < 4, the effect of water-solvent is unpredictable and at *log P* > 4, the solvent is very hydrophobic in nature. It was found that rate of lipase bio-catalyzed esterification in organic solvent is higher when solvents of higher *log P* are applied as the media [16], [37].

However, although lipase bio-catalyzed reaction (i.e., esterification) in organic solvents requires microaqueous environment, a critical amount of water activity (a_w) less than 0.02 is required in order to ensure the effectiveness of bio-catalysis in organic solvent [38]. Water, which behaves as a lubricant, favors an interaction between the enzyme's active sites and the ligands thus increases the amino acids mobility during catalysis.

3.2. Reaction of Lipase in Ionic Liquids (IL)

Ionic liquids are salts (ions) that do not crystallize at room temperature that have physical characteristic of low melting temperature *viz.* below 100°C [39]. The first work using ionic liquid was reported by Walden (1914) where he synthesized ethylammonium nitrate with melting temperature of 12°C. Ionic liquids are named as green solvents as they have no measurable vapor pressure which made them environmental friendlier than organic solvent [3], [39], [40]. ILs provide advantages over other reaction solvents due to their high thermal stability, high catalytic activity, low viscosity, high solvation power for wide range of organic and inorganic compounds and they are non-flammable in atmosphere [41]–[44].

The first successful bio-catalysed reaction in IL was done back in year 2000 as reported by Lauand his colleagues [45]. They applied *Candida antarctica* (CAL B) lipase as the driving force for reaction. The processes conducted were transesterification, ammoniolysis and epoxidation reactions. They claimed that the processes performed were better than those in organic solvents as enzyme showed higher substrate selectivity as well as greater stability [3].

However, despite advantages of ILs toward lipase bio-catalysed reactions, solvation of free lipase into the IL resulted in deactivation of the enzyme [40]. Apart from that, presence of counter ions in the system affects the stability of enzyme as well. In a particular study, Rantwijk and his colleagues reported that nitrate and lactate counter ions for substrate methylsulfate resulted in lipase inactivation [46]. Thus, to overcome the problems, immobilization of lipase onto solid carriers is crucial in ILs based bio-catalysis. Thus, simply said, a heterogenous form of reaction is required to ensure the feasibility of reactions in ILs [31]. Viability to recycle the enzyme lipase without affecting its integrity and efficiency in the downstream processing are among the biggest challenges that need to be tackled when upscaling the system [31].

3.3. Reaction of Lipase in Supercritical Fluids (SCF)

Supercritical fluid technology is said to be the emerging alternative method explored for biocatalysis in non-conventional media system. The system allows high substrate mass transfer, easy tunable of SCFs solvation power (by altering temperature and pressure of system) and provides high efficiency of downstream processing, thus making the technology to be one of the favorites nowadays [47], [48]. Change in pressure of SCF influences the parameters that affect the specificity and stability of enzyme. This is because the change in pressure will result in alteration of partition coefficient of solvent (log *P*), dielectric constant as well as the Hildebrant solubility [49].

SCF technology has been applied widely in enzymatic reactions (enzymatic oxidation, esterifications, hydrolysis, enantioselective synthesis) as the conditions assist functionalization of enzyme in the reactions. For example, the temperature required to run the reaction is relatively low and this does not affect the integrity of enzyme [47]. Apart from that, improvement in the product, substrate recovery processes were seen as the system provides medium for their recovery [50], [51]. From thermodynamic point of view, Knez and his colleagues claimed that the SCF assisted reaction require low energy input to run based on the calculated enthalpy, entropy and free energy [52].

An example for a widely exploited fluid in the system is carbon dioxide (CO_2). Supercritical carbon dioxide (scCO_2) is said to be an excellent choice as it is cheap, safe to environment, non-flammable, possesses a critical temperature of 31.1°C and a moderate critical pressure [53], [54]. It has the ability to extract hydrophobic compounds from some ILs and that very same ILs do not dissolve in scCO_2. The physical advantages improve the downstream processing whereby the organic substances can be extracted out from ILs without producing any cross contamination [55], [56]. Besides, one of the important criteria in producing successful reaction is an efficient mixing of the reaction mixture. ScCO_2 overcomes this difficulty by accelerating the mass transfer limited enzymatic reaction [57].

However, there were several cases in which lipase deactivation was observed in reaction supercritical media. Perhaps, explanations for this were firstly, rapid release of supercritical fluid (*eg*: CO_2) interacts with the bound water of the enzyme thus resulted in structural changes in enzyme which eventually cause inactivation [51], [58]. Secondly, the high pressure (>400 MPa) might contribute towards enzyme inactivation. However, the mentioned high pressure is rarely applied as normally, a pressure range between 10–40 MPa is suffice. Within this range, only reversible conformation may occur and does not affect the overall performance of enzyme [48]. Due to formation of covalent complexes between CO_2 and free amino groups on the surface of the lipase, resulted carbamates neutralized amino acid (lysine) residue thus affecting the integrity of enzyme [59].

4. Effect of Water Towards Lipase Bio-Catalysed Reaction in Non-Conventional Media

The thermodynamic water activity (a_w) is essential in describing the level of enzyme hydration in reaction [60]. Variations in thermodynamic water activity in the system also affect the rate of reaction and stereospecificity in chiral reactions [33], [61]. In native lipase structure, there are two types of enzyme water that are differentiated in organic solvents, *viz.* the 'inside' water that is bound to the interior of the protein thus forming hydrogen bonds, which affect conformation of the active site of enzyme, and 'contract' water that is weakly bonded to the surface of the enzyme and exchanges rapidly with non-enzyme associated water in the solvent [62].

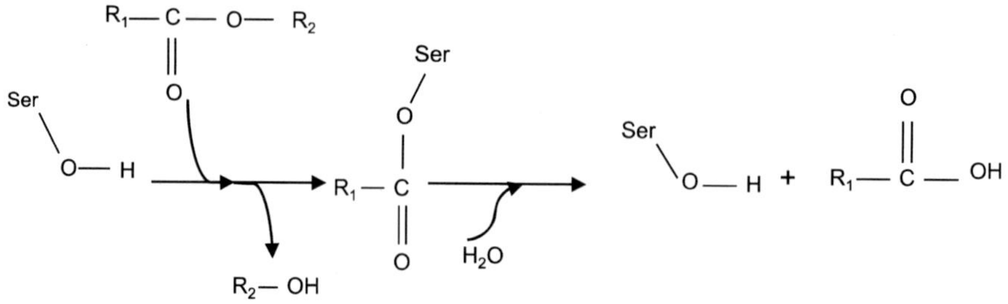

Figure 7. Hyrolysis of Acyl donor- Lipase complex.

Reversibility of lipase catalysed reaction is influenced by amount of water present in the system [63], [64]. In presence of large amount of water molecules, they behave as activators for lipase to hydrolyze the intermediate complex (substrate-enzyme complex). Here, the water acts as a competitive substrate thus enhancing the hydrolysis rate. This occurs in the environment with water activity of more than 0.1 [38]. The mechanism for the reverse reaction is shown in the reaction diagram as in Figure 7.

There are several methods for continuous removal of water in a reaction in forcing the forward reaction. The use of molecular sieves is one of the preferred ways to remove water generated during an esterification [28]. However, the amount of molecular sieve need to be increased with the amount of solvent as the reaction medium is applied. This may result in mass transfer limitation due to difficulties in mixing [65], [66]. Reaction in absence of organic solvent as the reaction medium was also attempted in the esterification process [66]. Azeotropic distillation, gas sparging as well as pervaporation were also studied and applied to systems in order to remove generated water from the reaction [67]–[69].

5. Improvement of Enzyme Lipase as Biocatalyst Non-Conventional Media

Mass exploitation of lipases as biocatalysts in various reactions has made the work development to be one of the crucial factors in ensuring the system effectiveness. Methods for enzyme lipase improvement have been studied and some even developed. Approaches in improving the versatility of enzyme can be divided into three ways *viz.* physicochemical methods, reaction engineering methods and molecular biology methods [34].

One of the easiest and yet effective methods in improving the lipase performance in non-conventional media is by immobilizing it onto a suitable carrier. This enhances the enzyme's activity, stabilizes the enzyme over many cycles, and even assists in the catalyst recovery post reaction [70]. The method is the key parameter to enhance the operational performance of enzymes as reported by several researchers [71]–[73]. As recently reported by Idris and Bukhari, the immobilization technique conditions employed during immobilization, type of support, nature of solvent and type of reactor intensively influenced the immobilized enzyme [74]. It is therefore important to monitor the immobilization technique in order to conserve the enzyme's integrity as the lipase needs to maintain its broad substrate specificity for organic synthesis [18].

To have fully functional lipases in non-conventional media, it is necessary to immobilize them in the open form (and does not cause protein denaturation). This can be achieved by applying solid carriers with surface activating groups and/or proper conjugation chemistry [75]. Here, the immobilization is performed through adsorption onto suitable hydrophobic matrices. Examples of hydrophobic supports for lipase immobilization are methylated silica, phenyl-Sepharose, poly-(ethylene glycol)-Sepharose, polypropylene particles, polypropylene hollow fibers as well as glass beads coated with hydrophobic material [16].

Examples for industrially successful immobilized lipases are Novozyme 435 (immobilized *Candida Antarctica* lipase B) and RM-IM (Immobilized *Rhizomucor miehei* lipase) [18]. Novozyme 435 is adsorbed onto a resin commercially known as Lewatit OC 160. The carrier is macroporous, divinlybenzene-crosslinked polymer based methacrylic ester. The enzyme loading is between 8.5 to 20% (w/w) at 50–100 μm wide front in the outer part of the beads [76], [77]. Novozyme 435 is mainly applied for the resolution of chiral intermediates and other high-price specialty chemicals where cost is of secondary importance compared to the performance and purity of catalyst. Many researchers have reported on the effectiveness of Novozyme 435 as an extraordinary catalyst for both small organic molecules and, more recently, for polymerization reactions [74].

Currently, the works of enzyme immobilization are enhanced using nanomaterials. These carriers highly assist in balancing the key factors for biocatalysis effectiveness *viz.* surface area, mass transfer limitations, and effective enzyme loading [70], [78]. Examples for nanomaterials applied into the systems were liquid- modified magnetic nanoparticles [79], hydrophobic magnetic nanoparticles [80], carbon nanotubes (CNTs) [81], [82] as well as magnetic multi-walled carbon nanotubes (M – MWNTs) [70]. Lipases immobilized onto these supports showed high stability at high temperature, low- mass transfer resistance, high stability, high activity, and high reusability. The lipases were able to undergo process for biodiesel production [2], resolution of racemic compounds, and others [22], [83].

Apart from improving the quality of enzyme through immobilization method, a technique named directed evolution is known to be powerful in improving the enzyme's properties [84]. The enzyme genes undergone mutagenesis and/or recombination before they were expressed in a host and screened for improved variants [85]. Resulted variants were found to have improved enzyme properties, for examples; high enzyme activity in organic solvents, high substrate specificity as well as high thermostability [84], [86]. For instance, work conducted by Liebeton and colleagues used extracellular lipase from *Pseudomonas aeruginosa* as the target enzyme [85]. The resulted variant showed high amount of glycine (gly) residues in the folding which in return increased the conformational flexibility of the enzyme. Increase in enzyme's flexibility has been identified as one of the key factors in determining the substrate selectivity for lipases [87].

CONCLUSION

In conclusion, lipases are enzymes of extremely high versatility that can be applied in industries to accommodate the world demand for green syntheses technologies. Several important considerations need to be taken into account in order to develop an efficient system optimization. Inadequate understanding of the enzymes' fundamental structure construction and folding will lead to process developmental loss as optimization in physical parameters in upscale reactions are greatly influenced by the nature of the enzyme themselves. Factors such as immobilization technique, microenvironment of enzyme, pH of immobilization, nature of solvent, properties of support and conditions of application play significant roles when formulating enzymes for industrial applications. The approach to utilise a combination of different immobilization techniques can contribute to progress in meeting the required performance of enzyme catalysis. The integrated use of ionic liquids and lipases immobilized on activated nanoparticles may be promising in green synthesis technologies.

Acknowledgments

The authors acknowledge the financial assistance by Ministry of Higher Education (MOHE) Malaysia for the HICOE (Vote No: 4J625) funding source.

References

[1] Poliakoff, M., Fitzpatrick, J., Farren, T., et al. (2002). Green chemistry: science and politics of change. *Science*, *297*: 807-810.

[2] Dussan, K., Cardona, C., Giraldo, O., et al. (2010). Analysis of a reactive extraction process for biodiesel production using a lipase immobilized on magnetic nanostructures. *Biores Technol.*,*101*: 9542-9549.

[3] Andreani, L. and Rocha, J. D. (2012). Use of ionic liquids in biodiesel production: a review. *Braz J. Chem. Eng.*, *29*: 1-13.

[4] Teo, C., Jamaluddin, H., Zain, N. A. and Idris, A. (2014). Biodiesel production via lipase catalysed transesterification of microalgae lipids from Tetraselmis sp. *Renew Energy*, *68*: 1-5.

[5] Idris, A., Misran, E., Hassan, N., et al. (2012). Modified PVA-alginate encapsulated photocatalyst ferro photo gels for Cr (VI) reduction. *J. Hazard.*, *227-228*: 309-316.

[6] Hasan, F., Shah, A., and Hameed, A. (2006). Industrial applications of microbial lipases. *Enz Microbial Technol.*, *39*: 235-251.

[7] Barros, M., Fleuri, L. and Macedo, G. A. (2010). Seed lipases: sources, applications and properties - A review. *Braz J. Chem. Eng.*, *27*: 15-29.

[8] Cherukuvada, S., Seshasayee, A., Raghunathan, K., et al. (2005). Evidence of a Double-Lid Movement in *Pseudomonas aeruginosa* Lipase: Insights from Molecular Dynamics Simulations. *PLoS Comput. Biol.*, *1*: 0182-0189..

[9] Magnusson, A., Rotticci-Mulder, J., Santagostino, A., et al. (2005). Creating Space for Large Secondary Alcohols by Rational Redesign of *Candida antarctica* Lipase B. *Chem BioChem.*, *6*: 1051-1056.

[10] Trodler, P. and Pleiss, J. (2008). Modeling structure and flexibility of *Candida antarctica* lipase B in organic solvents. *BMC Struct. Biol.*, *8*: 9-18.

[11] Ollis, D., Cheah, E., Cygler, M., et al. (1992). The α/β hydrolase fold. *Protein Eng.*, *5*: 197-211.

[12] Uppenberg, J., Niklas, O., Martin N., et al., (1995). Crystallographic and molecular-

modeling studies of lipase B from *Candida antarctica* reveal a stereospecificity pocket for secondary alcohols. *Biochem., 34*: 16838-16851.

[13] Tejo, B., Salleh, A. and Pleiss, J. (2004). Structure and dynamics of *Candida rugosa* lipase: the role of organic solvent. *J. Mol. Model.,10*: 358-366.

[14] Holmquist, M. (2000). Alpha Beta-Hydrolase Fold Enzymes Structures, Functions and Mechanisms. *Curr Protein Pept Sci., 1*: 209-235.

[15] Carlqvist, P., Eklund, R., Hult, K., et al. (2003). and Brinck, T., "Rational design of a lipase to accommodate catalysis of Baeyer?Villiger oxidation with hydrogen peroxide," *J. Mol. Model., 9*: 164-171.

[16] Verma, M., Azmi, W., and Kanwar, S. (2008). Microbial lipases: At the interface of aqueous and non-aqueous media. *Acta Microbiol Immunol Hung., 55*: 265-294.

[17] Jaeger, K-E. and Reetz, M. T. (1998). Microbial lipases form versatile tools for biotechnology. *Trends Biotechnol.,16*: 396-403.

[18] Petkar, M., Lali, A., Caimi, P., et al. (2006). Immobilization of lipases for non-aqueous synthesis. *J Mol Catal Enz., 39*: 83-90.

[19] Brzozowski, A., Derewenda, U., Derewenda, Z., et al. (1991). A model for interfacial activation in lipases from the structure of a fungal lipase-inhibitor complex. *Nature, 351*: 491-494.

[20] Derewenda, U., Brzozowski, A., Lawson, D., et al. (1992). Catalysis at the interface: the anatomy of a conformational change in a triglyceride lipase. *Biochem., 3*: 1532-1541.

[21] Yolanda C., Allan S., Girona, V., et al. et al. (1999). Interfacial Control of Lid Opening in Thermomyces lanuginosa Lipase. *Biochem., 39*: 413-423.

[22] Yilmaz, E. (2003). Orlistat-induced molecular bio-imprinting of microbial lipase," *World J. Microbiol. Biotechnol., 19*: 161-165.

[23] Martinelle, M., Holmquist, M. and Hult, K., (1995). On the interfacial activation of *Candida antarctica* lipase A and B as compared with *Humicola lanuginosa* lipase. *Biochim Biophys Acta - Lipids Lipid Metab., 1258*: 272-276.

[24] Vulfson, E. N. (1994). *Industrial applications of lipases*. ci.nii.ac.jp.

[25] McNeill, G., Shimizu, S. and Yamane, T. (1991). High-yield enzymatic glycerolysis of fats and oils. *J Am Oil Chem Soc., 68*: 1-5.

[26] Noureddini, H., Gao, X. and Philkana, R. (2005). Immobilized *Pseudomonas cepacia* lipase for biodiesel fuel production from soybean oil. *Biores Technol., 96*: 769-777.

[27] Shweta S., Sharma, S. and Gupta, M. (2004). Biodiesel Preparation by Lipase-Catalyzed Transesterification of Jatropha Oil. *Energy Fuels, 11*: 154-159.

[28] Ariffin, M., Annuar, M. and Heidelberg, T. (2014). Surfactant synthesis via lipase esterification of methyl α-d-glucopyranoside with selected aliphatic carboxylic acids," *J Surfactants Deterg., 17*: 683-692.

[29] Linko, Y-Y., Lamsa, M., Huhtala, A., et al. (1995). Lipase biocatalysis in the production of esters. *J Amr Oil Chem. Soc., 72*: 1293-1299.

[30] Turner, N. and Vulfson, E. (2000). At what temperature can enzymes maintain their catalytic activity? *Enz Microb Technol., 27*: 108-113.

[31] Krieger, N., Bhatnagar, T., BAratti, J., et al. (2004). Non-aqueous biocatalysis in heterogeneous solvent systems. *Food Technol Biotechnol., 42*: 279-286.

[32] Gandhi, N., Patil, N., Sawant, S., et al. (2000). Lipase-Catalyzed Esterification. *Catal Rev., 42*: 439-480.

[33] Klibanov, A. M. (2001). Improving enzymes by using them in organic solvents. *Nature,*

409: 241-246.

[34] Bornscheuer, U. T. (2010). Lipases, Synthesis of Chiral Compounds, Aqueous and Organic Solvents. In: *Encyclopedia of Industrial Biotechnology*. Hoboken, NJ, USA: John Wiley & Sons, Inc., pp. 1-12.

[35] Lortie, R. (1997). Enzyme catalyzed esterification. *Biotechnol Adv., 15*: 1-15.

[36] Lu, J., Nie, K., Wang, F., et al. (2008). Immobilized lipase *Candida* sp. 99-125 catalyzed methanolysis of glycerol trioleate: Solvent effect. *Biores Technol., 99*: 6070-6074.

[37] Li, Q., Li, G., Yu, S., et al. (2011). Ring-opening polymerization of ε-caprolactone catalyzed by a novel thermophilic lipase from *Fervidobacterium nodosum. Process Biochem.*, 46: 253-257.

[38] Leonard, V., Fransson, L., Lamare, S., et al. (2007). A Water Molecule in the Stereo specificity Pocket of *Candida antarctica* Lipase B Enhances Enantio selectivity towards Pentan-2-ol. *Chem Bio Chem., 8*: 662-667.

[39] Gordon, C. M. (2001). New developments in catalysis using ionic liquids. *Appl Catal Gen., 222*: 101-117.

[40] Park, S. and Kazlauskas, R. (2003). Biocatalysis in ionic liquids - advantages beyond green technology. *Curr Opin Biotechnol., 14*: 432-437.

[41] Earle, M. J. and Seddon, K. R. (2002). Ionic Liquids: Green Solvents for the Future. *Pure Appl Chem., 72*: 10-25.

[42] Welton, T. (2004). Ionic liquids in catalysis. *Coord Chem Rev., 248*: 2459-2477.

[43] Plechkova, N. V. and Seddon, K. R. (2008). Applications of ionic liquids in the chemical industry. *Chem Soc Rev.*, 37: 123-150.

[44] De Diego, T., Lozano, P., Abad, M., et al. (2009). On the nature of ionic liquids and their effects on lipases that catalyze ester synthesis. *J Biotechnol.*, 140: 234-241.

[45] Lau, M., Van Rantwijk, F., Seddon, K., et al. (2000). Lipase-Catalyzed Reactions in Ionic Liquids. *Org Lett., 2*: 4189-4191.

[46] van Rantwijk, F., Lau, M. and Sheldon, R. (2003). Biocatalytic transformations in ionic liquids. *Trends Biotechnol., 21*: 131-138.

[47] Knez, Z. and Habulin, M. (2002). Compressed gases as alternative enzymatic-reaction solvents: A short review. *J Supercrit Fluids, 23*: 29-42.

[48] Rezaei, K., Temelli, F. and Jenab, E. (2007). Effects of pressure and temperature on enzymatic reactions in supercritical fluids. *Biotechnol. Adv., 25*: 272-280.

[49] Nagesha, G., Manohar, B. and Udaya S. (2004). Enzymatic esterification of free fatty acids of hydrolyzed soy deodorizer distillate in supercritical carbon dioxide. *J Supercrit Fluids, 32*: 137-145.

[50] Hammond, D., Karel, M., Klibanov, A., et al. (1985). Enzymatic reactions in supercritical gases. *Appl Biochem Biotechnol., 11*: 393-400.

[51] Nakamura, K. (1990). Biochemical reactions in supercritical fluids. *Trends Biotechnol., 8*: 288-292.

[52] Knez, Z. (1998). Habulin, M., and Krmelj, V., "Enzyme catalyzed reactions in dense gases. *J. Supercrit. Fluids, 14*: 17-29.

[53] Dumont, T., Barth, D., Corbier, C., et al. (1992). Enzymatic reaction kinetic: Comparison in an organic solvent and in supercritical carbon dioxide. *Biotechno. Bioeng., 40*: 329-333.

[54] Srivastava, S., Madras, G. and Modak, J. (2003). Esterification of myristic acid in supercritical carbon dioxide. *J Supercrit Fluids, 27*: 55-64.

[55] Lynnette A., Zhiyong G. and Brennecke, J. (2001). High-Pressure Phase Behavior of Ionic Liquid/CO2 Systems. *J Phys Chem., 105*: 2437-2444.

[56] Blanchard, L. and Brennecke, J. (2001). Recovery of Organic Products from Ionic Liquids Using Supercritical Carbon Dioxide. *Ind Eng Chem Res., 40*: 287-292.

[57] Balcao, V., Paiva, A. and Xavier M. (1996). Bioreactors with immobilized lipases: State of the art. *Enz Microb Technol., 18*: 392-416.

[58] Lin, T-J., Chen, S-W. and Chang, A-C. (2006). Enrichment of n-3 PUFA contents on triglycerides of fish oil by lipase-catalyzed trans-esterification under supercritical conditions. *Biochem Eng J., 29*: 27-34.

[59] Kamat, S., Critchley, G., Beckman, E., et al. (1995). Biocatalytic synthesis of acrylates in organic solvents and supercritical fluids: III. Does carbon dioxide covalently modify enzymes? *Biotechnol Bioeng., 46*: 610-620.

[60] Bell, G., Janssen, A. and Halling, P. J. (1997). Water activity fails to predict critical hydration level for enzyme activity in polar organic solvents: Interconversion of water concentrations and activities. *Enz Microb Technol., 20*: 471-477.

[61] Halling, P. J. (1994). Thermodynamic predictions for biocatalysis in nonconventional media: Theory, tests, and recommendations for experimental design and analysis. *Enz Microb Technol., 16*: 178-206.

[62] Nakasako, M. (1994). Water-protein interactions from high-resolution protein crystallography. *Philos Trans R Soc Lond B Biol Sci., 359*: 1191-1206, 2004.

[63] Sang Yoo, I., Joon Park, S. and Hee Y. (2007). Enzymatic Synthesis of Sugar Fatty Acid Esters. *J Ind Eng Chem., 13*: 1-6.

[64] Hai, H., Kura, H., Takahashi, M., et al. (2010). Facile synthesis of Fe 3 O 4 nanoparticles by reduction phase transformation from γ-Fe Fe_2o_3 nanoparticles in organic solvent. *J colloid interface, 341*: 194-199.

[65] Yan, Y., Bornscheuer, U. and Schmid, R. (1999). Lipase-catalyzed synthesis of vitamin C fatty acid esters. *Biotechnol Lett., 21*:1051-1054.

[66] Zhang, Y., Dube, M. ., McLean, D., et al. (2003). Biodiesel production from waste cooking oil: 1. Process design and technological assessment. *Biores Technol., 89*: 1-16.

[67] Kwon, S., Han, J. and Rhee, J. (1995). Production and in situ separation of mono- or diacylglycerol catalyzed by lipases in n-hexane. *Enz Microb Technol., 17*: 700-704.

[68] Won, K. and Lee, S. (2001). Computer-Aided Control of Water Activity for Lipase-Catalyzed Esterification in Solvent-Free Systems. *Biotechnol Prog., 17*: 258-264.

[69] Belafi-Bako, K., Dormo, N., Ulbert, O., et al. (2002). Application of pervaporation for removal of water produced during enzymatic esterification in ionic liquids. *Desalination, 149*: 267-268.

[70] Tan, H., Feng, W. and Ji, P. (2012). Lipase immobilized on magnetic multi-walled carbon nanotubes. *Biores Technol., 115*: 172-176.

[71] Sheldon, R. A. (2007). Enzyme Immobilization: The Quest for Optimum Performance. *Adv Synth Catal.*, vol. *349*: 1289-1307.

[72] Idris, A., Zain, N. and Suhaimi, M. (2008). Immobilization of Baker's yeast invertase in PVA–alginate matrix using innovative immobilization technique. *Process Biochem., 43*: 331-338.

[73] Zain, N., Suardi, S. and Idris, A. (2010). Hydrolysis of liquid pineapple waste by invertase immobilized in PVA–alginate matrix. *Biochem Eng J., 50*: 83-89.

[74] Idris, A. and Bukhari, A. (2012). Immobilized *Candida antarctica* lipase B: Hydration,

stripping off and application in ring opening polyester synthesis. *Biotechnol. Adv.*, *30*: 550-563.

[75] Bastida, A., Sabuquillo, P., Armisen, P., et al. (1998). A single step purification, immobilization, and hyperactivation of lipases via interfacial adsorption on strongly hydrophobic supports. *Biotechnol Bioeng.*, 58: 486-493.

[76] Nakaoki, N., Miller, A., Miller, E., et al. (2005). *Candida antarctica* lipase B catalyzed polymerization of lactones: Effects of immobilization matrices on polymerization kinetics & molecular weight. *Ind Biotechnol.*, *1*: 126-134.

[77] Chen, B., Hu, J., Miller, E., et al. (2008). *Candida antarctica* Lipase B Chemically Immobilized on Epoxy-Activated Micro- and Nanobeads: Catalysts for Polyester Synthesis. *Biomacromol.*, *9*: 463-471.

[78] Youdan W., Pratixa P., Kevin L., et al. (2006). Nanostructured Biosensors Built by Layer-by-Layer Electrostatic Assembly of Enzyme-Coated Single-Walled Carbon Nanotubes and Redox Polymers. *Lanqmuir*, *22*: 9776-9783.

[79] Jiang, R., Zhu, H., Li, J., et al. (2016). Fabrication of novel magnetically separable BiOBr/CoFe2O4 microspheres and its application in the efficient removal of dye from aqueous phase by an environment-friendly and economic approach. *Appl Surf.*, *364*: 604-612.

[80] Lee, D-G., Ponvel, K., Kim, M., et al. (2009). Immobilization of lipase on hydrophobic nano-sized magnetite particles. *J Mol Catal B Enz.*, *57*: 62-66.

[81] Gao, Y. and Kyratzis, I. (2008). Covalent Immobilization of Proteins on Carbon Nanotubes Using the Cross-Linker 1-Ethyl-3-(3-dimethylaminopropyl)carbodiimide—a Critical Assessment. *Bioconjug Chem.*, *19*:1945-1950.

[82] Feng, W. and Ji, P. (2011). Enzymes immobilized on carbon nanotubes. *Biotechnol Adv.*, *29*: 889-895.

[83] Chaubey, A., Parshad, R., Taneja, S., et al. (2009). *Arthrobacter* sp. lipase immobilization on magnetic sol–gel composite supports for enantioselectivity improvement. *Process Biochem.*, *44*: 154-160.

[84] Arnold, F. H. and Volkov, A. A. (1999). Directed evolution of biocatalysts. *Curr Opin Chem Biol.*, *3*: 54-59.

[85] Liebeton, K., Zonta A., Schemossek, K., et al. (2000). Directed evolution of an enantioselective lipase. *Chem Biol.*, *9*: 709-718.

[86] Kuchner, O. and Arnold, F. (1997). Directed evolution of enzyme catalysts. *Trends Biotechnol.*, *15*: 523-530.

[87] Mats H., Daniel C. and Cygler, M. (1997). Identification of Residues Essential for Differential Fatty Acyl Specificity of Geotrichum candidum Lipases I and II. *Biochem.*, *36*: 15019-15025.

SECTION B. INDUSTRIAL MICROBIOLOGY

In: Microbial Catalysts. Volume 1
Editors: S. M. Abdel-Aziz, N. Garg, A. Aeron et al.
ISBN: 978-1-53614-679-0

Chapter 4

PRODUCTION AND APPLICATIONS OF MICROBIAL TANNASE

Nalan Yılmaz Sarıözlü*
Department of Biology, Faculty of Science,
Anadolu University, Eskişehir, Turkey

ABSTRACT

Tannin acyl hydrolase, commonly known as tannase, is a hydrolytic enzyme which catalyzes the hydrolysis of gallotannins, complex tannins, and gallic acid esters. Tannase is extensively used in food, feed, beverage, brewing, pharmaceutical, chemical, and leather industries. The most important commercial applications of tannase are production of gallic acid, manufacture of instant tea, and acorn wine. Although tannase can be obtained from some plants, animals, and microbial sources, it is mostly produced by fungi, mainly *Aspergillus* and *Penicillium* species, bacteria, and yeasts. This review describes sources, substrates, production, physicochemical properties, various industrial applications and potential uses of the microbial tannase.

Keywords: tannase, tannins, production, applications

1. INTRODUCTION

Tannase (tannin acyl hydrolase, EC 3.1.1.20) catalyzes the hydrolysis of ester and depside bonds in gallotannins, complex tannins, and gallic acid esters to release gallic acid and glucose [1, 2]. Scheele discovered gallic acid presence in an aqueous extract of gallnuts in 1786 [3]. Van Tieghem first reported that the formation of gallic acid during fermentation is due to the action of fungus organisms in 1867. Then, he stated that these organisms were *Penicillium glaucum* and a new organism which he named *Aspergillus niger* [4]. Fernbach and Pottevin extracted tannase from *Aspergillus niger* at the same time (1901) [3]. In 1943, Toth and Barsony

* Author's Email: nalany@anadolu.edu.tr.

reported that tannase contains two separate enzymes: an esterase and a depsidase [4]. Hatamoto et al. [5] cloned and sequenced the tannase gene from *Aspergillus oryzae*. They reported that the gene coded for 588 amino acid with a molecular weight of about 64000 and lacks introns. Many studies on sources, production, assay, properties, applications, purification processes, characterization, and immobilization of tannase were reported up to the present.

Tannase is widely used in the food, feed, beverages, brewing, cosmetic, pharmaceutical, chemical, and leather industries [6]. It is mainly used in the manufacture of instant tea, acorn wine and the production of gallic acid (3,4,5-trihydroxybenzoic acid) [7, 8]. Gallic acid is mostly used in the pharmaceutical industry for synthesis of trimethoprim, which is an antibacterial drug and also it is used as a substrate for the production of propyl gallate for a potent antioxidant in the food industry [9]. Tannase is used for the industrial processing and clarification of beer, some wines, fruit juices and coffee-flavored soft drinks. It is also used for the detannification of food, animal feeds, and for cleaning up extremely polluting tannin from the wastewater of the leather, pharmaceutical, and chemical industries [4, 10].

Tannase is commercialized by Biocon (India), Kikkoman (Japan), ASA Spezialenzyme GmbH (Germany), Julich Fine Chemicals (JFC) GmbH (Germany) [2, 9], Sigma-Aldrich Co. (USA), Wako Pure Chemical Industries Ltd. (Japan), Novo Nordisk (Denmark) [8], Winovazyme (China), etc. with different purity and catalytic units depending on presentation of the product. Tannases from Kikkoman and Biocon are produced by solid state fermentation (SSF) [2], while tannase (Nova Clear) from Winovazyme is produced by deep liquid fermentation (http://www.winovazyme.com).

2. Tannase Substrates: Tannins

Tannins are natural polyphenolic secondary metabolites of plants with different molecular weights ranging from 300 Da to 3000 Da, even up to 20000 Da [11]. Tannins are divided into four major groups according to their structural characteristics: gallotannins, ellagitannins, complex tannins, and condensed tannins [11]. Gallotannins are the simplest tannins formed by organic acids, such as gallic, digallic, and chebulic acids, esterified with the hydroxyl group of a polyol carbohydrate. Gallotannins can be readily hydrolyzed under mild acid or alkaline conditions, either in hot water or enzymatically [2]. Ellagitannins are esters of hexahydrodiphenic acid, which undergoes lactonization to produce ellagic acid. Ellagic acid is slowly hydrolyzed due to the further C–C coupling of the polyphenolic residue with the polyol unit [11]. Condensed tannins are oligomeric and polymeric proanthocyanidins composed of flavonoid units, and they are more difficult to be hydrolyzed [9]. Complex tannins consists of a gallotannin or an ellagitannin unit glycosidically bounded to a catechin or epicatechin unit.

The biodegradation of tannin molecules occur with different enzymes of some microorganisms. These enzymes refer to tannase, polyphenoloxidase, gallic acid decarboxylase [11], and ellagitannin acyl hydrolase [12]. Polyphenoloxidase can cause degradation of phenolic polymers or polymerization of phenolic substrates. It is an oxidative enzyme such as ligninase, manganese peroxidase, and laccase [11]. Gallic acid decarboxylase can catalyze the degradation of the tannic acid with the decarboxylation of gallic acid to pyrogallol. The enzyme is very unstable due to their oxygen sensitivity [6]. Ellagitannin acyl hydrolase acts on ellagitannin and releases ellagic acid.

Tannase specifically catalyzes the hydrolysis of gallotannins, complex tannins, and gallic acid esters such as tannic acid, digallic acid, methyl gallate, ethyl gallate, *n*-propylgallate, lauryl gallate, and isoamyl gallate [9, 13]. Tannase completely hydrolyzes tannic acid to gallic acid and glucose. Some microorganisms can only hydrolyze the galloyl residues of galloyl eaters in ellagitannins, some others from ellagitannin-rich soil, leaves and tannery effluents can produce highly active tannase to hydrolyze hexahydroxydiphenoyl and other residues of ellagitannins [11].

Tannin rich materials such as Turkish gall (*Quercus infectoria*), Chinese gall (*Rhus semialata*), sumac (*Rhus coriaria*) leaves, tara (*Caesalpinia spinosa*) pods, myrobalan nuts (*Terminalia chebula*), chestnut (*Castanea sativa*) [14], divi-divi (*Caesalpinia coriaria*) pods, dhawa (*Anogeissus latifolia*) leaves, babul (*Acacia arabica*), avarum (*Cassia auriculata*) [15], gilo (*Caesalpinia digyna*) fruits, mahua (*Madhuca indica*) barks [16], *Anacardium occidentale* barks, *Acacia auriculiformis* barks, konnam (*Cassia fistula*) barks, *Ficus benghalensis* barks, *Psidium guazava* barks [17], gobernadora (*Larrea tridentata*) leaves [18], jamun (*Syzygium cumini*) leaves [19, 20], amla (*Phyllanthus emblica*) leaves, keekar (*Acacia nilotica*) leaves [21], bahera (*Terminalia belerica*) fruit [22], and *Ficus nitida* leaves [23] are used for microbial tannase production. Different tannin-rich agro-industrial residues can potentially be used as substrates for tannase production under SSF or submerged fermentation (SmF). Some of these residues are coffee husk, cassava, carob bean, wine-grape, tea and coffee wastes [24], palm kernel cake, tamarind seed powder [25], redgram husk, groundnut shell [26], and olive-mill wastewater [27].

3. Microorganisms Producing Tannase

Tannase is obtained from plants, animals (ruminants, insects), and microorganisms. The main tannase producers are microorganisms such as bacteria, fungi, and yeast. Because their enzymes are more stable, in high yield, and in high quantities than obtained from other sources [28]. It is reported that tannase of plant origin from leaves of pedunculate oak (*Quercus ruber*) or fruit pods of divi-divi (*Caesalpinia coriaria*) closely resemble to the properties of fungal tannase [29]. Some of the tannase-producing microorganisms are listed in Table 1.

4. Microbial Production of Tannase

Tannase production was reported mostly from *Aspergillus* species with SmF and SSF techniques (Table 2). The SSF has advantages for higher tannase production when compared to SmF and liquid surface fermentation (LSF). The highest enzyme production in lesser fermentation time, extracellular nature of enzyme, increased productivity, higher activity titers, and better stability at higher temperatures and pH values are some of the advantages of SSF for industrial tannase production [131]. In addition, the SSF involves several economical and engineering advantages such as non-aseptic conditions with the use of cheap, simple and easily obtainable raw materials as substrates, low capital and energy costs, less expensive downstream processing, low water demand and less wastewater output, high productivity, higher product concentration, high reproducibility, lesser fermentation space, and easier control of contamination [132]. Moreover, biotechnological advantages of SSF are less catabolic repression, easy aeration, no anti-foam chemicals, cultivation of microorganisms requiring a solid support, and variability of microorganisms used, particularly the usage of mixed cultures [133]. Comparative studies were reported that tannase production by SSF is more advantageous over SmF [10, 131, 134, 135]. Lekha and Lonsane [131] compared the production of extracellular tannase by *Aspergillus niger* PKL 104 in the same medium in SSF, SmF, and LSF. They found that tannase production in SSF was 2.5 and 4.8 times higher at approximately half of the fermentation time when compared to those in SmF and LSF, respectively. Additionally, the extracellular tannase obtained from SSF showed good stability at higher temperatures and pH values. Aguilar et al. [10] evaluated the effects of different tannic acid and glucose concentrations on production of tannase by *A. niger* Aa-20 in SmF and SSF systems. Tannase activity and productivity were at least 2.5 times higher in SSF than in SmF and high tannic acid concentrations increased total tannase activity in SSF, while a decrease in SmF. They further observed that a strong catabolite repression of tannase synthesis was in SmF when an initial glucose concentration of 50 g/l was used, while in SSF, it was increased for 200 g/l. Aguilar et al. [134] reported similar results in another study. They determined at least 2 times higher tannase/biomass yield in SSF than SmF. Aguilar et al. [135] evaluated effect of protease production by *A. niger* Aa-20 in SmF and SSF using different tannic acid concentrations on tannase production. They determined up to 10 times higher expression of protease activity in SmF than obtained in SSF using same culture medium composition. Cruz-Hernández et al. [123] reported higher tannase activity on SSF (more than four times) as compared to SmF using by *A. niger* GH1 at 30ºC. Rana and Bhat [122] compared the physicochemical properties of intracellular tannase producing by *A. niger* van Tieghem MTCC 2425 in three fermentation processes and reported that the maximum production of tannase activity was observed at 120 h in LSF and SmF, while its activity was peaked at 96 h in SSF.

Tannase is induced and expressed with different levels of activity depending on strain and culture conditions [73]. Some factors such as substrates, carbon source, nitrogen source, incubation period, pH, moisture level, temperature, aeration, minerals, and metal ions affect tannase production. Bajpai and Patil [70] reported that phenolic compounds such as gallic acid, pyrogallol, methyl gallate, and tannic acid induced tannase synthesis. They found maximum induction ratios for *Aspergillus fischerii* (26.7), *Fusarium solani* (26.1), and *Trichoderma viride* (40.7) when the fungi were induced with gallotannin, gallic acid, and methyl gallate, respectively. *Aspergillus fischerii, Fusarium solani* and *Trichoderma viride* could not tolerate more than 4, 3, and 3% gallotannin concentration, respectively. Their tannase titer was considerably higher when compared to widely used *Aspergillus niger* which could tolerate up to 20% gallotannin concentration. Aguilar et al. [134] reported that increasing of tannic acid enhanced the expression of tannase activity in SSF. The addition of glucose (>20 g/l) resulted in strong catabolite repression in SmF system. Bradoo et al. [136] demonstrated that the adding of simple and complex sugars (glucose, arabinose, fructose, galactose, xylose, lactose, sucrose, carboxymethyl cellulose, starch, and mannitol) in the culture medium did not have a positive effect on tannase production by *Aspergillus japonicus* with the exception of glucose. They reported that 0.2% sodium nitrate was the preferred nitrogen source among all the nitrogen sources (sodium nitrate, potassium nitrate, ammonium nitrite, sodium nitrite, ammonium chloride, aspartic acid, glutamic acid, and urea) investigated for enzyme production.

Das Mohapatra et al. [137] reported effect of amino acids and their analogues on tannase production from *Bacillus licheniformis* KBR6. The maximum tannase (2.87-fold that of the control) was synthesized when alanine added to the culture medium. Other amino acids, such as DL-serine, L-cystine, glycine, L-ornithine, aspartic acid, L-glutamic acid, DL-valine, L-leucine and L-lysine, also induced tannase production. L-cysteine monohydrochloride and DL-threonine were the most potent inhibitors.

Sabu et al. [49] observed tannase production in SSF by *Lactobacillus* sp. ASR-S1 using tamarind seed powder, wheat bran, palm kernel cake and coffee husk. Maximum tannase production (0.85 U/gds) was obtained by using coffee husk, supplemented with 0.6% tannic acid and 50% (w/v) moisture.

Srivastava and Kar [92] reported production of tannase and gallic acid under SmF using agroresidue pomegranate rind from *A. niger*. They found that tannic acid (1%) stimulated tannase up to 245.9% while with 0.5% glucose, increase was negligible. The enzyme production was inhibited by gallic acid and nitrogen sources such as NH_4NO_3, NH_4Cl, KNO_3, asparatic acid, urea, and EDTA.

5. PROPERTIES OF TANNASE

Physicochemical properties of tannase vary in literature. These differences depend on the microorganisms and culture conditions. Most of the tannase from microorganisms have a molecular weight in the range of 50-320 kDa. It has been reported that tannase are formed by two or more subunits. Some properties of tannase from various micrororganisms are presented in Table 3.

Tannase activity is affected by various additives such as metal ions, cations, anions, surfactants, reducing agents, and chelators. Kar et al. [145] reported effect of different additives on tannase activity. According to their results; among all the metal ions studied, 1.0 mM Mg^{2+} or Hg^{+} activated tannase activity. Ba^{2+}, Ca^{2+}, Zn^{2+}, Hg^{2+} and Ag^{+} inhibited activity at same concentration. Fe^{3+} and Co^{2+} completely and Ag^{+}, Ba^{2+}, Zn^{2+} and Hg^{2+} competitively inhibited tannase activity. Among the anions studied, 1.0 mM Br^{-} or $S_2O_3^{2-}$ enhanced tannase activity. Tween 40 and Tween 80 enhanced tannase activity while Tween 60 inhibited. Sodium lauryl sulfate and Triton X-100 inhibited tannase activity. 1.0 mM EDTA or 1,10-*o*-phenanthrolein inhibited tannase activity within tested chelators. Urea stimulated tannase activity at a concentration of 1.5 M. Dimethyl sulphoxide, β-mercaptoethanol inhibited tannase activity at 1.0 mM concentration. Soybean extract inhibited activity at concentrations varying from 0.05 to 1.0% (w/v). Among the chosen inorganic nitrogen sources, ammonium ferrous sulfate, ammonium sulfate, ammonium nitrate and ammonium chloride enhanced tannase activity at 0.1% (w/v) concentration.

Srivastava and Kar [92] showed that Zn^{2+}, Ca^{+}, Mn^{2+}, Mg^{2+}, Ba^{2+} and Ag^{+} inhibited tannase activity from *A. niger* ITCC 6514.07. Kasieczka-Burnecka [120] reported that $Mg^{2+}$and Br^{-} ions and urea (0.5-2.0 M) activated two extracellular tannase (TAH I and TAH II) while other metal ions (Zn^{2+}, Cu^{2+}, K^{+}, Cd^{2+}, Ag^{+}, Fe^{3+}, Mn^{2+}, Co^{2+}, Hg^{2+}, Pb^{2+} and Sn^{2+}), CO_3^{2-} anions, Tween 20, Tween 60, Tween 80, Triton X-100, sodium dodecyl sulphate, β-mercaptoethanol, α-glutathione and 4- chloromercuribenzoate inhibited the activity. Qiu et al. [114] reported that Fe^{3+}, Zn^{2+}, dithiothrietol (DTT), β-mercaptoethanol, formaldehyde, and

Table 1. Some tannase producing microorganisms (modified from Chávez-González et al. [6])

Microorganisms			
Bacteria	Reference(s)	Bacteria	Reference(s)
Achromobacter sp.	[30]	*Leuconostoc mesenteroides*	[45]
Bacillus cereus	[31]	*Lonepinella koalarum*	[44, 50]
B. gottheilii	[32]	*Microbacterium terregens*	[51]
B. licheniformis	[7]	*Oenococcus oeni*	[43]
B. polymyxa	[33]	*Paenibacillus polymyxa*	[33]
B. pumilus	[33]	*Pantonea agglomerans*	[52]
B. sphaericus	[34]	*Pantonea* sp.	[53]
Citrobacter freundii	[9, 35]	*Pediococcus acidilactici*	[41]
Corynebacterium sp.	[33]	*Pediococcus pentosaceus*	[41]
Enterobacter asburiae	[36]	*P. pentosaceus*	[41, 43]
E. ludwigii	[37]	*Providencia rettgeri*	[51]
Enterobacter sp.	[38]	*Pseudomonas aeruginosa*	[21]
E. faecalis	[39, 40]	*P. solanacearum*	[33]
Klebisella planticola	[33]	*Rhodococcus* sp.	[54]
K. pneumoniae	[33]	*Selenomonas ruminantium*	[9, 55]
Lactobacillus acidophilus	[41]	*Serratia ficaria*	[51]
L. animalis	[41]	*S. marcescens*	[51]
L. apodemi	[42]	*Serratia* sp.	[53]
L. brevis	[43]	*Staphylococcus lugdunensis*	[56]
L. buchneri	[43]	*Streptococcus bovis*	[9]
L. case	[43]	*Streptococcus gallolyticus*	[57, 58]
L. fermentum	[43]	*Weissella paramesenteroides*	[45]
L. hilgardii	[43]	Yeasts	
L. murinus	[41]	*Arxula adeninivorans*	[59]
L. paraplantarum	[41, 44]	*Candida parapsilosis*	[36]
L. pentosus	[41, 45]	*Candida* sp.	[60, 61]
L. plantarum	[45-48]	*C. tropicalis*	[36]
Lactobacillus sp.	[49]	*C. utilis*	[62]
Leuconostoc fallax	[45]	*Debaryomyces hansenii*	[63]

Microorganisms			
Yeasts	Reference(s)	Fungi	Reference(s)
Kluyveromyces marxianus	[64]	*A. nivens*	[68]
Mycotorula japonica	[9]	*A. nomius*	[93]
Pichia kudriavzevii	[36]	*A. ochraceus*	[94]
Pichia spp.	[63]	*A. ornatus*	[81, 95]
Rhodotorula glutinis	[65]	*A. oryzae*	[71, 96]
Saccharomyces cerevisiae	[66]	*A. parasiticus*	[68]
Fungi		*A. penicilliformis*	[68]
Aspergillus aculeatus	[67]	*A. phoenicis*	[90, 97]
A. acolumaris	[68]	*A. recurriero*	[68]
A. amstellodemi	[68]	*A. ruber*	[19]
A. allahabadi	[69]	*A. rugulosa*	[81]
A. alliaceus	[68]	*A. rugulosus*	[71, 95]
A. aureus	[70]	*A. stellatus*	[68]
A. awamori	[71-73]	*A. striatus*	[68]
A. caespitosum	[68]	*A. sojae*	[77]
A. candidus	[74]	*A. tamarii*	[98]
A. carneus	[68]	*A. terreus*	[70, 99]
A. fischeri	[70]	*A. terricola*	[81]
A. flavus	[27, 68, 75-78]	*A. usamii*	[77]
A. foetidus	[79, 80]	*A. ustus*	[77]
A. fumigatus	[20, 68, 78, 81]	*A. variecol*	[68]
A. gallonyces	[9]	*A. versicolor*	[68]
A. glaucus	[78]	*Aureobasidium pullulans*	[100]
A. heteromorphus	[82]	*Chrysosporium* sp.	[78]
A. janus	[68]	*Cryphonectria parasitica*	[101]
A. japonicus	[71, 78, 83]	*Cunnighamella* sp.	[71]
A. javipes	[68]	*Cylindrocladiella peruviana*	[102]
A. melleus	[87]	*Doratomyces stemonitis*	[102]
A. nidulans	[68]	*Emericella nidulans*	[103]
A. niger	[10, 71, 78, 85-92]	*Fusarium oxysporum*	[71, 102]

Table 1. (Continued)

Microorganisms			
Fungi	Reference(s)	Fungi	Reference(s)
Fusarium solani	[71, 104]	*P. giganteum*	[68]
Fusarium subglutinans	[105]	*P. glabrum*	[24, 113]
Helicostylum sp.	[71]	*P. glaucum*	[4]
Hyalopus sp.	[106]	*P. granulatum*	[68]
Lenzites elegans	[107]	*P. herquei*	[114]
Mariannaea camptospora	[102]	*P. islandyicum*	[83]
Mucor sp.	[9]	*P. javanicum*	[77]
Neosartorya fischeri	[10]	*P. nigricans*	[68]
Neurospora crassa	[71]	*P. notatum*	[83]
Paecilomyces variotii	[78, 108, 109]	*P. oxalicum*	[77]
Penicillium aculeatum	[68]	*P. purpurogenum*	[78, 111]
P. acrellanum	[71]	*P. restrictum*	[68]
P. atramentosum	[110]	*P. rugulosum*	[68]
P. awrentiogiseum	[68]	*Penicillium* sp.	[68]
P. canescens	[111]	*P. spiculisporum*	[68]
P. capsulatum	[68]	*P. spinulosum*	[111]
P. carymbiferum	[68]	*P. variable*	[68, 115]
P. carylophilum	[71]	*P. zacinthae*	[111]
P. charlesii	[71, 68]	*Pycnoporus sanguineus*	[107]
P. chrysogenum	[71, 104, 112]	*Rhizopus oryzae*	[1, 79, 116-118]
P. citrinium	[71]	*Syncephalastrum racemosum*	[71]
P. commune	[81, 90]	*Trichoderma atroviride*	[102]
P. concentricum	[102]	*T. hamatum*	[71]
P. corylophilum	[68]	*T. harzianum*	[71]
P. crustosum	[68]	*T. viride*	[71, 78]
P. digitatum	[71]	*Trichophyton rubrum*	[119]
P. expansum	[77]	*Verticillium* sp.	[120]
P. frequentans	[90, 111]		

Table 2. Microorganisms, substrates, fermentation conditions, and maximum yields for tannase production in literature at last decades

Microorganisms	Substrate	Fermentation	Max. Yield	Reference
Aspergillus awamori ITCC 924	Czapek-Dox minimal medium with 1% (w/v) tannic acid	LSF: 37°C, pH 6.0, 48 h	19.25 U/ml	[71]
A. japonicus ITCC 2620	Czapek-Dox minimal medium with 1% (w/v) tannic acid	LSF: 37°C, pH 6.0, 48 h	25.60 U/ml	[71]
Bacillus licheniformis KBR 6	M_3 medium with 1% (w/v) tannic acid, basal salts and 0.1% (w/v) glucose	SmF: 35°C, pH 5.0, 18-21 h, 200 rpm	0.172 U/ml	[7]
B. licheniformis KBR 6	Basal medium with 1.5% (w/v) tannic acid	SmF: 30°C, pH 5.0, 18-21 h, 200 rpm	0.21 U/ml	[121]
Rhizopus oryzae	Modified Czapek-Dox medium with 2% (w/v) *Caesalpinia dygina* seed cover powder	SmF: 37°C, pH 5.0, 48 h (with shaking) SSF: 32°C, pH 4.5, 72 h, moisture 93% w/v MSSF: 32°C, pH 4.5, 72 h, moisture 93% w/v	23.86 U/ml 18.87 U/ml 32.76 U/ml	[15]
Aspergillus niger 11T25A5	Wheat bran medium with 0.8% (w/w) tannic acid	SSF: 32°C, 72 h, moisture 0.91% w/v $(NH_4)_2SO_4$ solution	67.5 U/g	[91]
Lactobacillus plantarum	Basal medium with 2% (w/v) tannic acid	LSF: 37°C, pH 6.0, 24 h	6 U/ml	[46]
Aspergillus flavus F2	Olive-mill wastewater	SmF: 30°C, pH 4.0, 72 h, 100 rpm	8000 U/l	[27]
A. niger van Tieghem MTCC 2425	MT-basal medium with 2% (w/v)tannic acid	SmF: 30°C, pH 5.0,6.0, 48 h, 120 rpm LSF: 30°C, pH 5.5, 6.5, 120 h SSF: 30°C, 72 h, pH 5.0, 6.5	19.2 U/g 14.2 U/g 30.9 U/g	[122]
A. niger GH1	Culture medium with 5% (w/v) tannic acid	SmF: 30°C, pH 5.5, 30 h, 250 rpm SSF: 30°C, pH 5.5, 20 h, moisture 70% (v/w)	537 U/l 2291 U/l	[123]

Table 2. (Continued)

Microorganisms	Substrate	Fermentation	Max. Yield	Reference
Bacillus licheniformis KBR6	*Anacardium occidentale* barks 0.5% (w/v)	SmF: 35°C, pH 5.0, 15-18 h, 200 rpm	0.62±0.04 U/ml	[17]
Rhizopus oryzae and *Aspergillus foetidus* (co-culture)	*Caesalpinia digyna* fruit powder, *Terminalia chebula* cover powder and *Madhuca indica* bark powder (1:1:1) (mixed substrates)	MSSF: 30°C, pH 5.0, 48 h, moisture 80% w/v	39.4 U/ml	[16]
Aspergillus aculeatus DBF9	Wheat bran with 5% (w/w) tannic acid	SSF: 30°C, pH 5.0, 72 h, moisture 80% w/v	8.16 U/g	[67]
A. niger MS101	Fermentation medium with 2% (w/v) tannic acid	SmF: 30°C, pH 5.5, 400 rpm	16.77 U/ml	[124]
A. ruber	Jamun leaves (*Syzygium cumini*)	SSF: 30°C, pH 5.5, 96 h, moisture level 1:2(w/v)	69 U/g	[19]
A. fumigatus MA	Jamun leaves (*Syzygium cumini*)	SSF: 25°C, pH 5.0, 96 h, moisture level 1:1(w/v)	174.32 U/g	[20]
A. niger ITCC 6514.07	Czapeck-Dox medium with 4% (w/v) pomegranate rind powder and 1% (w/v) tannic acid	SmF: 37°C, pH 5.0, 72 h, 220 rpm	101.2 U/ml	[92]
A. niger GH1	Culture medium with 5% (w/v) tannic acid	SSF: 30°C, 30 h, pH 5.5, moisture 70% (v/w)	390 U/l	[125]
Paecilomyces variotii	Tannic acid medium (TAM) with 1.5% (w/v) tannic acid, 1% (w/v) glucose and 0.2% (w/v) sodium nitrate	SmF: 35°C, pH 6.0, 72 h, 140 rpm LSF 35°C, pH 6.0, 96 h	123 ± 3.6 U/ml 102 ± 4.2 U/ml	[78]
	Wheat bran with 1.5% (w/v) tannic acid, 1% (w/v) glucose and 0.2% (w/v) sodium nitrate	SSF: 35°C, pH 6.0, 60 h, moisture 50-60% w/v	167 ± 3.6 U/ml	
Pseudomonas aeruginosa IIIB 8914	Minimal media with 2% (w/v) amla (*Phyllanthus emblica*) leaves Minimal media with 2% (w/v) keekar (*Acacia nilotica*) leaves	SmF: 37°C, pH 7.0, 24 h, 200 rpm SmF: 37°C, pH 5.5, 24 h, 200 rpm	13.65 U/ml 12.90 U/ml	[21]
Serratia ficaria DTC	Fermentation medium with 5.0 g/l tannic acid, ammonium nitrate 3 g/l, casein hydrolysate 3 g/l and yeast extract 10 g/l	SmF: 35°C, pH 6.0, 24 h, 150 rpm	0.56 U/l	[51]
Aspergillus awamori BTMFW032	Czapek-Dox minimal medium with 2.56% (w/v) tannic acid	SmF: 30°C, pH 5.6, 36 h, 93 rpm	5085 U/ml	[126]

Microorganisms	Substrate	Fermentation	Max. Yield	Reference
Aspergillus niger FETL FT3	Czapek-Dox medium with 1% (w/v) tannic	SmF: 30°C, pH 6.0, 96 h, 200 rpm	2.81U/ ml	[127]
A. niger GH1	Czapek-Dox broth with %5(w/v) tannic acid	SSF: 30°C, pH 4.0, 72 h	7955 ± 798 U/l	[128]
Bacillus sphaericus	Minimal medium with 2% (w/v) tannic acid	SmF, 37°C, pH 6.0, 36 h, 100 rpm	16.54 U/ml	[34]
Emericella nidulans	Khanna medium with 2% (w/v) tannic acid	SmF: 30°C, pH 6.0, 24 h, 100 rpm	0.031 U/mg	[103]
Klebsiella pneumoniae MTCC 7162	Minimal medium with 1% (w/v) tannic acid	SmF: 37°C, pH 6.0, 20-30 h, 100 rpm	3.4 U/ml	[129]
Lactobacillus plantarum MTCC 1407	M3 medium with 1% (w/v) tannic acid	SmF: 30°C, pH 5.5, 24 h, 120 rpm	5.22 U/ml	[48]
Aspergillus niger van Tiegh CBC 122722	Fermentation medium with 30 g/l *Ficus nitida* leaves tannin extract	LSF: 30°C, pH 5.0, 120 h, 200 rpm	451 U/50ml	[23]
	Fermentation medium with 30 g/l *Ficus nitida* leaves tannin extract	SSF: 30°C, pH 5.0, 120 h, 175 rpm	457.2 U/50ml	
	Fermentation medium (5ml) with 10 g of powdered *Ficus nitida* leaves	SlSF: 30°C, pH 5.0, 96 h, 175 rpm	324.3 U/50ml	
A.ochraceus	Khanna medium with 2% (w/v) tannic acid	SmF: 40°C, pH 6.0, 72 h, 100 rpm	1.1 ±0.05 U/mg	[94]
Bacillus massiliensis	MPSM medium with 0.5% (w/v) tannic acid	SmF: 30°C, pH 5.0, 12 h, 250 rpm	9.65 U/l	[130]
Aspergillus phoenicis FJ810504	Khanna medium with 2% (w/v) tannic acid	SmF: 30°C, pH 6.0, 72 h, 100 rpm	1.43 U/ml	[97]
A. nomius GWA5	Minimal medium with 2% (w/v) tannic acid	SmF: 30°C, pH 6.0, 72 h, static condition, immobilized cells	46.6 U/ml	[93]

SmF: Submerged fermentation, LSF: Liquid surface fermentation, SSF: Solid-state fermentation, MSSF: Modified solid-state fermentation, SlSF: Slurry state fermentation.

Table 3. Some physicochemical properties of tannases from various microorganisms

Microorganisms	Optimum temperature (°C)	Optimum pH	Molecular weight (kDa)	K_M (mM)	Isoelectric point (pI)	Reference(s)
Aspergillus flavus IFO 5839	50–60	5.0–5.5	192	0.05	4.0	[75, 76, 138]
Arxula adeninivorans	40	6.0	320	4.4	-	[59]
Aspergillus awamori BTMFW032	30	2.0, 8.0	230	1.9	4.4	[73]
Aspergillus niger LCF 8	35	6.0	186	-	4.3	[88]
A. niger Aa20	60–70	6.0	90,180	-	3.8	[139]
A. niger van Tieghem	60	4.5,6.0	-	0.20	-	[89]
A. niger ATCC 16620	30–40	6.0	168	1.03	-	[25]
A. niger van Tieghem MTCC 2425	30	5.0,6.5	185	0.60	-	[122, 140]
A. niger GH1	60	6.0	50,75,100	0.041	3.5	[125]
A. ochraceus	40	5.0	112	0.76	-	[94]
A. oryzae	-	-	310	-	-	[5]
A. oryzae	40	5.5	-	7.35	-	[141]
A. phoenicis FJ810504	60	5.0–6.5	218.8	1.7	-	[97]
A. tamarii (TAH I)	30	5.5	180	0.77	-	[142]
Bacillus sphaericus	50	5.0	-	-	-	[34]
Candida sp.	50	6.0	250	-	-	[60, 61]
Cryphonectria parasitica	30	5.5	240	0.95	-	[101]
Emericella nidulans (=*Aspergillus nidulans*)	45	5.0	302	14.01	-	[103]
Enterococcus faecalis	40	6.0	45	0.43×10^{-3}	-	[143]
Klebsiella pneumoniae MTCC 7162	40	5.5	46.5	-	-	[129]
Lactobacillus plantarum ATCC 14917 T	40	8.0	50	0.62	-	[144]
Lenzites elegans	60	5.5	163	5.5	-	[107]
Penicillium chrysogenum	30–40	5.0–6.0	-	0.048	-	[112]
P. herquei	30	6.0	90	0.62	-	[114]
P. variable IARI 2031	50	5.0	310	32	-	[115]
Paecilomyces variotii	40	6.0	149.8	-	-	[108]
Paecilomyces variotii	50	5.5	87.3, 71.5	0.61×10^{-3}	-	[109]
Selenomonas ruminantium subsp. *ruminantium*	30-40	7.0	59	1.6	-	[55]
Serratia ficaria DTC	35	8.9	-	-	-	[51]
Verticillium sp. P9 (TAH I)	25	5.5	154.5	1.05	5.8	[120]
Verticillium sp. P9 (TAH II)	20	5.5	154.5	1.05	6.2	[120]

ethanol inhibited tannase obtained from *Penicillium herquei* while K^+, Mn^{2+}, Tween 80, and Triton X-100 induced. Tannase from *Emericella nidulans* (=*Aspergillus nidulans*) showed hypertolerance to temperature and organic solvents. Zn^{2+}, Hg^{2+}, Co^{2+}, and the detergents SDS and Triton X-100 stimulated tannase activity. Organic solvents particularly isopropanol, acetonitrile and ethanol (about 50%, v/v) increased activity [103]. A thermostable tannase from *A. ochraceus* was enhanced by Mn^{+2} (33-39%) and NH_4^+ (15%) [94]. A salt-tolerant tannase from *Bacillus cereus* KBR9 was stable up to 2 M of NaCl and retains 82% original activity in 3 M [31].

6. Tannase Applications and Potential Uses

Tannase has broad range of applications in food, feed, beverage, brewing, chemical, pharmaceutical, and leather industries. The main commercial applications of the tannase are production of gallic acid and in the manufacture of instant tea, acorn wine. Tannase is used in the clarification of beer and fruit juices, stabilization of malt polyphenol, manufacture of coffee flavoured soft drinks, improvement in the flavour of grape wine and as a sensitive analytical probe for determining the structures of naturally occurring gallic acid esters [146, 147]. It is also used for the cleavage of polyphenolics such as dehydrodimer crosslinks exist in cell wall of plants which are essential for plant cell wall digestibility [148].

The enzymatic product, gallic acid is a phenolic compound and finds applications in many fields. Its major use area is in the pharmaceutical industries for manufacture of trimethoprim (TMP), an antibacterial agent that is usually given along with sulfonamide; together they have a broad spectrum of action [15, 72]. Gallic acid is used in enzymatic synthesis of gallic acid esters such as propyl gallate, which is mainly used as an antioxidant in fats and oils, as well as in beverages [15]. It is also used in the manufacture of pyrogallol, in leather industry, as photosensitive resin in semiconductor production, and in chemical industry as a substrate for cosmetics, hair products, adhesives, and lubricants [6, 7].

During production of tea beverages mainly ready-to-drink tea products, hot and clear tea infusion tends to produce turbid precipitates after cooling, named as tea cream. The tea cream is water-insoluble precipitates formed by a complex mixture of polyphenols [149, 150]. Instability due to formation of tea cream is a common quality problem in manufacture of tea beverages. Therefore, decreaming is a significant step in the process to meet the cold stability requirements of the product. Enzymatic decreaming overcome some of the disadvantages associated with other conventional decreaming processes such as cold water extraction, chill decreaming, chemical stabilization, and chemical solubilization [151]. Tannase is the most commonly employed enzyme aiming at improving cold water extractability/solubility and decreasing tea cream formation as well as improving the clarity. Tannase is hydrolyzes esters of phenolic acids, including the gallated polyphenols found in tea. Epigallocatechin-3-gallate (EGCG) and epicatechin- 3-gallate (ECG) are the most abundant catechins in fresh tea leaves. Tannase treatment hydrolyzes EGCG to yield EGC and gallic acid, and ECG to yield EC and gallic acid by cleaving their ester bonds [150, 151]. Enzymatic treatment of tea beverages causes a better color and taste, less cream formation, good mouth feeling, and increasing antioxidant activity [150].

High tannin content presence in fruits such as blueberries, pomegranate, cranberry, and raspberry is responsible for haze and sediment formation, as well as for color, bitterness, and astringency of the juice upon storage. The enzymatic treatment of fruit juices to reduce the bitterness has got advantages such as the higher quality of juice due to the lower haze and non-deterioration of juice quality [2, 16]. Rout and Banerjee [16] studied debittering of pomegranate juice using tannase. While tannase hydrolyzed tannin 25%, tannase and gelatin (1:1) combination hydrolyzed 49%. Tannase treatment has no negative impact on the biochemical and quality attributes of the fruit juice. Srivastava and Kar [92] reported that immobilized tannase from *Aspergillus niger* ITCC 6514.07 removed 68.8% tannin from juice of aonla/myrobalan/Indian gooseberry (*Phyllanthus emblica*), a tropical fruit, rich essential nutrients and vitamin C. Treatment of aonla juice with tannase caused minimum reduction in vitamin C content (2%), which was considerably low compared to other conventional methods. Hamdy and Fawzy [23] demonstrated that the partially purified tannase from *Aspergillus niger* van Tiegh CBC 122722 removed up to 55% and 45% of tannin after 120 minute from tea extract and pomegranate juice respectively. Tannase is used as a clarifying agent for haze reduction in the manufacture of wine and beer. Tannase has also been utilized with laccase for the treatment of grape juice and grape musts to remove phenolic compounds for stabilizing and increasing the quality of wines [4]. The tannins caused 50% of the color of the wine and if these compounds are oxidized to quinones by contact with the air, they could form an undesirable turbidity, which causes quality problems. This problem has been best resolved with the use of tannase [28]. Selwal et al. [110] reported that the tannase from *Penicillium atramentosum* KM reduced tannic acid content 38.05% of jamum wine, 43.59% of grape wine and 74% of tea extract after 3 h at 35°C.

Tannase is mostly used for decreasing the tannin content in manufacture of beer. High quantities of beer proteins cause an undesirable turbidity due to the tannins [28]. Tannase treatment could be used to resolve this problem and to improve the quality of beer.

Garcia-Conesa et al. [148] reported that tannase from *Aspergillus oryzae* exhibits some esterase activity on several synthetic ethyl esterified diferulates. Tannases may contribute to plant cell wall degradation by cleaving some of the cross-links existing between polymers.

Using tannase in feed preparations increases the bioavaibility of nutrients by hydrolyzing phenolic (tannins) anti-nutritional factors [152]. Sorghum (*Sorghum bicolor*) is generally used as an animal feed complement and the fifth most produced cereal worldwide; however, some varieties of sorghum present a high content of tannins. If this cereal is first treated with tannase, tannin content could be decreased and this type of sorghum could then be used as complement in animal diet [28, 153].

CONCLUSION

Tannase has great significance in biotechnological area with wide-range applications in pharmaceutical, food, chemical, and beverages industries. But large-scale use is still limited, due to high cost of the enzyme production. Thus, most of the reports on tannase are focused on production, purification and characterization of tannase from new microbial sources using alternative techniques, cheap and easily obtainable substrates. The use of molecular biology techniques to study the biochemical, regulatory, and molecular aspects of tannase could have significant impact on its use on current and potential industrial applications.

REFERENCES

[1] Kar B., Banerjee R., and Bhattacharyya B. C. (1999). Microbial production of gallic acid modified solid state fermentation. *J Ind Microbiol Biotechnol.*, *23*:173-177.

[2] Aguilar C. N., Rodríguez R., Gutiérrez-Sánchez G., et al. (2007). Microbial tannases: advances and perspectives. *Appl Microbiol Biotechnol.*, *76*:47-59.

[3] Knudson L. (1913). Tannic acid fermentation-I. *J Biol Chem.*, *14*:159-184.

[4] Lekha P. K. and Lonsane B. K. (1997). Production and application of tannin acyl hydrolase: state of the art. *Adv Appl Microbiol.*, *44*:215-260.

[5] Hatamoto O., Watarai T., Kikuchi M., et al. (1996). Cloning and sequencing of the gene encoding tannase and a structural study of the tannase subunit from *Aspergillus oryzae*. *Gene*, *175*:215-221.

[6] Chávez-González M., Rodríguez-Durán L. V., Balagurusamy N., et al. (2012). Biotechnological advances and challenges of tannase: an overview. *Food Bioprocess Technol.*, *5*:445-459.

[7] Mondal K. C. and Pati B. R. (2000). Studies on the extracellular tannase from newly isolated *Bacillus licheniformis* KBR 6. *J Basic Microbiol.*, *40*:223-232.

[8] Govindarajan R. K., Revathi S., Rameshkumar N., et al. (2016). Microbial tannase: current perspectives and biotechnological advances. *Biocatal Agric Biotechnol.*, *6*:168-175.

[9] Belmares R., Contreras-Esquivel J. C., Rodríguez-Herrera R., et al. (2004). Microbial production of tannase: an enzyme with potential use in food industry. *LWT-Food Sci Technol.*, *37*:857-864.

[10] Aguilar C. N., Augur C., Favela-Torres E., et al. (2001). Production of tannase by *Aspergillus niger* Aa-20 in submerged and solid state fermentation: influence of glucose and tannic acid. *J Ind Microbiol Biotechnol.*, *26*:296-302.

[11] Mingshu L., Kai Y., Qiang H., et al. (2006). Biodegradation of gallotannins and ellagitannins. *J Basic Microbiol.*, *46*:68-84.

[12] Huang W., Niu H., Gong G. H., et al. (2007). Individual and combined effects of physicochemical parameters on ellagitannin acyl hydrolase and ellagic acid production from ellagitannin by *Aspergillus oryzae*. *Bioproc Biosyst Eng.*, *30*:281-288.

[13] Curiel J. A., Rodríguez H., Acebrón I., et al. (2009). Production and physicochemical properties of recombinant *Lactobacillus plantarum* tannase. *J Agric Food Chem.*, *57*:6224-6230.

[14] Bhat T. K., Singh B., and Sharma, O. P. (1998). Microbial degradation of tannins- a current perspective. *Biodegradation*, *9*: 343-357.

[15] Kar B. and Banerjee R. (2000). Biosynthesis of tannin acyl hydrolase from tannin-rich forest residue under different fermentation conditions. *J Ind Microbiol Biotechnol.*, *25*:29-38.

[16] Rout S. and Banerjee R. (2006). Production of tannase under mSSF and its application in fruit juice debittering. *Indian J Biotechnol.*, *5*:346-350.

[17] Das Mohapatra P. K., Mondal K. C., and Pati B. R. (2006). Production of tannase through submerged fermentation of tannin-containing plant extracts by *Bacillus licheniformis* KBR6. *Pol J Microbiol., 55*:297-301.

[18] Treviño-Cueto B., Luis M., Contreras-Esquivel J. C., et al. (2007). Gallic acid and tannase accumulation during fungal solid state culture of a tannin-rich desert plant (*Larrea tridentata* Cov.). *Bioresour Technol., 98*:721-724.

[19] Kumar R., Sharma J., and Singh R. (2007). Production of tannase from *Aspergillus ruber* under solid-state fermentation using jamun (*Syzygium cumini*) leaves. *Microbiol Res., 162*:384-390.

[20] Manjit A. Y., Aggarwal N. K., Kumar K., et al. (2008). Tannase production by *Aspergillus fumigatus* MA under solid-state fermentation. *World J Microbiol Biotechnol., 24*:3023-3030.

[21] Selwal M. K., Yadav A., Selwal K. K., et al. (2010). Optimization of cultural conditions for tannase production by *Pseudomonas aeruginosa* IIIB 8914 under submerged fermentation. *World J Microbiol Biotechnol., 26*:599-605.

[22] Prasad D., Gupta R. K., Venkataratnam G. S., et al. (2011). Utilization of *Bahera* fruits for production of tannase and gallic acid by *Aspergillus heteromorphus* MTCC 5466 and synthesis of propyl gallate thereof. *Global J Biotech Biochem., 6*:119-128.

[23] Hamdy H. S. and Fawzy E. M. (2012). Economic production of tannase by *Aspergillus niger van* Tiegh adopting different fermentation protocols and possible applications. *Rom Biotech Lett., 17*:7441-7457.

[24] Van de Lagemaat J. and Pyle D. L. (2001). Solid-state fermentation and bioremediation: development of a continuous process for the production of fungal tannases. *Chem Eng J., 84*:115-123.

[25] Sabu A., Pandey A., Jaafar Daud M., et al. (2005). Tamarind seed powder and palm kernel cake: two novel agro residues for the production of tannase under solid state fermentation by *Aspergillus niger* ATCC 16620. *Bioresour Technol., 96*:1223-1228.

[26] Kuppusamy M., Thangavelu V., and Chokalingam A. (2012). Optimization and production of tannase by *Aspergillus foetidus* through submerged fermentation. *Asian J Biochem Pharm Res., 2*:22-29.

[27] Kachouri S., Halaouli S., Lomascolo A., et al. (2005). Decolourization of black oxidized olive-mill wastewater by a new tannase-producing *Aspergillus flavus* strain isolated from soil. *World J Microbiol Biotechnol., 21*:1465-1470.

[28] Aguilar C. N. and Gutiérrez-Sánchez G. (2001). Review: sources, properties, applications and potential uses of tannin acyl hydrolase. *Food Sci Tech Int., 7*:373-382.

[29] Niehaus J. U. and Gross G. G. (1997). A gallotannin degrading esterase from leaves of pedunculate oak. *Phytochem., 45*:1555-1560.

[30] Lewis J. A. and Starkey R. L. (1969). Decomposition of plant tannins by some soil microorganisms. *Soil Sci., 107*:235-241.

[31] Mondal K., Banerjee D., Banerjee R., et al. (2001). Production and characterization of tannase from *Bacillus cereus* KBR9. *J Gen Appl Microbiol., 47*:263-267.

[32] Subbalaxmi S. and Vytla R. M. (2016). Process optimization for tannase production by *Bacillus gottheilii* M2S2 on inert polyurethane foam support. *Biocatal Agric Biotechnol., 7*:48-55.

[33] Deschamps A. M., Otuk G., and Lebeault J. M. (1983). Production of tannase and degradation of chestnut tannin by bacteria. *J Ferment Technol., 61*:55-59.

[34] Raghuwanshi S., Dutt K., Gupta P., et al. (2011). *Bacillus sphaericus*: the highest bacterial tannase producer with potential for gallic acid synthesis. *J Biosci Bioeng.*, *111*:635-640.

[35] Murugan S, Uma Devi P., Mahesh P., et al. (2008). Production of tannase by *Citrobacter freundii* under solid state fermentation and its application in fruit juices debittering. *Biosci Biotechnol Res Asia*, *5*:301-306.

[36] Mandal S. and Ghosh K. (2013). Isolation of tannase-producing microbiota from the gastrointestinal tracts of some freshwater fish. *J App Ichthyol.*, 29:145-153.

[37] Singh B., Bhat T. K., Sharma O. P., et al. (2012). Isolation of tannase-producing *Enterobacter ludwigii* GRT-1 from the rumen of migratory goats. *Small Rumin Res.*, *102*:172-176.

[38] Sharma K. P. and John P. J. (2011). Purification and characterization of tannase and tannase gene from *Enterobacter* sp. *Process Biochem.*, *46*:240-244.

[39] Goel G., Puniya A. K., Aguilar C. N., et al. (2005). Interaction of gut microflora with tannins in feeds. *Naturwissenschaften, 92*:497-503.

[40] Goel G., Puniya A. K., and Singh K. (2007). Phenotypic characterization of tannin–protein complex degrading bacteria from faeces of goat. *Small Rumin Res.*, *69*:217-220.

[41] Nishitani Y., Sasaki E., Fujisawa T., et al. (2004). Genotypic analyses of *Lactobacilli* with a range of tannase activities isolated from human feces and fermented foods. *Syst Appl Microbiol.*, *27*:109-117.

[42] Osawa R., Fujisawa T., and Pukall R. (2006). *Lactobacillus apodemi* sp. nov., a tannase-producing species isolated from wild mouse faeces. *Int J Syst Evol Microbiol.*, *56*:1693-1696.

[43] Matthews A., Grbin P. R., and Jiranek V. (2006). A survey of lactic acid bacteria for enzymes of interest to oenology. *Aust J Grape Wine Res.*, *12*:235-244.

[44] Nishitani Y. and Osawa R. (2003). A novel colorimetric method to quantify tannase activity of viable bacteria. *J Microbiol Methods*, *54*:281-284.

[45] Kostinek M., Specht I., Edward V. A., et al. (2007). Characterisation and biochemical properties of predominant lactic acid bacteria from fermenting cassava for selection as starter cultures. *Int J Food Microbiol.*, *114*:342-351.

[46] Ayed L. and Hamdi M. (2002). Culture conditions of tannase production by *Lactobacillus plantarum*. *Biotechnol Lett.*, *24*:1763-1765.

[47] Rodríguez H., de las Rivas B., Gómez-Cordovés C., et al. (2008). Characterization of tannase activity in cell-free extracts of *Lactobacillus plantarum* CECT 748T. *Int J Food Microbiol.*, *121*:92-98.

[48] Kannan N., Aravindan N., and Viruthagiri T. (2011). Effect of culture conditions and kinetic studies on extracelluler tannase production by *Lactobacillus plantarum* MTCC 1407. *Indian J Biotechnol.*, *10*:321-328.

[49] Sabu A., Augur C., Swati C., et al. (2006). Tannase production by *Lactobacillus* sp. ASR-S1 under solid-state fermentation. *Process Biochem.*, *41*:575-580.

[50] Osawa R., Rainey F., Fujisawa T., et al. (1995). *Lonepinella koalarum* gen. nov., sp. nov., a new tannin-protein complex degrading bacterium. *Syst Appl Microbiol.*, *18*:368-373.

[51] Belur P. D., Gopal M., Nirmala K. R., et al. (2010). Production of novel cell-associated tannase from newly isolated *Serratia ficaria* DTC. *J Microbiol Biotechnol.*, *20*: 732-736.

[52] Zeida M., Wieser M., Yoshida T., et al. (1998). Purification and characterization of gallic acid decarboxylase from *Pantoea agglomerans* T71. *Appl Environ Microbiol., 64*:4743-4747.

[53] Pepi M., Lampariello L. R., Altieri R., et al. (2010). Tannic acid degradation by bacterial strains *Serratia* spp. and *Pantoea* sp. isolated from olive mill waste mixtures. *Int Biodeterior Biodegradation, 64*:73-80.

[54] Nadaf N. H. and Ghosh J. S. (2011). Production, purification and characterization of tannase from *Rhodococcus* NCIM 2891. *Curr Res J Biol Sci., 3*:246-253.

[55] Skene I. K. and Brooker J. D. (1995). Characterization of tannin acylhydrolase activity in the ruminal bacterium *Selenomonas ruminantium. Anaerobe, 1*:321-327.

[56] Noguchi N., Ohashi T., Shiratori T., et al. (2007). Association of tannase-producing *Staphylococcus lugdunensis* with colon cancer and characterization of a novel tannase gene. *J Gastroenterol., 42*:346-351.

[57] Osawa R., Fujisawa T., and Sly L. I. (1995). *Streptococcus gallolyticus* sp. nov.; gallate degrading organisms formerly assigned to *Streptococcus bovis. Syst Appl Microbiol., 18*:74-78.

[58] Sasaki E., Shimada T., Osawa R., et al. (2005). Isolation of tannin-degrading bacteria isolated from feces of the Japanese large wood mouse, *Apodemus speciosus*, feeding on tannin-rich acorns. *Syst Appl Microbiol., 28*:358-365.

[59] Böer E., Bode R., Mock H. P., et al. (2009). Atan1p-an extracellular tannase from the dimorphic yeast *Arxula adeninivorans*: molecular cloning of the *ATAN1* gene and characterization of the recombinant enzyme. *Yeast, 26*:323-337.

[60] Aoki K., Shinke R., and Nishira H. (1976). Purification and some properties of yeast tannase. *Agric Biol Chem., 40*:79-85.

[61] Aoki K., Shinke R., and Nishira H. (1976). Chemical composition and molecular weight of yeast tannase. *Agric Biol Chem., 40*:297-302.

[62] Shi B., He Q., Yao K., et al. (2005). Production of ellagic acid from degradation of valonea tannins by *Aspergillus niger* and *Candida utilis. J Chem Technol Biotechnol., 80*:1154-1159.

[63] Deschamps A. M. (1989). Microbial degradation of tannins and related compounds. In *Plant Cell Wall Polymers Biogenesis and Biodegradation*, edited by Lewis N. G. and Paice M. G., 559-566. Washington, DC: American Chemical Society.

[64] Mahmoud A. E., Fathy S. A., Rashad M. M., et al. (2018). Purification and characterization of a novel tannase produced by *Kluyveromyces marxianus* using olive pomace as solid support, and its promising role in gallic acid production. *Int J Biol Macromol., 107*:2342-2350.

[65] Taskin M. (2013). Co-production of tannase and pectinase by free and immobilized cells of the yeast *Rhodotorula glutinis* MP-10 isolated from tannin-rich persimmon (*Diospyros kaki* L.) fruits. *Bioprocess Biosyst Eng., 36*:165-172.

[66] Zhong X., Peng L., Zheng S., et al. (2004). Secretion, purification, and characterization of a recombinant *Aspergillus oryzae* tannase in *Pichia pastoris. Protein Expr Purif., 36*:165-169.

[67] Banerjee D., Mondal K. C., and Pati B. R. (2007). Tannase production by *Aspergillus aculeatus* DBF9 through solid-state fermentation. *Acta Microbiol Immunol Hung., 54*:159-166.

[68] Batra A. and Saxena R. K. (2005). Potential tannase producers from the genera *Aspergillus* and *Penicillium*. *Process Biochem.*, *40*:1553-1557.

[69] Gaikaiwari R. P., Wagh S. A., and Kulkarni B. D. (2012). Extraction and purification of tannase by reverse micelle system. *Sep Purif Technol.*, *89*:288-296.

[70] Bajpai B. and Patil S. (1997). Induction of tannin acyl hydrolase (EC 3.1.1.20) activity in some members of *fungi imperfecti*. *Enzyme Microb Technol.*, *20*:612-614.

[71] Bradoo S., Gupta R., and Saxena R. K. (1996). Screening of extracellular tannase-producing fungi: development of a rapid simple plate assay. *J Gen Appl Microbiol.*, *42*:325-329.

[72] Seth M. and Chand S. (2000). Biosynthesis of tannase and hydrolysis of tannins to gallic acid by *Aspergillus awamori*- optimisation of process parameters. *Process Biochem.*, *36*:39-44.

[73] Beena P. S., Soorej M. B., Elyas K. K., et al. (2010). Acidophilic tannase from marine *Aspergillus awamori* BTMFW032. *J Microbiol Biotechnol.*, *20*:1403-1414.

[74] Murugan K. and Al-Sohaibani S. A. (2010). Biocompatible removal of tannin and associated color from tannery effluent using the biomass and tannin acyl hydrolase (E.C.3.1.1.20) enzymes of mango industry solid waste isolate *Aspergillus candidus* MTTC 9628. *Res J Microbiol.*, *5*:262-271.

[75] Yamada H., Adachi O., Watanabe M., et al. (1968). Tannase (tannin acyl hydrolase), a typical serine esterase. *Agric Biol Chem.*, *32*:257-258.

[76] Yamada H., Adachi O., Watanabe M., et al. (1968). Studies on fungal tannase Part I. Formation, purification and catalytic properties of tannase of *Aspergillus flavus*. *Agric Biol Chem.*, *32*:1070-1078.

[77] Yamada K., Iibuchi S., and Minoda Y. (1968). Studies on tannin acyl hydrolase of microorganisms. Isolation and identification of producing molds and studies on the conditions of cultivation. *Agric Biol Chem.*, *45*:233-240.

[78] Raaman N., Mahendran B., Jaganathan C., et al. (2010). Optimisation of extracellular tannase production from *Paecilomyces variotii*. *World J Microbiol Biotechnol.*, *26*:1033-1039.

[79] Mukherjee G. and Banerjee R. (2004). Biosynthesis of tannase and gallic acid from tannin rich substrates by *Rhizopus oryzae* and *Aspergillus foetidus*. *J Basic Microbiol.*, *44*:42-48.

[80] Naidu R. B., Saisubramanian N, Selvakumar D., et al. (2008). Partial purification of tannase from *Aspergillus foetidus* by aqueous two phase extraction and its characterization. *Curr Trends Biotechnol Pharm.*, *2*:201-207.

[81] Cruz-Hernández M., Contreras-Esquivel J. C., Lara F., et al. (2005). Isolation and evaluation of tannin-degrading fungal strains from the Mexican desert. *Z Naturforsch C.*, *60*:844-848.

[82] Chhokar V., Seema B. V., Salar R., et al. (2010). Purification and characterization of extracellular tannin acyl hydrolase from *Aspergillus heteromorphus* MTCC 8818. *Biotechnol Bioprocess Eng.*, *15*:793-799.

[83] Ganga P. S., Nandy S. C., and Santappa M. (1977). Effect of environmental factors on the production of fungal tannase. *Leather Sci.*, *24*:8-16.

[84] Lekha P. K., Chand N., and Lonsane B. K. (1994). Computerized study of interactions among factors and their optimization through response surface methodology for the production of tannin acyl hydrolase by *Aspergillus niger* PKL 104 under solid state fermentation. *Bioprocess Eng.*, *11*:7-15.

[85] Pourrat H., Regerat F., Pourrat A., et al. (1985). Production of gallic acid from tara tannin by a strain of *A. niger*. *J Ferment Technol.*, 63:401-403.

[86] Regerat F., Pourrat H., and Pourrat A. (1989). Hydrolysis by fermentation of tannins from gall nuts. *Jalca*, *84*:323-328.

[87] Liu T. P. S. L., Costa R. M. P. B., Freitas D. J. V., et al. (2017). Tannase from *Aspergillus melleus* improves the antioxidant activity of green tea: purification and biochemical characterisation. *Int J Food Sci Technol.*, *52*:652-661.

[88] Barthomeuf C., Regerat F., and Pourrat H. (1994). Production, purification and characterization of tannase from *Aspergillus niger* LCF8. *J Ferment Bioeng.*, *77*:320-323.

[89] Sharma S., Bhat T. K., and Dawra R. K. (1999). Isolation, purification and properties of tannase from *Aspergillus niger* van Tieghem. *World J Microbiol Biotechnol.*, *15*:673-677.

[90] Van de Lagemaat J., Augur C., and Pyle D. L. (2000). Screening of filamentous fungi for the production of extra-cellular tannase in solid state fermentation (SSF). In *Coffee Biotechnology and Quality*, edited by Sera T., Soccol C. R., Pandey A., and Roussos S., 455-460. Dordrecht: Kluwer Academic Publisher.

[91] Pinto G. A. S., Leite S. G. F., Terzi S. C., et al. (2001). Selection of tannase-producing *Aspergillus niger* strains. *Braz J Microbiol.*, *32*:24-26.

[92] Srivastava A. and Kar R. (2009). Characterization and application of tannase produced by *Aspergillus niger* ITCC 6514.07 on pomegranate rind. *Braz J Microbiol.*, *40*:782-789.

[93] Hassan S. W. M, Ghanem K. M., El-Shahat A. F., et al. (2018). Production of tannase by free and immobilized marine *Aspergillus nomius* under optimized culture conditions. *Res J Pharm Biol Chem Sci.*, *9*:30-44.

[94] Gonçalves H. B., Riul A. J., Quiapim A. C., et al. (2012). Characterization of a thermostable extracellular tannase produced under submerged fermentation by *Aspergillus ochraceus*. *Electron J Biotechnol.*, (published online: September 15, 2012) doi: 10.2225/vol15-issue5-fulltext-5.

[95] De la Cerda Gómez A. A., Reyes Valdés M. H., Meléndez Rentería N. P., et al. (2011). Tannase production under solid and submerged culture by xerophilic strains of *Aspergillus* and their genetic relationships. *Micol Apl Int.*, *23*:21-27.

[96] Beverini M. and Metche M. (1990). Identification, purification, physiochemical properties of tannase from *Aspergillus oryzae*. *Sci Aliment.*, *10*:807-816.

[97] Riul A. J., Gonçalves H. B., Jorge J. A., et al. (2013). Characterization of a glucose- and solvent-tolerant extracellular tannase from *Aspergillus phoenicis*. *J Mol Catal B Enzym.*, *85-86*:26-133.

[98] Costa A. M., Ribeiro W. X., Kato E., et al. (2008). Production of tannase by *Aspergillus tamarii* in submerged cultures. *Braz Arch Biol Technol.*, *51*:399-404.

[99] Kawakubo J., Nishira H., Aoki K., et al. (1993). Isolation of a gallic acid-producing microorganisms with sake cake medium and production of gallic acid. *Biosci Biotech Biochem.*, *57*:1360-1361.

[100] Banerjee D. and Pati B. R. (2007). Optimization of tannase production by *Aureobasidium pullulans* DBS 66. *J Microbiol Biotechnol.*, *17*:1049-1053.
[101] Farias G. M., Gorbea C., Elkins J. R., et al. (1994). Purification, characterization, and substrate relationships of the tannase from *Cryphonectria parasitica*. *Physiol Mol Plant Pathol.*, *44*:51-63.
[102] Peterson R. A., Bradner J. R., Roberts T. H., et al. (2009). Fungi from koala (*Phascolarctos cinereus*) faeces exhibit a broad range of enzyme activities against recalcitrant substrates. *Lett Appl Microbiol.*, *48*:218-225.
[103] Gonçalves H. B., Riul A. J., Terenzi H. F., et al. (2011). Extracellular tannase from *Emericella nidulans* showing hypertolerance to temperature and organic solvents. *J Mol Catal B Enzym.*, *71*:29-35.
[104] Bajpai B. and Patil S. (1996). Tannin acyl hydrolase (EC 3.1.1.20) activity of *Aspergillus*, *Penicillium*, *Fusarium* and *Trichoderma*. *World J Microbiol Biotechnol.*, *12*:217-220.
[105] Hamdy H. S. (2008). Purification and characterisation of a newly isolated stable long-life tannase produced by *F. subglutinans* (Wollenweber and Reinking) Nelson et al. *J Pharm Innov.*, *3*:142-151.
[106] Mahapatra S. and Banerjee D. (2009). Extracellular tannase production by endophyitc *Hyalopus* sp. *J Gen Appl Microbiol.*, *55*:255-259.
[107] Ordoñez R. M., Colombo I., Alberto M. R., et al. (2011). Production of tannase from wood-degrading fungus using as substrate plant residues: purification and characterization. *World J Microbiol Biotechnol.*, *27*:2325-2333.
[108] Mahendran B., Raman N., and Kim D. J. (2006). Purification and characterization of tannase from *Paecilomyces variotii*: hydrolysis of tannic acid using immobilized tannase. *Appl Microbiol Biotechnol.*, *70*:444-450.
[109] Battestin V. and Macedo G. A. (2007). Tannase production by *Paecilomyces variotii*. *Bioresour Technol.*, *98*:1832-1837.
[110] Selwal M. K., Yadav A., Selwal K. K., et al. (2011). Tannase production by *Penicillium atramentosum* KM under SSF and its applications in wine clarification and tea cream solubilization. *Braz J Microbiol.*, *42*:374-387.
[111] Yılmaz Sarıözlü N. and Kıvanç M. (2009). Isolation of gallic acid-producing microorganisms and their use in the production of gallic acid from gall nuts and sumac. *Afr J Biotechnol.*, *8*:1110-1115.
[112] Suseela Rajakumar G. and Nandy S. C. (1983). Isolation, purification, and some properties of *Penicillium chrysogenum* tannase. *Appl Environ Microbiol.*, *46*:525-527.
[113] Van de Lagemaat J. and Pyle D. L. (2005). Modelling the uptake and growth kinetics of *Penicillium glabrum* in a tannic acid-containing solid-state fermentation for tannase production. *Process Biochem.*, *40*:1773-1782.
[114] Qiu Y., Niu H., Huang W., et al. (2011). Properties and secondary structure of tannase from *Penicillium herquei*. *Biotechnol Bioprocess Eng.*, *16*:858-866.
[115] Sharma S., Agarwal L., and Saxena R. K. (2008). Purification, immobilization and characterization of tannase from *Penicillium variable*. *Bioresour Technol.*, *99*:2544-2551.
[116] Hadi T. A., Banerjee R., and Bhattarcharyya B. C. (1994). Optimization of tannase biosynthesis by a newly isolated *Rhizopus oryzae*. *Bioprocess Eng.*, *11*:239-243.

[117] Misro S. K., Kumar M. R., Banerjee R., et al. (1997). Production of gallic acid by immobilization of *Rhizopus oryzae*. *Bioprocess Eng.*, *16*:257-260.
[118] Purohit J. S., Dutta J. R., Nanda R. K., et al. (2006). Strain improvement for tannase production from coculture of *Aspergillus foetidus* and *Rhizopus oryzae*. *Bioresour Technol.*, *97*:795-801.
[119] Krishnasamy L., Jayanthi S., and Ponmurugan P. (2009). Effect of various biotic and abiotic factors on the activity of tannase. *Eur J Appl Sci.*, *1*:1-5.
[120] Kasieczka-Burnecka M., Kuc K., Kalinowska H., et al. (2007). Purification and characterization of two cold-adapted extracellular tannin acyl hydrolases from an Antarctic strain *Verticillium* sp. P9. *Appl Microbiol Biotechnol.*, *77*:77-89.
[121] Mondal K. C., Banerjee R., and Pati B. R. (2000). Tannase production by *Bacillus licheniformis*. *Biotechnol Lett.*, 0:767-769.
[122] Rana N. K. and Bhat T. K. (2005). Effect of fermentation system on the production and properties of tannase of *Aspergillus niger* van Tieghem MTCC 2425. *J Gen Appl Microbiol.*, *51*:203-212.
[123] Cruz-Hernández M., Augur C., Rodríguez R., et al. (2006). Evaluation of culture conditions for tannase production by *Aspergillus niger* GH1. *Food Technol Biotechnol.*, *44*:541-544.
[124] Murugan K., Saravanababu S., and Arunachalam M. (2007). Screening of tannin acyl hydrolase (E.C.3.1.1.20) producing tannery effluent fungal isolates using simple agar plate and SmF process. *Bioresour Technol.*, *98*:946-949.
[125] Mata-Gómez M., Rodríguez L.V., Ramos E.L., et al. (2009). A novel tannase from the xerophilic fungus *Aspergillus niger* GH1. *J Microbiol Biotechnol.*, *19*:987-996.
[126] Beena P. S., Basheer S. M., Bhat S. G., et al. (2011). Propyl gallate synthesis using acidophilic tannase and simultaneous production of tannase and gallic acid by marine *Aspergillus awamori* BTMFW032. *Appl Biochem Biotechnol.*, *164*:612-628.
[127] Darah I., Sumathi G., Jain K., et al. (2011). Tannase enzyme production by entrapped cells of *Aspergillus niger* FETL FT3 in submerged culture system. *Bioprocess Biosyst Eng.*, *34*:795-801.
[128] Rodríguez-Durán L. V., Contreras-Esquivel J. C., Rodríguez R., et al. (2011). Optimization of tannase production by *Aspergillus niger* in solid-state packed-bed bioreactor. *J Microbiol Biotechnol.*, *21*:960-967.
[129] Sivashanmugam K. and Jayaraman G. (2011). Production and partial purification of extracellular tannase by *Klebsiella pneumoniae* MTCC 7162 isolated from tannery effluent. *Afr J Biotechnol.*, *10*:1364-1374.
[130] Belur P. D., Goud R., and Goudar D. C. (2012). Optimization of culture medium for novel cell-associated tannase production from *Bacillus massiliensis* using response surface methodology. *J Microbiol Biotechnol.*, *22*:199-206.
[131] Lekha P. K. and Lonsane B. K. (1994). Comparative titres, location and properties of tannin acyl hydrolase produced by *Aspergillus niger* PKL 104 in solid-state, liquid surface and submerged fermentations. *Process Biochem.*, *29*:497-503.
[132] Krishna C. (2005). Solid-state fermentation systems-an overview. *Crit Rev Biotechnol.*, *25*:1-30.
[133] Hölker U., Höfer M., and Lenz J. (2004). Biotechnological advantages of laboratory-scale solid-state fermentation with fungi. *Appl Microbiol Biotechnol.*, *64:*175-186.

[134] Aguilar C. N., Augur C., Favela-Torres E., et al. (2001). Induction and repression patterns of fungal tannase in solid-state and submerged cultures. *Process Biochem.*, *36*:565-570.

[135] Aguilar C. N., Favela-Torres E., Viniegra-González G., et al. (2002). Culture conditions dictate protease and tannase production in submerged and solid-state cultures by *Aspergillus niger* Aa-20. *Appl Biochem Biotechnol.*, *102–103*:407–414.

[136] Bradoo S., Gupta R., and Saxena R. K. (1997). Parametric optimization and biochemical regulation of extracellular tannase from *Aspergillus japonicus*. *Process Biochem.*, *32*:135–139.

[137] Das Mohapatra P. K., Pati B. R., and Mondal K. C. (2009). Effect of amino acids on tannase biosynthesis by *Bacillus licheniformis* KBR6. *J Microbiol Immunol Infect.*, *42*:172–175.

[138] Adachi O., Watanabe M., and Yamada H. (1968). Studies on fungal tannase part II. Physicochemical properties of tannase of *Aspergillus flavus*. *Agric Biol Chem.*, *32*:1079–1085.

[139] Ramirez-Coronel M. A., Viniegra-Gonzalez G., Darvill A., et al. (2003). A novel tannase from *Aspergillus niger* with β-glucosidase activity. *Microbiol-SGM.*, *149*:2941–2946.

[140] Bhardwaj R., Singh B., and Bhat T. K. (2003). Purification and characterization of tannin acyl hydrolase from *Aspergillus niger* MTCC 2425. *J Basic Microbiol.*, *43*:449–461.

[141] Abdel-Naby M. A., Sherif A. A., El-Tanash A. B., et al. (1999). Immobilization of *Aspergillus oryzae* tannase and properties of the immobilized enzyme. *J Appl Microbiol.*, *87*:108–114.

[142] Costa A. M., Kadowaki M. K., Minozzo M. C., et al. (2012). Production, purification and characterization of tannase from *Aspergillus tamarii*. *Afr J Biotechnol.*, *11*:391–398.

[143] Goel G., Kumar A., Beniwal V., et al. (2011). Degradation of tannic acid and purification and characterization of tannase from *Enterococcus faecalis*. *Int Biodeterior Biodegradation*, *65*:1061–1065.

[144] Iwamoto K., Tsuruta H., Nishitaini Y., et al. (2008). Identification and cloning of a gene encoding tannase (tannin acyl hydrolase) from *Lactobacillus plantarum* ATCC 14917 T. *Syst Appl Microbiol.*, *31*:269–277.

[145] Kar B., Banerjee R., and Bhattacharyya B. C. (2003). Effect of additives on the behavioural properties of tannin acyl hydrolase. *Process Biochem.*, *38*:1285–1293.

[146] Canteralli C., Brenna O., Giovanelli G., et al. (1989). Beverage stabilization through enzymic removal of phenolics. *Food Biotechnol.*, *3*:203–213.

[147] Giovanelli G. (1989). Enzymic treatment of malt polyphenols for stabilization. *Ind Bevenda (Ital)*, *18*:497–502.

[148] García-Conesa M. T., Ostergaard P., Kauppinen S., et al. (2001). Hydrolysis of diethyl diferulates by a tannase from *Aspergillus oryzae*. *Carbohydr Polym.*, *44*:319–324.

[149] Lu M. J., Chu S. C., Yan L., et al. (2009). Effect of tannase treatment on protein-tannin aggregation and sensory attributes of green tea infusion. *LWT-Food Sci Technol.*, *42*:338–342.

[150] Rodríguez-Durán L. V., Valdivia-Urdiales B., Contreras-Esquivel J. C., et al. (2011). Novel strategies for upstream and downstream processing of tannin acyl hydrolase. *Enzyme Res.*, Vol 2011, Article ID 823619, pp. 20.

[151] Kumar C. S., Subramanian R., and Rao L. J. (2013). Application of enzymes in the production of RTD black tea beverages: a review. *Crit Rev Food Sci Nutr.*, *53*:180–197.

[152] Madeira J. V. Jr., Macedo J. A., and Macedo G. A. (2012). A new process for simultaneous production of tannase and phytase by *Paecilomyces variotii* in solid-state fermentation of orange pomace. *Bioprocess Biosyst Eng.*, *35*:477–482.

[153] Schons P. F., Ries E. F., Battestin V., et al. (2011). Effect of enzymatic treatment on tannins and phytate in sorghum (*Sorghum bicolor*) and its nutritional study in rats. *Int J Food Sci Technol.*, *46*:1253–1258.

In: Microbial Catalysts. Volume 1
Editors: S. M. Abdel-Aziz, N. Garg, A. Aeron et al.
ISBN: 978-1-53614-679-0

Chapter 5

XYLANASE: AN INDUSTRIALLY IMPORTANT ENZYME

Sharad Agrawal[+] and Ritu Mahajan[*]
[+]Department of Biotechnology, Sharda University,
Greater Noida, India
[*]Department of Biotechnology, Kurukshetra University,
Kurukshetra, India

ABSTRACT

This chapter encompasses various microbial sources of xylanase enzyme as well as its production conditions, yield, production processes, optimization parameters, characterization with respect to pH and temperature optima, molecular weight, kinetic parameters and also various industrial aspects of the enzyme. Xylanolytic enzymes are the upcoming enzymes of commercial sector. Production of extracellular xylanase through industrially suitable microbial sources is highly advantageous because of quick synthesis in copious quantity and easy product recovery. Also, since biocatalysts are highly specific for their substrates and the microbial cells are eminently biodegradable, they offer complete environment friendly solutions for various industrial processes. Various production parameters viz. inoculum age and size, pH of fermentation media, incubation temperature, incubation time, carbon and nitrogen sources etc. need to be modeled for obtaining maximum yield of enzyme with least economic input. Characterization of enzymes with respect to Km and Vmax values, pH and temperature optima as well as their stability data at different pH and temperatures is necessary to reveal the suitability of enzymes for industrial application purposes. Lot of research is being done to explore industrially suitable xylanase producing microorganisms. The importance of this enzyme can be judged from the quantum of work done during last two decades, as reported in this review.

Keywords: xylanase, solid state fermentation, submerged fermentation, purification, industrial applications

[*] Corresponding Author's Email: ritupanipat@rediffmail.com.

INTRODUCTION

Global market for industrial enzymes is projected to reach from $1.06 billion in 1995 [1] to almost $6.3 billion in 2022 [2] and is expected to increase further, as new applications are coming up. Xylanolytic enzymes are one of the most important enzymes of commercial sector, as they are widely applicable in various industrial processes viz. removal of xylan prior to bleaching of wood pulp in paper and pulp industries [3-7], recycling of waste paper [8-13], improving quality of dough and baked products [14-18], for extraction and clarification of fruit juices [19-20], extraction of oils [21-23], degumming of plant fibres [24-28], animal feed [29-33], biofuel production [34-35] etc. Alarming level of environmental pollution, caused by increased industrialization since last few decades, has promoted an awareness and switched the focus of researchers towards various eco-friendly approaches and thus in return, has encouraged a great deal of research for xylanolytic enzymes. Incorporation of microbial biocatalysts, into various industrial processes, makes them less energy demanding and less polluting. Enzymes, exhibiting high activity as well as stability over broad range of temperature and pH, are capable of meeting the industrial requirements more precisely.

Mode of Action

The xylanolytic enzyme concoction consists of a range of hydrolytic enzymes, including endo-1,4-β-xylanase (E.C.3.2.1.8), xylan-1,4-β-xylosidase (E.C.3.2.1.37), α-glucuronidase (E.C.3.2.1.139), α-L-arabinofuranosidase (E.C.3.2.1.55) and acetylxylan esterase (E.C.3.1.1.72), which act synergistically and specifically on their respective bond cleaving sites, for complete degradation of xylan polysaccharide into its constituent sugars.

Microbial Sources of Xylanase

Microorganisms have been considered as an excellent source for extracellular production of enzymes, since they are capable of multiplying at higher rates and are manufacturing machines for biologically active products. In recent years, there has been a phenomenal uplift in the use of bio-catalysts at industrial level. Several researchers have reported the production of xylanase, using a variety of microbial species. Among bacteria, *Bacillus* has been reported as one of the most prominent source of xylanase and several different species such as *B. altitudinis* DHN8 [8], *B. amyloliquefaciens* [36, 37], *B. circulans* [38, 39], *B. thermantarcticus* [40], *B. pumilus* [41-43], *Geobacillus stearothermophilus* [44], *B. subtilis* [45, 46], *B. aerophilus* [47] and *B. halodurans* [48] have been reported to exhibit considerable xylanolytic activity. Apart from bacteria, several fungal species namely, *Aspergillus* [49-52] and *Trichoderma* [53, 54] have been found to be potent producers of xylanase (Table 1 and Table 2).

Table 1. Production of xylanase under solid state fermentation conditions by various microorganisms

Microorganism	Substrate	Fermentation conditions		Activity (IU/g)	Reference
		Temperature (°C)	pH		
Bacillus pumilus	Wheat bran	35	9.0	21431	Poorna and Prema, [166]
Bacillus sp. GRE7	Wheat bran	55	7.0-7.5	3950	Kiddinamoorthy et al. [125]
Burkholderia sp. DMAX	Wheat bran	37	8.6	5650	Mohana et al. [65]
Bacillus subtilis ASH	Wheat bran	37	7.0	8964	Sanghi et al. [84]
Aspergillus niger	Corn cob and wheat bran	30	-	794	Betini et al. [167]
A. niveus	Wheat bran	30	-	927	Betini et al. [167]
A. ochraceus	Corn cob and wheat bran	30	-	752	Betini et al. [167]
Streptomyces sp. 7b	Wheat bran	50	7.0-8.0	2180	Bajaj and Singh, [103]
Aspergillus foetidus	Wheat bran	30	5.3	8450	Chapla et al. [67]
Bacillus pumilus	Wheat bran	37	8.0	25,103	Garg et al. [62]
Scytalidium thermophilum	Deoiled *Jatropha curcas* seed cake	45	9.0	1455	Joshi and Khare, [168]
Bacillus pumilus	Wheat bran	30	8.0	73,000	Nagar et al. [63]
Penicillium citrinum	Sugarcane baggase	30	7.0	1645	Ghoshal et al. [99]
Cellulosimicrobium	Wheat bran	40	7.0	4962	Kamble and Jadhav, [66]
Aspergillus niger	Barley bran	35	5.5	42.5	Soliman et al. [152]
Fusarium solani SyrN7	Wheat bran	25	8.0	1593	Bakri et al. [105]
Trichoderma reesei SAF3	Wheat bran	30	6.0	291	Kar et al. [128]
Rhizopus stolonifer JS1008	Corn cob	32	7.5	13.9	Zhang et al. [134]
Thermoascus aurantiacus RCKK	Wheat bran	45	5.0	6543	Jain et al. [169]
Aspergillus tubingensis FDHN1	Sorghum straw	35	5.0	4105	Adhyaru et al. [49]
Cellulosimicrobium cellulans CKMX1	Apple pomace	35	8.0	1027.65	Walia et al. [111]
Penicillium citrinum HZN13	Sweet sorghum bagasse	40	4.0	30,144	Bagewadi et al. [69]
Aspergillus niger USM SD2	Oil-palm-empty-bunches	25	6.3	3246	Ajijolakewu et al. [93]
Schizophyllum commune ARC-11	Rice straw	30	7.0	8591.4	Gautam et al. [98]
Aspergillus niger CCUG33991	Wheat bran	28-35	6.7	2919	Khanahmadi et al. [110]

Table 2. Production of xylanase under submerged fermentation conditions by various microorganisms

Microorganism	Substrate	Fermentation conditions		Activity (IU/mL)	Reference
		Temperature (°C)	pH		
Bacillus subtilis	Oat spelt xylan	55	9	128	Annamalai et al. [129]
Streptomyces sp. SU9	Wheat bran	45	8-10	5200	Bajaj et al. [70]
Bacillus pumilus	Oat	30	4.913	2.38	Mullai et al. [170]
B. pumilus SV-85S	Wheat bran	37	6.0	2995	Nagar et al. [43]
B. halodurans	Wheat bran	45	9.0	40	Kumar and Satyanarayana, [129]
B. mojavensis AG137	Oat bran	37	8.0	302.466	Sepahy et al. [88]
Paenibacillus macerans	Corn cob chips	50	6.0	120	Dheeran et al. [171]
Bacillus subtilis BS05	Sugarcane baggase	37	5.0	439	Irfan et al. [145]
Bacillus sp. SV-34S	Wheat bran	37	7.0	3454	Mittal et al. [172]
B. pumilus SV-205	Wheat bran	60	10.0	7382.7	Nagar et al. [71]
Stenotrophomonas maltophilia	Wheat bran	35	8.0	26.4	Raj et al. [130]
Geobacillus sp. WSUCF1	Birchwood xylan	60	7.0	23.8	Bhalla et al. [173]
Penicillium citrinum	Czapek yeast extract	30	6.0	138.73	Ghoshal et al. [96]
Bacillus mojavensis UEB-FK	Wheat bran	37	7.0	41.9	Kallel et al. [100]
B. cereus BSA1	Xylan	35	6.0	7.40	Mandal et al. [76]
Aspergillus niger DR02	Sugarcane bagasse liquor	29	5.0	458.1	Robl et al. [174]
Bacillus pumilus	Wheat bran	33.5	9.0	372	Thomas et al. [175]
Aspergillus niger DX-23	Corn Cob Waste	30	5.0	59.5	Desai and Iyer, [74]
Bacillus pumilus 3GAH	Wheat bran	37	7.0	1430	Kaur et al. [86]
B. tequilensis	Birchwood xylan	35	7.0	86.62	Khusro et al. [56]
Aspergillus niger	Beechwood xylan	28	7.0	2524	Uday et al. [77]
Bacillus altitudinis DHN8	Sorghum straw	32	8.0	492	Adhyaru et al. [8]
Aspergillus fumigatus SD5A	Corn cob	30	6.0-6.7	50.55	Elegbede and Lateef, [133]
Fusarium sp. BVKT R2	Birchwood xylan	30	5.0	4200	Ramanjaneyulu and Reddy, [97]
Streptomyces sp. ER1	Beechwood xylan	28	7.0	10,220	Rosmine et al. [90]

Xylanase in Combination with Other Enzymes

Although the individual production of xylanolytic enzymes using different sources has been reported by several researchers; few reports are also available, showing their production in combination with other industrially significant enzymes, in the same fermentation medium by a single microbial isolate. Extracellular release of xylanase in combination with other important enzymes such as pectinase, cellulase etc., makes the enzyme production process economically more beneficial from industrial point of view. Researchers and Research & Development (R&D) sectors in industries, nowadays are putting their efforts for screening of microbes, capable of producing a combination of two or more enzymes concurrently in high quantity along with wide range of thermo-pH stability and suitable to the need of industries. Table 3 summarizes various reports showing different catalytic combinations produced by various microbial isolates.

Table 3. Production of xylanase in combination with different enzymes by various microbial isolates

Microorganism	Enzymes produced	Activity	Units	Reference
Deuteromyces	Xylanase	83	IU/mL	Gomes et al. [176]
	Pectinase	121	IU/mL	
Streptomyces sp. QG-11-3	Xylanase	96	IU/mL	Beg et al. [177]
	Pectinase	46	IU/mL	
Aspergillus niger 3T5B8	Xylanase	30.62	IU/mL	Couri et al. [178]
	Protease	5.27	IU/mL	
	Polygalacturonase	30.75	IU/mL	
Bacillus circulans BL53	Xylanase	0.953	IU/mg	Heck el al. [179]
	Mannanase	0.540	IU/mg	
Thermomyces lanuginosus	Xylanase	5846	IU/mL	Mendoza et al. [180]
	Pectinase	840	IU/mL	
Aspergillus awamori	Xylanase	30 -40	IU/g	Botella et al. [181]
	Exopolygalacturonase	30 -40	IU/g	
A. niger F3	Xylanase	65	IU/g	Rodríguez-Fernández et al. [182]
	Pectinase	265	IU/g	
A. awamori 2B.361 U2/1	Xylanase	43.7	IU/g	Diaz et al. [183]
	CMCase	2.8	IU/g	
	Exopolygalacturonase	16.9	IU/g	
A. niger FGSCA733	Xylanase	6087	IU/g	Thembekile et al. [107]
	Cellulase	3974	IU/g	
A. fumigatus ABK9	Xylanase	1130	IU/g	Das et al. [184]
	Endoglucanase	826	IU/g	
	ß-glucosidase	255	IU/g	
	FPase	102	IU/g	
Cerrena unicolor VKM F-3196	Xylanase	6.5	IU/mL	Belova et al. [185]
	Cellulase	12.7	IU/mL	
Stenocarpella maydis	Xylanase	19266	IU/L	Harnandes-Dominguez et al. [186]
	Cellulase	9439	IU/L	
	Protease	806	IU/L	

Table 3. (Continued)

Microorganism	Enzymes produced	Activity	Units	Reference
Trichoderma virens	Xylase	348	IU/g	El-Shishtawy et al. [119]
	CMCase	123.26	IU/g	
	FPase	49.3	IU/g	
	Pectinase	499.9	IU/g	
Aspergillus japonicus PJ01	Xylanase	12.07	IU/mL	Li et al. [187]
	CMCase	2.69	IU/mL	
	Pectinase	53.48	IU/mL	
Penicillium chrysogenum QML-2	Xylanase	19613.25	IU/g	Zhang and Sang, [68]
	Mannanase	8479.82	IU/g	
Microorganism	Enzymes produced	Activity	Units	Reference
Trichoderma longibrachiatum MDU-6	Xylanase	3811	IU/g	Chutani and Sharma, [188]
	Mannanase	9.9	IU/g	
Trichoderma citrinoviride AUKAR04	Xylanase	55000	IU/g	Periyasamy et al. [189]
	CMCase	375	IU/g	
	β-1,3-glucanase	695	IU/g	
Talaromyces stipitatus	Xylanase	582	IU/g	Bharti et al. [95]
	FPase	2.86	IU/g	
	Endoglucanase	30	IU/g	
Bacillus pumilus	Acidic Xylanase	315	IU/mL	Kaur et al. [41]
	Alkaline Xylanase	290	IU/mL	
	Pectinase	88	IU/mL	
Trichorderma reesei CCT-2768	Xylanase	99.5	IU/g	Silva et al. [190]
	CMCase	13	IU/g	
*Trichoderma reesei*QM9414	Xylanase	93.08	IU/mL	Silva et al. [191]
	Endoglucanase	3.14	CMC/mL	
	Exoglucanase	1.25	IU/mL	
	Total Cellulase	0.85	FPU/mL	

CMCase – carboxymethylcellulase, FPase – filter paperase

Screening of Xylanolytic Enzyme Producing Isolates

For isolation of xylanase enzyme producing microorganisms, xylan enriched screening media is required, for observing zone of substrate hydrolysis. Reports are available in the literature showing the utilization of commercial xylan for screening purpose [55-57]. However, employment of commercial xylan for qualitative estimation of enzyme activity makes the methodology expensive, since the cost of 100 g commercial xylan is US $294.39. Hence, to formulate an economically viable screening process, few researchers have reported the utilization of xylan rich agro-residues such as wheat bran as an alternate source to xylan [58-61].

Production of Xylanase Enzyme

Submerged liquid culture (SmF) and solid substrate fermentation (SSF) are the two main processes through which enzymes can be produced. Table 1. and Table 2. enlists several reports on xylanase production using various bacterial and fungal species through SSF and SmF conditions respectively. Production of nearly 25,103 IU [62] and 73,000 IU [63] of xylanase per g of solid substrate has been reported under solid state fermentation conditions using *Bacillus pumilus* strain. Researchers have also obtained 8964 IU [64], 5650 IU [65] and 4962 IU [66] of xylanase using *B. subtilis* ASH, *Burkholderia* sp. DMAX and *Cellulosimicrobium* strain respectively, under solid substrate fermentation conditions. Using fungal strains also, under SSF conditions, 8450 [67], 19,613 [68] and 30,144 [69] IU of xylanase per g of solid substrate was obtained using *Aspergillus foetidus*, *Penicillium chrysogenum* QML-2 and *P. citrinum* HZN13 strain respectively, whereas, under SmF conditions using *Streptomyces* sp. SU9 and *Bacillus pumilus* strain respectively, 5200 IU [70] and 7382.7 IU [71] of xylanase titer was obtained per mL of liquid medium.

Optimization of Process Parameters for Xylanolytic Enzymes Production

Optimization of fermentation media components and other parameters such as pH, incubation temperature, incubation time, inoculum age and size, carbon and nitrogen sources etc. play a critical role in enhancing the yield of industrially important enzymes.

Inoculum Size and Age

Microbial enzymes titer produced in the fermentation medium is largely dependent on inoculum size. The number of microbial colony forming units in production media must be sufficient enough to promote ample growth as well as optimum enzyme titer [65]. Inoculum concentration above or below optimal, results in poor enzyme production due to disruption in mass: substrate ratio. An insufficient inoculum size results in delayed preparatory phase and hence enzyme synthesis is also retarded, whereas, an extortionate microbial concentration leads to a conflict among cells for the availability of nutrients present in production medium [72].

Several researchers have reported an optimum inoculum concentration between 1.0–5.0% (v/v) for adequate xylanase synthesis under submerged fermentation conditions [8, 73-77]. Researchers have also experimentally determined 10% [78-81] and 12.5% [82] concentration of inoculum to be optimum for maximum xylanase synthesis under SmF conditions using different microorganisms.

Under solid state fermentation conditions, 5.0% - 10% concentration of inoculum was found to be best for maximum xylanase production [66, 83]. Inoculum concentration of 11.6% [68] and 15% [63, 84] has also been reported as optimum by workers.

An inoculum age of 16-18 h has been reported to be optimum by several researchers for obtaining maximum xylanase titer by different microorganisms viz. *Bacillus altitudinis* DHN8 [85], *B. pumilus* [86], *Streptomyces* RCK-2010 [87], *B. pumilus* SV-85S [43], *B. mojavensis* AG137 [88] etc. While optimizing the production conditions through statistical design approach, Gupta et al. [89] determined 36 h of inoculum age to be optimum for obtaining maximum xylanase production under submerged fermentation conditions. Kumar et al. [79] observed 8 h of inoculum age to be optimum for maximum xylanase production from *B. amyloliquefaciens* SH8, whereas Rosmine et al. [90] determined 21.05 h of inoculum age to be significant for getting maximum xylanase activity from *Streptomyces* sp.

Under SSF conditions, 16 h [91] and 18 h [66] old inoculum of *B. megaterium* and *Cellulosimicrobium* respectively, yielded maximum xylanase production.

pH of Fermentation Media

Productivity of enzymes by a microorganism is largely dependent on the concentration of hydrogen ions in fermentation media. The growth rate of microorganism as well as the enzyme yield changes significantly with change in pH. The optimum pH for obtaining maximum enzyme activity may thus vary from acidic to alkaline range, depending upon the microorganism involved and the pH required for its growth and enzyme productivity.

Under acidic conditions at pH 4.0, Gowdhaman et al. [92] reported maximum xylanase synthesis from *B. aerophilus* KGJ2. Between pH 5.0 – 6.5, *A. tubingensis* FDHN1 [49], *Trichoderma harzianum* PPDDN10 NFCCI-2925 [53], *A. niger* DX-23 [74], *B. cereus* BSA1 [76], *A. niger* USM SD2 [93], *A. fumigatus* AR1 [94], *Talaromyces stipitatus* [95], *Pencillium citrinum* [96] and *Fusarium* sp. BVKT R2 [97] gave maximum xylanase activity.

Kamble and Jadhav [66], Gautam et al. [98], Ghoshal et al. [99], Kallel et al. [100], Kuancha and Apiraksakorn [101] and Sharma & Chand [102] reported pH 7.0 to be optimum for maximum xylanase synthesis from *Cellulosimicrobium* sp., *Schizophyllum commune* ARC-11, *P. citrinum*, *B. mojavensis* UEB-FK, *B. subtilis* GN156 and *Pseudomonas* sp. XPB-6, respectively. Bajaj and Singh [103] also obtained highest xylanase titer between pH 7 to 8 from *Streptomyces* sp. 7b strain. Under alkaline conditions between pH 8.0 – 9.0, maximum xylanase yield was obtained from *Streptomyces* sp. AOA40 [14], *B. subtilis* [75], *B. megaterium* [78] and *A. caespitosus* [104].

Incubation Temperature

The growth rate of microorganisms as well as the production of enzymes is largely dependent on the temperature of fermentation medium. The microorganisms exhibit maximum synthesis of enzyme at an optimized temperature condition. Different xylanases from fungus, *A. foetidus* MTCC 4898 [67], *A. niger* USM SD2 [93], *Talaromyces stipitatus* [95], *Schizophyllum commune* ARC-11 [98], *Penicillium citrinum* [99], *Fusarium solani* [105], *A. tubingensis* JP-1 [106] and *A. niger* FGSCA733 [107] have been found to be maximally produced in temperature ranging from 20°C to 30°C. Several researchers have reported the production of xylanase enzymes from different microbial sp. viz. *Candida tropicalis* HNMA-1 [34], *B. pumilus* SV-85S [43], *T. harzianum* PPDDN10 NFCCI-2925 [53], *B. tequilensis* strain ARMATI [56], *B. subtilis* [75], *B. cereus* BSA1 [76], *B. subtilis* GN156 [101], *Pseudomonas* sp. XPB-6 [102], *Aspergillus niger* LCBT-14 [108], *A. tubingensis* FDHN1 [109], *A. niger* CCUG33991 [110] and *Cellulosimicrobium cellulans* CKMX1 [111] in a temperature range of 30 to 40°C. Optimum temperature range for production of xylanase from *Mucor indicus*, *Mucor hiemalis* and *Rhizopus oryzae* was observed in between 40–45°C [112]. Some thermophillic microbes like *Paenibacillus* sp. N1 [82] and *Bacillus subtilis* [113] have also been reported to produce the xylanase enzyme in a temperature ranging from 50°C to 60°C.

Incubation Time

Fermentation time for obtaining maximum enzyme titer varies from one species to another and is also dependent upon the environmental conditions as well as genetic makeup of the microbe. Different researchers have reported different cultivation periods for xylanase production viz. under submerged fermentation conditions, 18 h from *Bacillus subtilis* [113]; 24 h from *B. pumilus* MK001 [58], *Pseudomonas* sp. XPB-6 [102] and *Paenibacillus campinasensis* BL11 [114]; 48 h from *B. aerophilus* KGJ2 [47], *B. subtilis* [75], *A. awamori* [115] and *B. halodurans* PPKS-2 [116]; 72 h from *B. licheniformis* P11(C) [117] and 96 h from *Penicillium* sp. SS1 [118]. Desai and Iyer [74] reported 84 h of incubation interval to be optimum for obtaining maximum production of xylanase from *Aspergillus niger* DX-23.

Under SSF conditions also, various incubation periods have been reported by different researchers. Thembekile et al. [107] and Khanahmadi et al. [110] employed *A. niger* FGSCA733 and *A. niger* CCUG33991 respectively, for obtaining maximum xylanase activity after 48 h of incubation time. A cultivation time of 50 – 55 h was found to be optimum for xylanase production from *M. indicus*, *M. hiemalis* and *R. oryzae* [112]. Abdullah et al. [108], Walia et al. [111] and El-Shishtawy et al. [119] obtained maximum xylanase activity from *A. niger* LCBT-14, *Cellulosimicrobium cellulans* CKMX1 and *Trichoderma virens*, respectively after 72 h of incubation time. *T. harzianum* PPDDN10 NFCCI-2925 [53] and *A. foetidus* [120] gave maximum xylanase activity after 96 h of fermentation interval. Researchers have also reported 5-8 days of incubation time to be optimum for obtaining maximum xylanase activity using different microorganisms viz. *A. tubingensis* FDHN1 [49], *P. chrysogenum* QML-2 [68], *Streptomyces* ER1 [90], *Talaromyces stipitatus* [95], *Schizophyllum commune* ARC-11 [98], *A. tubingensis* JP-1 [106], *Coprinellus disseminatus* SW-1 [121], *P. canescens* [122] and *Paecilomyces themophila* J18 [123].

Carbon Sources

Among nutritional parameters, carbon sources are one of the most crucial factors for enzyme production. Commercially available xylan could be an outstanding source of carbon for xylanase production because the low molecular weight compounds obtained, can act as chief xylanase inducers. Implementation of such substrates results in increased yield of xylanase enzyme [124-126].

However, due to high cost of commercial xylan, researchers have shifted their focus towards utilization of abundantly available high xylan content containing plant waste materials as an alternative source of substrate. Various agro-waste residues, such as wheat bran, corn cob, rice bran, rice husk and other xylan rich similar substrates are readily available, commercially viable alternatives and hence can be used to achieve higher xylanase yield with significant reduction in overall production cost of enzymes. Several researchers have reported the suitability of wheat bran as a cost effective and economically preferrable carbon source for the production of xylanase enzyme [41, 62, 66, 110, 119, 127-130].

Apart from wheat bran, various other agro-residues such as rice bran [131], corn cob [74, 132-134], wheat straw [135, 136], rice straw [98, 137], melon peel [138], maize straw [139], paddy husk [140], cassava bagasse [141], apple pomace [142, 143], banana waste [144] and sugarcane pulp or baggase [115, 145] have also been used for xylanase production by different microorganisms.

Nitrogen Sources

The nitrogen source used in the fermentation medium is one of the major factors influencing enzyme production. Several researchers have utilized organic nitrogen sources viz. beef extract [47], peptone [95], meat extract [102], soybean flour [113], soya bean meal [121, 140], yeast extract [56, 93, 96, 97, 141], soybean bran [146] etc. for xylanase production. Utilization of inorganic nitrogen sources such as sodium nitrate [74], ammonium nitrate [72, 76], ammonium sulfate [110, 147] etc. have also been reported.

Researchers have also reported a combination of nitrogen sources for maximum enzyme production. A combination of yeast extract and peptone was found to be most appropriate nitrogen source for obtaining 410 IU/mL of xylanase yield from *B. subtilis* ASH [64]. Maximum activity of xylanase enzyme (9.61 IU/mL) was obtained with a combination of yeast extract and tryptone [149] as chief nitrogen source. Xylanase production from *B. pumilus* VLK-1 (7,295 IU/ml) and *B. pumilus* 3GAH (1430 IU/mL) was higher using a combination of peptone, yeast extract and KNO_3 as compared to individual nitrogen sources [86, 148]. Using a combination of ammonium sulfate and peptone, Irfan et al. [75] obtained maximum xylanase activity from *B. subtilis*. Zhang and Sang [68] observed that xylanase production (19613.25 IU/mL) from *Penicillium chrysogenum* QML-2 was promoted more efficiently using a mixture of ammonium sulphate and yeast extract, as compared to other nitrogen sources.

Additives

Various metal ions, chelators, detergents, surfactants and amino acids have positive and negative effects on enzymes production and activity. Optimized salts concentration creates an isotonic environment which supports an ideal growth of microbial cells and also enhances the yield of enzymes [44]. Hyper or hypotonic production media results in either shrinkage or swelling and bursting of microbial cells, thereby decrease the overall enzyme synthesis phenomenon.

Addition of proline, thiamine and Tween-40 in production media at optimum concentration resulted in nearly two fold enhancement in xylanase synthesis from *Gracilibacillus* sp. TSCPVG [149]. Also, the presence of either NaCl or Na_2SO_4 at 3.5% concentration resulted in maximum xylanase production (3.5 and 3.68 IU/mL, respectively) from *Gracilibacillus* sp. TSCPVG, while all other metal ions viz. Mn^{2+}, Ca^{2+}, Mg^{2+}, Ni^{2+}, Co^{2+}, Fe^{2+} and Zn^{2+} resulted in reduced enzyme yield [149]. Presence of metal ions such as Na^{+}, Mg^{2+}, Mn^{2+}, Fe^{3+}, Zn^{2+}, Cu^{2+}, Co^{2+} and also, Tween 80 resulted in an increase in xylanase production from *Penicillium citrinum*, while Hg^{2+}, Ca^{2+} and EDTA inhibited enzyme activity [99]. Soliman et al. [150] obtained nearly 1.45 and 2.0 fold enhanced xylanase synthesis from *A. niger* and *T. viride* respectively under SSF conditions in the presence of KCl. Using statistical design approach, Gowdhman et al. [151] determined $MgSO_4$ as a significant variable, affecting xylanase yield from *Bacillus aerophilus* KGJ2 and a concentration of 1.0 g/l was found to be optimum for maximum enzyme production. Bibi et al. [44] reported that addition of salts such as K_2HPO_4 (0.25%), KH_2PO_4 (0.05%) and $CaCl_2$ (0.01%) in production media, at their respective concentration resulted in maximum production of endo-1,4-ß-xylanase enzyme. Goluguri et al. [152] observed stimulatory effect of metal ions Mn^{2+} and K^{+} on xylanase production from *Thielaviopsis basicola* by 8% and 12% respectively, while Cu^{2+} and Zn^{2+} inhibited the enzyme activity by about 50%. Guan et al. [153] observed that supplementing media with metal ions such as Zn^{2+}, Fe^{2+}, Mg^{2+}, Mn^{2+} or Ca^{2+} resulted in an increase in enzyme activity with Mg^{2+} ions exerting maximum positive impact on xylanase production (55.92 IU/mL) from *Cladosporium oxysporum* GQ-3. Similarly, addition of Tween 20 (0.10% w/v) to production media enhanced the xylanase yield from *Schizophyllum commune* ARC-11 [98]. Xylanase yield from *B. amyloliquifaciens* SH8 was enhanced in the presence of metal ions Zn^{2+}, Ca^{2+} and Mn^{2+}, whereas Cu^{2+}, Mg^{2+} and Na^{+} strongly inhibited the enzyme productivity [79]. Rosmine et al. [90] used olive oil as an additive at a concentration of 33.10 mg/mL for obtaining enhanced production of xylanase enzyme (10220 IU/mL) from *Streptomyces* ER1 strain.

Xylanase activity yield from *Chromohalobacter* sp. TPSV 101 was not enhanced in the presence of divalent cations such as Ca^{2+}, Cu^{2+}, Mn^{2+}, and Fe^{2+}, whereas Zn^{2+} and Hg^{2+} completely inhibited the bacterial growth and enzyme production [154].

Optimization of various physico-chemical and nutritional factors is indispensable in order to determine the best conditions for maximum level of enzyme production. This can be done either by using a conventional 'one variable at a time' approach or multi-interactive statistical design approach. Traditional "one factor at a time" approach for optimization of media components is time consuming and laborious and often does not yield reliable results, as the interactive outcome among various media components is neglected. Response surface methodology (RSM), on the other hand, is advantageous in terms of rapid sifting of the factors, significantly affecting the production of primary and secondary metabolites [155]. Researchers have reported an increase in xylanase production, upto 3.7 fold using *Bacillus tequilensis* ARMATI [56], 2.19 fold using *B. pumilus* [62], 10.31 fold using *P. chrysogenum* [68], 3.14 fold using *Penicillium citrinum* HZN13 [69], 1.3 fold using *Bacillus* sp. PKD9 [83], 1.56 fold from *Streptomyces* ER1 [90], 3.1 fold using *Cellulosimicrobium cellulans* [111], 8.5 fold using *Fusarium verticillioides* [156] and 1.9-fold using *A. fischeri* [157] under optimized conditions, using statistical design approach.

Purification and Characterization of Xylanases

Purification and characterization of enzymes is significant for understanding the most suitable conditions required for unexcelled performance of an enzyme and their applicability from industrial point of view. Purification of xylanase requires a sequence of steps, commonly involving precipitation of proteins using ammonium sulfate followed by various combinations of column chromatography techniques as reported by different workers. Utilization of both, anion exchangers viz. QAE-Sephadex, DEAE-Toyopearl, DEAE-Sepharose, Q-Sepharose and cation exchangers viz. SP-Sephadex, CM-Sepharose and SE-52 have been reported for xylanase purification. Commonly used gel filteration chromatography media for xylanase purification are Biogel, Sephacryl, Sephadex and Utrogel.

Nearly 11 fold purified xylanase was obtained from *Aspergillus oryzae* DSM 1863 with 28% recovery yield using Sephadex G-200 and DEAE-Sephadex column [52]. Kiddinamoorthy et al. [125] purified xylanase, upto 28.5 fold, from *Bacillus* sp. GRE7 using DEAE Cellulose and Sephadex G-75 column with 27% recovery yield. Xylanase obtained from *Cladosporium oxysporum* was purified to 20.98 fold along with 19.85% yield recovery using ammonium sulfate precipitation (0-40%) and DEAE-Sepharose column [153]. Approximately 7.93 fold purified xylanase was obtained with 48.63% yield recovery, from *Aspergillus niger* [158] after precipitating with ammonium sulfate and passing through Sephadex G-75 column. Nearly 38.35 fold purified xylanase was obtained from *Caldicoprobacter algeriensis* sp. TH7C1T using Sephacryl S-200 and Mono-S Sepharose column [159]. Adigüzel and Tuncer [14] and Muthusamy et al. [160] purified xylanase upto 3.35 and 4.27 fold respectively, from *Streptomyces* sp. (strain AOA40 and MSU3, respectively) using different purification steps and obtained 26.99 and 15.57% recovery yield. Table 4a and Table 4b enlists steps of purification and characterization of various xylanolytic enzymes and isozymes respectively.

Table 4a. Purification and characterization of xylanase enzyme from various microbial isolates

Microorganism (xylanase source)	Purification Steps	Fold Purification	Yield %	pH optima	Temperature optima (°C)	Molecular weight (kDa)	Km (mg/mL)	Vmax (IU/mg)	Reference
Arthrobacter sp.	$(NH_4)_2SO_4$ (0-80%), Sephadex G-200 DEAE-Sepharose FF CM-Sepharose FF	21	14	9.0	100	20	0.9	3571	Khandeparkar and Bhosle, [192]
Bacillus sp. GRE7	$(NH_4)_2SO_4$ (40-80%), DEAE Cellulose Sephadex G-75	28.5	27	7.0	70	42	2.23	296.8	Kiddinamoorthy et al. [125]
Burkholderia sp. DMAX	$(NH_4)_2SO_4$ (30-70%)	3.21	48	8.6	50	-	12.75	165	Mohana et al. [65]
Aspergillus flavus MTCC 9390	$(NH_4)_2SO_4$ (30-70%) Sephadex G-100	12.86	22.3	5.0	55	35	-	200 IU/mL	Bhushan et al. [193]
Bacillus mojavensis UEB-FK	$(NH_4)_2SO_4$ (0-60%) Biogel P100 S-Sepharose	2.8	2.08	4.0	50	14	3.85	250.02	Kallel et al. [194]
Gracilibacillus sp. TSCPVG	Acetone fraction Phenyl Sepharose Sephadex G75 Electroelution	12	22	7.5	60	42	21	1667	Poosarla and Chandra, [195]
Cellulosimicrobium cellulans CKMX1	$(NH_4)_2SO_4$ (0-80%) Sephadex G-100 DEAECellulose Ultrafiltration (10kDa)	5.92	20.28	8.0	55	29	2.64	2,000	Walia et al. [196]
Paenibacillus sp. NF1	Octyl-Sepharose Sephadex G75 Q-Sepharose	14.55	10.63	6.0	60	37	6.32	3,216.44	Zheng et al. [197]
Scytalidium thermophilum	Sephacryl S-200 Anion exchanger	21.8	9.6	6.5	65	21	2.4	168.6 IU/mL	Kocabas et al. [199]

Microorganism (xylanase source)	Purification Steps	Fold Purification	Yield %	pH optima	Temperature optima (°C)	Molecular weight (kDa)	Km (mg/mL)	Vmax (IU/mg)	Reference
Streptomyces sp. AOA40	DEAE-Sepharose Sephadex G-100	3.35	26.99	6.0	60	-	2.4	165.4	Adigüzel and Tuncer, [14]
Caldicoprobacter algeriensis sp. TH7C1^T	Heat treatment (1 h at 70°C) Sephacryl S-200 Mono-S Sepharose	38.35	27	11	70	35	1.33	595	Amel et al. [161]
Streptomyces thermovulgaris TISTR1948	DEAE-Toyopearl Toyopearl HW-55	15	13	6.5	65	46.2	37.6 μM	303	Boonchuay et al. [200]
Thielaviopsis basicola	$(NH_4)_2SO_4$ (30-60%) DEAE Sephadex Sephacryl S-200	19.7	33	5.5	60	32	1.447	60.04 IU/mL	Goluguri et al. [154]
Streptomyces althioticus LMZM	$(NH_4)_2SO_4$ (0-60%) Sephadex G-25 DEAE cellulose Sephadex G–100	12.65	5.39	8.0	60	31.75	43.03	312.5	Luo et al. [201]
S. olivaceus	DEAE cellulose Sephadex G–75	4.27	15.57	8.0	40	42	8.16	250.01	Muthusamy et al. [162]
Bacillus methylotrophicus CSB40	$(NH_4)_2SO_4$ (30–80%) DEAE sepharose fast flow	8.12	15.88	6.0	50	27	0.08	794.63	Panthi et al. [202]
Trichoderma citrinoviride AUKAR04	$(NH_4)_2SO_4$ (0–55%) Three phase partitioning	5.7	99.8	5.0	50	29.8	0.713	24.28 IU/mL	Periyasamy et al. [190]
Bacillus altitudinis DHN8	$(NH_4)_2SO_4$ (0–70%) DEAE Cellulose	4.9	57.0	8.0	45-55	98.0	**ND**	**ND**	Adhyaru et al. [8]

ND – No data available

Table 4b. Purification and characterization of xylanase isozymes from various microbial isolates

Microorganisms	Purification Steps	Isozyme	Fold Purification	Yield %	pH optima	Temperature optima (°C)	Molecular weight (kDa)	Km (mg/mL)	Vmax (IU/mg)	Reference
Thermotoga maritime	Q-Sepharose Phenyl Sepharose (3-0 M NaCl) Phenyl Sepharose (0.5-0 M NaCl) Ultrafiltration	Xyn A	31	4.4	6.3	92	120	0.11%	374	Winterhalter and Liebl, [203]
	Q-Sepharose Phenyl Sepharose Mono P Gel filtration	Xyn B	288	3.5	5.4	105	40	0.029%	4600	
Cephalosporium sp. RYM-202	$(NH_4)_2SO_4$ (30–80%) DEAE Sephadex A-50 Sephacryl S-200 HR Superose 12 HR	CX-I	17.3	9.9	7.5	50	35	3.18	140.2	Kang et al. [204]
		CX-II	22.9	15.0	8.0	50	24	2.20	175.6	
Melanocarpus albomyces	$(NH_4)_2SO_4$ DEAE Sephadex A-50 Ultrogel AcA-54 QAE-Sephadex A-50	IA	1.91	1.6	6.6	65	38	0.33	311	Prabhu and Maheshwari, [205]
		IIIA	8.69	27.02	5.6	65	24	1.69	500	
Myceliophthora sp. IMI 387099	Ultrafiltration DEAE Sepharose Phenyl Sepharose	Xyl Ia	13.23	2.61	6.0	75	53	1.63	55.5	Chadha et al. [206]
		Xyl Ib	8.54	2.36	6.0	75	53	2.85	33.3	

Microorganisms	Purification Steps	Isozyme	Fold Purification	Yield %	pH optima	Temperature optima (°C)	Molecular weight (kDa)	Km (mg/mL)	Vmax (IU/mg)	Reference
Penicillium sclerotiorum	DEAE Sephadex A-50 Sephadex G-75	I	2.49	27.1	2.5	50	23.9	6.5	189.70	Knob and Carmona, [207]
		II	2.41	9.8	4.5	55	33.1	26.61	90.25	
Malbranchea flava	DEAE Sepharose Phenyl Sepharose	MFX I	21.30	4.26	9.0	60	25.2	1.25	1666	Sharma et al. [208]
		MF II	11.39	0.37	9.0	60	30	3.7	1923	
Bacillus pumilus	Ultrafiltration $(NH_4)_2SO_4$ (20-70%) DEAE Sepharose	Xyl 1	3.12	4.2	7.0	50	14	4.0	5000	Poorna, [209]
		Xyl 2	1.06	3.1	7.0	50	35	3.5	3448	
		Xyl 3	2.68	1.46	7.0	55	60	ND	ND	
Trichoderma reesei F418	$(NH_4)_2SO_4$ (20-85%) Sephacryl S-300 DEAE-Sepharose	X_I	6.2	14.7	5.5	50	50	1.33	34.83	Soroor et al. [210]
		X_{II}	5.5	13.6	5.5	50	31	2.50	295.43	
		X_{III}	8.6	31.6	5.5	50	30	1.12	448.4	
Bacillus pumilus	Ultrafiltration $(NH_4)_2SO_4$ (0-50%) Sephadex G-100	Xylanase I	19.0	67.0	6.0	55	24.5	4.7	500 IU/mL	Kaur et al. [211]
		Xylanase II			8.5	55	13.0			
Trichoderma stromaticum AM7	$(NH_4)_2SO_4$ (0–80%) Desalting Sephadex G-25 Ion exchange Mono Q Sepharose Sephacryl S-100	Xyl 1	84.6	3.25	5.0	50	i. 26.6 ii. 31.0 iii. 34.2	ND	ND	Almeida et al. [15]
		Xyl 2	329.1	6.70	5.0	60	22.0	9.6	28.57	
Penicillium janczewskii	DEAE Sephadex A-50 Sephadex G-75	Xylanase	5.4	28.4	6.0	65	23.6	7.5	100 IU/mL	Terrasan et al. [212]
	$(NH_4)_2SO_4$ (0–55%) CM Sephadex C-50 Phenyl agarose Sephadex G-200	ß-xylosidase	557.9	6.2	5.0	75	200	3.4	33.2 IU/mL	

ND – No data available

Industrial Applications of Xylanase

Xylanase enzymes, either alone or in combination with several other enzymes viz. pectinases, cellulases, amylases etc. find tremendous applications in several different industrial sectors such as biobleaching and deinking of pulp in paper industry, bio-scouring and bleaching in textile industry, extraction and clarification of fruit and vegetable juices and alcoholic beverages, reducing viscosity of animal feed etc. Depending upon the applicationand the type of ligno-cellulosic material to be treated, the applicability of xylanase enzymes is broadly divided under two major categories.

Acidic Xylanases

Juices obtained by mechanical crushing of fruits and vegetables are cloudy, turbid and highly viscous in appearance and thus do not meet the consumer expectations. The turbidity and viscosity of juice is mainly due to the presence of several polysaccharides such as starch, pectin, cellulose, hemicelluloses and lignin [161]. Degradation of hemicellulosic polysaccharides using xylanolytic synergism is an efficient alternative for reducing the turbidity, viscosity and improving the clarity and nutritive value of juices. Enzymatic treatment enhances the juice yield in addition to improvement in various other quality characteristics [19, 20, 117, 133, 162-165].

Since fruit and vegetable juices are acidic in pH, xylanases showing considerable enzyme activity under acidic conditions can be suitably applied for extraction and clarification of such beverages at industrial scale.

Alkaline Xylanases

Chemical based bleaching steps in paper manufacturing and paper recycling process chiefly employ chlorine, chlorine dioxide agents, NaOH, Na_2SiO_3, Na_2CO_3, H_2O_2, chelating agent and surfactants etc. Extensive use of these chemicals has resulted in accumulation of highly toxic by-products. In textile industries also, chemical scouring and bleaching has resulted in an increase in the accumulation of highly toxic and environment damaging waste products and led to larger BOD and COD values of water bodies. Treatment with xylanase causes breakdown of xylan into its constituent monosaccharides, creates pores and hence allows easy access of bleaching chemicals [3, 5-7]. Since these processes are carried out at high pH, xylanases showing pH optima and stability under alkaline conditions are the most appropriate alternatives for commercial usage in such applications.

Conclusion

With an increase in the emission of toxic chemicals from various industrial processes, regulatory bodies have nowadays imposed strict environmental regulations in order to cut down the pollution level. Scientific awareness has also added an incentive for future research in this direction. Employing bio-catalysts such as xylanase either alone or in combination with other significant enzymes such as pectinase, laccase etc. in paper-pulp and textile industries helps in bringing down the usage of chemicals in various industrial processes and thus in return results in decreased BOD and COD values of water bodies. However, to make the enzymatic process economically feasible at industrial level, it is indispensable to isolate microorganisms possessing high enzyme yielding potential with improved pH and thermo-stability. Thus, applicability of xylanase in different industrial sectors has stimulated the research and development for hunting of more potent xylanase producing strains. Enhancement in the yield as well as stability of enzymes can be achieved by inducing mutations in the genetic make-up of source organisms. But the search for natural producers of xylanase with high titer along with increased stability over broad range of pH and temperature conditions is still on. Optimization of various physico-chemical as well as nutritional parameters must also be carried out critically in order to harness maximum enzyme synthesizing capacity of the organism. Industries and scientific organizations are continuously putting their efforts in the direction of making the enzymatic processes more cost-effective. Availability of industrially suitable xylanases in combination with other important enzyme/s along with their high titer will make the process eco-friendly and economically feasible and will promote various industries to switch over to enzymatic processes, hence will reduce toxic chemicals' load on the environment.

Acknowledgments

The authors thankfully acknowledge the financial support provided by Department of Biotechnology, Ministry of Science & Technology, Government of India (Grant number: BT/PR 20438/BCE/8/1220/2016 for 3 years).

References

[1] Godfrey, T. and West, S. (1996). Textiles, In *Industrial Enzymology*, Macmillan Press, London, UK, 2nd edition; 360–371.

[2] Nabrdalik, M. (2016). Production and optimization of exocellular lipases produced by *Proteus mirabilis*. *Proceedings of ECOpole*; *10* (2).

[3] Agrawal, S., Yadav, R. and Mahajan, R. (2016). Synergistic effect of xylano-pectinolytic enzymes produced by a bacterial isolate in bleaching of plywood industrial waste. *J Cleaner Production*, *118*: 229–233.

[4] Kaur, A., Mahajan, R., Singh A., et al. (2010). Application of cellulase-free xylano-pectinolytic enzymes from the same bacterial isolate in biobleaching of kraft pulp. *Biores Technol.*, *101* (23): 9150-9155.

[5] Lin, XQ., Han, SY., Zhang, N., et al. (2013). Bleach boosting effect of xylanase A from *Bacillus halodurans* C-125 in ECF bleaching of wheat straw pulp. *Enz Microbial Technol.*, *52* (2): 91-98.

[6] Sharma, D., Agrawal, S., Yadav, R. D., et al. (2017). Improved efficacy of ultrafiltered xylanase–pectinase concoction in biobleaching of plywood waste soda pulp. 3- *Biotech.*, *7*: 1-7.

[7] Singh, S., Dutt, D., Tyagi, CH., et al. (2011). Bio-conventional bleaching of wheat straw soda-AQ pulp with crude xylanases from SH-1 NTCC-1163 and SH-2 NTCC-1164 strains of *Coprinellus disseminatus* to mitigate AOX generation. *New Biotechnol.*, *28* (1): 47-57.

[8] Adhyaru, D., Bhatt, N., Modi, H., et al. (2017). Cellulase-free-thermo-alkali-solvent-stable xylanase from *Bacillus altitudinis* DHN8: Over-production through statistical approach, purification and bio-deinking/bio-bleaching potential. *Biocatal Agric Biotechnol., 12*: 220-227.

[9] Chutani, P. and Sharma, K. (2015). Biochemical evaluation of xylanases from various filamentous fungi and their application for the deinking of ozone treated newspaper pulp. *Carbohydr Poly., 127*: 54–63.

[10] Desai, D. and Iyer, B. (2015). Biodeinking of old newspaper pulp using a cellulase-free xylanase preparation of *Aspergillus niger* DX-23. *Biocatal Agric Biotechnol., 5*: 78-85.

[11] Kumar, N.V., Rani, M. E., Gunaseeli, R., et al. (2017*a*). Paper pulp modification and deinking efficiency of cellulase-xylanase complex from *Escherichia coli* SD5. *Intr J Biolog Macromol., 111*: 289-295.

[12] Singh, A., Yadav, R., Kaur, A., et al. (2012). An ecofriendly cost effective enzymatic methodology for deinking of school waste paper. *Biores Technol, 120*: 322–327.

[13] Xu, Q. H., Wang, Y., Qin, M., et al. (2011). Fibre surface characterization of old newprint pulp deinked by combining hemicellulase with laccase-mediated system. *Biores Technol., 102* (11): 6536–6540.

[14] Adigüzel, A. O. and Tuncer, M. (2016). Production, characterization and application of a xylanase from *Streptomyces* sp. AOA40 in fruit juice and bakery industries. *Food Biotechnol., 30* (3): 189-218.

[15] Almeida, C. E., Mendes, S. G., Uetanabaro, A. P., (2016). Thermoresistant xylanases from *Trichoderma stromaticum*: application in bread making and manufacturing xylo-oligosaccharides. *Food Chem., 221*: 1499-1506.

[16] Butt, M. S., Tahir-Nadeem, M., Ahmad, Z., et al. (2008). Xylanases and their applications in baking industry. *Food Technol Biotechnol., 46* (1): 22-31.

[17] Jaekel, L. Z., Camila, B. S., Steel, C. J., et al. (2012). Influence of xylanase addition on the characteristics of loaf bread prepared with white flour or whole grain wheat flour. *Ciência e Tecnologia de Alimentos, 32* (4): 844-849.

[18] Liu, W., Brennan, M., Serventi, L., et al. (2017). Effect of cellulase, xylanase and α-amylase combinations on the rheological properties of Chinese steamed bread dough enriched in wheat bran. *Food Chem., 234*: 93-102.

[19] Kumar, L., Nagar, S., Mittal, A., et al. (2014*b*). Immobilization of xylanase purified from *Bacillus pumilus* VLK-1 and its application in enrichment of orange and grape juices. *J Food Sci Technol., 51* (9): 1737–1749.

[20] Shahrestani, H., Asghar, T., Soozanipour, A., et al. (2016). Enzymatic clarification of fruit juices using xylanase immobilized on 1,3,5-triazine-functionalized silica-encapsulated magnetic nanoparticles. *Biochem Eng J., 109*: 51–58.

[21] Baby, K. C. and Ranganathan, T. V. (2016). Effect of enzyme pre-treatment on extraction yield and quality of cardamom (*Elettaria cardamomum* maton.) volatile oil. *Ind Crops Prod., 89*: 200–206.

[22] Chandran, J., Amma, K., Menon, N., et al. (2012). Effect of enzyme assisted extraction on quality and yield of volatile oil from black pepper and cardamom. *Food Sci Biotechnol., 21* (6): 1611-1617.

[23] Shah, S., Sharma, A. and Gupta, M. (2005). Extraction of oil from *Jatropha curcas* L. seed kernels by combination of ultrasonication and aqueous enzymatic oil extraction. *Biores Technol., 96*: 121–123.

[24] Kaur, A., Singh, A., Patra, A. K., et al. (2016*a*). Cost-effective scouring of flax fibers using cellulase-free xylano-pectinolytic synergism from a bacterial isolate. *J Clean Prod., 131*: 107–111.

[25] Singh, A., Kaur, A., Patra, A., et al. (2016). Xylano-pectinolytic synergism from a bacterial isolate for ecofriendly processing of jute fabric. *J Clean Prod., 124*:120–125.

[26] Vardhini, K. J. and Murugan, R. (2016). Effect of laccase and xylanase enzyme treatment on chemical and mechanical properties of banana fiber. *J Nat Fibers, 14* (2): 217-227.

[27] Wang, R., Liu, Z., Cheng, L., et al. (2017). A novel endo-ß-1,4-xylanase GH30 from *Dickeya dadantii* DCE-01: Clone, expression, characterization, and ramie biological degumming function. *Text Res J., 0* (00): 1–10.

[28] Zhao, D., Liu, P., Pan, C., et al. (2016). Flax retting by degumming composite enzyme produced by *Bacillus licheniformis* HDYM-04 and effect on fiber properties. *The J Text., 108* (4): 507-510.

[29] Cowieson, A. J., Hruby, M. and Isaksen, M. F. (2005). The effect of conditioning temperature and exogenous xylanase addition on the viscosity of wheat-based diets and the performance of broiler chickens. *Br Poult Sci., 46* (6): 717-724.

[30] Gemma, G., Krzysztof, K., Aleksandra, D., et al. (2017). Response of turkeys fed wheat-barley-rye based diets to xylanase supplementation. *Anim Feed Sci Technol., 229*: 117-123

[31] Romero, J., Macias, E., Ma, Z., et al. (2016). Improving the performance of dairy cattle with a xylanase-rich exogenous enzyme preparation. *J Dairy Sci., 99*: 3486–3496.

[32] Vallejo, L. H., Salem, A. Z., M., Camacho, L. M., et al. (2016). Effects of xylanase supplementation on feed intake, digestibility and ruminal fermentation in Rambouillet sheep. *J Agric Sci, 154*: 1110–1117.

[33] Wu, Y. B., Ravindran, V., Thomas, D. G., et al. (2004). Influence of phytase and xylanase, individually or in combination, on performance, apparent metabolisable energy, digestive tract measurements and gut morphology in broilers fed wheat-based diets containing adequate level of phosphorus. *Br Poul Sci., 45*: 76–84.

[34] Nouri, H., Azin, M. and Mousavi, M. L. (2017). Xylan-hydrolyzing thermotolerant *Candida tropicalis* HNMA-1 for bioethanol production from sugarcane bagasse hydrolysate. *Ann Microbiol., 67*: 633–641.

[35] Uday, U. S., Choudhury, P., Bandopadhyay, T., et al. (2016b). Classification, mode of action and production strategy of xylanase and its application for biofuel production from water hyacinth. *Int J Biol Macromo.l, 82*: 1041–1054.

[36] Amore, A., Parameswaran, B., Kumar, R., et al. (2014). Application of a new xylanase activity from *Bacillus amyloliquefaciens* XR44A in brewer's spent grain saccharification. *Wiley Online Library, 90* (3): 573–581.

[37] Breccia, J. D., Sineriz, F., Baigori, M. D., et al. (1998), Purification and characterization of a thermostable xylanase from *Bacillus amyloliquefaciens. Enzyme Microbial Technol., 22* (1): 42-49.

[38] Bocchini, D. A., Alves-Prado, H. F., Baida, L. C., et al. (2002). Optimization of xylanase production by *Bacillus circulans* D1 in submerged fermentation using response surface methodology. *Proc Biochem., 38*: 727-731.

[39] Dhillon, A., Gupta, J., Jauhari, B., et al. (2000). A cellulase-poor, thermostable, alkalitolerant xylanase produced by *Bacillus circulans* AB16 grown on rice straw and its application in biobleaching of eucalyptus pulp. *Biores Technol.*, 73 (3): 273-277.

[40] Lama, L., Valeria, C., Agata, G., et al. (2004). Purification and characterization of thermostable xylanase and β-xylosidase by the thermophilic bacterium *Bacillus thermantarcticus*. *Res Microbiol.*, *155* (4): 283–289.

[41] Kaur, A., Singh, A., Dua, A., et al. (2017). Cost- effective and concurrent production of industrially valuable xylano-pectinolytic enzymes by a bacterial isolate *Bacillus pumilus* AJK. *Prep Biochem Biotechnol.*, *47* (1): 8-18.

[42] Monisha, R., Uma, M. V., Murthy, V. K. (2009). Partial purification and characterization of *Bacillus pumilus* xylanase from soil source. *J Sci Eng Technol.*, 5(II): 137-148.

[43] Nagar, S., Gupta, V. K., Kumar, D., et al. (2010). Production and optimization of cellulose-free, alkali-stable xylanase by *Bacillus pumilus* SV-85S in submerged fermentation. *J Ind Microbiol Biotechnol.*, 37(1): 71–83.

[44] Bibi, Z., Ansari, A., Zohra, R. R., et al. (2014). Production of xylan degrading endo-1, 4-ß-xylanase from thermophilic *Geobacillus stearothermophilus* KIBGE-IB29. *J Rad Res Appl Sci.*, 7: 478-485.

[45] Banka, A. L., Guralp, S. A., Gulari, E. (2014). Secretory expression and characterization of two hemicellulases, xylanase and ß-xylosidase, isolated from *Bacillus subtilis* M015. *Appl Biochem Biotechnol.*, *174* (8): 2702–2710.

[46] Chang, S., Guo, Y., Wu, B., et al. (2016) Extracellular expression of alkali tolerant xylanase from *Bacillus subtilis* Lucky9 in *E. coli* and application for xylooligosaccharides production from agro-industrial waste. *Int J Biol Macromol.*, *96*: 249-256.

[47] Gowdhaman, D., Jeyalakshmi, G., Sugumaran, K., et al. (2014a). Optimization of the xylanase production with the newly isolated *Bacillus aerophilus* KGJ2. *Turk Biyokim Derg.*, *39* (1): 70–77.

[48] Gupta, V., Garg, S., Capalash, N., et al. (2015). Production of thermo-alkali-stable laccase and xylanase by co-culturing of *Bacillus* sp. and *B. halodurans* for biobleaching of kraft pulp and deinking of waste paper. *Bioproc Biosyst Eng.*, *38* (5): 947–956.

[49] Adhyaru, D., Bhatt, N. and Modi, H. (2015). Optimization of upstream and downstream process parameters for cellulose poor thermosolvent-stable xylanase production and extraction by *Aspergillus tubingensis* FDHN1. *Biores Bioprocess*, *2* (3): 1-14.

[50] Maitan-Alfenas, G. P., Oliveira, M. B., Nagem, R. A., et al. (2016). Characterization and biotechnological application of recombinant xylanases from *Aspergillus nidulans*. *Int J Biol Macromol.*, *91*: 60–67.

[51] Thomas, L., Parameswaran, B., Pandey, A. (2016). Hydrolysis of pretreated rice straw by an enzyme cocktail comprising acidic xylanase from *Aspergillus* sp. for bioethanol production. *Renew Energy*, *98*: 9–15.

[52] Tuyen, D. T., Nguyen, T. S. L., Nguyen, T. T. (2016). Purification and evaluation for effects of temperature on extracellular xylanase activity from *Aspergillus oryzae* DSM 1863. *J Viet Env.*, 8 (1): 9-13.

[53] Pathak, P., Bhardwaj, N., Singh, A. (2014). Production of crude cellulase and xylanase from *Trichoderma harzianum* PPDDN10 NFCCI-2925 and its application in photocopier waste paper recycling. *Appl Biochem Biotechnol.*, *172* (8): 3776-3797.

[54] Silva, L. A., Terrasan, C. R., Carmona, E. C. (2015). Purification and characterization of xylanases from *Trichoderma inhamatum*. *Electron J Biotechnol.*, *18* (4): 307-313.

[55] Gupta, V. K., Gaur, R., Gautam, N. et al. (2009). Optimization of xylanase production from *Fusarium solani* F7. *Amr J Food Technol.*, *4* (1): 20-29.

[56] Khusro, A., Kaliyan, B. K., Al-Dhabi, N. A., et al. (2016). Statistical optimization of thermo-alkali stable xylanase production from *Bacillus tequilensis* strain ARMATI. *Electron J Biotechnol.*, *22*: 16–25.

[57] Nair, S., Sindhu, R. and Shashidhar, S. (2008). Fungal xylanase production under solid state and submerged fermentation conditions. *Afr J Microbiol Res.*, *2*: 082-086.

[58] Kaur, A., Mahajan, R., Singh, A., et al. (2011). A novel and cost effective methodology for qualitative screening of alkalo-thermophilic cellulase free xylano-pectinolytic microorganisms using agricultural wastes. *World J Microbiol Biotechnol.*, *27* (2): 459–463.

[59] Nagar, S., Mittal, A. and Gupta, V. (2012a). A cost effective method for screening and isolation of xylan degrading bacteria using agro waste material. *Asian J Biol Sci.*, *5*(8): 384-394.

[60] Singh, A., Kaur, A., Dua, A., et al. (2015). An efficient and improved methodology for the screening of industrially valuable xylano-pectino-cellulolytic microbes. *Enz Res.*, Article ID 725281. DOI: 10.1155/2015/725281.

[61] Varghese, L. M., Agrawal, S., Sharma, D., et al. (2017). Cost-effective screening and isolation of xylano-cellulolytic positive microbes from termite gut and termitarium. 3 *Biotech.*, *7*:108.

[62] Garg, G., Mahajan, R., Kaur, A., et al. (2011). Xylanase production using agro-residue in solid-state fermentation from *Bacillus pumilus* ASH for biodelignification of wheat straw pulp. *Biodegradation*, *22* (6):1143-1154.

[63] Nagar, S., Mittal, A., Kumar, D., et al. (2011). Hyper production of alkali stable xylanase in lesser duration by *Bacillus pumilus* SV-85S using wheat bran under solid state fermentation. *New Biotechnol.*, *28* (6): 581-587.

[64] Sanghi, A., Garg, N., Kuhar, K., et al. (2009). Enhanced production of cellulase-free xylanase by alkalophilic *Bacillus subtilis* ASH and its application in biobleaching of kraft pulp. *BioRes.*, *4* (3): 1109-1129.

[65] Mohana, S., Shah, A., Divecha, J., et al. (2008). Xylanase production by *Burkholderia* sp. DMAX strain under solid state fermentation using distillery spent wash. *Biores Technol.*, *99* (16): 7553-7564.

[66] Kamble, R. D. and Jadhav, A. R. (2012). Optimization and scale up of cellulase-free xylanase production in solid state fermentation on wheat bran by *Cellulosimicrobium* sp. MTCC 10645. *Jordan J Biol Sci.*, *5*(4): 289-294.

[67] Chapla, D., Divecha, J., Madamwar, D., et al. (2010). Utilization of agro-industrial waste for xylanase production by *Aspergillus foetidus* MTCC 4898 under solid state fermentation and its application in saccharification. *Biochem Eng J.*, *49*(3): 361–369.

[68] Zhang, H. and Sang, Q. (2015). Production and extraction optimization of xylanase and mannanase by *Penicillium chrysogenum* QML-2 and primary application in saccharification of corn cob. *Biochem Eng J.*, *97*: 101-110.

[69] Bagewadi, Z. K., Mulla, S. I., Shouche, Y., et al. (2016). Xylanase production from *Penicillium citrinum* isolate HZN13 using response surface methodology and characterization of immobilized xylanase on glutaraldehyde-activated calcium-alginate beads. *3 Biotech.*, *6*(2): 164.

[70] Bajaj, K., Razdan, K. and Sharma, A. (2010). Thermoactive alkalistable xylanase production from a newly isolated *Streptomyces* sp. SU9. *Ind J Chem Technol.*, *17*: 375-380.

[71] Nagar, S., Mittal, A., Kumar, D., et al. (2012b). Production of alkali tolerant cellulase free xylanase in high levels by *Bacillus pumilus* SV-205. *Int J Biol Macromol., 50*: 414-420.

[72] Fang, T. J., Liao, B. C. and Lee, S. C. (2010). Enhanced production of xylanase by *Aspergillus carneus* M34 in solid-state fermentation with agricultural waste using statistical approach. *New Biotechnol., 27*(1): 25-32.

[73] Battan, B., Sharma, J., Dhiman, S., et al. (2007). Enhanced production of cellulase-free thermostable xylanase by *Bacillus pumilus* ASH and its potential application in paper industry. *Enz Microbial Technol., 41*: 733-739.

[74] Desai, D. and Iyer, B. (2016). Utilization of corn cob waste for cellulase-free xylanase production by *Aspergillus niger* DX-23: Medium optimization and strain improvement. *Waste Biomass Valor*. DOI: 10.1007/s12649-016-9567-4.

[75] Irfan, M., Asghar, U., Nadeem, M., et al. (2015). Optimization of process parameters for xylanase production by *Bacillus* sp. in submerged fermentation. *J Rad Res Appl Sci., 9*(2): 139-147.

[76] Mandal, A., Kar, S., Dutta, T., et al. (2015). Parametric optimization of submerged fermentation conditions for xylanase production by *Bacillus cereus* BSA1 through Taguchi Methodology. *Acta Biologica Szegediensis, 59*(2): 189-195.

[77] Uday, U. S. P., Bandyopadhyay, T. K., Goswami, S., et al. (2016a). Optimization of physical and morphological regime for improved cellulose free xylanase production by fed batch fermentation using *Aspergillus niger* (KP874102.1) and its application in bio-bleaching. *Bioeng., 0*(0): 1–10.

[78] Kareem, R. A. W. A., Khushk, I., Bhutto, M. A., et al. (2014). Xylanase production using fruit waste as cost effective carbon source from thermotolerant *Bacillus megaterium. Afr J Microbiol Res., 8*(38): 3463-3470.

[79] Kumar, S., Sharma, N. and Pathania, S. (2017b). Optimization of process parameters and scale-up of xylanase production from *Bacillus amyloliquifaciens*SH8 in a stirred tank bioreactor. *Cellul Chem Technol., 51*: 403-415.

[80] Sharma, N., Burgohain, P., Kaushal, R., et al. (2012). Use of microwave pretreated *Cedrus deodara* wood residue as a substrate for enhanced production of cellulase free xylanase from *Geotrichum* sp. F3 isolated from rural compost. *J Microbiol Biotechnol Res., 2*(4): 621-631.

[81] Wu, Y. R. and He, J. (2015). Characterization of a xylanase producing *Cellvibrio mixtus* strain J3-8 and its genome analysis. *Sci Rep., 5*: 10521.

[82] Pathania, S., Sharma, N. and Verma, S. (2012). Optimization of cellulase-free xylanase produced by a potential thermoalkalophilic *Paenibacillus* sp. N1 isolated from hot springs of Northern Himalayas in India. *J Microbiol Biotechnol Food Sci., 2*: 1-24.

[83] Panwar, D., Srivastava, P. and Kapoor, M. (2013). Production, extraction and characterization of alkaline xylanase from *Bacillus* sp. PKD-9 with potential for poultry feed. *Biocatal Agric Biotechnol., 3*(2): 118-125.

[84] Sanghi, A., Garg, N., Sharma, J., et al. (2008). Optimization of xylanase production using inexpensive agro-residues by alkalophilic *Bacillus subtilis* ASH in solid state fermentation. *World J Microbiol Biotechnol., 24*: 633-640.

[85] Adhyaru, D. N., Bhatt, N. S. and Modi, H. A. (2013). Enhanced production of cellulase-free, thermo-alkali-solvent-stable xylanase from *Bacillus altitudinis* DHN8, its

characterization and application in sorghum straw saccharification. *Biocatal Agric Biotechnol.*, *3*(2): 182-190.

[86] Kaur, P., Bhardwaj, N. and Sharma, J. (2016b). Process optimization for hyper production of xylanase via statistical methodology from isolated *Bacillus pumilus* 3GAH using lignocellulosic waste. *Biocatal Agric Biotechnol.*, *6*: 159-167.

[87] Kumar, A., Gupta, R., Shrivastava, B., et al. (2012). Xylanase production from an alkalophilic actinomycete isolate *Streptomyces* sp. RCK-2010, its characterization and application in saccharification of second generation biomass. *J Mol Catal B: Enzymatic*, *74*: 170–177.

[88] Sepahy, A., Ghazi, S. and Sepahy, M. (2011). Cost-effective production and optimization of alkaline xylanase by indigenous *Bacillus mojavensis* AG137 fermented on agricultural waste. *Enzyme Res*, Article ID 593624. DOI: 10.4061/2011/593624.

[89] Gupta, G., Sahai, V. and Gupta, R. K. (2013) Optimization of xylanase production from *Melanocarpus albomyces* using wheat straw extract and its scale up in stirred tank bioreactor. *Ind J Chem Technol.*, *20*: 282-289.

[90] Rosmine, E., Changanedassery, N. S., Reshma, S., et al. (2017). Statistical optimization of xylanase production by estuarine *Streptomyces* sp. and its application in clarification of fruit juice. *J Genetic Eng Biotechnol.*, *15*: 393–401.

[91] Sindhu, I., Chhibber, S., Capalash, N., et al. (2006). Production of cellulase-free xylanase from *Bacillus megaterium* by solid state fermentation for biobleaching of pulp. *Curr Microbiol.*, *53*(2): 167-172.

[92] Gowdhaman, D., Manaswini, V. S., Jayanthi, V., et al. (2014b). Xylanase production from *Bacillus aerophilus* KGJ2 and its application in xylooligosaccharides preparation. *Int J Biol Macromol.*, *64*: 90–98.

[93] Ajijolakewu, A., Peng, L., Abdullah, W., et al. (2017). Optimization of production conditions for xylanase production by newly isolated strain *Aspergillus niger* through solid state fermentation of oil palm empty fruit bunches. *Biocatal Agric Biotechnol.*, *11*: 239-247.

[94] Anthony, T., Chandra, R. K., Rajendran, A., et al. (2003). High molecular weight cellulase-free xylanase from alkali-tolerant *Aspergillus fumigatus* AR1. *Enzyme Microbial Technol.*, *32*: 647–654.

[95] Bharti, A. K., Kumar, A., Kumar, A., et al. (2017). Exploitation of *Parthenium hysterophorous* biomass as low-cost substrate for cellulase and xylanase production under solid-state fermentation using *Talaromyces stipitatus* MTCC 12687. *J Rad Res Appl Sci.*, doi: 10.1016/j.jrras.2018.01.003.

[96] Ghoshal, G., Banerjee, U. C., Shivhare, U. S. (2015). Isolation, screening and optimization of xylanase production in submerged fermentation using *P. citrinum*. *J Sci Ind Res.*, *74*: 400-405.

[97] Ramanjaneyulu, G., Sridevi, A., Seshapani, P., et al. (2017). Enhanced production of xylanase by *Fusarium* sp. BVKT R2 and evaluation of its biomass saccharification efficiency. *3 Biotech.*, *7*:351.

[98] Gautam, A., Kumar, A., Bharti, A. K., et al. (2018). Rice straw fermentation by *Schizophyllum commune* ARC-11 to produce high level of xylanase for its application in pre-bleaching. *J Genetic Eng Biotechnol.*, doi: 10.1016/j.jgeb.2018.02.006.

[99] Ghoshal, G., Banerjee, U. C., Chisti, Y., et al. (2012). Optimization of xylanase production from *Penicillium citrinum* in solid-state fermentation. *Chem Biochem Eng Q.*, *26*(1): 61-69.

[100] Kallel, F., Driss, D., Chaari, F., et al. (2015). Statistical optimization of low-cost production of an acidic xylanase by *Bacillus mojavensis* UEB-FK: its potential applications. *Biocatal Agric Biotechnol.*, *5*: 1-10.

[101] Kuancha, C. and Apiraksakorn, J. (2012). Cultural condition improvement for xylanase production by *Bacillus subtilis* GN156. *KKU Res J.*, *17*(6): 933-938.

[102] Sharma, P. K. and Chand, D. (2012). Production of cellulase free thermostable xylanase from *Pseudomonas* sp. XPB-6. *Int Res J Biol Sci.*, *1*(5): 31-41.

[103] Bajaj, B. K. and Singh, N. P. (2010). Production of xylanase from an alkalitolerant *Streptomyces* sp. 7b under solid-state fermentation, its purification, and characterization. *Appl Biochem Biotechnol.*, *162*: 1804–1818.

[104] Sandrim, V., Rizzatti, A., Terenzi, H., et al. (2005). Purification and biochemical characterization of two xylanases produced by *Aspergillus caespitosus* and their potential for kraft pulp bleaching. *Proc Biochem.*, *40*: 1823–1828.

[105] Bakri, Y., Jawhar, M., Arabi, M. I. (2013). Xylanase production by *Fusarium Solani* in solid state fermentation. *Res Biotechnol.*, *4*(1): 31-37.

[106] Pandya, J. J. and Gupte, A. (2012). Production of xylanase under solid-state fermentation by *Aspergillus tubingensis* JP-1 and its application. *Bioproc Biosyst Eng.*, *35*:769–779.

[107] Thembekile, N., Rachmond, L. and Emil, K. (2012). *Jatropha curcas* seed cake as substrate for production of xylanase and cellulase by *Aspergillus niger* FGSCA733 in solid-state fermentation. *Ind Crops Prod.*, *37*: 118–123.

[108] Abdullah, R., Nisar, K., Aslam, A., et al. (2014). Enhanced production of xylanase from locally isolated fungal strain using agro-industrial residues under solid-state fermentation. *Nat Prod Res.*, *29*: 1006-1011.

[109] Adhyaru, D. N., Bhatt, N. S., Modi, H. A., et al. (2016). Insight on xylanase from *Aspergillus tubingensis* FDHN1: Production, high yielding recovery optimization through statistical approach and application. *Biocatal Agric Biotechnol.*, *6*: 51-57.

[110] Khanahmadi, M., Arezi, I., Amiri, M., et al. (2018). Bioprocessing of agro-industrial residues for optimization of xylanase production by solid- state fermentation in flask and tray bioreactor. *Biocatal Agric Biotechnol.*, *13*: 272-282.

[111] Walia, A., Mehta, P., Guleria, S., et al. (2015). Improvement for enhanced xylanase production by *Cellulosimicrobium cellulans* CKMX1 using central composite design of response surface methodology. 3 *Biotech.*, *5*(6): 1053–1066.

[112] Behnam, S., Karimi, K., Khanahmadi, M., et al. (2016). Optimization of xylanase production by *Mucor indicus*, *Mucor hiemalis,* and *Rhizopus oryzae* through solid state fermentation. *Biol J Microorganism*, *4*(16): 1-10.

[113] Sa-Pereira, P., Mesquita, A., Duarte, J. C., et al. (2002). Rapid production of thermostable cellulase-free xylanase by a strain of *Bacillus subtilis* and its properties. *Enz Microbial Technol*; *30*: 924-933.

[114] Ko, C. H., Lin, Z. P., Tu, J., et al. (2010). Xylanase production by *Paenibacillus campinasensis* BL11 and its pretreatment for hardwood kraft pulp bleaching. *Int Biodeterior Biodegradation*, *64*: 13-19.

[115] Paredes, R., Barros, R., Inoue, H., et al. (2015). Production of xylanase, α-L-arabinofuranosidase, β-xylosidase, and β-glucosidase by *Aspergillus awamori* using the

liquid stream from hot-compressed water treatment of sugarcane bagasse. *Biomass Convers Biorefin.*, *5*(3): 299-307.

[116] Prakash, P., Jayalakshmi, S., Prakash, B., et al. (2012). Production of alkaliphilic, halotolerent, thermostable cellulase free xylanase by *Bacillus halodurans* PPKS-2 using agro waste: single step purification and characterization. *World J Microbiol Biotechnol.*, *28*: 183-192.

[117] Bajaj, B. and Manhas, K. (2012). Production and characterization of xylanase from *Bacillus licheniformis* P11(C) with potential for fruit juice and bakery industry. *Biocatal Agric Biotechnol.*, *1* :330–337.

[118] Bajaj, B., Sharma, M., Sharma, S. (2011). Alkalistable endo-ß-1,4-xylanase production from a newly isolated alkalitolerant *Penicillium* sp. SS1 using agro-residues. *3 Biotech.*, *1*: 83–90.

[119] El-Shishtawy, R., Mohamed, S., Asiri, A., et al. (2015). Saccharification and hydrolytic enzyme production of alkali pre-treated wheat bran by *Trichoderma virens* under solid state fermentation. *BMC Biotechnol.*, *15*: 37.

[120] Shah, A. and Madamwar, D. (2005). Xylanase production by a newly isolated *Aspergillus foetidus* strain and its characterization. *Proc Biochem.*, *40*: 1763-1771.

[121] Agnihotri, S., Dutt, D., Tyagi, C., et al. (2010). Production, biochemical characterization of a novel cellulase-poor alkali-thermo-tolerant xylanase from *Coprinellus disseminatus* SW-1 NTCC1165. *World J Microbiol Biotechnol.*, *26*: 1349–1359.

[122] Antoine, A., Jacqueline, D. and Thonart, P. (2010). Xylanase production by *Penicillium canescens* on Soya Oil Cake in solid-state fermentation. *Appl Biochem Biotechnol.*, *160*: 50–62.

[123] Yang, S. Q., Yan, Q. J., Jiang, Z. Q., et al. (2006). High-level of xylanase production by the thermophilic *Paecilomyces themophilis* J18 on wheat straw in solid state fermentation. *Biores Technol.*, *97*: 1794-1800.

[124] Annamalai, N., Thavasi, R., Jayalakshmi, S., et al. (2009). Thermostable and alkaline tolerant xylanase production by *Bacillus subtilis* isolated from marine environment. *Ind J Biotechno.l*, *8*: 291–297.

[125] Kiddinamoorthy, J., Anceno, A., Haki, G., et al. (2008). Production, purification and characterization of *Bacillus* sp. GRE7 xylanase and its application in eucalyptus Kraft pulp biobleaching. *World J Microbiol Biotechnol.*, *24*: 605-612.

[126] Subramaniyan, S. (2012). Isolation, purification and characterisation of low molecular weight xylanase from *Bacillus pumilus* SSP-34. *Appl Biochem Biotechnol.*, *166*: 1831–1842.

[127] Garai, D. and Kumar, V. (2013). A Box–Behnken design approach for the production of xylanase by *Aspergillus candidus* under solid state fermentation and its application in saccharification of agro residues and *Parthenium hysterophorus* L. *Ind Crops Prod.*, *44*: 352–363.

[128] Kar, S., Gauri, S., Das, A., et al. (2013) Process optimization of xylanase production using cheap solid substrate by *Trichoderma reesei* SAF3 and study on the alteration of behavioral properties of enzyme obtained from SSF and SmF. *Bioproc Biosyst Eng.*, *36*: 57–68.

[129] Kumar, V. and Satyanarayana, T. (2011). Applicability of thermo-alkali-stable and cellulase-free xylanase from a novel thermo-halo-alkaliphilic *Bacillus halodurans* in producing xylooligosaccharides. *Biotechnol Lett.*, *33*: 2279–2285.

[130] Raj, A., Kumar, S. and Singh, S. K. (2013). A highly thermostable xylanase from *Stenotrophomonas maltophilia*: purification and partial characterization. *Enzyme Res.*, Article ID 429305. DOI: 10.1155/2013/429305.

[131] Virupakshi, S., Babu, G., Gaikwad, S., et al. (2005). Production of a xylanolytic enzyme by a thermoalkaliphilic *Bacillus* sp. JB-99 in solid state fermentation. *Proc Biochem.*, *40*: 431-435.

[132] Ali, U., Ibrahim, Z. ands Isaac, G. (2013). Ethanol and xylitol production from xylanase broth of *Thermomyces lanuginosus* grown on some lignocellulosic wastes using *Candida tropicalis* EMCC2. *Life Sci J.*, *10*: 968-978.

[133] Elegbede, J. A. and Lateef, A. (2017). Valorization of corn-cob by fungal isolates for production of xylanase in submerged and solid state fermentation media and potential biotechnological applications. *Waste Biomass Valo.r*, *9* (8): 1273-1287.

[134] Zhang, Z., Li, J., Feng, F., et al. (2013). Optimization of nutrition constituents for xylanase activity by *Rhizopus stolonifer* under solid state fermentation on corncob. *BioRes.*, *8* (2): 2018-2032.

[135] Bakri, Y., Jawhar, M. and Arabi, M. I. (2008). Improvement of xylanase production by *Cochliobolus sativus* in solid state fermentation. *Braz J Microbiol*, *39*: 602-604.

[136] Jin, N., Ma, S., Liu, Y., et al. (2012). Thermophilic xylanase production by *Aspergillus niger* in solid state fermentation using wheat straw and corn cob. *Afr J Microbiol Res.*, *6* (2): 2387-2394.

[137] Riswan, A., Muthuvelayudham, R., Viruthagiri, T. (2012). Statistical optimization of nutrients for production of cellulase and hemicellulase from rice straw. *Asian J Biochem Pharm Res.*, *2*: 154-174.

[138] Seyis, I. and Aksoz, N. (2005). Xylanase production from *Trichoderma harzianum* 1073 D3 with alternative carbon and nitrogen sources. *Food Technol Biotechnol.*, *43*: 37-40.

[139] Goyal, M., Kalra, K.L., Sareen, V., et al. (2008). Xylanase production with xylan rich lignocelluosic wastes by a local soil isolate of *Trichoderma viride*. *Braz J Microbiol.*, *39*: 535–541.

[140] Kapilan, R. and Arasaratnam, V. (2011). Paddy husk as support for solid state fermentation to produce xylanase from *Bacillus pumilus*. *Rice Sci.*, *18*: 36-45.

[141] Sugumaran, K., Kumar, B., Mahalakshmi, M., et al. (2013). Cassava bagasse - Low cost substrate for thermotolerant xylanase production using *Bacillus subtilis*. *Int J ChemTech Res.*, *5* (1): 394-400.

[142] Villas-Boas, S., Esposito, E., de Mendonca, M. (2002). Novel lignocellulolytic ability of *Candida utilis* during solid-substrate cultivation on apple pomace. *World J Microbiol Biotechnol.*, *18*: 541-545.

[143] Walia, A., Mehta, P., Chauhan, A., et al. (2013). Optimization of cellulase-free xylanase production by alkalophilic *Cellulosimicrobium* sp. CKMX1 in solid-state fermentation of apple pomace using central composite design and response surface methodology. *Ann Microbiol.*, *63*: 187-198.

[144] Reddy, G., Babu, P., Komaraih, P., et al. (2003). Utilization of banana waste for the production of lignolytic and cellulolytic enzymes by solid substrate fermentation using two Pleurotus species (*P. ostreatus* and *P. sajorcaju*). *Proc Biochem.*, *38*: 1457-1462.

[145] Irfan, M., Nadeem, M., Syed, Q., et al. (2012). Effect of medium composition on xylanase production by *Bacillus subtilis* using various agricultural wastes. *Am Eur J Agric Env Sci.*, *12*: 561-565.

[146] Alberton, L., Vandenberghe, L., Assmann, R., et al. (2009). Xylanase production by *Streptomyces viridosporus* T7A in submerged and solid state fermentation using agro-industrial residues. *Braz Arch Biol Technol.*, *52*: 171–180.

[147] Zimbardi, A., Sehn, C., Meleiro, L., et al. (2013). Optimization of β-glucosidase, β-xylosidase and xylanase production by *Colletotrichum graminicola* under solid state fermentation and application in raw sugarcane trash saccharification. *Int J Mol Sci.*, *14*: 2875-2902.

[148] Kumar, L., Kumar, D., Nagar, S., et al. (2014a). Modulation of xylanase production from alkaliphilic *Bacillus pumilus* VLK-1 through process optimization and temperature shift operation. 3 *Biotech.*, *4*: 345–356.

[149] Giridhar, P. V. and Chandra, T. S. (2010). Production of novel halo-alkali-thermo-stable xylanase by a newly isolated moderately halophilic and alkali-tolerant *Gracilibacillus* sp. TSCPVG. *Proc Biochem.*, *45*: 1730–1737.

[150] Soliman, H., Sherief, A., EL-Tanash, A. (2012). Production of xylanase by *Aspergillus niger* and *Trichoderma viride* using some agriculture residues. *Int J Agric Res.*, *7* (1): 46-57.

[151] Gowdhaman, D., Manaswini, V., Jayanthi, V., et al. (2013). Xylanase production from *Bacillus aerophilus* KGJ2 and its application in xylooligosaccharides preparation. *Int J Biol Macromol.*, *64*: 90-98.

[152] Goluguri, B., Chiranjeevi, T., Uma, A., et al. (2016). Novel alkali-thermostable xylanase from *Thielaviopsis basicola* (MTCC1467): Purification and kinetic characterization. *Int J Biol Macromo.l*, *82*: 823–829.

[153] Guan, G., Zhao, P., Zhao, J., et al. (2016). Production and partial characterization of an alkaline xylanase from a novel fungus *Cladosporium oxysporum*. *BioMed Res Int.*, Article ID 4575024. DOI: 10.1155/2016/4575024.

[154] Prakash, S., Veeranagouda, Y., Kyoung, L., et al. (2009). Xylanase production sing inexpensive agricultural wastes and its partial characterization from a halophilic *Chromohalobacter* sp. TPSV 101. *World J Microbiol Biotechnol.*, *25*: 197–204.

[155] Dai, X. J., Liu, M. Q., Jin, H. X., et al. (2011). Optimisation of solid-state fermentation of *Aspergillus niger* JL-15 for xylanase production and xylooligosaccharides preparation. *Czech J Food Sci.*, *29*: 557–567.

[156] Almeida, M., Guimarães, V., Falkoski, D., et al. (2014). Optimization of endoglucanase and xylanase activities from *Fusarium verticillioides* for simultaneous saccharification and fermentation of sugarcane bagasse. *Appl Biochem Biotechnol.*, *172*: 1332–1346.

[157] Senthilkumar, S., Ashokkumar, B., Chandra, R., et al. (2005). Optimization of medium composition for alkali-stable xylanase production by *Aspergillus fischeri* Fxn 1 in solid-state fermentation using central composite rotary design. *Biores Technol.*, *96*: 1380-1386.

[158] Ahmad, Z., Butt, M., Riaz, M. (2013). Partial purification and characterization of xylanase produced from *Aspergillus niger* using wheat bran. *Pak J Agric Sci.*, *50*: 433-437.

[159] Amel, B., Nawel, B., Khelifa, B., et al. (2016). Characterization of a purified thermostable xylanase from *Caldicoprobacter algeriensis* sp. nov. strain TH7C1T. *Carbohydr Res.*, *419*: 60–68.

[160] Muthusamy, S., Silambarasan, T., Palavesam, A., et al. (2016). Biosynthesis, purification and characterization of β-1,4-xylanase from a novel mangrove associated

actinobacterium *Streptomyces olivaceus* (MSU3) and its applications. *Protein Expr Purif., 130*: 1-12.

[161] Lee, W., Yusof, S., Hamid, N., et al. (2006). Optimizing conditions for enzymatic clarification of banana juice using response surface methodology (RSM). *J Food Eng.*, 73: 55–63.

[162] Bhat, M. K. (2000). Cellulases and related enzymes in biotechnology. *Biotechnol Adv.*, *18*: 355-383.

[163] Mishra, A., Mishra, V., Akhter, P., et al. (2017). Industrial application of glutaraldehyde activated immobilized xylanase for clarification of tomato juice. *Int J Adv Res Biol Sci.*, *4* (9): 71-82.

[164] Pal, A. and Khanum, F. (2011). Efficacy of xylanase purified from *Aspergillus niger* DFR-5 and in combination with pectinase and cellulase to improve yield and clarity of pineapple juice. *J Food Sci Technol.*, *48*: 560–568.

[165] Shah, S. (2007). Optimization of an enzyme assisted process for juice extraction and clarification from litchis (*Litchi chinensis* Sonn.). *Int J Food Eng*, *3*(3):1–17.

[166] Poorna, C. A. and Prema, P. (2007). Production of cellulase-free endoxylanase from novel alkalophilic thermotolerent *Bacillus pumilus* by solid-state fermentation and its application in wastepaper recycling. *Biores Technol.*, *98*: 485–490.

[167] Betini, J., Michelin, M., Peixoto-Nogueira, S., et al. (2009). Xylanase from *Aspergillus niger*, *Aspergillus niveus* and *Aspergillus ochraceus* produced under solid-state fermentation and their application in cellulose pulp bleaching. *Bioproc Biosyst Eng.*, *32*: 819-824.

[168] Jain, K. K., Dey, T. B., Kumar, S., et al. (2014). Production of thermostable hydrolases (cellulases and xylanase) from *Thermoascus aurantiacus* RCKK: a potential fungus. *Bioproc Biosyst Eng.*, *38*: 787-796.

[169] Joshi, C. and Khare, S. K. (2011). Utilization of deoiled *Jatropha curcas* seed cake for production of xylanase from thermophilic *Scytalidium thermophilum*. *Biores Technol.*, *102*: 1722–1726.

[170] Mullai, P., Fathima, N. and Rene, E. R. (2010). Statistical analysis of main and interaction effects to optimize xylanase production under submerged cultivation conditions. *J Agric Sci.*, *2* (1): 144-153.

[171] Dheeran, P., Nandhagopal, N., Kumar, S., et al. (2012). A novel thermostable xylanase of *Paenibacillus macerans* IIPSP3 isolated from the termite gut. *J Ind Microbiol Biotechnol.*, *39*: 851–860.

[172] Mittal, A., Nagar, S., Gupta, V. (2012). Production and purification of high levels of cellulase-free bacterial xylanase by *Bacillus* sp. SV-34S using agro-residue. *Ann Microbiol.*, *63*: 1157–1167.

[173] Bhalla, A., Bischoff, K. and Sani, R. (2015). Highly thermostable xylanase production from a thermophilic *Geobacillus* sp. strain WSUCF1 utilizing lignocellulosic biomass. *Front Bioeng Biotechnol.*, *3*: 84-91.

[174] Robl, D., Priscila, S., Patrícia, S., et al. (2015). Xylanase production by endophytic *Aspergillus niger* using pentose-rich hydrothermal liquor from sugarcane bagasse. *Biocatal Biotransformation*, *33*: 175-187.

[175] Thomas, L., Sindhu, R., Parameswaran, B., et al. (2015). Production of a cellulose-free alkaline xylanase from *Bacillus pumilus* MTCC 5015 by submerged fermentation and its application in biobleaching. *Ind J Exp Biol.*, *53*: 356-363.

[176] Gomes, I., Saha, R., Mohiuddin, G., et al. (1992). Isolation and characterization of a cellulase-free pectinolytic and hemicellulolytic thermophilic fungus. *World J Microbiol Biotechnol.*, *8*: 589-592.

[177] Beg, Q. K., Bhushan, B., Kapoor, M., et al. (2000). Production and characterization of thermostable xylanase and pectinase from *Streptomyces* sp. QG-11-3. *J Ind Microbiol Biotechnol*, *24*: 396-402.

[178] Couri, S., Terzi, S. C., Pinto, G. A. S., et al. (2000) Hydrolytic enzyme production in solid-state fermentation by *Aspergillus niger* 3T5B8. *Proc Biochem.*, *36*: 255-261.

[179] Heck, J., Soares, L. and Ayub, M. (2005). Optimization of xylanase and mannanase production by *Bacillus circulans* strain BL53 on solid-state cultivation. *Enz Microbial Technol.*, *37*: 417–423.

[180] Mendoza, N., Caliwara, M. and Unciano, N. (2006) A local isolate of non-cellulolytic, xylanolytic and pectinolytic thermophilic fungus *Thermomyces lanuginosus* C1a. *Philipp J Sci.*, *135*: 113-119.

[181] Botella, C., Diaz, A., Ory, I., et al. (2007) Xylanase and pectinase production by *Aspergillus awamori* on grape pomace in solid state fermentation. *Proc Biochem.*, *42*: 98–101.

[182] Rodriguez-Fernandez, D., Rodriguez-Leon, J., Carvalho, J., et al. (2011). The behaviour of kinetic parameters in production of pectinase and xylanase by solid state fermentation. *Biores Technol.*, *102*: 10657-10662.

[183] Díaz, A., Ory, I., Caro, I., et al. (2012). Enhance hydrolytic enzymes production by *Aspergillus awamori* on supplemented grape pomace. *Food Bioprod Proc.*, *90*: 72–78.

[184] Das, A., Paul, T., Halder, S., et al. (2013). Production of cellulolytic enzymes by *Aspergillus fumigatus* ABK9 in wheat bran-rice straw mixed substrate and use of cocktail enzymes for deinking of waste office paper pulp. *Biores Technol.*, *128*: 290–296.

[185] Belova, O., Lisov, A., Vinokurova, N., et al. (2014). Xylanase and cellulase of fungus *Cerrena unicolor* VKM F 3196: Production, properties and applications for the saccharification of plant material. *Appl Biochem Microbiol.*, *50*: 148–153.

[186] Hernandez-Dominguez, E., Rios-Latorre, R., Alvarez-Cervantes, J., et al. (2014). Xylanases, cellulases, and acid protease produced by *Stenocarpella maydis* grown in solid state and submerged fermentation. *BioRes.*, *9* (2): 2341-2358.

[187] Li, P., Xia, J., Shan, Y., et al. (2015). Effects of surfactants and microwave-assisted pretreatment of orange peel on extracellular enzymes production by *Aspergillus japonicus* PJ01. *Appl Biochem Biotechnol.*, *176*: 758-771.

[188] Chutani, P. and Sharma, K. K. (2016). Concomitant production of xylanases and cellulases from *Trichoderma longibrachiatum* MDU-6 selected for the deinking of paper waste. *Bioproc Biosyst Eng.*, *39*: 747-758.

[189] Periyasamy, K., Santhalembi, L., Mortha, G., et al. (2016). Production, partial purification and characterization of enzyme cocktail from *Trichoderma citrinoviride* AUKAR04 through solid-state fermentation. *Arab J Sci Eng.*, *42*: 53–63.

[190] Silva, da F., de Oliveira, C., dos Santos, D., et al. (2017). Pretreatments of Carnauba (*Copernicia prunifera*) straw residue for production of cellulolytic enzymes by *Trichorderma reesei* CCT-2768 by solid state fermentation. *Renew Energ.*, *116*: 299-308.

[191] Silva, D., Hergesel, L., Campioni, T., et al. (2018). Evaluation of different biological and chemical treatments in agroindustrial residues for the production of fungal glucanases and xylanases. *Proc Biochem.*, *67*: 29-37.

[192] Khandeparkar, R. and Bhosle, N. (2006). Isolation, purification and characterization of the xylanase produced by *Arthrobacter* sp. MTCC 5214 when grown in solid state fermentation. *Enz Microbial Technol.*, *39*: 732-742.

[193] Bhushan, B., Pal, A., Kumar, S., et al. (2013). Biochemical characterization and kinetic comparison of encapsulated haze removing acidophilic xylanase with partially purified free xylanase isolated from *Aspergillus flavus* MTCC 9390. *J Food Sci Technol.*, *52*: 191-200.

[194] Kallel, F., Driss, D., Bouaziz, F., et al. (2014). Production of xylooligosaccharides from garlic straw xylan by purified xylanase from *Bacillus mojavensis* UEB-FK and their in vitro evaluation as prebiotics. *Food Bioprod Process*, *94*: 536-546.

[195] Poosarla, V. G. and Chandra, T. S. (2014). Purification and characterization of novel halo-acid-alkali-thermo-stable xylanase from *Gracilibacillus* sp. TSCPVG. *Appl Biochem Biotechnol.*, *173*: 1375–1390.

[196] Walia, A., Mehta, P., Chauhan, A., et al. (2014). Purification and characterization of cellulase-free low molecular weight endo ß-1,4 xylanase from an alkalophilic *Cellulosimicrobium cellulans* CKMX1 isolated from mushroom compost. *World J Microbiol Biotechnol.*, *30*: 2597–2608.

[197] Zheng, H., Sun, M., Meng, L., et al. (2014). Purification and characterization of a thermostable xylanase from *Paenibacillus* sp. NF1 and its application in xylooligosaccharides production. *J Microbiol Biotechnol.*, *24*: 489–496.

[198] Chang, L., Arai, T., Ibrahim, D., et al. (2015). Purification and characterization of xylanase from newly isolated *Penicillium rolfsii* c3-2(1) IBRL. *Biores.*, *10*: 1627-1643.

[199] Kocabas, D., Guder, S., Ozben, N. (2015). Purification strategies and properties of a low-molecular weight xylanase and its application in agricultural waste biomass hydrolysis. *J Mol Catal B: Enzymatic*, *115*: 66-75.

[200] Boonchuay, P., Takenaka, S., Kuntiya, A., et al. (2016). Purification, characterization, and molecular cloning of the xylanase from *Streptomyces thermovulgaris* TISTR1948 and its application to xylooligosaccharide production. *J Mol Catal B: Enzymatic*, *129*: 61–68.

[201] Luo, L., Cai, J., Wang, C., et al. (2016). Purification and characterization of an alkaliphilic endo-xylanase from *Streptomyces althioticus* LMZM and utilization in the pulp paper industry. *J Chem Technol Biotechnol.*, *91*: 1093–1098.

[202] Panthi, S., Choi, Y., Choi, Y., et al. (2016). Biochemical and thermodynamic characterization of a novel, low molecular weight xylanase from *Bacillus methylotrophicus* CSB40 isolated from traditional Korean food. *Appl Biochem Biotechnol.*, *179*: 126-42.

[203] Winterhalter, C. and Liebl, W. (1995). Two extremely thermostable xylanases of the hyperthermophilic bacterium *Thermotoga maritima* MSB8. *Appl Env Microbiol.*, *61*: 1810–1815.

[204] Kang, M., Maeng, P. and Rhee, Y. (1996). Purification and characterization of two xylanases from alkalophilic *Cephalosporium* sp. strain RYM-202. *Appl Env Microbiol.*, *62*: 3480-3482.

[205] Prabhu, A. and Maheshwari, R. (1999). Biochemical properties of xylanases from a thermophilic fungus, *Melanocarpus albomyces*, and their action on plant cell walls. *J Biosci.*, *24*: 461-470.

[206] Chadha, B., Ajay, B., Mellon, F., et al. (2004). Two endoxylanases active and stable at alkaline pH from the newly isolated thermophilic fungus, *Myceliophthora* sp. IMI 387099. *J Biotechnol.*, *109*: 227-237.

[207] Knob, A. and Carmona, E. (2010). Purification and characterization of two extracellular xylanases from *Penicillium sclerotiorum*: A novel acidophilic xylanase. *Appl Biochem Biotechnol.*, *162*: 429–443.

[208] Sharma, M., Chadha, B. and Saini, H. S. (2010). Purification and characterization of two thermostable xylanases from *Malbranchea flava* active under alkaline conditions. *Biores Technol.*, *101*: 8834–8842.

[209] Poorna, C. A. (2011). Purification and biochemical characterization of xylanases from *Bacillus pumilus* and their potential for hydrolysis of polysaccharides. *Ferment Technol.*, *1*:101.

[210] Soroor, M., Abd-El-Hady, M., El-Sayed, M., et al. (2013). Purification and characterization of cellulase-poor xylanases from *Trichoderma reesei* F418 grown on rice straw by solid-state fermentation. *J Appl Sci Res.*, *9*: 1702-1713.

[211] Kaur, A., Singh, A. and Mahajan, R. (2014). Characterization of industrially-valuable xylano-pectinolytic enzymes produced concurrently by a novel isolate of *Bacillus pumilus*. *Biotechnol Lett.*, *36*: 2229–2237.

[212] Terrasan, C., Guisan, J. and Carmona, E. (2016). Xylanase and β-xylosidase from *Penicillium janczewskii*: Purification, characterization and hydrolysis of substrates. *Electron J Biotechnol.*, *23*: 54-62.

In: Microbial Catalysts. Volume 1
Editors: S. M. Abdel-Aziz, N. Garg, A. Aeron et al.
ISBN: 978-1-53614-679-0

Chapter 6

MICROBIAL ALKALINE PROTEASES: PRODUCTION AND APPLICATION IN THE LEATHER INDUSTRY

Hamid Mukhtar*
Institute of Industrial Biotechnology,
Government College University, Lahore, Pakistan

ABSTRACT

Proteases are hydrolytic enzymes that cleave peptide bonds in proteins. Produced by a number of microbial sources these enzymes may be categorized based upon their pH optima. Alkaline proteases are used widely in various industries, however, the most important application of these enzymes is in the leather industry whereby the enzymes are utilized in bating and dehairing processes. The use of these enzymes can significantly lower the need of toxic chemicals that are otherwise, utilized in the process. Therefore, proteases tend to increase the efficiency of the process without the production of any harmful chemicals.

Keywords: hydrolases, bating, dehairing, enzyme production, fermentation

Proteases (EC 3.4.21-24 and 99; peptidyl-peptide hydrolases) are the hydrolytic enzymes, which act upon native proteins and catalyze the breakdown of peptide bonds [1]. They also catalyze the peptide synthesis in organic solvents and in solvents with low water content [2]. They refer to a mixture of enzymes including proteinases, peptidases and amidases. The proteinases hydrolyze intact protein molecules to proteoses, peptides and some amino acids. The oligopeptide products of proteinase action are further hydrolyzed by peptidases to amino acids. The amino acids are in turn acted upon by amidases, which release ammonia from amino compounds however, it is not necessary that all the proteases contain these three enzymes. The hydrolysis of peptide bonds by proteases as shown in Figure 1 is termed as proteolysis.

* Author's Email: hamidmukhtar@gcu.edu.pk.

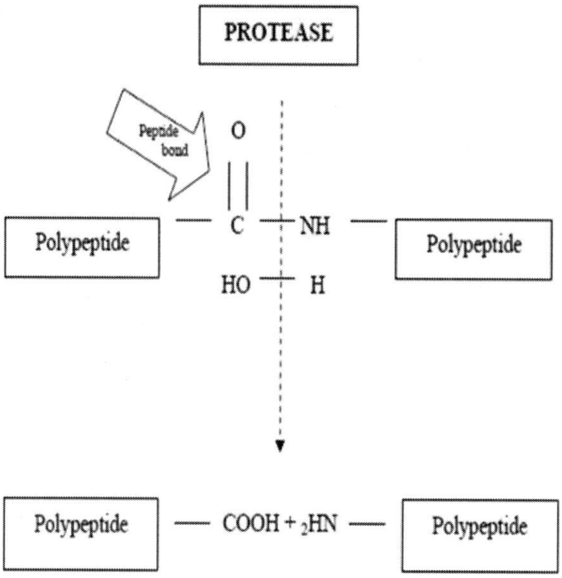

Figure 1. Protease catalysis of peptide bonds (Proteolysis).

Proteases possess some characteristics of biotechnological interest due to which these have become the most important industrial enzymes [3]. Almost all proteases are heat resistant, maintain their activity at temperature 60-70ºC and possess some milk clotting activity. The proteases vary widely in their specific activities, optimum pH, pH stability range (2-13), heat sensitivity, substrate specificity, active site, catalytic mechanism and stability profiles [4].

On the basis of their acid-base behavior, proteases have been classified in to three categories i.e., acid, neutral and alkaline proteases. The acid proteases are those which have pH optima in the range of 2.0-5.0 and are mainly fungal in origin. Important examples of acid proteases are Pinguinain, Tabernamontanain, proteolytic enzymes of yeasts, *Aspergillus, Rhizopus* and *Penicillium* species [5, 6]. Proteases having pH optima in the range of 7.0 or around are called neutral proteases. They are mainly of plant origin however, some bacteria and fungi also produce neutral proteases e.g., euphorbain, pomiferain, papain, ficin, bromelain and soyin etc. [7]. While proteases that have pH optima in the range of 8.0-11.0 are grouped under the category of alkaline proteases. Some of the important alkaline proteases are solanain, hurain and proteolytic enzymes of *Bacillus* and *Streptomyces* species [8, 9].

Bergmann and Frankel-Conrat (1937) [10] classified the proteases into two categories on the basis of their mode of action i.e., endopeptidases and exopepetidases. Endopeptidases are those which can attack and break both centrally located as well as terminal peptide bonds. Common examples of endopeptidases are trypsin, chymotrypsin and pepsin. They can be inhibited by organophosphorus compounds such as DFP or PMSF [11]. Exopeptidases are proteases, which attack the peptide linkage adjacent to terminal (-NH) or carboxylic (-COOH) groups e.g., amino peptidases, di- and tripeptidases. Their action is dependent upon bound metal ions such as Mg^{+2}, Fe^{+2} and Co^{+2} which act as cofactors. Proteases, like all other enzymes, have also been classified as intracellular (endo-proteases) and extracellular (exo-proteases). Intracellular proteases are more difficult to isolate as compared to extracellular proteases. This is due to their great heat lability and low concentration. They are usually present in higher animals and plants. Extracellular proteases are widely distributed in bacteria and fungi, which secrete enzymes on their substrates in order to break down their complex organic compounds.

Most of the alkaline proteases found in nature belong to the class of serine proteases. These enzymes are characterized by the presence of a nucleophilic serine residue in the active site along with histidine and aspartate. Serine alkaline proteases are active in the pH range of 7 to 11, though the optimum pH varies according to the source of the enzyme and its characteristic features. The isoelectric points of these enzymes often lies near their pH optimum i.e., pH 8 to 11. They exhibit temperature optima between 50 and 70°C [12]. Serine alkaline proteases demonstrate broad substrate specificity and generally have a low molecular mass ranging from 18 to 35kDa [13]. They are mostly stable in the presence of calcium ions and are mostly inhibited by serine active-site inhibitors, the examples of which are phenylmethylsulphonyl fluoride (PMSF) and diisopropyl-fluorophosphate (DFP).

There are various sub-classes of serine alkaline proteases. Chymotrypsin-like proteases are those that show a pH optima of 8 and have a molecular weight of nearly 20 kDa. These are usually found in animals and are produced by some species of *Streptomyces* such as *S. erythreus, S. fradiae and S. griseus* [14]. These are specific for basic amino acids and are mostly inhibited in the presence of diisopropyl-fluorophosphate, N-tosyl-L-lysine chloromethyl ketone and soyabean trypsin inhibitor [4].

Subtilisin-like proteases are those that show substrate specificity for hydrophobic and aromatic residues and have a pH optima of 10. The molecular weight of this sub-class of serine alkaline proteases is in the range of 15 to 30kDa. Produced mostly by bacteria including *Flavobacterium, Bacillus amyloliquifaciens* and *Bacillus subtilis*, these show inhibition in the presence of diisopropyl-fluorophosphate, phenyl methyl suphonyl fluoride and potato inhibitor [13]. Myxobacter α-lytic proteases show maximum activity at pH 9 and have specificity for the carboxyl group of aliphatic and neutral amino acids. Diisopropyl-fluorophosphate is a well-known inhibitor of the enzyme. Various species of *Sorangium* are reported as producers of this type of serine alkaline protease. Prolyloligopetidase-like serine proteases have not been studied much but homology studies show that these belong to the class of serine alkaline proteases [15].

Since proteases are physiologically necessary for living organisms, they occur ubiquitously in a wide diversity of sources such as higher plants, animal tissues and microorganism. In the past, the enzyme preparations were made from classical sources such as animal and higher plant tissues but now, the increasing demands for proteases have stimulated the enzyme production from microbial world [16]. Microbial enzymes are often more versatile than enzymes derived from plants or animals, because of the high yields, availability of great variety of catalytic activities, ease of genetic manipulation, regular supply due to absence of seasonal fluctuations and rapid growth of the microorganisms on inexpensive media. Microbial enzymes are also more stable than their corresponding plant and animal enzymes with more convenient and safer production [4, 17].

Microorganisms represent an excellent source of proteolytic enzymes owing to their broad biochemical diversity and their susceptibility to genetic manipulation. Microbial **proteases** account for approximately 40% of the total worldwide enzyme sales [18]. **Proteases** from microbial sources are preferred to the enzymes from plant and animal sources, since they possess almost all the characteristics desired for their biotechnological applications. The proteases can be obtained economically and profitably from microorganisms by the process of fermentation. Many of the organisms excrete more than one kind of proteases and the type of proteolytic enzyme formed may depend upon the composition of the medium [19].

Among fungi, *Conidiobolus* sp., *Rhizopus* sp. *Aspergilli* and *Agaricus bisporus* are well known producers of serine alkaline proteases [20]. *Candida* sp. has been the most studied amongst yeast as a serine alkaline protease producer [21].

Alkaline proteases are produced by a large number of bacterial species [22]. The bacterial proteases usually diffuse into the medium (extracellular) and can be concentrated by filtration or by precipitation. Bacteria form proteolytic enzymes, when grown in protein medium however, very few bacteria produce proteases on protein- free media [19]. *Bacillus* species do possess remarkable biotechnological value due to their non-pathogenicity and ability to produce extracellular proteases in large amounts [23]. Different species of *Bacillus* producing high titers of proteases include *B. subtilis* [24, Hameed et al., 1996; 25], *B. licheniformis* [26, 27], *B. pseudofirmus* [28], *B. cereus, B. pumilus* [29], *B. stearothermophilus* [30, 31], *B. intermedius* [32], *B. amyloliquefaciens* [33], *B. thermoproteolyticus* [34] and *Bacillus mycoids* [35]. Proteases are also formed by *B. prodigiosus, B. proteus, B. pychyaneus, B. diphtherae* and *B. polymyxa* [36]. Protease production has also been reported by alkalophilic *Streptomyces* spp. [37], *Pseudomonas* spp. [38, 39], *Photorhabdus* spp. [40], *Lactobacillus* spp. [41], *Alcaligenes faecalis* [42] and *Xanthmonas maltophila* [43].

Bacterial alkaline proteases are characterized by their high activity at alkaline pH (9-11), their broad substrate specificity and optimal temperature around 60°C. The most important group of bacterial alkaline proteases is serine alkaline proteases. They are generally produced extracellularly for the purpose of scavenging nutrients [44]. They are most active in the range of pH 9-10, with a molecular weight range of 15-30 KDa and an isoelectric point near pH 9. They are highly sensitive towards phenyl methyl sulphonyl fluoride (PMSF), diisopropyl-flourophosphate (DFP) and potato inhibitor [12]. These properties of bacterial alkaline proteases make them suitable for use in a number of industries. This class of proteases is well represented by various species of *Bacillus* such as *B. amyloliquifaciens, B. licheniformis* and *B. subtilis* [13, 45]. However, bacterial neutral proteases are active in a narrow pH range (pH 5 to 8) and have relatively low thermotolerance. Due to their intermediate rate of reaction, neutral proteases generate less bitterness in hydrolyzed food proteins than do the animal proteinases and hence, are valuable for use in the food industry. Some of the neutral proteases belong to the metalloprotease type and require divalent metal ions for their activity, while others are serine proteinases, which are not affected by chelating agents [46].

The production of extracellular alkaline proteases and other commercial enzymes by fermentation mainly depends on the bacterial strain, composition of the medium, cell growth, nutrient and metal ions requirement, pH of the medium, incubation temperature and period and method of cultivation of the microorganism [19, 47]. It is also strongly influenced by variation in C/N ratio, presence of some easily metabolizable sugars (glucose) and metal ions [48, 49]. Protease synthesis is also affected by rapidly metabolizable nitrogen sources (amino acids). In addition, several other physical factors such as aeration, agitation, inoculum density, pH, temperature and incubation also affect the amount of protease produced [50, 51]. The culture medium that is used in fermentation must not only meet the requirements of the microorganism, but also those of the industrial processes. Thus, such factors as cost, efficiency of utilization, rheology and its effect on downstream processing are all important factors in the design of a fermentation process. All microorganisms require water, sources of carbon, nitrogen, mineral elements and possibly specific requirements of vitamins and amino acids for growth and metabolism [52] which should also be fulfilled.

Bacterial alkaline proteases can be produced both by solid-state as well as submerged fermentation. For solid-state fermentation, different species of *Bacillus* including *B. subtilis* and *B. amyloliquefaciens* [33, 53, 54] have been used for high yields of proteases. Different agricultural by-products such as wheat bran, other cereal brans and different defatted oil seed cakes are being successfully used for hyper production of alkaline proteases [55]. However, submerged fermentation in shake flasks or fermentors is a preferred method for cultivation of bacteria and production of proteases. This is due to several inherent advantages such as easy control of environmental conditions (pH, temperature, oxygen level etc), good homogenization of fermentation medium components and easy inoculation etc. Different strategies have been used in order to scale up the production of alkaline proteases from different species of bacteria. Controlled batch, fed-batch fermentations [56, 57] and chemostat cultures [58] have been successfully used for improving protease yields. In recent years, a great amount of research and development efforts have been carried out to use the statistical approach methods with the aim of obtaining high yields of alkaline proteases in the fermentation medium [51, 59]. The application of properly designed approaches with multi-factor models allows the biochemical engineers to design scale up strategies for increased enzyme production.

Many of the organisms excrete more than one kind of protease and the type of the protease formed may depend on the composition of the medium. For example, *Bacillus* NRRL B-3411 produces preferably neutral protease when grown on grain medium, but mainly alkaline protease when cultured on a fishmeal-enzose-cerelose medium [60]. The biosynthesis of protease is often correlated with the particular growth phases of the microbial culture. Under most growth conditions, *Bacillus* species produce extracellular protease during the post-exponential growth phase [61]. Mandelstam (1958) [62] attributed this behavior to an increased need for turnover of cell proteins at the slower growth rate. Other bacilli synthesize proteases during the exponential growth phase. However, these kinetics depend on the composition of the medium [63].

The commercial production of extracellular alkaline proteases commonly involves the fermentation process carried out in fermentors/bioreactors containing suitable medium for the growth of the microorganisms and induction of the enzyme synthesis. Production of the enzymes can be increased by genetic manipulation or providing right environment to the microorganism for the growth and product formation. The provision of right environment requires the physiological studies to determine the physical and nutritional requirements for maximum growth and high yield of the desired product [64].

Some of the most important mechanisms that control primary and secondary metabolism are substrate induction, catabolite repression and inhibition, feedback repression and inhibition and ATP regulation [65, 66]. In substrate induction, constitutive and inducive enzymes appear to be involved. Constitutive enzymes are controlled by gene expression, while the activity of inducible ones depends on the concentration of substrate. Feedback regulation can be controlled by two basic mechanisms, feedback regulation at the genetic level to control the synthesis of the enzymes and feedback inhibition at the molecular level to control the activity of enzymes. In both cases, regulation is dependent upon the concentration of the end product. Sometimes, the synthesis of enzyme proteins can be repressed through nutrient effect such as the suppressive action of the substrates like glucose that promote rapid growth and this type of regulation has been termed as catabolite repression [67].

Secondary metabolite production is often suppressed in exponential phase of growth by nutrients like carbon, nitrogen and phosphorus [68]. Glucose is an excellent source of carbon and energy for microbial growth; however it is not often a major source in secondary metabolism fermentation because of its suppressive effect on metabolite production. Malathi and Dhar (1987) [69] observed that production of extracellular alkaline protease in *Aspergillus flavus* is induced in the presence of glucose; however the catabolite repression was observed, when its percentage was increased up to 3%. Nitrogen catabolite regulation of primary metabolism involves repression of enzymes such as protease, amylase and urease that act on nitrogen containing substrates, especially ammonium containing compounds. Baig (1982) [70] studied the effect of different nitrogen sources on the production of protease from *Bacillus subtilis* and found the maximum production with NH_4NO_3. Amino acid mixture containing proline, aspartic acid, glutamine and aspargine also produced maximum repression of protease production. Pinotti et al., (2000) [71] showed a strong correlation between the nitrogen source and protease yield and the presence of glucose repression. They achieved highest enzyme concentration using casein hydrolyzate and cheese whey as nitrogen sources. Several others found soybean meal [72, 73], keratin [43], poultry feathers [74] and casein [48, 75] as best nitrogen sources for the production of alkaline proteases from bacteria.

A suitable concentration of specific metal ions in the fermentation medium is very important for the microbial growth and subsequent production of primary and secondary metabolites [76]. A lower concentration of these ions may result in a decreased microbial growth and enzyme production, while some metal ions in higher concentration often inhibit microbial growth and enzyme production. Proom and Knight (1955) [77] have found that Ca^{+2}, Zn^{+2}, Mn^{+2}, Mg^{+2}, Na^{+2}, Fe^{+2} and Fe^{+3} ions have a positive effect on the growth of the microorganisms and production of enzymes at an optimum concentration of 0.05%, 0.0005%, 0.01%, 0.2%, 0.1%, 0.002% and 0.001%, respectively. They also found that the maximum inhibitory effect was exerted by Fe^{+2} and Fe^{+3} ions above their optimum concentration. The activation of protease in the medium in the presence of Ca^{+2} and Zn^{+2} has also been reported [78]. Zn^{+2} has also been found very important for the production and stability of protease from *Bacillus mycoids* and *Streptomyces naraensis* [35, 37]. Presence of Mn^{+2}, Fe^{+2} and Co^{+} ions in the culture medium has been found to exert an adverse effect on the enzyme production and activity [35, 79].

The cultural factors especially temperature and pH are also very important parameters to consider while obtaining higher yields of proteases by bacteria. Temperature affects all the reactions in a living cell, so it is an important environmental factor in controlling the growth and metabolic activities of microbial populations. Most organisms have optimal growth temperature between 25-40°C and are termed as mesophiles however, many thermophilic bacteria have been reported to grow and produce proteases above 40°C such as *Bacillus* sp. JB-99 has been found with 50°C as optimum temperature for growth and production of alkaline protease [80]. Several thermophilic anaerobic bacteria with proteolytic activity growing at temperatures between 50°C to 90°C were isolated from the island of Sao Miguel in the Azores (Portugal) by Riessen and Antranikian (2001). Similarly, hydrogen ion concentration (pH) of the fermentation medium is also an equally important factor. The rate and extent of growth and metabolism of a microorganism depends on pH of its environment. Most bacteria are capable of growth and protease production between pH 5.0 to 9.0 [81], however, many alkalophilic bacteria have been reported to grow at high pH values and produce alkaline proteases. For example, the protease produced by an alkalophilic and thermophilic *Bacillus* sp. strain B18 exhibited optimum pH of 12 to 13 [82].

Catabolic enzymes are frequently inducible as they are only produced in the presence of an inducer, which is normally the substrate of the enzyme or a structurally similar compound. Thus, the medium for commercial production of a catabolic enzyme must include the relevant inducer. In case of protease production, various proteins such as soybean meal, casein, sunflower meal and gelatin etc. are included in the medium as an inducer of proteases. Similarly, the production of catabolic enzymes is frequently repressed by the presence of certain compounds in the fermentation medium. For example, protease production in many bacilli is repressed in certain amino acid containing media and protease production in *A. niger* is apparently sensitive to repression by sulfur containing amino acids [19]. Rapidly utilizable carbon sources have been demonstrated to repress the formation of a wide range of secondary metabolites [83]. Nitrogen sources have also been reported to have repressive effects on secondary metabolite production. The fact that soybean meal and proline are the best nitrogen sources for metabolite production is probably due to their slow utilization, thus avoiding nitrogen metabolite repression [83].

In fermentation biotechnology, as soon as one has a microorganism of his choice for the production of certain useful product, there exists the aim to enhance its yield. This can be done either by improving the external conditions (process optimization) or by genetic manipulation in which random mutagenesis by using physical or chemical mutagen is one of the successful methods. Some of the mutagens principally induce base substitutions such as hydroxylamine, *N*-methyl-*N*-nitro-*N*-nitrosoguanidine (MNNG), bromouracil, ethyl methyl sulphonate (EMS) and 2-aminopurine, while others such as ultraviolet radiations (UV) and nitrous acid are efficient deletion mutagens [84]. The mutagenic spectrum of these mutagens has been well established to induce favorable genetic changes in industrially important microorganisms [85]. Various successful attempts have been made to produce positive mutants with several fold higher production of proteases. Bierbaum et al., (1994) [86] selected two mutant strains of *Bacillus licheniformis* insensitive to catabolite repression through classical mutagenesis in connection with the development of a fed-batch procedure for protease production. Yoneda and Maruo (1975) [87] successfully induced mutation in *Bacillus subtilis* by *N*-methyl-*N*-nitro-*N*-nitrosoguanidine treatment. The mutant produced 16 times more protease than the parent. They also observed changes in the morphology of cells as the mutant cells became filamentous due to the formation of chains of cells. The autolysis of cells was also decreased due to mutation. In addition to higher yields, mutagenesis has also been carried out to improve the characteristics of proteases. For example, more stable proteases towards higher temperatures and alkaline conditions were created by introducing a point mutation in the *Bacillus* sp. [82].

LEATHER INDUSTRY AND PROTEASES

Alkaline proteases are envisaged to have extensive applications in leather industry. In a tannery, a rawhide is subjected to a series of chemical treatments prior to tanning and finally converted to finished leather. Proteases may play a vital role in these treatments by replacing these hazardous chemicals especially involved in soaking, dehairing and bating [88, 89] Increased usage of enzymes for dehairing and bating not only prevents pollution problems, but also is effective in saving energy [90].

Soaking involves removal of blood, dung and dirt from hides and some structural changes. In advanced tanneries, soaking is usually performed with combination of proteolytic enzymes that are optimally active in the alkaline or neutral pH. Alkaline proteases play an important role in solubilization of albumin and globulin, opening up of contracted fibrous proteins and washing dirt and excessive fat at soaking stage [18, 91].

In leather processing, the most constrained operation from environmental point of view is the dehairing of skin/hide. The conventional method used for depilation (dehairing) involves the use of sodium sulfide and hydrated lime and is described as lime-sulfide process. A paste made by sodium sulfide and hydrated lime is applied on the flesh side of the skin. The objective of this treatment is the removal of hair by hydrolysis of mucoids, swelling of collagen fibers and elastin digestion. This process has now been clearly recognized to be environmentally objectionable accounting for the discharge of 100% of sulfide and 80% of the suspended solids in the tannery effluent [92].

The use of enzymes for unhairing skin and hides has been widely discussed since the beginning of this century and many microbial sources and methods of application have been suggested [89]. The enzymes catalyze the breakdown of the protein keratin in the hair and allow the hair to be easily removed. The pH at which these enzymes are active ranges from acidic, in which pepsin and papain can be used, to neutral and alkaline for enzymes such as from *Bacillus subtilis*. The advantage of using proteases for dehairing of skins are the reduction of the sulfide contents in the effluent, recovery of the hair/wool which is of good quality, an increased yield of leather area, easy handling of the pelts by workmen, simplification of the pretreatment, the elimination of the bate in the deliming stage and finally the production of a good quality pelts/leather [93]. However, there are some disadvantages such as the enzymatic process is more expensive than the conventional chemical process; it requires careful control and lakes swelling of the collagen during the enzymatic dehairing [89].

Proteases, which are mostly produced by bacteria are stable in alkaline range, have become more and more widespread for the depilation and opening up of the hide. These proteases may be used in conjunction with alkalies such as lime, sodium carbonate and with reducing agents such as sulfides or mercaptants at pH 10-14. Depending on the nature of finally prepared leather, the treatment can last 6-24 hours. Enzymatically aided unhairing processes help to increase the surface and facilitate the cleaning and the dying of the leather [94].

The bating following the dehairing process involves the degradation of elastin and keratin, removal of hair residues, and the deswelling of collagen, which produces a good and soft leather mainly used for making leather clothes and goods. Traditionally, bating is an enzymatic process mainly involving the use of pancreatic enzyme preparations containing trypsin. Today, microbial proteases have replaced the use of trypsin, as they are more economical. Quality of the finished leather mainly depends upon the bating efficiency [95].

Leather shavings and dust produced during the finishing and preparation of end products mainly consists of collagen protein. It may be degraded with collagenase (a type of protease) to get amino acids and oligopeptidases [96].

REFERENCES

[1] Raju, K., Jaya, R. and Ayyanna, C. (1994). Hydrolysis of casein by bajara protease importance. *Biotechnol Coming Decades, 181*: 55-70.

[2] Beg, Q. K., Sahai, V. and Gupta, R. (2003). Statistical media optimization and alkaline protease production from *Bacillus mojavensis* in a bioreactor. *Process Biochem., 39*: 203-209.

[3] Barindra, S., Debashish, G., Malay, S., et al. (2006). Purification and characterization of a salt, solvent, detergent and bleach tolerant protease from a new gamma *Proteobacterium* isolated from the marine environment of the Sundarbans. *Process Biochem., 41*: 208-215.

[4] Ward, O. P. (1985). Proteolytic enzymes. In: *Comprehensive Biotechnology,* Pergmon Press Ltd., New York, pp: 789-818.

[5] Keay, L. (1971). *Bacillus megaterium* neutral proteases. A Zinc containing metals enzyme. *Process Biochem., 6*: 20.

[6] Haq, I. and Mukhtar, H. (2004). Biosynthesis of protease by *Rhizopus oligosporous* IHS_{13} by solid-state fermentation in a low-cost medium. *J Basic Microbiol., 44*: 280-287.

[7] Seong, C. N., Jo, J. S., Choi, S. K., et al. (2004). Production, purification, and characterization of a novel thermostable serine protease from soil isolate *Streptomyces tendae. Biotechnol Lett., 26*: 907-909.

[8] Hameed, A., Natt, M. A. and Evans, C. S. 1996. Production of alkaline proteases by a new *Bacillus subtilis* isolate for use as a bating enzyme in leather treatment. *World J Microbiol Biotechnol., 12*: 289-291.

[9] Tang, X. M., Shen, W., Lakay, F. M., et al. (2004). Cloning and over-expression of an alkaline protease from *Bacillus licheniformis. Biotechnol Lett., 26*: 975-979.

[10] Bergmann M. and Frankel-Conrat H. (1937). The role of specificity in the enzymic synthesis of proteins: synthesis with intracellular enzymes. *J Biol Chem, 119*: 707–720.

[11] Feng, Y. Y., Yang, W. B., Ong, S. L., et al. H. (2001). Fermentation of starch for enhanced alkaline protease production by constructing an alkalophilic *Bacillus pumilus* strain. *Appl_Microbiol Biotechnol., 57*:153-160.

[12] Gupta, R., Beg, Q. and Lorenz, P. (2002). Bacterial alkaline proteases: molecular approaches and industrial applications. *Appl Microbiol Biotechnol., 59*: 15-32.

[13] Rao, M. B., Tanksale, A. M., Ghatge, M. S., et al. (1998). Molecular and biotechnological aspects of microbial proteases. *Microbiol Mol Biol Rev., 62*: 597-635.

[14] Kalisz, H. M., (1988). Microbial proteinases. In: *Advances in Biochemical Engineering/Biotechnology*, Fiechter, A. (Ed.), Vol. 36. Springer, Berlin, pp. 1–65.

[15] Rawlings, N. and Barrett, A. (1997). Structure of membrane glutamate carboxypeptidase. *Biochim Biophys Acta, 1339*: 247–252.

[16] Imshenetskii, A., Kasatkina, I. and Zheltova, E. (1971). Repression in protease synthesis in *A. terricola* by exogenic amino acids. *Microbiol., 40*: 382-386.

[17] Wiseman, A. (1995). *Handbook of enzyme Biotechnology*, Ed. III, 95-96, Ellis Horwood.

[18] Godfrey, T. (1996). Leather, In: *Industrial Enzymology*, 2nd Ed., (Godfrey, T., and S. West, Edt.) Stocton Press, New York, pp. 285-291.

[19] Reed, G. (2004). *Prescott and Dunn's Industrial Microbiology,* 4th Ed., CBS Publishers & Distributors, New Delhi, India, pp: 684-691.

[20] Siezen R. J. and Leunissen J. A. (1997). Subtilases: The superfamily of subtilisin-like serine proteases. *Protein Sci.*, *6*: 501–523.

[21] Poza, M., Sieiro, C., Carreira, L., et al. (2003). Production and characterization of the milk-clotting protease of *Myxococcus xanthus* strain 422. *J Ind Microbiol Biotechnol.*, *30*: 691-698.

[22] Sizer, I. W. (1972). *Advances in Applied Microbiology*, 1st Edition, Elsevier, vol. 68. pp15-18.

[23] Chandi, C. R. and Subramanyam, V R. (2002). Enhanced protease and β-Lactamase activity by immobilization of a thermophilic *Bacillus* sp. Isolated from a local hot spring in Orissa, India, In: *Recent trends in Biotechnology*, (Kumar, V. S., Agrobios, J. eds.). India, pp:93-100.

[24] Pavlova, N., Dobrev, I. and Aleksiev, B. (1992). Regioselective peptide synthesis by the alkaline protease from *Bacillus subtilis* strain *Dy. Proc Jur P Symp.*, *22nd*: 405-406.

[25] Lee, H. Suh, D. B., Hwang, J. H. et al. (2002). Characterization of a keratinolytic metalloprotease from *Bacillus* sp. SCB-3. *Appl Biochem Biotechnol.*, *97*: 123-133.

[26] Calik, P., Calik, G. and Ozdamar, T. H. (1998). Oxygen transfer effects in serine alkaline protease fermentation by *Bacillus licheniformis*: Use of citric acid as the carbon source. *Enz Microbial Technol.*, *23*: 451-461.

[27] Serkina, A. V., Shevelev, A. B. and Chestukhina, G. G. (2001). Structure and functions of bacterial proteinase precursors. *Bioorg Khim.*, *27*: 323–346.

[28] Pedersen, N., Wimmer, R., Matthiesen, R., et al. (2003). Synthesis of sucrose laurate using a new alkaline protease. In: *Tetrahedron:Asymmetry*, vol. *14*: 667-673.

[29] Kim, J. M., Lim, W. J. and Suh, H. J. (2001). Feather degrading *Bacillus* species from poultry waste. *Process Biochem.*, *37*: 287-291.

[30] Rehman, N. Z., Raja, C., Razak, K., et al. (1994). Purification and characterization of a heat-stable alkaline porotease from *Bacillus stearothermophilus* F1. *Appl Microbiol Biotech.*, *40*: 822-827.

[31] Peter, D., Johanna, M. and Renate, U. H. (2001). Differentiation between conformational and autoproteolytic stability of the neutral protease from *Bacillus stearothermophilus* containing an engineered disulfide bond. *Eur J Biochem.*, *268*: 3612-3618.

[32] Balban, N., Sharipova, M., Usmanova, A., et al. (1993). Extracellular alkaline protease from *Bacillus intermedius;* isolation and some properties. *Biokhimya*, *58*: 1923-1928.

[33] George, S., Raju, M., Krishnan, V., et al. (1995). Production of protease by *Bacillus amyloiquefaciens* in solid-state fermentation and its application in the unhairing of hides and skins. *Process Biochem.*, *30*:457-462.

[34] Voordouw, G., Gaucher, G. and Roche, R. (1974). Physicochemical properties of thermomycolase, the thermostable, extracellular, serine protease of the fungus *Malbranchea pulchella*. *Can J Biochem.*, *52*: 981-990.

[35] Mohamed, A. N., Ismail, A., Ahmed, S. et al. (1998). Production and Immobilization of alkaline protease from *Bacillus Mycoides Biores Technol.*, *64*: 205-210.

[36] Ruechle, G. L. A. 1923. Enzymatic contents of bacterial spores. *J. Bacteriol.*, 8:487-491.

[37] Hiramatsu, A. and Ouchi, T. (1978). Zinc of neutral proteinases from *Streptomyces naraensis*. *J Agric Biol Chem.*, *42*: 1309-1313.

[38] Ko, J., Park, S., Kim, E., et al. (2002). Thermostability of an alkaline protease, AprP, is enhanced by replacement of Ser307 and Ser331 at the cleavage sites. *Biotechnol Lett.*, *24*: 1749-1755.

[39] Ravaud, S., Gouet, P., Haser, R., et al. (2003). Probing the role of divalent metal ions in a bacterial psychrophilic metalloprotease: Binding studies of an enzyme in the crystalline state by X-ray crystallography. *J Bacteriol.*, *185*: 4195-4203.

[40] Cabral, C., Cherqui, A., Pereira, A., et al. (2004). Purification and Characterization of Two Distinct Metalloproteases Secreted by the Entomopathogenic Bacterium *Photorhabdus* sp. Strain Az29. *Appl Env Microbiol.*, *70*: 3831-3838.

[41] Bintsis, T., Vafopoulou-Mastrojiannaki, A., Litopoulou-Tzanetaki, E., et al. (2003). Protease, peptidase and estrase activities by *lactobacilli* and yeast isolates from Feta cheese brine. *J Appl Microbiol.*, 95: 68-77.

[42] Thangam, E. B. and Rajkumar, G. S. (2002). Purifcation and characterization of alkaline protease from *Alcaligenes faecalis*. *Biotechnol Appl Biochem.*, *35*: 149-154.

[43] Toni, C., Richter, M., Chagas, J., et al. (2002). Purification and characterization of an alkaline serine endopeptidase from a feather degrading *Xanthomonas maltophila* strain. *Can J Microbiol.*, *48*: 342-348.

[44] Graycar, T. P. (1999). Proteolytic cleavage, reaction mechanism. In: *Bioprocess technology: fermentation, biocatalysis and bioseparation*. Flickinger M.C., Drew SW (eds.) Wiley, New York, pp 2214–2222.

[45] Singh, J., Vohra, R. M. and Sahoo, D. K. (2001). Purification and characterization of two extracellularalkaline proteases from a newly isolated obligate alkalophilic *Bacillus sphaericus*. *J Ind Microbiol., Biotechnol.*, *26*: 387-393.

[46] Uehara, H., Yoneda, Y., Yamane, K., et al. (1974). Regulation of neutral protease productivity in *Bacillus subtilis*: Transformation of high protease productivity. *J Bacteriol.*, *119*: 82-91.

[47] Stanbury, P., Whitaker, A. and Hall, S. (1995). *Principles of fermentation Technology*, 2nd Ed., Elsevier Science Ltd., Oxford.

[48] Beg, Q., Saxena, R. and Gupta, R. (2002). De-repression and subsequent induction of protease synthesis by *Bacillus mojavensis* under fed-batch operations. *Process Biochem.*, *37*: 1103-1109.

[49] Varela, H., Ferrari, M., Belobradjic, L., et al. (1996). Effect of medium composition on the production by a new *Bacillus subtilis* isolate of protease with promising unhairing activity. *World J Microbiol Biotechnol.*, *12*: 643-645.

[50] Hameed, A., Keshavarz, T. and Evans, C. (1999). Effect of oxygen tension and pH on the production of extracellular protease from a new isolate of *Bacillus subtilis* K2, for use in leather processing. *J Chem Technol Biotechnol.*, *74*: 5-8.

[51] Puri S., Beg, Q. K. and Gupta, R. (2002). Optimization of alkaline protease production from *Bacillus* sp. using response surface methodology. *Curr Microbiol.*, *44*: 286-290.

[52] Trevan, M., Boffey, S., Goulding, K. et al. (2004). The Biological Principles: In *Biotechnology*, Tata McGraw Hill, New Delhi, pp. 85-96.

[53] Haque, H., Nadeem, B. and Qadeer, M. (1990). Biosynthesis of enzyme by solid state fermentation III. Production of protease by *Bacillus subtilis*. *Sci Int.*, *2*: 31-34.

[54] Pandey, A., Selvakumar, P., Soccol, C., et al. (2001). *Solid-State Fermentation in Biotechnology: Fundamentals and Applications*, Asia Tech Publishers Inc., New Delhi, P. 221.

[55] Arroteia, E., Mitchell, D., Meien, O., et al. (2001). The implications of temperature instability of enzymes for their production in solid-state fermentation processes. *Electronic J. Biotechnol.*, *26*: 271-275.

[56] Hubner, U., Bock, U. and Schugerl, K. (1993). Production of alkaline serine protease subtilisin Carlsberg by *Bacillus licheniformis* on complex medium in a stirred tank reactor. *Appl Microbiol Biotechnol.*, *40*: 182–188.

[57] Van, P., Spitzenberger, F., Kretzmer, G., et al. (1996). Improvement of the production of subtilisin Carlsberg alkaline protease by *Bacillus licheniformis* by on-line monitoring and control in a stirred tank reactor. *J Biotechnol.*, *49*: 83-93.

[58] Frankena J., Koningstein C., Verseveld H., et al. (1986). Effect of different limitations in chemostat cultures on growth and production of exocellular protease by *Bacillus licheniformis. Appl Microbiol Biotechnol.*, *4*: 106-112.

[59] De Coninck J., Bouquelet, S., Dumortier, V., et al. (2000). Industrial media and fermentation processes for improved growth and protease production by *Tetrahymena thermophila*. *J Ind Microbiol Biotechnol.*, *24*: 285-290.

[60] Keay, L., Moseley, M., Anderson, R., et al. (1972). Production and isolation of microbial proteases. *Biotechnol Bioeng Symp.*, *3*: 63-92.

[61] Schaeffer, P. (1969). Sporulation and production of antibiotics, exoenzymes and exotoxins. *Bacteriol Rev.*, *33*: 48-71.

[62] Mandelstam, J. (1958). Turnover of protein in growing and nongrowing population of *Escherichia coli. Biochem J.*, *69*: 110-119.

[63] Chaloupka, J. (1969). Dual control of megateriopeptidase synthesis. *Ann Ins Pasteur, Paris, 117*: 631-636.

[64] Michael, D. and George, J. (1987). Regulation of secondary metabolite biosynthesis in *Gibberella fujikuroi. Trans Brit Mycol Soc.*, *62*: 377-389.

[65] Drew, S. W. and Demain, A. L. (1977). Effect of primary metabolites on secondary metabolism. *Ann Rev Microbiol.*, *31*: 343-356.

[66] Rose, A. H. (1979). Production and industrial importance of secondary metabolites in economic. *Microbiol.*, Vol. 3. Academic Press.

[67] Gallo, M. and Katz, E. (1972). Regulation of secondary metabolite biosynthesis: catabolite repression of phenoxazine synthase and actinomycine formation by glucose. *J Bacteriol.*, *109*: 659-667.

[68] Demain, A. L., Kennel, Y. and Aharonowitz, Y. (1979). *Microbial technology- current state future prospects*. Bull, A. T., Ellwood, D. C., C. Retledge. (Eds). London Society, Cambridge Univ. Press.

[69] Malathi, S. and Dhar, S. (1987). Production of extracellular protease by an *Aspergillus flavus* isolate and its application in the depletion of skin. *Leather Sci.*, *34*: 67-76.

[70] Baig, M. A. (1982). Nutritional requirements of *Bacillus subtilis* for the production of protease - effect of nitrogen source. *Pak J Biochem.*, *15*: 56-61.

[71] Pinotti, L., Silva, A., Silva, R., et al. (2000). Study of different media for production of penicillin G acylase from *Bacillus megaterium* ATCC 4945. *Appl Biochem Biotechnol.*, *84*: 655-663.

[72] Mabrouk, S., Hashem, A., El-Shayeb, N., et al. (1999). Optimization of alkaline protease productivity by *Bacillus lichniformis* ATCC 21415. *Biores Technol.*, *69*: 155-159.

[73] Moriera, K., Cavalcanti, M., Duarte, H., et al. (2001). Partial characterization of proteases from *streptomyces clavuligerus* using an inexpensive medium. *Braz J Microbiol.*, *32*: 215-220.

[74] Gessesse A., Hatti-Kaul, R., Gashe B., et al. (2003). Novel alkaline proteases from alkaliphilic bacteria grown on chicken feather. *Enzy Microb Technol.*, *32*: 519-524.

[75] Haq, I. and Mukhtar, H. (2006). Fuzzy logic control of bioreactor for enhanced biosynthesis of alkaline protease by an alkalophilic strain of *Bacillus subtilis, Curr Microbiol., 52*: 149-152.

[76] Linden, A., Mayans, O., Meyer-Klaucke, W., et al. (2003). Differential regulation of a hyperthermophilic alpha amylase with a novel (Ca, Zn) two-metal centre by zinc. *J Biol Chem., 278*: 9875-9884.

[77] Proom, H. and Knight, B. (1955). The minimal requirements of some species in the genus *Bacillus subtilis. J Gen Microbiol., 13*: 474-480.

[78] Chessa, J., Petrescu, I., Bentahir, M., et al. (2000). Purification, physico-chemical characterization and sequence of a heat labile alkaline mettaloprotease isolated from a psychrophilic *Pseudomonas* species. *Biochim Biophys Acta, 1479*: 265-274.

[79] Macfarlane, G. and Gibson, G. (1993). Characteristics of protease in *Bacteroids splanchnicus* NCTC 10825. *J Appl Microbiol Biotechnol., 39*: 506-511.

[80] Johnvesly, B. and Naik, G. (2001). Studies on production of thermostable alkaline protease from thermophilic and alkalophilic *Bacillus sp.* JB-99 in chemically defined medium. *Process Biochem., 37*: 139-144.

[81] Riessen, S. and Antranikian, G. (2001). Isolation of thermoanaerobacter *keratinophilus* sp., a novel thermophilic, anaerobic bacterium with keratinolytic activity. *Extremophiles, 5*: 399-408.

[82] Masui, A., Yasuda, M., Fujiwara, N., et al. (2004). Enzymatic hydrolysis of gelatin layers on used lith film using thermostable alkaline protease for recovery of silver and PET film. *Biotechnol Prog., 20*: 1267-9.

[83] Demain, A. L. (1984). *Biotechnology of Industrial Antibiotics*. Vandamme, E. J., (Ed.), New York, Marcel Dekker, pp: 33-44.

[84] Miller, J. H. (1982). Carcinogens induce targeted mutations in *Escherichia coli. Cell, 31*: 5-7.

[85] Stanier, R., Doudoroff, M. and Adelberg, E. (1970). *General Microbiology*, 3rd Ed., p. 418. Macmillan and Co.

[86] Bierbaum, G., Karutz, M., Weusterbotz, D., et al. (1994). Production of protease with *Bacillus lichniformis* mutants insensitive to repression of Exoenzyme biosynthesis. *Appl Microbiol Biotechnol., 40*: 611-617.

[87] Yoneda, Y. and Maruo, B. (1975). Mutation of *Bacillus subtilis* causing hyperproduction of alpha amylase and protease, and its synergistic effect, *J Bacteriol., 124*: 48-54.

[88] Puvankrishnan, R. and Dhar, S. (1986). Recent advances in the enzymatic depilation of hides and skins. *Leather Sci., 33*: 177-191.

[89] Mukhtar, H. and Haq, I. (2008). Production of alkaline protease by *Bacillus subtilis* and Its application as a depilating agent in leather processing, *Pak J Bot., 40*: 1673-1679.

[90] Sachidanandham, R. Kumar, I., Krishnan, M., et al. (1999). Mathematical model-based estimation of volumetric oxygen transfer coefficient(s) in the production of proteolytic enzymes in *Bacillus amyloliquefaciens. Bioprocess Eng., 21*: 319-322.

[91] Christner, J. (1995). Modern enzyme applications in the beamhouse. *World Leather, 10*: 60-64.

[92] Peper, K. W. and Wyatt, K. G. (1989). Enzymatic unhairing of heavy hides. *J Ind Leather Technol Asso., 36*: 214-233.

[93] Mitra, P. and Chakraverty, R. (1998). Proteolytic enzymes as depilatory agents in the tannery. *Proceedings of National Symposium on Frontiers in Applied and Environmental Microbiology*, pp. 195-197.
[94] Cecil, A., Christner, J. and Wick, G. (1995). Liming skin and hide with high concentration of protease. *European Patent No. 2282148.*
[95] Haines, B. M. and Barlow, J. R. (1975). Review-The anatomy of leather, *J Material Sci.*, *10*: 525-538.
[96] Steinmann, B., Royce, P. and Furga, A. (1993). The Ehlers Danlos syndrome; In *Connective tissue and its heritable disorders: Molecular, genetic and Medical aspects*, pp. 351-407, Willey-Liss Inc. N.Y.

SECTION C. AGRICULTURAL MICROBIOLOGY

In: Microbial Catalysts. Volume 1
Editors: S. M. Abdel-Aziz, N. Garg, A. Aeron et al.
ISBN: 978-1-53614-679-0

Chapter 7

THE BENEFICIAL ROLE OF MICROBIAL CATALYST FOR SUSTAINABLE ENVIRONMENT

Sadia Sharmeen, Taslim U. Rashid, Abul K. Mallik, PhD, Md. Shahruzzaman, PhD, Shanta Biswas, Md. Minhajul Islam, Asad Uzzaman, Papia Haque, PhD and Mohammed Mizanur Rahman*, PhD

Department of Applied Chemistry and Chemical Engineering,
Faculty of Engineering and Technology, University of Dhaka, Dhaka, Bangladesh

ABSTRACT

Microorganisms have been using over the last 70 years for their uniqueness in improving agricultural productivity and environmental conditions, removing toxic chemicals from the environment, desulfurizing many fossil fuels, modeling of the ecosystem, controlling oil spillage, and saving of energy resources. Often for their unpredictable characteristics and biosynthetic capabilities to produce a certain set of environmental and cultural conditions lead them for solving particularly difficult problems in the life sciences and other fields as well in an economically effective and environmentally sustainable manner. Microorganisms achieved significant results in eliminating problems associated with the use of chemical fertilizers and pesticides; nowadays they are widely used in nature farming and organic agriculture. Numerous oleaginous yeasts, microalgae, bacteria, in particular, those of the actinomycete group offer great potential as a sustainable feedstock for the production of third-generation biofuels, such as biodiesel and bioethanol. Biological treatment methods involving microorganisms have been widely used to degrade chloro-organic pollutants and aromatic compounds, to remove heavy metals, wide variety chemicals of a complex structure like dyes having azo, aromatic and heterocyclic groups, surfactants, toxicants, and salts are being released in the effluent of the textile industry. Recently, bio-electrochemical systems (BES) such as microbial fuel cells (MFCs) and microbial electrolysis cells (MECs) have emerged as a promising technology for simultaneous energy recovery and wastewater treatment.

* Corresponding Author's Email: mizanur.rahman@du.ac.bd.

1. INTRODUCTION

The environment is what we are surrounded by through our life. The quality of our lives depends on how is the environment and at the same time, our daily life governs the excellence or vulnerability of the environment. Our activities today will make the future of our next generation tomorrow. From ancient age, human activities are targeting the well-being of human; however, nowadays people have the understanding that even for the sake of humanity we cannot come to any compromise to destroy our nature. Our life will only survive when the environment can sustain. To keep a sustainable environment human activities should be controlled and make friendly to it. The sustainable environment can be achieved while we can keep it pollution free.

Biological different processes naturally correct many environmental problems. An enormous number of synthetic chemicals have continuously been produced and used for many purposes worldwide. These synthetic chemicals can pollute our environment directly or indirectly. However, the application of several biological processes can reduce or eliminate the potential toxicity of chemical contaminants by different means like; degradation, transformation, immobilization, etc. These bioremediation processes include the use of microbes, plants and/or animals. Biodegradation is the most commonly practiced biological means to remove many chemical contaminants from the environment. In this process, living microorganisms enzymatically breaks down chemical compounds into completely non-toxic or less toxic compounds [1]. Human and animal waste is another role player for environmental pollution. Many chemical or physical methods have so far been used in the treatment of this type of wastes. Nevertheless, the use of microbes has been proved to be the best method for the management of these wastes [2].

The microbial system can not only rectify the environment, it can also be advantageous for the production of many items. The use of the microbial system is better than that of the conventional chemical system in terms of several aspects, like the microbial system needs less energy, little cost, produces fewer wastes and for some microbes pollutants sometimes act as its growth promoters [3]. Effective microorganism technology has become popular in the last two decades in various fields (agriculture, food processing, biotechnology, genetic engineering, odor control, human health, animal husbandry, industrial catalysts, etc.) [4]. The use of microbial system already showed very satisfactory performance in water decontamination process, production of biofuel, removal of color from textile effluent, desulfurization of fossil fuel, degradation of lignin and control of oil spillage, etc. [5-7]. In this chapter, the beneficial effects of the use of microbial catalysts in the above-mentioned sectors will be discussed.

2. MICROBIAL CATALYSTS

Due to their omnipresence, microorganisms impact the entire biosphere i.e., plays an immense role in regulating the biogeochemical systems. Microorganisms produce different enzymes that are required for the numerous industrial and therapeutic purposes. These microbial enzymes are preferred due to their economic feasibility, high efficiency, consistency, availability, inexpensive formulation and so on. However, one of the mentionable properties of the microbial enzyme is its catalytic activity which helps to accelerate the biological/biochemical reactions inside as well as outside the cell.

Table 1. List of some of the important enzymes and their industrial applications are listed below

Source	Microorganism	Enzyme	Application
Bacterial	*Bacillus subtilis*	Proteases	Detergent Industry
Fungal	*T. lanuginosus*	Lipases	Detergent Industry; Pulp and paper industry
Fungal	*Aspergillus voryzae*	Amylase	Textile industry
Bacterial	*Pseudomonas stutzeri*	azoreductase	Textile dye decolorization
Bacterial	*Bacillus subtilis*	Amylases	Pulp and Paper Industry
Fungal and Bacterial	Different Microorganisms	Cellulases	Oil and Petroleum Industry; Pulp and Paper Industry
Fungal	*A. niger*	catalase	Textile Industry

Microbes, especially bacteria, often engage in symbiotic relationships (either positive or negative) with other organisms, and these relationships affect the ecosystem. In the year 1812 Gottlieb Sigismund Kirchhoff enlightened the application of biomolecules as a catalyst. French chemist Anselme Payen discovered the first enzyme, 'diastase' in 1833. Two years later in a catalytic reaction of hydrolysis of starch, diastase was first used. From then with industrial expansion and population growth, various enzymes from different microbes were used as a catalyst in different important chemical and biochemical reaction [Table 1].

3. ENZYMES AS WATER DECONTAMINATING AGENTS

A vital resource for human health and societal development is water. For a sustainable society, it is critical to use the proper treatment of water and wastewater at a reasonable cost due to the lack of sufficient freshwater source for increasing population. And the pollution of natural water bodies is also increasing with time. Extensive soil and water pollution are increasing due to the discharge of an enormous volume of nitrogen-containing wastewater from domestic and industrial origins. From an environmental and human health point of view, it is the earnest necessity to eliminate these pollutants from the effluents in industrialized countries. Distillery wastewater plays a detrimental role when it is disposed into the environment. When distillery water is discharged into a water body such as rivers, lake or lagoons, it reduces sunlight penetration through the water which lowers the photosynthetic activity and dissolved oxygen concentrations and thus ultimately is harmful to aquatic life. However, disposal on land is also unsafe as it reduces alkalinity and Manganese availability in the soil thus inhibiting seed germination and vegetation growth. So, appropriate treatment of this waste stream is necessary. Recently, textile industry effluents are a huge problem as it generates a high volume of waste products most of those are discharged into water bodies which affect the quality of water. The threat of climate change and energy crisis have been provoked the researchers to develop energy-efficient alternatives to conventional activated sludge wastewater treatment processes. Nowadays, for the production of energy from wastewater, a potential, as well as interesting technology named bioelectrochemical wastewater treatment, has appeared.

3.1. Bioelectrochemical Wastewater Treatment

The bioelectrochemical wastewater treatment is dependent on the use of electrochemically active microorganisms [8, 9]. It uses extracellular electron transfer (EET) [10, 11] mechanism to transfer electrons to an anode while microorganisms are used to catalyze the electrochemical oxidation of the organic materials in wastewaters (Figure 1) [12]. The electrode used in this process is therefore called a microbial bioanode.

Bioelectrochemical systems (BES) is not a new concept, and it can be detected back to one century ago when the bioelectricity was first described [13], which led to the investigation of microbial fuel cells (MFCs). MFCs is advantageous for microbial metabolism of organic compounds with a solid electron acceptor (an electrode) and can convert organic wastes directly into electric energy [14]. MFCs can efficiently recover bioenergy from wastewater; unlike anaerobic digestion. The direct electricity generation in MFCs avoids the collection, treatment, and conversion of biogas into electricity, thereby potentially saving the cost associated with biogas generators. Further development of MFCs recognizes that with an externally applied potential, a modified MFC can produce hydrogen gas, instead of electricity, and this device is called microbial electrolysis cell (MEC) [15]. When an additional chamber is installed between the anode and the cathode, an MFC is turned into a microbial desalination cell (MDC) that achieves desalination without much requirement of external energy [16]. The schematics of MFCs, MECs, and MDCs are shown in Figure 1.

3.1.1. Mechanism of Bioelectrochemical Wastewater Treatment

Bioelectrochemical wastewater treatment can be achieved by electrical connection between a cathode and microbial bioanode. The electrical connection between the electrodes makes a reduction reaction which ultimately determines the electron flow from the anode to the cathode. Overall reaction is dependent on the Gibbs free energy change which governs the operation theorem of bioelectrochemical system (BES) [17] i.e., when the change of the Gibbs free energy for overall reaction is negative then electricity can be generated which makes the BES work as a microbial fuel cell (MFC) [8] and when the change in Gibbs free energy for overall reaction is positive, it requires electrical energy from outside and the BES is operated as a microbial electrolysis cell (MEC) [18, 19].

At the anode, electrons are produced by the oxidation reaction which occurs due to the presence of organic materials and electrochemically active microorganisms [Figure 2]. Then these electrons are transferred to the cathode via an electrical circuit through the extracellular electron transfer (EET) by the help of the microorganisms which further utilized for oxygen reduction (in case of MFCs) or to form product like hydrogen (in case of MECs). The cathodic reaction occurred in both cases can be occurred either through biocatalysis (in the presence of a microbial biocathode) or through direct chemical catalysis (e.g., in presence of platinum). Thus an electro-neutrality is maintained in the system by the flow of ions in between the electrodes.

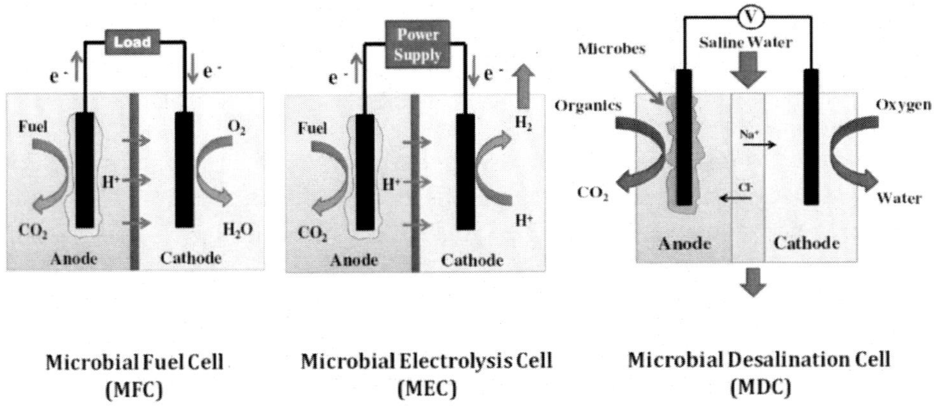

Figure 1. The schematics of MFCs, MECs, and MDCs [20].

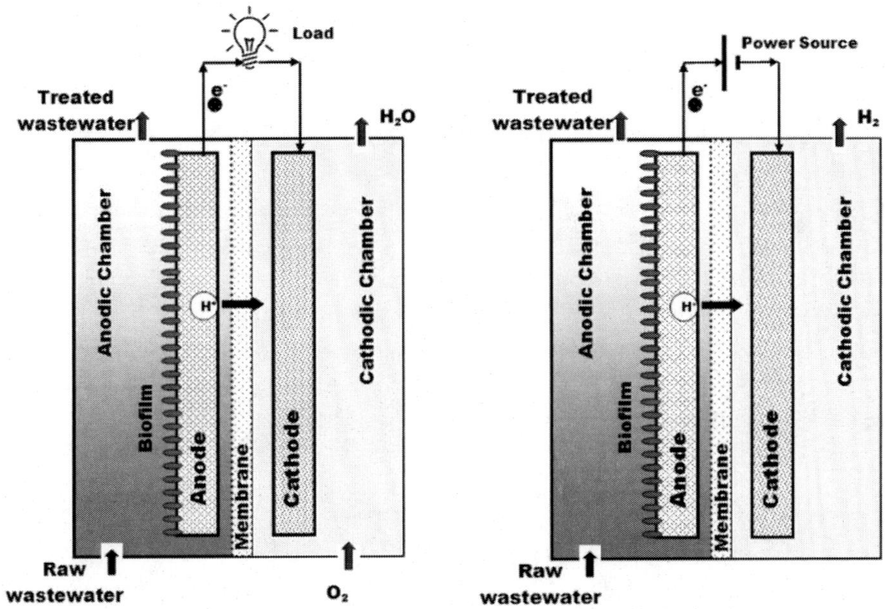

Figure 2. Schematic representation of the two most common bioelectrochemical wastewater treatment systems: (a) the microbial fuel cell (MFC) for electricity production and (b) the microbial electrolysis cell (MEC) for hydrogen production [19]].

3.1.2. Organic Wastewater Streams Treatment Using a Rotatable Bioelectrochemical Contactor (RBEC)

Rotatable bioelectrochemical contractor (RBEC) is a new approach of BES which is evaluated for its ability to convert the soluble organics directly into electrical current without the use of the conventional pH control methods or without the need of the wastewater re-circulation which requires more electricity. The RBEC can operate as an MFC, achieving net electricity recovery from wastewater organics. Here cathode is exposed in air, so, no ion exchange membrane or wastewater recirculation is required [21].

3.2. Microbial Treatment Processes for the Removal of Nitrogen from Wastewater

Due to some shortcomings of conventional nitrogen removal method (litho autotrophic nitrification and denitrification), novel and cost-effective biological nitrogen removal processes have been developed including simultaneous nitrification and denitrification system, anaerobic ammonium oxidation system or the combined system.

3.2.1. Types of Nitrogen-Containing Wastewaters

3.2.1.1. Nitrogen Present in Agricultural Wastewater

The recovery of nitrogen compounds from agricultural manures can be possible. Nitrogen concentration varies with the original source of the respective manure, for example, poultry manure in a farm near Istanbul, Turkey, contained 1580 mg/l Kjeldhal nitrogen and1318 mg/l NH_4^+-N [22]. Wastewater composition, volume and the rate of formation depend on the livestock species, their type and age, the nature of their feed, how it is fed, whether or not the livestock is housed, climate, etc. [23].

3.2.1.2. Nitrogen Present in Domestic Wastewater

Domestic wastewater effluent contains a high concentration of inorganic nitrogen leading to the eutrophication of the receiving water bodies [24]. Basically, the presence of ammonia and nitrates in the effluent causes eutrophication. In some areas, where the sewer system is not present for the disposal of domestic wastewater, is managed by use of septic tanks and seepage fields which generally shows high concentrations of ammonia and nitrates. In Rajasthan, India, the raw sewage received at the activated sludge plant has a BOD of 600–800 mg/l and a NH_4^+-N concentration of 80–110 mg/l during summer [25]. Zeng et al. reported that the ammonia concentration in real domestic wastewater from one septic tank in China was 54-74 mg/l [26].

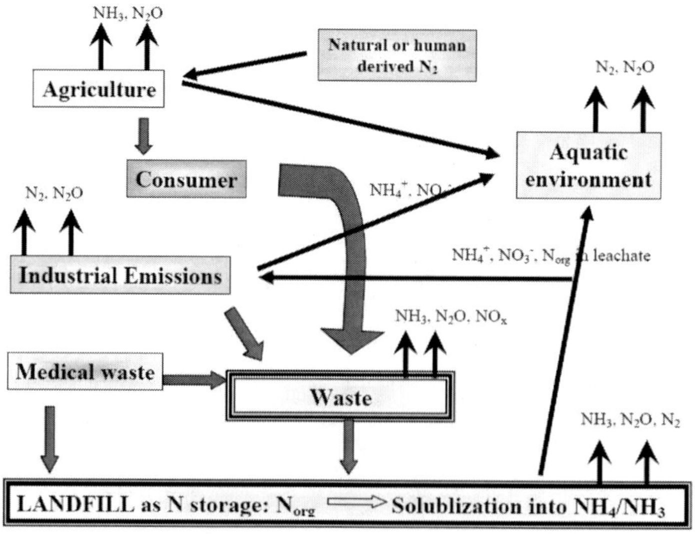

Figure 3. Nitrogen flow in waste management stream [27].

3.2.1.3. Nitrogen Present in Agro-Industrial and Industrial Wastewaters

Presence of nitrogen in different industrial and agro-industrial wastewaters is higher than the other wastewaters. Several authors reported that the stream contains more than 200 mg/l nitrates [28] and some contain even higher nitrate levels. For instance, the wastewater from glasshouses contained 325 mg/l nitrate$^-$-N [29]. Several industries generate wastewater with varying amounts of nitrate, such as 222 mg/l in a tannery wastewater of Pisa [30], 2320 mg/l in wastewater from the cochineal insects are usually processed to produce natural carmine used principally as a coloring agent in cosmetics, beverages and products with low pH [31] in Italy. 3600 mg/l usually generated from an initiating explosive factory in China [32] and 4000-6000 mg/l are produced during the frosting process of bottles in a winery [33]. Figure 3 shows the source of nitrogen flow in waste management stream.

3.2.2. Processes for Nitrogen Removal from Wastewater

3.2.2.1. Nitrification/Denitrification

Activated sludge system is the most popular removal process for biologically available nitrogen, where nitrification and denitrification both steps are done consequently [34]. Wastewater from antibiotics production, for example, usually contains large amounts of fermentation products, some residual antibiotic activity, and a high concentration of ammonia which is treated by activated sludge method [35].

Under strict aerobic conditions, complete nitrification of wastewater is a two-step process; where the first step includes the conversion of ammonia to nitrite by *Nitrosomonas*:

$$NH_4^+ + 1.5\,O_2 \rightarrow NO_2^- + H_2O + 2H^+ \quad (1)$$

while the second step indicates the further oxidation of nitrite to nitrate, which is commonly carried out by Nitrobacter [36]:

$$NO_2^- + 0.5\,O_2 \rightarrow NO_3^- \quad (2)$$

Both of the oxidation stages use different bacteria as an energy source, while CO_2 is used as a carbon source and molecular oxygen is used as an electron acceptor. Bacterial genera usually include autotrophs, where, *Nitrobacter* is not a restrain autotrophy and hence can grow using organic carbon. The whole process of nitrification and growth is balanced by two types of nitrifiers - autotrophic and heterotrophic which are inhibited by high concentrations of their own substrates and thus have a very little energy for high-affinity uptake systems. Several calculations have been proved that 80% of the energy generated by autotrophs is used to fix carbon dioxide [37]. On the contrary, heterotrophic nitrification is not associated with cellular growth like autotrophic nitrification [38]. However, *Nitrosococcus, Nitrosopira, Nitrosovibrio,* and *Nitrosolobus* can also be used to oxidize ammonium to nitrite [39].

Subsequently, denitrification is carried out by heterophic denitrifiers under anoxic conditions where instead of oxygen, oxidized nitrogen compounds i.e., NO_2^- and NO_3^- are reduced to gaseous nitrogen as electron acceptors along with the organic matter as carbon and energy source. Denitrifiers are common in gram-negative bacteria like *Pseudomonas, Alcaligenes, Paracoccus,* and *Thiobacillus*. Some gram-positive bacteria (such as *Bacillus*) and a few *halophilic archaeal* microorganisms (e.g., *Haloferax denitrificans*) are capable to denitrify [40].

3.2.2.2. Simultaneous Nitrification and Denitrification (SND)

In the SND process, a partial nitrification of NH_4^+ to nitrite occurs first which continues with a direct reduction of nitrite to N_2 gas [41]. Nitrification and denitrification take place concurrently in the same reactor vessel under an identical operating condition. An advantage of this process is it reduces the relatively large reactor volumes and energy costs for recirculation that are required for a separated aerobic and anoxic system. Several types of treatment units have been suggested in which SND can be run [42].

3.2.2.3. Autotrophic Denitrification

The rate of heterotrophic denitrification is strongly dependent on the type and concentration of carbon source and the C/N ratio [43]. This might also differ for different microorganisms, water streams, and environmental conditions [44].

3.2.2.4. Anaerobic Ammonium Oxidation (Anammox) Process

Especially effective biological alternative to conventional nitrogen removal from wastewaters is anaerobic ammonium oxidation (Anammox) process, where, ammonium is oxidized to nitrogen gas with nitrite being the electron acceptor under anaerobic conditions. Here, carbon dioxide is used for the growth of the anammox microorganisms required. This autotrophic method consumes 100% less biodegradable organic carbon and at least 50% less oxygen compared to the traditional nitrification-denitrification process [45]. Therefore, a lower operating cost is observed.

$$NH_4^+ + NO_2^- \rightarrow N_2 + 2\,H_2O \tag{3}$$

An approximate ratio of one to one ammonium and nitrite is needed in the Anammox process. However, for sludge digester effluents, this ratio can be attained without control, because these effluents contain bicarbonate as the anion for ammonium. The alkalinity of the water is exhausted leading to a drop in pH and preventing further nitrification when the conversion of ammonium is almost half done [46]:

$$NH_4^+ + HCO_3^- + 0.75\,O_2 \rightarrow 0.5\,NH_4^+ + 0.5\,NO_2^- + CO_2 + 1.5\,H_2O \tag{4}$$

The updated nitrogen cycle with ANAMMOX is represented in Figure 4 [47]. In putrescibles, a large amount of readily soluble nitrogen is present. The solubilization can be influenced by the characteristics of substrate and mode of reactor operation and hence, the amount of readily soluble nitrogen present in putrescibles can be desired to vary between 2 and 4 g NH_4^+-N/ Kg vs. the TKN (total Kjeldahl Nitrogen) value for MSW value is often reported, and thus the NH_4^+-N/TKN ratio may be a more suitable ratio to evaluate the amount of readily available dissolved nitrogen. Gallert and Winter (1997) reported a NH_4^+ -N/TKN ratio of 0.11 for MSW, whereas Held et al. (2002) reported 0.084 [48, 49]. Thus, almost 10% of the total nitrogen of putrescible is easily solubilized.

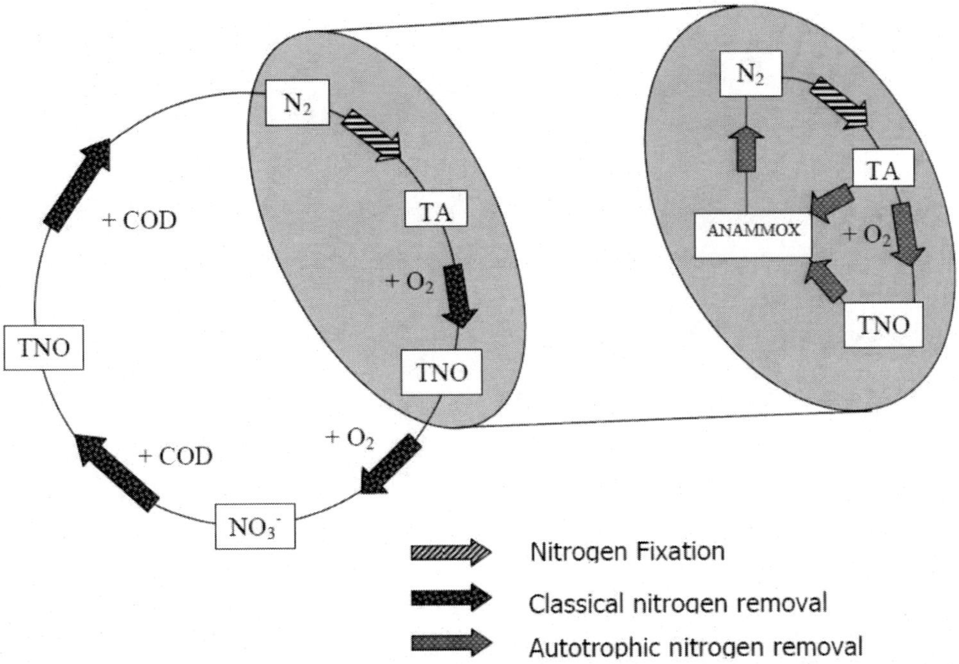

Figure 4. ANAMMOX process for removal of nitrogen [27].

The physiology of anaerobic ammonium-oxidizing aggregates produced in a sequencing batch reactor was reported by Strouset al. [50] where the maximum specific substrate conversion rate was measured as a function of temperature and pH and the activation energy of ANAMMOX was estimated to be 70 kJ/mol. Strouset al. also reported that the affinity constants for the substrates, ammonium, and nitrite, are less than 0.1 mg N/L inhibited the ANAMMOX process completely [51]. In another study, they had shown that the ANAMMOX process was reversibly inhibited by the presence of oxygen [52].

3.2.2.5. Partial Nitrification-Anammox

Another novel microbial process for wastewater treatment is the combination of Annamox reaction with partial nitrification in a reactor named as completely autotrophic nitrogen removal over nitrite (CANON) [53].

3.2.2.6. Bioreactor Landfill

Burton and Watson-Craik proposed that during the nitrification and the denitrification stages of leachate recirculation nitrogen transforms ammonia and nitrogen fluxes in a landfill [Figure 5] [54]. Lema et al. reported on the characteristics of nitrogen and it's removal from landfill leachate [55]. Afterward, in solid waste as well as in wastewater, anthropogenic nitrogen and nitrous oxide emissions also have been reported by Barton and Atwater [56]. Studies showed that the total Kjeldahl Nitrogen (TKN) content of various municipal wastes varies from 1.2 to 4% of total solids in different anaerobic digestion and landfill.

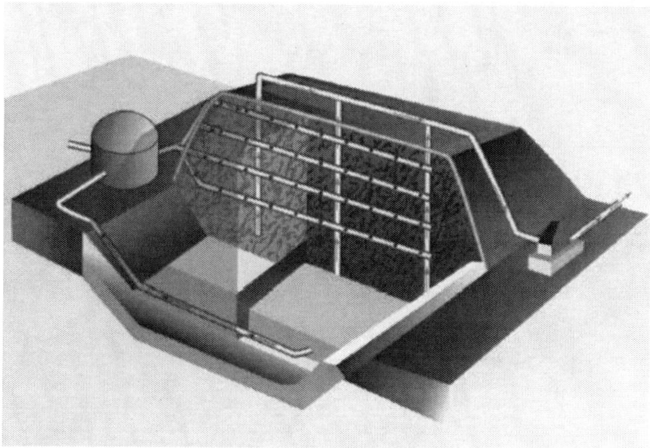

Figure 5. Schematic of bioreactor landfill design [27].

3.3. Biological Approaches for Treatment of Distillery Wastewater

Distillery wastewater usually contains a large amount of dissolved organic and inorganic materials. Several treatment methods have proposed but they are not authenticated much because of the wide variation in the characteristics of distillery wastewater based on different raw materials, agricultural practices, and distillery operating techniques. Color, chemical composition, turbidity and temperature of the effluents also influence the required treatment techniques. The treatment techniques should be implemented in such a way that it gives dual/triple benefits to the industry i.e., they reduce the pollution load as well as generate some useful byproducts and energy.

3.3.1. Different Biological Methods for Distillery Wastewater Treatment

Biological treatments are satisfactory when the ratio BOD/COD is greater than 0.25. They are most suitable as they produce very useful byproducts. Different biological treatments for distillery wastewater are as follows:

3.3.1.1. Aerobic Degradation-Pretreatment

Aerobic systems such as aerated lagoons or activated sludge plants are commonly practiced for biological treatment of distillery wastewater to remove the COD. But these techniques have high operating costs and produce large quantities of waste sludge, which need to be disposed of [57].

Benitez et al. reported that the wastewater generated from wine distillery was purified by combined processes i.e., an aerobic digestion followed by anaerobic digestion. The first stage is a pre-treatment stage along with the second stage named as a performance stage. The pre-treatment stage of the wine distillery wastewater resulted in a significant increase in the methane yield of the following, anaerobic stage. The results of this research also reviewed that single aerobic treatment achieves a potential reduction of the substrate (± 90%) and 66 to 79% removal of the total phenolic compounds [58]. These techniques are also used to reduce the BOD [59].

3.3.1.2. Anaerobic Digestion

Anaerobic digestion is a very typical technique. If the organic content is high (i.e., high BOD) in the effluent, a low-cost treatment adopted in industries is anaerobic digestion. Still, it is the best in treating the generated raw spent wash after alcohol separation [60]. Anaerobic digestion treatment efficiently removes 90% of the COD, 80–90% of the BOD along with the recovery of 85–90% of biogas generated [61].

3.3.1.3. Phycoremediation-Post Treatment

The quality of anaerobically treated spent wash is enhanced by phycoremediation post treatment which further treated the effluent utilizing microalgae. Nowadays microalgae treatment is of great importance.

3.3.2. Decolorization of Distillery Wastewater

In distilleries for ethanol production, molasses is used as a raw material for producing the dark brown color spent wash with a high BOD, COD, low pH and toxic substances as phenols [62]. The source of colors of the distillery wastewater is mainly caramels from overheated sugars, phenolic i.e., tannic and humic acids from the feedstock, furfural from acid hydrolysis and mellanoids from maillard reaction of carbohydrates i.e., sugars with proteins i.e., amino groups [63]. The dark color needs a pretreatment before dumping into the environment. Distilleries spent wash (DS) contains about 2% (w/w) of a dark brown recalcitrant pigment, melanoidin [64].

Several studies reviewed that decolorization of distillery spent wash can be achieved by physico-chemical treatment. Melanoids can be removed by physico-chemical treatment but they need a high dosage of reagents and produce a large amount of sludge [65]. Biological methods are preferable for decolorization and bioremediation of spent wash as they are a competitively cheap alternative to chemical decomposition processes, environment-friendly, and publicly acceptable treatment [65].

A wide variety of aerobic microorganisms are suitable to decolorize spent wash include bacteria, fungi, cyanobacteria, and yeasts.

3.3.2.1. Bacterial Treatment

In the years, microbial treatment with pure bacterial culture has been reported [Table 2]. Isolation of bacterial strains from sewage and acclimatization on increasing concentrations of distillery wastes has been reported by Kumar and Viswanathan. All of these strains reduced COD by 80% in 4-5 days without any aeration. Biomass, carbon dioxide and volatile acids were the major products after the treatment [66].

Table 2. Bacteria employed for the decolorization of distillery wastewater

Name	Color removal (%)	References
Xanthomonas fragariae	76	[67]
Bacillus smithii	35.5	[68]
Lactobacillus hilgardii	40	[69]
Acetobacter acetii	76.4	[70]
Pseudomonas fluorescens	94	[71]
Bacillus thuringiensis	22	[72]
Pseudomonas aeruginosa	67	[65]

3.3.2.2. Fungal Treatment

Currently, a few varieties of *basidiomycetes* and *ascomycetes* type fungi are used in the decolorization of natural and synthetic melanoidin present in distillery wastewaters [Table 3]. They purify the effluent by consuming the organic substances; hence reduce its COD and BOD. Consequently, some valuable products like fungal biomass for protein-rich animal feed or some specific fungal metabolites are also obtained.

Table 3. Fungi employed for the decolorization of distillery wastewater

Name	Color Removal (%)	References
Phanerochaete chrysosporium	85 (free) 59 (immobilized)	[73]
P. chrysosporium	53.5	[74]
Trametes versicolour	80	[75]
Penicillium sp.	30	[76]
Aspergillus niger UM2	80	[77]
Corilus hirsutus	80	[78]
Flavodon flavus	80	[79]
Coriolu sversicolour	63	[80]
Mycelia sterilia	93	[81]
Aspergillus niveus	56	[82]
Coriolus sp. no. 20	80	[83]
Aspergillus - UB2	75	[84]

3.4. Removal of Colors and Toxic Elements from Textile Wastewater Using Microorganisms

Textile wastewater effluent generally contains wide varieties of chemicals with complex structures having aromatic and heterocyclic groups, toxicants, surfactants, salts and different types of dyes like azo dyes which are carcinogenic in nature. Also, these chemicals have high COD and BOD value which is responsible for the depletion of the dissolved oxygen in the reservoir.

Textile processing frequently requires heavy metals and often they are found in textile dyeing wastewater as free ions or in a complex form which contributes to environmental pollution. Heavy elements like chromium are highly toxic for aquatic life. Different chemical, physical and biological processes are being used for the treatment of textile effluents. Chemical and physical methods are not widely popular due to their high costs and pollutant to the environment where the biological approach is environment-friendly as well as make a complete mineralization of the organic pollutants. Biological approach is also attractive due to its diverse metabolic pathways, the versatility of microorganisms and cost-effectiveness. Different microorganisms have the different natural capability to metabolize the harmful pollutants and convert them to more acceptable form which is less toxic to the environment.

3.4.1. Decolorization of Dyes Using Microorganisms

Textile dyes were classified as azo, diazo, cationic, basic, anthraquinone and metal complex group, depending on the nature of their chemical structure [85]. More than 7×10^5

tons of dyestuff produced annually from 100,000 commercially available dyes [86]. On exposure to light, water, and many chemicals the chemical structure of the dyes was remain unchanged [87].

The azo dye is the most extensively applied dye in the textile industries. Due to the emergence of the aromatic ring in the molecule, the azo dye is considered recalcitrant. It causes environmental damage, like the interference in the absorption and reflection of light in the aquatic environment, and can be carcinogenic and mutagenic, when it is inserted into the environment [88].

Beside azo dye, many other dyes were known as carcinogens, such as benzidine and other aromatic components all of which might be reformed as a result of microbial metabolism. It has already been well documented that azo and nitro-compounds were decreased in the sediment and intestinal environment, resulting in the regeneration of the parent toxic amines, though, this compound was not readily removed by typical microbial based waste treatment processes. The highest rates of toxicity were found for basic, diazo and direct dyes [89].

A wide variety of microorganisms (bacteria, fungi, yeasts, algae etc.) are capable of decolorization of a wide range of dyes and competent to completely mineralize many azo dyes under certain environmental conditions.

3.4.1.1. Decolorization with Fungi

Fungus is capable of degrading several dyes, including azo dyes; also used for oxidizing different chemical compounds and mineralizing persistent aromatic pollutants. The power of fungus to demote a diverse group of compounds depends upon the nonspecific ligninolytic enzymatic system containing manganese peroxides, lignin peroxides, and laccase. The degradation and mineralization of dyes by white rot fungi was widely studied. The fungus *Geotrichum* sp. CCMI 1019 has the ability to decolorize large amounts of dyes like reactive red 158, reactive black 5, and reactive yellow 27 [90, 91].

In the case of *Phanaerochaete chrysosporium* a ligninolytic fungus pre-adapted to the medium and degrade the azobenzene constitute of azo dye for aromatic ring cleavage [92]. Textile industry wastewater has been decolorized by the production of, lignin peroxidase (Lip), manganese-dependent peroxidase (MnP) and laccase. These are the enzymes of white rot fungus *Irpex lacteus* in stationary cultures without adding any chemicals [93].

Most scientific studies on an azo dye biodegradation have focused on the fungal cultures mainly belonging to white rot fungi which were used to develop bioprocesses for mineralization of azo dyes. The *Phanerochaete chrysosporium* is the most widely studied of white-rot fungi, as well as *Bjerkandera adusta*, *Pleurotus*, *Trametes* (*Coriolus*) *versicolor*, *Aspergillus* and *Phlebia* species, and variety of other isolates, are also studied. *Penicillium chrysosporium* lignin peroxidase is also responsible for decolorization of dyes and Laccase activity for the decolorization is found in *T. versicolor*. However, in most cases, enzymatic degradation is the major mechanism.

For the removal of a dye such as Sulfur black from textile effluent, a fungal isolated from textile industry *Aspergillus terreus* SA3 was efficiently used. The removal of azo dyes from aqueous solution by *Schizophyllum commune* (50 μm concentration) showed 97.57% in Methyl orange, 96.86% in Congo red and 97.40% in Erichrome black-T color removal on the fourth day respectively. *Lenzites eximia* decolorized methyl orange by 94.79%, congo red by 95.50%, and Erichrome black-T by 95.36% at 50 μm concentration respectively on the fourth day [94].

3.4.1.2. Decolorization with Yeast

A few works were followed to exploring the decolorization ability of yeast, mainly studied for the biosorption. Yeasts, such as *Klyveronyces marxianus* have the capacity to decolorize dyes [95]. Actively growing *Kluyveromyces marxianus* IMB3 yeast cells almost completely decolorize remazol black B within 24 hours at various acidic pH values under aerobic conditions [96]. Two novel dye decolorizing yeast strains of *Candida krusei* G-1 and *Pseudo zymarugulosa* Y-48 exhibited excellent color removal of reactive azo dye. *Saccharomyces cerevisiae* MTCC463 effectively decolorize methyl red due to biodegradation and it shows a different rate of metabolism at different pH with the involvement of azo reductase [97].

3.4.1.3. Decolorization with Algae

Color removal by algae is due to three intrinsically different mechanisms i.e., CO_2 and H_2O transformation of colored molecules to non-colored molecules, assimilative usage of chromophores for the production of algal biomass and adsorption of chromophores on algal biomass.

Several species of *Oscillatoria and Chlorella* were capable of degrading azo dyes to their aromatic amines and further metabolize the aromatic amines to simpler organic compounds, some were even capable of utilizing azo dyes as their sole carbon and nitrogen source [98]. The biodegradation of azo dyes by the algae (*C. vulgaris, Chlorella pyrenoidosa,* and *Oscillatoria tenuis*) has also been assessed. Moreover, the algae can play an outright role in the degradation of azo dyes [99]. *Chlorella vulgaris* have biosorption capacity for several reactive dyes [100]. Dried *Spirogyra rhizopus* have the ability to decolorize acid red 274 dye by both biosorption and bio coagulation process and the removal amounts decreased while the removed concentration of AR 274 dye increased with increasing *S. rhizopus* concentration [101].

For biological treatment of triphenylmethane dye and malachite green, the potential of *Cosmarium* sp. belonging to green algae was investigated as a viable biomaterial [102]. Immobilized thermophilic cyanobacterial strain *Phormidium* sp. has good decolorization activity under thermophilic condition [103].

3.4.1.4. Decolorization with Bacteria

Bacterial biodegradation of textile dyes is a winsome and inexpensive method. Adaptation of existing strain or isolation of new strains for decomposition of dyes will increase the efficiency of bioremediation. The mechanism involved in color revocation by bacteria was absorption or adsorption. The effectiveness of a biological treatment system was greatly affected by the operational parameters such as temperature, pH, dye concentration and nutrients. The influence of each parameter on the color revocation process must be optimized to increase dye reduction.

Citrobacter sp. decolorizes several recalcitrant triphenylmethane and azo dyes by adsorption mechanism [104]. Reactive orange 16 was effectively degraded by *Enterococcus faecalis YZ* 66. Biodegradation of acid orange 7 is accelerated by the availability of organism in the surfaces that have expedient pores to permit microbial growth [105]. In recent years the biodegradation/decolorization potential of bacteria has come into the spotlight. An effluent adapted bacterium was found to be more capable in decolorizing the effluent containing reactive black B dye than the effluent non adapted bacteria [106].

Two bacterial strains *Bacillus pumilus* (KEB-10) and *Bacillus cereus* (KEB-7) decolorize indigo dye was reported by Khelifi et al. [107]. Degradation and decolorization of orange II containing textile effluent with biodigester sludge from an endemic municipality waste treatment plant using sulfate-reducing bacteria was a cheaper alternative way [108]. Decolourization of textile azo dyes by *Staphylococcus arlettae* strain VN-11 by sequential microaerophilic/aerobic process was shown to be very useful [109]. Under aerobic and anaerobic conditions bacterial strains are able to decolorize azo dyes [110-112]. Many researchers applied sulfate-reducing bacteria (SRB) for the textile effluent bioremediation for decreasing sulfate [113].

Researchers were able to successfully decolorize four different azo dyes under microaerophilic conditions by 97% by using a facultative *Staphylococcus arlettae* bacterium [109]. Effluent adapted strains of *Acinetobacter, Bacillus,* and *Legionella* was found to have potentials to remove color. However, in chemical oxygen demand (COD) removal activities strains of *Acinetobacter, Bacillus,* and *Pseudomonas* have found to present strong activity [114].

The ability of *Brevibacterium* sp. (strain VN-15) to decolorize four azo dyes C.I. Reactive Red 198 (RR198), C.I. Reactive Yellow 107 (RY107), C.I. Direct Blue 71 (DB71) and C.I. Reactive Black 5 (RB5) was evaluated in a static/agitated sequential batch process [115]. The bacterial strain *Pseudomonas stutzeri* ETL-4 (isolated from Common Effluent Treatment Plant) has been reported to decolorize the dye (Congo red) at a concentration of 100 mg/L up to 94% within 24 h in static conditions. The pH and temperature for optimum activity and growth of the isolate were found as 7.0 and 37 °C, respectively [116]. Near to a textile industrial outlet, a novel strain was separated having a high capacity for rapid decolorization of Remazol Black-B as an azo dye. The isolate was recognized as Pseudomonas putida ETL-7. The bacterium exhibited a remarkable color removal capability over a wide range of dye concentration (50-200 mg/L). Colorless cells of *P. putida* and UV Visible spectroscopic analyses suggested that the decolorizing activity only through biodegradation, not by inactive surface adsorption [117]. A reactive group of azo dye named C.I: Reactive Red was established to be degraded using Pseudomonas spp. strain MPS-2 that can degrade azo dyes to yellowish colored intermediates of aromatic amines and/or even to partially mineralize them into the simpler compound. This is a common metabolic in an organism, which is safe and less toxic to the environment [118]. A useful bacterial strain, *Paracoccus* sp. GSM 2, exhibited complete decolorization of Reactive Violet 5 (100mg/L) within 16 h. However, under the static condition, it could maximally decolorize 800 mg/L of dye within 38 h with 73% removal [119].

3.4.1.5. Decolorization with Actinomycetes

Decolorization of reactive dyes, including anthraquinone, azo and phthalocyanine by using *actinomycete* strains has been reported by many researchers. In this process, adsorption of dyes is done without any degradation to the cellular biomass. Some Cu based azo dyes like formazon-copper based complex dyes are completely decolorized through degradation by same actinomycete strains [120]. *Streptomyces ipomoea* decolorized azo dyes orange II by the production of salt-resistant laccase under versatile pH [121].

3.4.2. Mechanism of Decoloration of Textile Effluents Using Enzymes [122]

Azo compounds can biologically degrade under both aerobic and anaerobic conditions. The color removal process under anaerobic conditions is also named as dye reduction where the biochemistry of azo dye reduction is investigated by most literature. Theoretically, in anaerobic conditions, decolorization of azo dyes is a simple and non-specific process where fission of azo bond produces some simple aromatic amines. The potency of various anaerobic treatments for degradation of many varieties of synthetic dyes has been denoted many times in many kinds of literature.

The actual mechanism of degradation under anaerobic conditions begins with the reduction in the cleavage of the azo bonds (–N=N–) using azo reductase. Consequently, four-electrons i.e., reducing equivalents transfer proceeds through two stages (i.e., in each stage two electrons are transferred) at the cleavage of the azo dyes which act as a final electron acceptor. This results in dye decolorization and the forms the colorless solutions.

$$(R_1 - N = N - R_2) + 2\,e^- + 2\,H^+ \rightarrow R_1 - HN - NH - R_2 \quad (5)$$

$$(R_1 - HN - NH - R_2) + 2\,e^- + 2\,H^+ \rightarrow R_1 - NH_2 + R_2 - NH_2 \quad (6)$$

3.4.3. Degradation of Toxic Elements from Textile Effluent

Textile effluents are complex and contain different varieties of pigments, colorants, dyes and other products such as a dispersant, acids, alkalis, salts, humectants, oxidants, heavy metals like Ni, Zn, Cr; detergents etc. All of these colored effluents are discharged into the rivers and lakes, lowering the concentration of dissolved oxygen, thus creating an anoxic environment that is lethal to resident organisms.

Major contaminants present in the effluents from the textile industry are phenol and chromium and these can simultaneously be degraded by using indigenous microorganisms such as *P. putida* MTCC (1194) and *E. coli* NCIM (5051). These microorganisms also help to decrease the content of the other heavy metals present in the effluent to become below their permissible limits.

4. Soil Decontamination Using Enzymes

Several highly polluting substances are present in the environment and affect the soil and sediments. Different organic pollutants possibly encountered in the soil are pesticides, dyes, phenols, chlorophenols, nitro compounds, N and P derivatives, choroanilines, plastics, biopolymers, etc. and inorganic pollutants are arsenic, copper, lead, cyanide, chromium, etc. [123].

4.1. Advantages of Enzymatic Approach for Soil Decontamination over the Conventional Methods

Conventional methods of soil decontamination are expensive and show limited efficiency along with some unwanted side-effects such as demand for prolonged treatment time, the formation of hazardous by-products, and unhealthy working conditions for people. Current remediation approaches usually involve the removal of masses of polluted soil from the area by mechanical means which is unmanageable, transportation of this unmanageable mass to a treatment area and incubation for days or weeks in biological reactors for a prolonged time or treated by the physicochemical techniques or incineration. Due to many expensive steps, it is expected to use enzymes where the remediation process could be conducted *in situ*, within a few hours, and at a considerably reduced cost [124].

However, due to the oxido reductase activity [125] of numerous plants, they were subjected as a source of crude enzymes in the remediation process [126]. Plant materials like peroxidase (horseradish peroxidase is mostly studied) have been used in remediation of contaminated soil. Recently as solid crude, the peroxidase is extracted from the potato tubers showed significant enzymatic activity [124]. Here plants are cut into small pieces in order to maximize the enzyme contact area with contaminated soil and further enzyme isolation step is not necessary [124].

An important group of enzymes is formed by oxidative enzymes [127, 128] which plays a predominant role in soil decontamination. Currently, another alternative approach to decontaminate the soil is proposed with the polymerized chlorinated phenols and anilines binding to humus [129]. Binding and polymerization both can be attained by the reaction of oxidative coupling which is further controlled by applying different types of enzymes. Due to the controlled release of the detoxified pollutants from the polymers and humus by means of decontamination, the oxidative coupling process is considered safe [130].

4.2. Mechanisms of Soil Decontamination Using Oxidative Enzymes

The detoxification of polluted environment using oxidative enzymes can be done by (Figure 6) several steps starting with the formation of humus materials in the soil, in the space between plants and soil through degradation and by synthetic mechanisms.

They can also be involved in the reactions between humus constituents and xenobiotic molecules and form bound residues in water systems and humus materials in soil (Figure 6a). Their efficiency in the detoxification relies on their ability to catalyze the polymerization reaction of toxic compounds (alone or, by cross-reaction) with other phenolics or with co-substrates (toxic/harmless); and production of polymeric products (dimers, trimers, hybrid oligomers), which readily accumulate in soil and/or in water systems (Figure 6b).

4.3. Limitations of Using Enzymes in Soil Decontamination [123]

- If the product of the enzyme-mediated reaction is less toxic than the substrate i.e., pollutant then the use of enzyme is profitable.
- Only specific microorganisms are suitable for decontamination where many enzymes needed to act consequently i.e., the detoxification process demands a multistep process.
- The use of cofactors with the enzyme may be problematic unless it was prepared correctly.
- The rapid degradation of the free enzyme by proteases released by soil microorganisms may occur during using enzymes to detoxify organic-polluted soil.

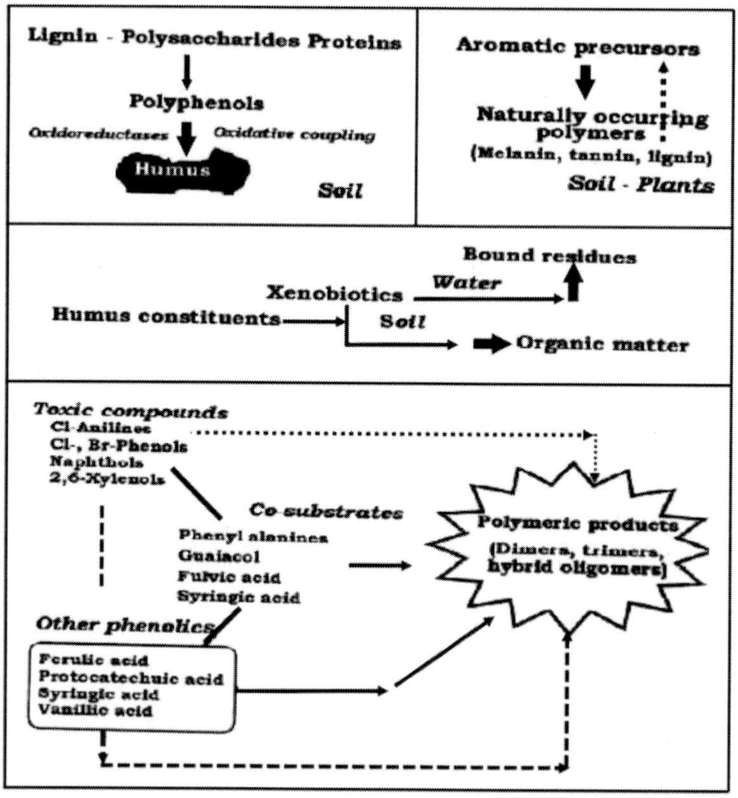

Figure 6. Role of oxidative enzymes in soil decontamination a) Involvement of oxidative enzymes in synthetic and degradation mechanisms occurring in soil, plant, and water; b) detoxification by oxidative enzymes by polymerization and cross-reaction of different pollutants [123].

5. Removal of Toxic Chemicals from the Environment Using Microbial Catalyst

Bioremediation is a procedure in which pollutants or contaminants are converted into less hazardous or nonhazardous compounds with the aid of various microorganisms. A number of enzymes secreted from bacteria, fungi, algae take part in the bioremediation of toxic organic pollutants. Bioremediation is an economically feasible and eco-friendly process. Although most bioremediation process occurs under aerobic conditions, some anaerobic environments also help bioremediation to occur [Figure 7] [131].

5.1. Microbial Enzymes Used in Bioremediation

Various microbial enzymes take part in effective bioremediation process leading to the formation of innocuous compounds and thereby help to develop a sustainable environment.

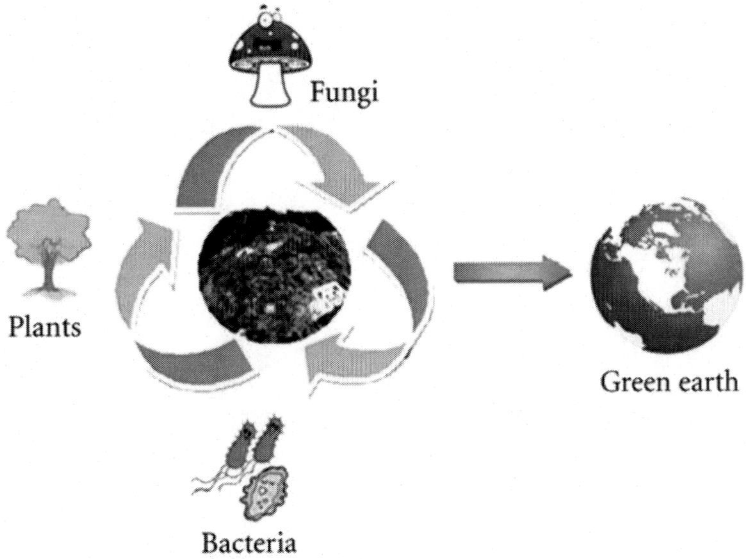

Figure 7. The process of waste bioremediation [132].

5.1.1. Microbial Oxidoreductases

Microbial oxidoreductase enzymes take part in the rupture of chemical bonds in toxic organic compounds [133].

5.1.1.1. Microbial Oxygenases

There are two types of oxygenases, monooxygenases, and dioxygenases. A wide range of compounds can be ruptured by oxygenases. Aromatic rings undergo cleavage when O_2 atoms are introduced into organic molecules [Figure 8] [131].

5.1.1.2. Monooxygenases

Monooxygenases catalyze various reactions e.g., ammonification, dehalogenation, desulfurization, denitrification, biotransformation, and biodegradation of various organic compounds. In oxidative dehalogenation reactions monooxygenase catalyzes under oxygen-rich conditions and in reductive dechlorination, it works under low oxygen levels. On the other hand, methane monooxygenase enzyme degrades various hydrocarbons like substituted alkanes, cycloalkanes, heterocyclic hydrocarbons etc. [134, 135].

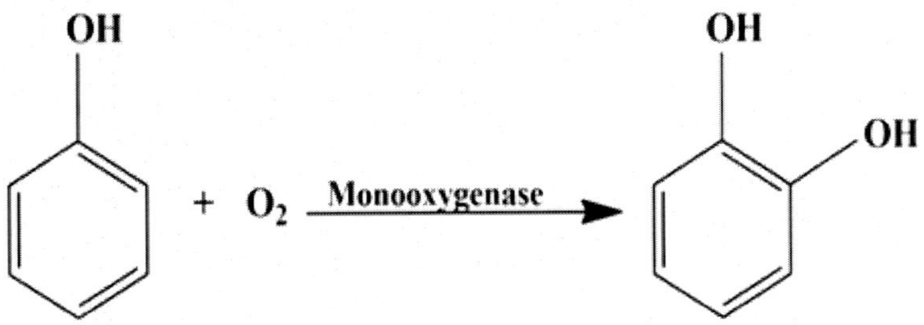

Figure 8. Degradation of the aromatic compound by monooxygenase [136].

5.1.1.3. Microbial Dioxygenases

Dioxygenases molecular oxygen into their substrate catalyzes enantio specifically the oxygenation of a wide range of substrates [Figure 9]. While the extradiol cleaving enzymes utilize Fe (II) and Mn (II), the intradiol cleaving enzymes utilize Fe (III).

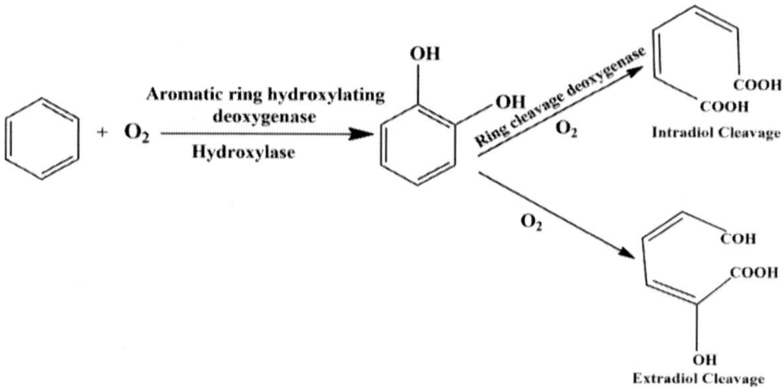

Figure 9. Degradation of aromatic compounds by dioxygenase [131].

5.1.2. Microbial Laccases

Intra and extracellular laccases are produced by various microorganisms and these enzymes are capable of catalyzing various compounds such as aminophenols, ortho and para diphenols, polyphenols, lignins, polyamines, and sometimes inorganic ions [128, 137].

Phenolic and methoxyphenolic acids are not only oxidized by laccases but also decarboxylated [Figure 10]. Oxidoreductase enzymes represented by laccases have potential biotechnological and bioremediation applications [133].

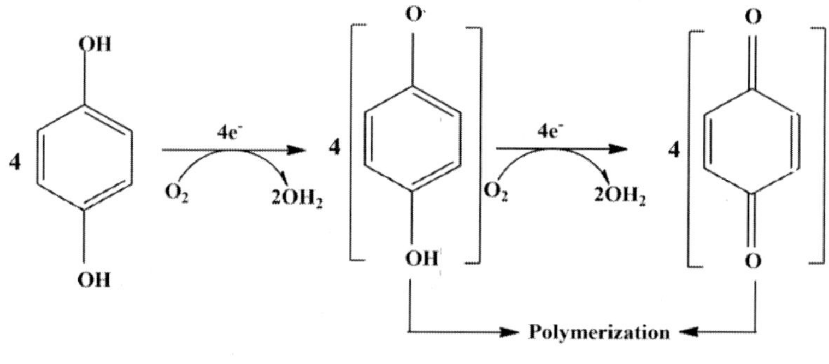

Figure 10. General reaction mechanism for phenol oxidation by laccase [138].

5.1.3. Microbial Peroxidases

Lignin and other phenolic compounds are oxidized by peroxidases in the presence of a mediator by using hydrogen peroxide (H_2O_2). Peroxidases can be of two types: haem and nonhaem proteins. Different types of microbial peroxide enzymes are as follows:

5.1.3.1. Microbial Lignin Peroxidases

White rot fungus secrete lignin peroxidases during secondary metabolism. Lignin peroxidase (LiP) degrades lignin and other phenolic compounds in the presence of co-substrate

H_2O_2 and mediator like veratryl alcohol. Aromatic compounds which have redox potentials higher than 1.4V(NHE) can be oxidized by LiP by single electron abstraction [139].

5.1.3.2. Microbial Manganese Peroxidases

Lignin degrading basidiomycete's fungus can secrete MnP, an extracellular heme enzyme that oxidizes Mn^{2+} to the oxidant Mn^{3+} in a multistep reaction. This Mn^{3+} can act as a mediator in the oxidation process of various phenolic compounds.

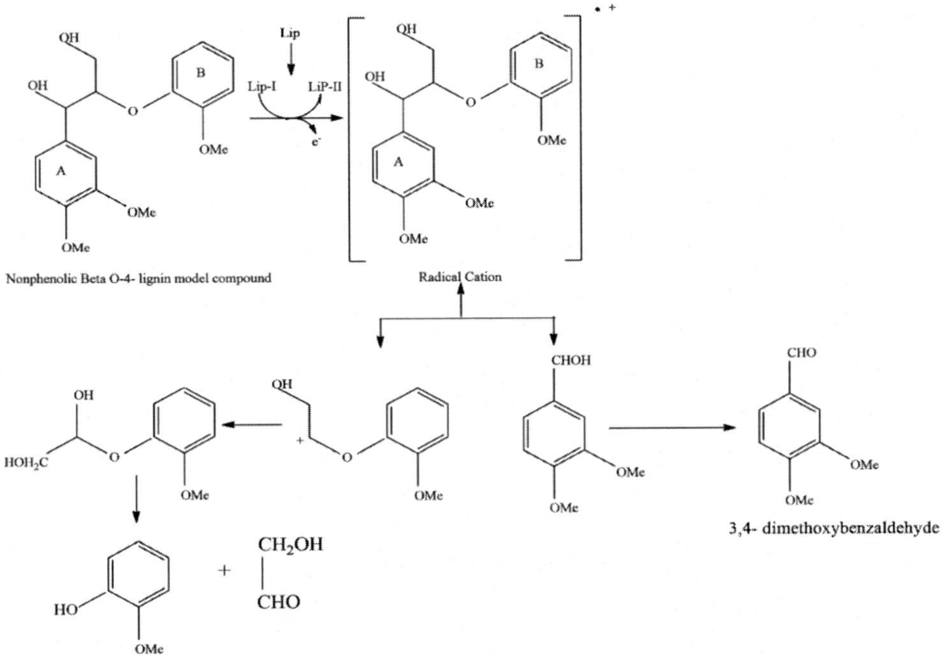

Figure 11. Lignin peroxidase (LiP)-catalyzed oxidation of non-phenolic β -O-4 lignin model compound [140].

5.1.3.3. Microbial Versatile Peroxidases

Like MnP, LiP, and horseradish peroxidase VP can oxidize Mn^{2+}, methoxybenzenes, phenolic aromatic substrates and nonphenolic lignin model dimers directly in the absence of manganese [Figure 11]. Therefore, in the industry for biotechnological applications, a highly efficient VP overproduction system is desirable [140, 141].

5.1.4. Microbial Hydrolytic Enzymes

Chemical bonds in the oil spill and organophosphate and carbamate insecticides can be effectively disrupted by hydrolytic enzymes. Condensations and alcoholysis reactions are also catalyzed by hydrolases. Extracellular hydrolytic enzymes like amylases, proteases, lipases, DNAses, pullulanases, and xylanases have quite diverse potential applications in different areas such as food industry, feed additive, biomedical sciences, and chemical industries [142]. The hemicellulase, cellulase, and glycosidase are also very important for their application in biomass degradation [143].

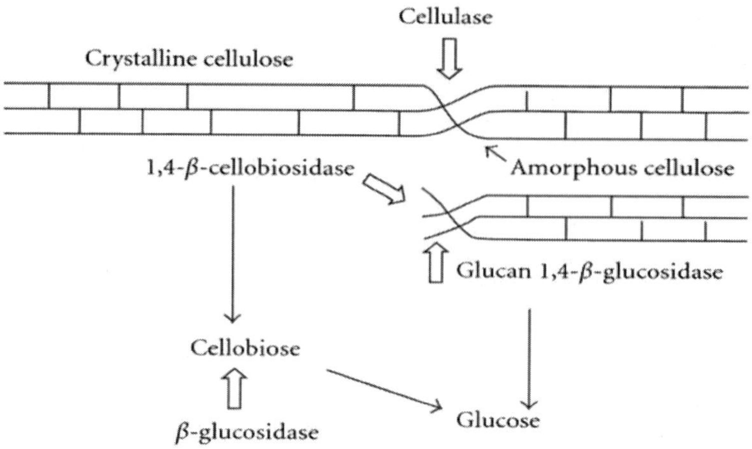

Figure 12. Proposed mechanism for the hydrolysis of cellulose by the fungal cellulose enzyme system [143].

5.1.4.1. Microbial Lipases

Lipases can degrade lipids obtained from a large variety of microorganisms, animals, and plants. They also catalyze many reactions like hydrolysis, interesterification, esterification, aminolysis, and alcoholysis [144].

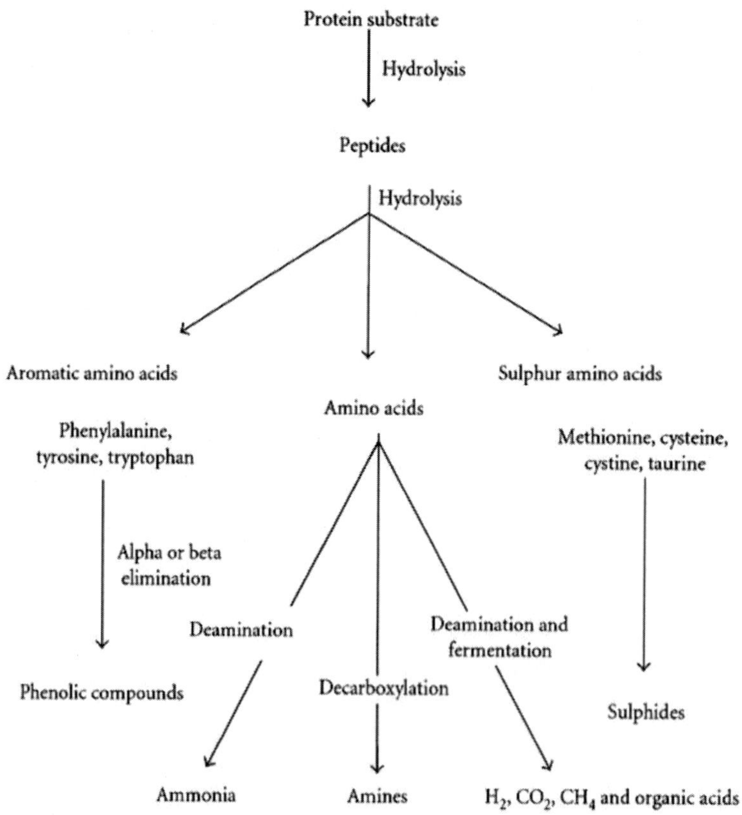

Figure 13. Proposed pathway for protease hydrolysis [132].

5.1.4.2. Microbial Cellulases

In the hydrolysis process three major groups of cellulases are involved (1) endoglucanase (EG, endo-1, 4-Dglucanohydrolase), (2) exoglucanase or cellobiohydrolase (CBH, 1, 4-b-D-glucan cellobiohydrolase) and (3) β-glucosidase [Figure 12]. Cellulases degrade cellulose into reducing sugars during the enzymatic hydrolysis process. The reducing sugars are further fermented by yeasts or bacteria to ethanol [145].

5.1.4.3. Microbial Proteases

The breakdowns of the protein based substance which enter atmosphere are hydrolyzed by proteases [Figure 13]. In an aqueous environment, proteases hydrolyze peptide bonds and in non-aqueous environment synthesizes them.

6. Enzyme Treatment to Decrease Solids and Improve Digestion of Primary Sewage Sludge

Enzymatic pre-treatment process can lead to an overall decrease in solids and improvement of digestion of primary sewage sludge [146]. An experiment carried out at mesophilic temperature (37 °C) showed that there was 11.5% and 10.6% decrease in volatile solids (VS) with 50% digested sludge and 0.1% enzyme dosage of carezyme and the mixed enzymes.

In the activated sludge process microbial cells undergo death or lysis. During this process, the cell contents (substrates and nutrients) are usually released into the medium and deliver an autochthonous substrate. This substrate is finally used in microbial metabolism. A portion of the carbon was released as products of respiration and microbial metabolism results in reduced overall biomass production [147]. Some enzymes that can accelerate sludge hydrolysis are explained below:

6.1. Lysozyme

Lysozyme is also known as muramidase or N-acetylmuramide glycanhydrolase which can dissolve polysaccharides as substrates in the bacterial cell wall. It breaks β-1,4 glycosidic bond between N-acetyl glucosamine and N-acetylmuramic acid in the bacterial cell wall. The dissolution of the insoluble mucopolysaccharide in the bacterial cell wall occurs into soluble glycopeptides. This enzyme also degrades organic molecules.

6.2. Protease

Proteases, abundant in activated sludge, exist in the intracellular space and the external EPS of living microorganisms. Protease has the ability to catalyze the hydrolysis of peptide bonds in proteins. As a result, peptide chains and amino acids are produced. In this way, the protease can significantly and efficiently reduce total sludge [148-150].

6.3. Lipase

Usually oil refineries and some restaurants waste have extremely high lipid content. Lipase, widely distributed in microorganisms, can catalyze these type of triacylglycerols and other water insoluble esters of hydrolysis, esterification, and alcoholysis, as well as trans-esterification [151, 152].

6.4. Cellulase

Large amounts of cellulose are highly present in the activated sludge obtained from the paper and textile industries. Cellulase is mainly composed of β-glucosidase, endoglucanase, and exoglucanase (CBH). Filamentous fungi are the main source of cellulose [148, 152-159].

7. Contributions of Microbial Enzymes to Improve Agricultural Productivity

In recent years, there is a big concern about the excessive use and misuse of chemical fertilizers and pesticides, especially in the developing countries. Because these are harmfully affecting the environment and creating many problems related to food safety and quality and also human and animal health. Therefore, there has been a recent interest in growing natural farming and organic agriculture using organic fertilizer and pesticides as well as applying the principles of natural ecosystems, both in developed and developing countries. At the same time, agricultural researchers have developed beneficial microorganisms for alternative agriculture and sustainable agriculture system [160-162].

As an alternative agrochemicals the demand for biological based organic fertilizers are increasing [163]. The use of organic fertilizer improve nutrient supply to the soil and consequently preserve the field management [164]. Generally, fertile and prolific soils have large populations of various microorganisms (i.e., both species and genetic diversity) and a high content of organic matter as well as a wide ratio of beneficial to detrimental microorganisms [165].

7.1. Functions of Beneficial Microorganisms in Agricultural Sectors

Soil microorganisms have an important influence on soil fertility and plant health [166]. Organic/natural farming includes the concepts and practices comprise the use of beneficial microorganisms to improve soil health [167]. For example, the natural microflora of the soil containing all kinds of useful bacteria and fungi (i.e., *arbuscular mycorrhiza* fungi (AMF) or plant growth promoting rhizobacteria (PGPR)) [168]. In organic farming, biofertilizers are commonly used which can recover and improve the soil environment rich in various types of micro and macro-nutrients. Soil environment is improved by biodegradation of organic matter in the soil, nitrogen fixation, phosphate and potassium mineralization, the release of plant growth regulating substances, and production of antibiotics [169]. Moreover, if biofertilizers are used as seed or soil inoculants, they multiply and participate in nutrient cycling and assistance in crop productivity [170]. Generally, plants take up 10 to 40% of the total used fertilizer and the remaining 60 to 90% is lost. Therefore, microbial inoculants have vital importance in integrated nutrient controlling systems to sustain agricultural productivity as well as the healthy environment [171]. Again, organic farming confirms food safety and biodiversity of cultivable soil [172]. Furthermore, biofertilizers increase shelf life causing no adverse effects to the ecosystem [173]. To summarize, the main functions of beneficial microorganisms in agricultural sectors are fixation of atmospheric nitrogen; decomposition of organic wastes and residues; recycling and increased availability of plant nutrients; degradation of toxicants including pesticides; production of antibiotics and other bioactive compounds; solubilization of insoluble nutrient sources; production of polysaccharides to improve soil aggregation and so on.

7.2. Mechanisms of Bacterial Plant Growth Promotion

The potential mechanisms of bacterial plant growth promotion are to provide nutrients to plants which they lack like fixed nitrogen, iron, and phosphorus.

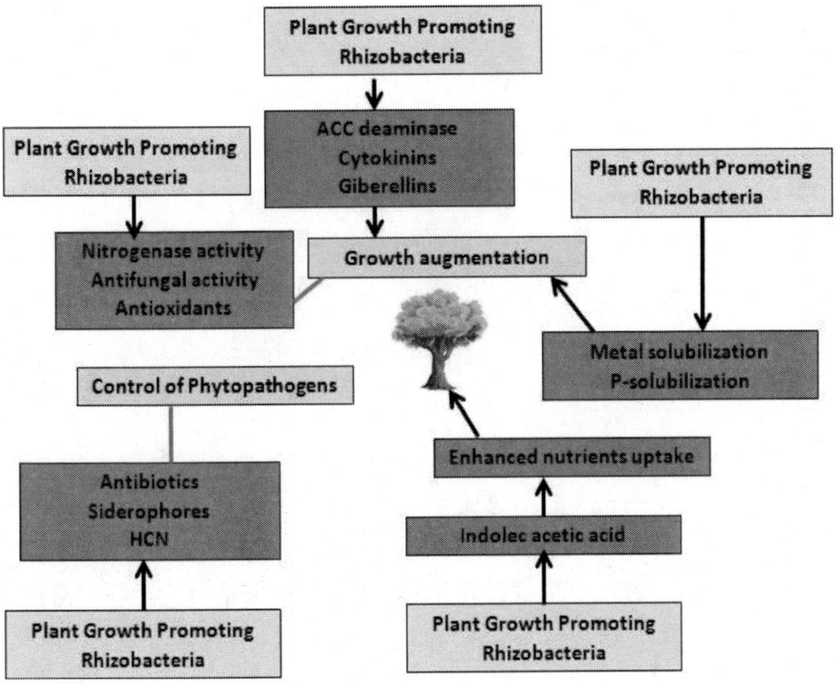

Figure 14. Mechanism of plant growth promotion by rhizobacteria [174].

7.2.1. Nitrogen Fixation

Some free living bacteria (*Azospirillum spp.*, *Rhizobia spp.* etc.) [Table 4, Figure 4 [175] can provide a little amount of fixed nitrogen which required by the bacterially associated host plant [176]. However, more nitrogen fixation can be achieved by nitrogenase genes such as structural genes (for activation of the Fe protein, iron-molybdenum cofactor biosynthesis, and electron donation) and regulatory genes (for the synthesis and function of the enzyme). In nitrogen-fixing bacteria (diazotrophic), nitrogenase genes are usually found in a cluster of about 20–24 kb with seven operons encoding 20 dissimilar proteins.

7.2.2. Phosphate Solubilization

The limited bioavailability of phosphorus from the soil often limits plant growth. Therefore, solubilization and mineralization of phosphorus using phosphate solubilizing bacteria is an important characteristic in plant growth promoting bacteria (PGPB) and in plant growth promoting fungi (e.g., mychorrizae) [177, 178]. In general, the solubilization of inorganic phosphorus happens by the action of low molecular weight organic acids (gluconic and citric acid), which are synthesized by different types of soil bacteria [Figure 15] [178-180]. On the other hand, the synthesis of various phosphatases, catalyzing the hydrolysis of phosphoric esters can create mineralization of organic phosphorus [178]. Interestingly, in the same bacterial strain, phosphate solubilization and mineralization can coexist [181].

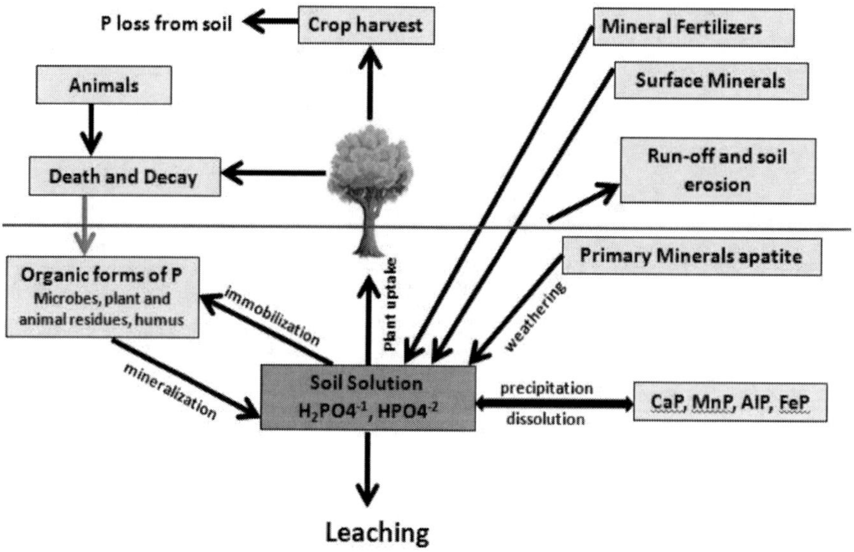

Figure 15. Movement of phosphorus in soil [174].

7.2.3. Sequestering Iron

Iron is very important for both microorganisms and plants and they need a high level of it. In the rhizosphere, plant, bacteria, and fungi compete for iron and getting enough iron is problematic for them [182, 183]. Therefore, bacteria synthesize low molecular mass siderophores (400–1500 Da), molecules with a very high affinity for Fe^{3+} (K_a ranging from $10^{23}$to 10^{52}) as well as membrane receptors able to bind the Fe-siderophore complex, to persist with such an inadequate supply of iron. By this way, iron uptake to plant is facilitating by microorganisms [184, 185]. The chemical structures of 250 siderophores have been determined over 500 known of it [184].

Table 4. Different plant growth promoting rhizobacteria tested for various plant types

Plant growth promoting Rhizobacteria	Plant	Results after the addition of bacteria	References
Pseudomonas putida, Azospirilium, Azotobacter	Artichoke *(Cynara scolymus)*	Phosphate solubilizing bacteria together with nitrogen-fixing bacteria can increase radicle and shoot length, shoot weight, the coefficient of the velocity of germination, seedling vigority index significantly, and a decrease in mean time of germination.	[186]
Pseudomonas sp. A3R3	*Alyssum serpyllifolium, Brassica juncea*	Significantly increased the biomass (*B. juncea*) and Ni content (*A. serpyllifolium*) in plants, which grow in Ni-stressed soil.	[187]
Psychrobacter sp. SRS8	*Ricinus communis, Helianthus annuus*	Increased plant biomass, chlorophyll, and protein content as well as stimulated plant growth and Ni accumulation in plant species.	[188]
Rhizobium phaseoli	*Vignaradiata L.*	*Rhizobium* significantly mitigated the adverse effects of salinity and enhanced the plant height, number of nodules per plant, plant biomass, grain yield, and grain nitrogen concentration in the presence of tryptophan.	[189]
Bacillus species PSB10	Chickpea *(Cicer arietinum)*	Significantly improved growth, nodulation, chlorophyll, leg hemoglobin, seed yield and grain protein and at the same time reduction of the uptake of chromium in roots, shoots and grains occurs.	[190]
Bradyrhizobium sp. 750, Ochrobactrum cytisi, Pseudomonas sp.	*Lupinus luteus*	Increased biomasses, nitrogen content, and accumulation of metals.	[191]
Paenibacillus Polymyxa	Pepper	enhanced the biomass of plants and elicited induced systemic resistance against bacterial spot pathogen *Xanthomonas axonopodis pv. Vesicatoria* untreated plants.	[192]
Pseudomonas fluorescensstrain R-93, P. putida strain *R-168*	Maize *(Zea mays L.)*	Enhancement of plant height, seed weight, number of seed per ear and leaf area, shoot dry weight occurs sufficiently.	[193]
Azospirillum amazonense	Rice *(Oryza sativa L.)*	Grain dry matter accumulation (7–11.6%), the number of panicles (3–18.6%), and nitrogen accumulation at grain maturation (3.5–18.5%) enhanced.	[194]
Pseudomonas species	Rice *(Oryza sativa)*, maize *(Zea mays)*	Pseudomonad isolated from rice showed better ability to control bacterial and fungal root pathogens than that of maize.	[195]

Table 4. (Continued)

Plant growth promoting Rhizobacteria	Plant	Results after the addition of bacteria	References
Pseudomonas aeruginosa strain MKRh3	Black gram	It showed enhanced plant growth, lessened cadmium accumulation, and extensive rooting.	[196]
P. aeruginosa	Indian mustard and pumpkin	Stimulated plant growth, reduced Cd uptake	[197]
Azospirillum brasilense Sp245	Common bean	Root growth increased	[198]
Bacillus sp., Paenibacillus sp.	Rice	The root and shoot growth increased significantly.	[199]
Azospirillum lipoferum, Azotobacter chroococcum	Cotton *(Gossypium hirsutum L.)*	Seed yield (21%), plant height (5%) and microbial population In soil (41%) enhanced over their respective controls while boll weight and staple length.	[200]
Rhizobium sp. RP5	Pea *(Pisum sativum)*	The dry matter, nodule numbers, root N, shoot N, leg hemoglobin, seed yield and grain protein increased by 19%, 23%, 26%, 47%, 112%, 26%, and 8%, respectively, at 290 mg Ni/kg.	[201]
Xanthomona ssp. RJ3, Azomona ssp. RJ4, Pseudomonas sp. RJ10, Bacillus sp. RJ31	*Brassica napus*	Plant growth stimulated and cadmium accumulation enhanced.	[202]
Ochrobactrum, Bacillus cereus	Mungbean	By reducing Cr (VI) to Cr (III), the toxicity of chromium to seedlings decreased.	[203]
Pseudomonas jessenii PS06, Mesorhizobium ciceri C-2/2	Chickpea *(Cicer arietinum)*	The seed yield (52%) enhanced by the co-inoculation treatment compared to the uninoculated control treatment and nodule fresh weight.	[204]
Pseudomas fluorescens	Soybean	Increased plant growth	[205]
Unidentified PGPR isolate	Wheat	In inoculated wheat seedlings, enhancement in root elongation (up to 17.3%), root dry weight (up to 13.5%), shoot elongation (up to 37.7%) and shoot dry weight (up to 36.3%) occurs.	[206]

8. The Removal of Heavy Metal Pollution from the Environment using Microorganisms

Metals are indispensable for biological systems and must be present within a certain concentration range. Because of their indisputable nature, at high concentrations metals can act in a deleterious manner by blocking essential functional groups, displacing other metal ions, or modifying the active conformation of biological molecules. For that reason, heavy metal pollution in soils and aquatic environments is considered a severe ecological problem.

Heavy metal contamination means the excessive deposition of toxic heavy metals in the soil instigated by human activities. Heavy metals in the soil include mercury (Hg), cadmium (Cd), lead (Pb), chromium (Cr), arsenic (As) etc. Other heavy metals of certain biological toxicity include zinc (Zn), copper (Cu), nickel (Ni), stannum (Sn), vanadium (V) and so on. There is a number of applications such as basic engineering works, paper and pulp industries, leather tanning, petrochemicals, fertilizers, etc. which result in heavy metal contamination.

Conventional physicochemical methods such as electrochemical treatment, ion exchange, precipitation, reverse osmosis, evaporation, and sorption for heavy metal removal from waste streams are cost effective. The biological approach has also been considered as another remediation for heavy metal contamination that is environment friendly too. Due to their better performance, low cost and large availability, a huge array of biological materials, especially bacteria, algae, yeasts, and fungi have received increasing attention for heavy metal removal and recovery.

8.1. Removal of Heavy Metals by Biosorption

Biosorption is a physicochemical process that is a property of certain types of inactive, dead, microbial biomass to bind and concentrate metal or metalloid species, compounds and particulates from very dilute aqueous solution. It is particularly the cell wall structure of certain algae, fungi, and bacteria which was found responsible for this phenomenon. There are so many advantages of biosorption process such as low cost, high efficiency, minimization of chemical and biological sludge, no additional nutrient requirement, ability to regenerate biosorbent and possibility of metal recovery. For the extensive removal of metals, a large number of microorganisms have been investigated as biosorbents. The tested biosorbents can be classified into the following categories:

- Bacteria (e.g., *Bacillus subtillis*)
- Fungi (e.g., *Rhizopus arrhizus*)
- Yeast (e.g., *Saccharomyces cerevisiae*)
- Algae

8.1.1. Bacterial Biosorbents

Because of the small size, ubiquity, ability to grow under controlled conditions and resilience to a wide range of environmental situations, bacteria were used as biosorbents [207]. Bacteria species such as *Bacillus, Pseudomonas, Streptomyces, Escherichia, Micrococcus,* etc. have been tested for uptake of heavy metals such as Pb, Zn, Cu, Cd, Fe (III), Cr, Pt, U, and Th.

8.1.2. Fungal Biosorbents

Three groups of fungi (molds, yeasts, and mushrooms) among the large and diverse group of eukaryotic microorganisms have major practical importance. Yeasts and filamentous fungi have been detected in many cases to fix metallic elements. Heavy metal ions and radio nuclides are successfully removed by using filamentous fungi (such as *Penicillium spp., Aspergillus spp., Rhizopus spp.* and white rot fungi) from aqueous solutions. Different species of *Penicillium*, as well as *Aspergillus*, have been reported as good biosorbents of metal ions. The genus *Rhizopus*, such as *Rhizopus Arrhizus* and *Rhizopus javanicus*, has been discovered to owe the relatively good sequestering properties [208].

8.1.2.1. Penicillium

A variety of heavy metal ions such as Cu, Au, Zn, Cd, Mn, U, and Th can be removed by *Penicillium* from aqueous solutions [209]. *Penicillium spinulosum, P. oxalicum* [210], *P. austurianum* [211], *P. verrucosum* [212], *P. purpurogenum* [213], *P. canescens* [214], *P. chrysogenum*, etc. were reported to adsorb various metals. As *P. chrysogenum* can extract gold from a cyanide solution, the biosorption capacity was not promising [215]. *P. spinulosum* was reported to be capable of removing Cu, Au, Zn, Cd, Mn [216].

8.1.2.2. Aspergillus

Aspergillus niger showed a good affinity for binding Cu^{2+}, Zn^{2+} and Ni^{2+} ions in a single composition system. On the other hand, only copper and zinc captured in the multi-component system [217]. Moreover, there are many toxic metal ions such as Ni^{2+}, Ca^{2+}, Fe^{2+}, Fe^{3+}, and Cr^{2+} etc. that can be removed efficiently by using a waste fungal biomass containing killed cells of *A. niger*.

8.1.2.3. Other Fungi

Cell wall of *Rhizopus arrhizus* involves a high content of chitin that has the ability to remove complex metal ions [218]. Growing *Rhizopus arrhizus* and *Aspergillus niger* did a comparative investigation on the bioaccumulation of heavy metal ions where viable *R. arrhizus* and *A. niger* could remove Cu (II). Biosorption of metal ions, such as Li^{+}, Ag^{+}, Pb^{2+}, Cd^{2+}, Ni^{2+}, Zn^{2+}, Cu^{2+}, Sr^{2+}, Fe^{2+}, Fe^{3+}, Al^{3+} etc. by *R. nigricans* biomass was reviewed. Furthermore, *R. arrhizus* were also reported to adsorb Th, Pb, Ph, Cd, Ni, and Cr [219].

8.1.2.4. Yeast

Most yeast can absorb a wide range of metal ions or be strictly specific in respect of only one metal ion. The genus of yeasts such as *Saccharomyces, Candida, Pichia* is proficient biosorbents for heavy metal ions. Among them, *Saccharomyces cerevisiae* is one of the most special types of biosorbent [220]. Moreover, Ag, Au, Cd, Co, Cr, Cu, Ni, Pb, U, Th, and Zn have been successfully removed from the aqueous solution using yeast biomass.

8.1.3. Marine Algae as Biosorbents

Due to high sorption capacity and availability in unlimited quantities in the seas and oceans, Algae are possessing special interest amongst scientists for the development of new biosorbents materials [221]. The *cyanophyceae, L. taylorii* exhibited a high uptake capacity for four metal ions Pb, Ni, Zn, and Cd. There are some marine micro and macro algae species that can act as biosorbents for heavy metal ions [222]. Having good metal uptake capacity, some algae were reported to be able to absorb one or more heavy metal ions, including K, Mg, Ca, Fe, Sr, Co, Cu, Mn, Ni, V, Zn, As, Cd, Mo, Pb, Se, and Al [222, 223].

8.2. Mechanism of Biosorption

The abundance of functional groups in the structure of microorganisms denotes that there are many ways for the metal to be taken up by the microbial cell. The biosorption mechanisms are very complex and sometimes are not fully understood. They may be classified according to various criteria.

According to the dependence on the cell's metabolism, biosorption mechanisms can be divided into:

- Metabolism dependent
- Non-metabolism dependent

According to the location where the metal removed from the solution, biosorption can be classified as:

- Extra cellular accumulation/ precipitation
- Cell surface sorption/ precipitation
- Intracellular accumulation

During metabolism dependent biosorption, transport of the metal across the cell membrane yields intracellular accumulation. This implies that this type of biosorption may take place only with viable cells and associated with an active defense system of the microorganism, which reacts in the presence of toxic metal.

On the other hand, metal uptake is occurred by the physicochemical interaction between the metal and the functional groups present on the microbial cell surface in case of non-metabolism dependent biosorption. This is centered on physical adsorption, ion exchange and chemical sorption, which is not dependent on the cells' metabolism. As we know that polysaccharides, proteins, and lipids constitute cell walls of microbial biomass and these cell walls have abundant metal binding groups such as carboxyl, sulfate, phosphate and amino groups. For that reason, non-metabolism dependent is relatively rapid and can be reversible [224].

In case of precipitation, the metal uptake may take place both in the solution and on the cell surface [225]. Precipitation may be dependent on the cell's' metabolism if the microorganism produces compounds in the presence of toxic metals. On the contrary, precipitation may not be dependent on the cells' metabolism, if it occurs after a chemical interaction between the metal and cell surface [226].

Metals play a fundamental role in the life processes of microorganisms. Many of metals, such as calcium, chromium, cobalt, copper, iron, magnesium, manganese, nickel, potassium, sodium and zinc, are actually act as micronutrients. These metals are used for the following purposes-

- For redox-processes
- To stabilize molecules through electrostatic interactions
- As components of various enzymes and
- For regulation of osmotic pressure

In complex biochemical reactions, metal ions play vital roles as trace elements. There are many other metals, such as silver, aluminium, cadmium, gold, lead, and mercury that have no biological role as well as potentially toxic to microorganisms [227].

These heavy metal ions form unspecific complex compounds within the cell at higher concentrations which leads to toxic effects as well as making them too threatening for any physiological function. Toxicity of nonessential metals occurs through the displacement of essential metals from their native binding sites or through ligand interactions [227, 228]. Enzymes which are essential for microbial metabolism such as Hg^{2+}, Cd^{2+} and Ag^{2+} tend to bind to sulfhydryl (–SH) groups to inhibit the activity of sensitive enzymes [228]. Most metal ions have to enter the microbial cell for a physiological or toxic effect. Many divalent heavy metal cations (e.g., Mn^{2+}, Fe^{2+}, Co^{2+}, Ni^{2+}, $Cu^{2+,}$ and Zn^{2+}) are structurally very similar. Moreover, the structure of oxy anions such as chromate resembles that of sulfate, and the same is true for arsenate and phosphate. In such cases, for rendering the enzyme non-functional, these toxic metal cations may substitute for physiological essential cations within an enzyme. Therefore, the microbial uptake systems have to be tightly regulated to be able to differentiate between structurally very similar metal ions.

This problem has been solved by most microorganisms using two types of uptake mechanisms for heavy metal ions. The first one is non-specific, fast, constitutively expressed and driven by the chemiosmotic gradient across the cytoplasmic membrane of bacteria [228].

The second type of uptake system is inducible and slower, has high substrate specificity. This uptake mechanism often uses ATP hydrolysis as the energy source and is only produced by the cell in times of need, starvation or during a special metabolic situation [229]. Even though microorganisms have specific uptake systems, high concentrations of nonessential metals may be transported into the cell by a constitutively expressed unspecific system [228]. In addition, both essential and nonessential metals can damage cell membranes, alter enzyme specificity, disrupt cellular functions and damage the structure of DNA at high levels [227]. Also, oxidative stress on microorganisms is imposed by concentrations of elevated levels of heavy metals [230]. As a result, some microorganisms have been forced to develop metal-ion homeostasis factors and metal resistance determinants [227, 228] as metal ions cannot be degraded or modified like toxic organic compounds. There is six possible metal resistance mechanisms exist: (i) exclusion by permeability barrier, (ii) intra-cellular sequestration, (iii) extra-cellular sequestration, (iv) active transport efflux pumps, (v) enzymatic detoxification and (vi) reduction in the sensitivity of cellular targets to metal *ions* [227-229, 231]. One or more of these resistance mechanisms permits microorganisms to function in metal co-contaminated environments [232].

8.3. Advantages of Microbial Remediation

Microbial remediation has several advantages while comparing with chemical remediation and phytoremediation processes. Some of the advantages of microbial remediation include [233]:

- It requires moderate capital investment as it uses relatively low-cost, cheap-technology. For that reason, it is fairly inexpensive while comparing with other remediation processes.

- This process usually perceived by the public as an acceptable treatment process because it is an environmental safe as well as a natural process.
- As it does not generate waste, so produce less wasted groundwater.
- It is self sustaining as the microbes able to degrade the contaminant.
- The residues for the treatment are usually undamaging products that include carbon dioxide, water, and cell biomass.
- It can be used alongside other technologies.
- It can be carried out on site, often without causing a major disruption of normal life.

9. Microbial Desulfurization of Fossil Fuels

The availability of low sulfur crude has decreased over the last decade as a consequence of the increasing reserves of heavy crude. Terrestrial oil or petroleum deposits are being increasingly employed for the production of fuels which contain high levels of sulfurous hydrocarbons. Combustion of sulfur containing fossil fuels results in acidic deposition posing a multitude of threats to the environment [234-236]. Physical and chemical methods are expensive to prevent the release of sulfur compounds as atmospheric pollutants from coal and crude oil, though the methods are used widely.

9.1. Different Sulfur Compounds Present in Fossil Fuels

Sulfur is the most abundant, multivalent and nonmetallic element in petroleum that can vary from 0.05% to 10% of the composition. Other than elemental sulfur, sulfate, sulfite, thiosulphate, and sulfide, more than 200 sulfur containing organic compounds are recognized in crude oils. These include sulfides, thiols, thiophenes, substituted benzo and dibenzothiophenes, benzonaphthothiophene and many more complex molecules [237]. The condensed thiophenes are a very common form of various fossil fuels in which sulfur is present [238]. Dibenzothiophene (DBT), benzothiophene (BT), and their substitutes are the major sulfur containing aromatic compounds in fuels, consisting up to 70% of the sulfur content [239]. They have boiling points more than 200 °C and it is difficult to remove them from outlet streams to the atmosphere [240, 241]. Typical thiophenic compounds such as benzothiophene (BT), non-b, single-b, and di-b-substituted benzothiophenes (Boiling points 219 °C) are found up to 30% in crude oils [242]. Inorganic sulfur compounds such as pyrites (FeS_2) along with elemental sulfur and hydrogen sulfide can also be present in fossil fuels [243].

9.2. Hydrodesulphurization Process (Traditional Method)

A traditional method for desulfurization of oil compounds is hydrodesulfurization (HDS). Hydrodesulfurization is a catalytic reaction between hydrogen and organic sulfur to remove it at the pressure between 5 up to 10 MPa and in the range of 300 to 350 °C temperature [244]. Hydrodesulfurization is not able to remove and get deep desulfurization of the oils that contain some other molecules especially the polyaromatic sulfur heterocyclies (PASHs) remaining in the heavier fractions [245]. Moreover, the high costs of operation and building, unwanted secondary effects, inherent chemical limitations are associated with hydrodesulphurization, discharging the high amount of CO_2 in the environment [246, 247]. Moreover, deep hydrodesulfurization may reduce the caloric value of treated oil in the high temperature during hydrodesulfurization, produce carbonaceous coke on the catalysts by extreme conditions that are applied in hydrodesulfurization, generate H_2S during hydrodesulfurization is a poison for catalysts and truncate their functional life [248]. The fact that hydrodesulfurization can remove the bulk of inorganic sulfur and simple organic sulfur but inadequate to produce low sulfur fuels due to the fact that the complex polycyclic sulfur compound present in petroleum and coal is not removed by the process (HDS) [249].

9.3. Biodesulfurization Process

For removing organic sulfur by desulfurization from fossil fuels, which have lower sulfur content, require new strict economic regulations and more efficient methods. Recently, microbial degradation of organic sulfur compounds has attracted attention for its potential application of desulfurization of coal and petroleum oil. Microbial desulfurization may provide an improved technique for the removal of organic sulfur from fossil fuels.

Biodesulfurization (BDS) is a process that is being developed based on bacterial potential that can remove organosulfur from oil fractions through their metabolism [250]. The biodesulfurization process uses microorganisms for the pre-combustion removal of sulfur compounds from fossil fuels. Enzymes can be applied to remove sulfur compounds more safely under mild conditions. Moreover, bioprocessing generally is specific for substrates. Such a high substrate specificity of biological reactions is more convenient for treatment of petroleum fractions containing a limited number of organic sulfur compounds [251, 252].

The main advantage of biodesulfurization compared to hydrodesulfurization is that this process does not need high reaction conditions such as high temperature and pressure and can happen under normal pressure and temperature condition. In addition, biodesulfurization has some other merits such as capital and operating cost saving, capability and flexibility to treat a wide variety of petroleum streams, more rapid engineering and construction time, more energy efficient process, safer and milder condition and no generation of undesirable side-products [253, 254]. DBT and its derivatives or a range of sulfur compounds in several crude oil fractions can selectively desulfurize by using different bacterial strains such as:

- *Rhodococcus rhodochrous* IGTS8 [235, 255]
- *Rhodococcus erythropolis* H-2 [256]
- *Corynebacterium* SY-1 [257]
- *Agrobacterium* sp. [258]

9.3.1. Types of Biodesulfurization

Biodesulfurization can occur in aerobic and anaerobic conditions. Biodesulfurizing microorganisms are able to use and remove sulfuric compounds into three ways, such as oxidative C-S cleavage, oxidative C-C cleavage, and reductive C-S cleavage [259]. The aerobic and anaerobic mechanisms can break down the carbon-sulfur bonds in sulfur heterocyclic compounds such as benzothiophene, thiophene, and dibenzothiophene [260]. Anaerobic desulfurization is performed in the presence of hydrogen or nitrogen gases by anaerobic bacteria. H_2S gas and organic sulfur compounds production can be accompanied by anaerobic biodesulfurization [261]. The first anaerobic biodesulfurization of oil was observed in 1953 by recruitment of *D. desulfuricans*. One of the advantages of anaerobic biodesulfurization is that it is more compatible with traditional desulfurizing methods [262]. The rate of biodesulfurization under the anaerobic condition is very slower than that in aerobics biodesulfurization [263]. At present, there is no substantial anaerobic biodesulfurization process in industrial scale because of its drawback and difficulties such as the cost of required hydrogen [259].

9.4. Desulfurization of Dibenzothiophene (DBT)

Desulfurization of DBT can be performed in two pathways by aerobic bacteria Kodama pathways and 4S pathways.

9.4.1. Kodama Pathway (Oxidative C-C Cleavage)

Kodama pathway is the first reported aerobic pathway of DBT biodesulfurization [259, 264]. This pathway involves the ring cleavage of one aromatic ring in sulfur compounds because of initial oxidative reaction. Kodama pathway is done into three main steps; i) hydroxylation, ii) ring cleavage and iii) hydrolysis. In fact, Kodama pathway is an incomplete ring destructive of DBT without removal of sulfur. *Pseudomonas jijani* and *Pseudomonas abikonesis* are able to follow Kodama pathway to degrade DBT [259].

9.4.2. 4S Pathway (Oxidative C-S Cleavage)

The 4S pathway was suggested firstly by Kilbane in 1990, in which microorganisms are able to oxidize sulfur atoms in DBT without any breakage of C-C bonds [260]. In this method, sulfur atoms are released from DBT. The hydrocarbon parts of DBT are protected. This pathway is a sulfur specific DBT desulfurization. It is called 4S because it involves four intermediate sulfur compounds. This pathway of desulfurization is more considerable in biodesulfurization compare to Kodama pathway because of its neutral influence in removing sulfur on hydrocarbon structure and caloric value of fuels. The 4S pathway is a DBT sulfur specific desulfurization pathway that contains consecutive sulfur oxidation of DBT by breaking down the S-C bonds leading to 2-hydroxybiphenyl (2-HBP) and sulfate formation and leaving hydrocarbon skeleton unchanged. This pathway involves four steps of enzymatic reactions. These enzymes attack the sulfur part of DBT selectively and then release the sulfur from hydrocarbon parts [265].

10. Biodiesel Production by Microalgal Biotechnology

Biodiesel is a renewable and alternative fuel that can be used instead of diesel fuel made from petroleum. Biodiesel can be made from vegetable oils and animal fats and is becoming increasingly important due to diminishing petroleum reserves and adverse environmental consequences of exhaust gases from petroleum fueled engines. The term "biodiesel" commonly refers to an oxygenated diesel fuel made from various feedstocks by conversion of the triglyceride fats to methyl or ethyl esters via transesterification for reducing the high viscosity of the oil closer to that of petrodiesel. Biodiesel fuel is renewable, biodegradable, non-toxic, and essentially free of sulfur and aromatics which are better compared to petrodiesel fuel.

Because of several favorable environmental properties biodiesel have no net increase in the release of carbon dioxide and very low sulfur content. The gas generated during combustion could be reduced using biodiesel as an energy source. Also, biodiesel has high oxygen content which decreases carbon monoxide emission. Moreover, the harmful effect of biodiesel on the environment is less as biodiesel contains no aromatic compounds.

Recent investigation has directed that air toxicity and cancers can be decreased using biodiesel compared to common diesel source. As a non-renewable fuel, biodiesel appears to have significant economic potential because fossil fuel prices will increase inexorability further in the future. Ultimately, in terms of flash point and biodegradability biodiesel is better than diesel fuel.

10.1. Microorganisms Available for Biodiesel Production

Oleaginous microorganisms may be defined as microbial with the content of microbial lipid excess of 20%. Biodiesel, named single cell oil (SCO) is produced using microbial lipids, has attracted great interest among the scientists all over the world. Although there is all kinds of microorganism such as microalgae, bacillus, fungi, and yeast are used for storing oils, not all of them are available for biodiesel production (Table 5).

Table 5. Oil content of some microorganisms [272]

Microorganisms	Oil content (% dry wt.)	Microorganisms	Oil content (% dry wt.)
Microalgae		Yeast	
Botryococcus braunii	25-75	*Candida curvata*	58
Cylindrothica sp.	16-37	*Crptococcus albidus*	65
Nidzschia sp.	45-47	*Lipomyces starkeyi*	64
Schizochytrium sp.	50-77	*Rhodotorla glutinis*	72
Bacterium	-	Fungi	-
Arthrobacter spp.	>40	*Aspergillus oryzae*	57
Acinetobacter calcoaceticus	28-38	*Mortierella isabellina*	86
Rhodococcus opacus	24-25	*Humicola lanuginosa*	75
Bacillus alcalophilus	18-24	*Mortierella vinacea*	66

Numerous oleaginous yeasts and microalgae's have been reported to grow and accumulate significant amounts of lipids similar to vegetable oil [266], methyl-esters and soaps [267], used as sole carbon and energy sources. In recent years, as substrates for SCO, a large number of hydrophobic lipid accumulating microorganisms have been studied, especially for the production of biodiesel. The extent of lipid accumulation in microorganisms is determined by the genetic constitution. The maximum attainable lipid contents can differ immensely among species and even among individual strains. Microalgae species mostly used for biodiesel production and are sunlight driven cell factories that convert carbon dioxide to potential biofuels [268], It can provide several different types of renewable biofuels, including methane which is produced by anaerobic digestion using algal biomass, biodiesel [269], and biohydrogen [270]. The study showed that most microalgae biomasses are a rich source of ω3 and ω6 fatty acids, essential amino acids, like leucine, isoleucine, valine, etc. [271]. The accumulated oil in almost all microalgae is mainly triglyceride (>80%).

Microalgae grow extremely fast and many are very rich in oil, and double their biomass within 24h. Though the average lipid content of algal cells varies between 1% and 70%, it can reach up to 90% of dry weight under certain conditions [268]. It is believed as one of the good candidates for biodiesel production, based on higher photosynthetic efficiency, higher biomass production, and faster growth compared to other energy crops.

Table 5 tells that nearly all bacteria accumulate lower lipid than microalgae. The average oil content of dry biomass is about 20–40%. For example, the oil content of *Arthrobacter sp.* and *Acinetobacter calcoaceticus* is about 40% and 38%, respectively. For the production of biodiesel with the highest growth rate, bacterial reach the superiority among other microorganisms.

Besides, storage lipid accumulating bacteria containing *actinomycete* group may be used for synthesizing remarkably high amounts of fatty acids (up to 70 wt% of the cellular dry weight) from simple carbon sources like glucose under growth restricted conditions TAGs [273].

Since 1980, yeasts and fungi (especially molds) are considered as favorable oleaginous microorganisms [274]. There are some yeast strains, such as *Rhodosporidium sp.*, *Rhodotorula sp.* and *Lipomyces sp.*, which can gather intracellular lipids as high as 70% of their biomass dry weight. The most effective oleaginous yeast *Cryptococcus curvatus* can accumulate storage lipid up to >60% on a dry weight basis. Further, these lipids usually consist of SOC-90% w/w triacylglycerol while grows under N-limiting conditions, with a percentage of saturated fatty acids (% SFA) of about 44% which is similar to many plant seed oils [275]. Oleaginous yeasts and molds which have specific structure accumulate triacylglycerols rich in polyunsaturated fatty acids [276], and nearly all of them are unsaturated fatty acid. For example, *Rhodosporidium toruloides* Y4 which is a well-known microbial lipid producer contained mainly long chain fatty acids with 16 and 18 carbon atoms. The extensive characterization of the oleaginous profile of *R. toruloides* CBS 14 has been reported [277]. A recent study showed that in flask fed batch cultures the lipid content of dry biomass and cellular can reach 151.5 g/L and 48.0% (w/w), respectively [278]. A filamentous fungus – *Mortierella alliacea* Strain YN-15, accumulated arachidonic acid (AA, C20: 4n-6) mainly in the form of triglyceride in its mycelia by 7-d cultivation in a 50 L jar fermenter, which yielded 46.1 g/L dry cell weight, 19.5 g/L total fatty acid, and 7.1 g/L AA [279]. Based on these records, oleaginous yeasts and fungi are used as alternative oil resources for biodiesel production.

10.2. Microalgal Biotechnology for Lipids Production

Compared to other oil crops microalgae have high potentials in biodiesel production. First of all, the cultivation of microalgae does not need much land while comparing with terraneous plants [280]. Microalgae produce biodiesel and that biodiesel will not compromise the production of food and other products derived from crops. Secondly, the growing ability of microalgae is extremely rapid and many algal species are rich in oils. For instance, the heterotrophic growth of *Chlorella protothecoides* in fermenters can accumulate lipids as high as 55% of the cell dry weight after 144 h of cultivation with the feeding of corn powder hydrolysate [281]. In microalgae, oil levels of 20–50% are common [280].

10.2.1. Biosynthesis of Lipids/Fatty Acids in Microalgae

It is recognized that for lipids production based on both inorganic carbon (CO_2) and organic carbon sources (glucose, acetate, etc.) can be utilized by microalgae [Figure 16]. The components and contents of lipids in microalgal cells vary from species to species. Neutral lipids (e.g., triglycerides, cholesterol) and polar lipids (e.g., phospholipids, galactolipids) are the two basic classes of lipid. Triglycerides as neutral lipids are used as the main materials for the production of biodiesel. The synthesis routes of triglycerides in microalgae may entail the following three steps [Figure 17]:

- The formation of acetyl coenzyme A (acetyl-coA) in the cytoplasm;
- The elongation and desaturation of the carbon chain of fatty acids; and
- The biosynthesis of triglycerides in microalgae.

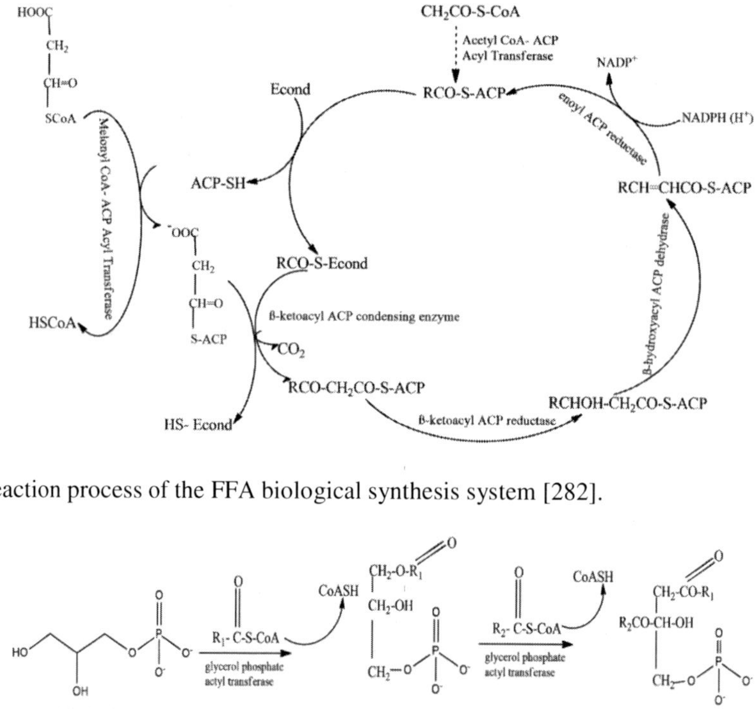

Figure 16. Reaction process of the FFA biological synthesis system [282].

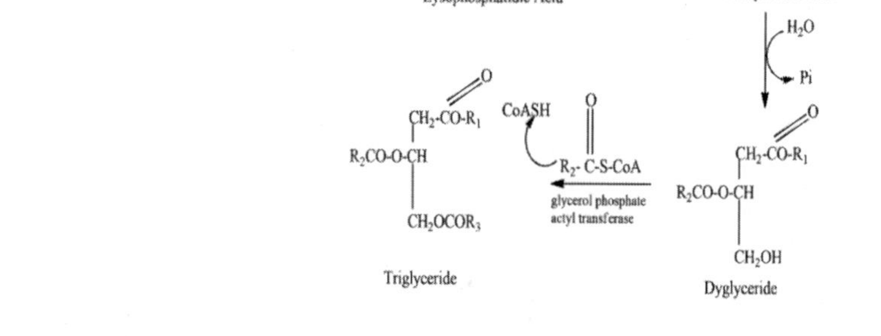

Figure 17. The biosynthesis of triglycerides in microalgae [282].

10.2.2. Phototrophic Cultivation of Microalgae for Lipids Production

In phototrophic culture, open ponds and enclosed photobioreactors are the two systems where microalgae can be grown up. Enclosed photobioreactor system is more suitable for some microalgae that are readily contaminated by other microbes. On the other hand, in the open pond system, some special microalgae can survive well in extreme environments such as high pH (e.g., *Spirulina*) and high salinity (e.g., *Dunaliella*) or can grow very rapidly (e.g., *Chlorella*).

Enclosed photobioreactor system has been suggested for the production of high value long chain fatty acids (e.g., DHA, EPA) because the environmental control is much better than the open pond system. Instead, because of its relatively cheap operating cost open pond system is perhaps more suitable for cultivating microalgae for biodiesel compared to the enclosed photobioreactor system. Carbon dioxide, light, macro, and micronutrients are the basic requirements for microalgal phototrophic growth. Power plants release large amounts of waste gases daily which are the main sources of carbon dioxide. For example, up to 13% CO_2 is released from fuel gas that is emitted by typical coal fired power plants. This high concentration of CO_2 enhances transfer and uptake of CO_2 in the pond system. The combination of a coal fired power plant with algae cultivation provides a feasible approach and it is considered as a good concept to recycle CO_2 from coal combustion into usable liquid fuel.

Microalgae could use CO_2 from fuel gas directly while grown in large outdoor ponds [283]. Besides, the wastewater is rich in inorganic irons (nutrients) which are essential for microalgal growth. With the aim of developing an economically feasible system to remove ammonia from wastewater and CO_2 from flue gas, *Chlorella vulgaris* was grown in wastewater from a steelmaking plant [284].

10.2.3. Heterotrophic Cultivation of Microalgae for Lipids Production

Though microalgae can effectively use light, the phototrophic growth of microalgae is often slow because of the limitation of light at high cell densities on a large scale [285] or due to excessive light especially in sunny days which is called "photo-inhibition" [286]. Heterotrophic growth of microalgae in conventional fermenters should be favorably considered because of these disadvantages which are associated with photoautotrophic cultivation [287]. The cultivation of microalgae by the heterotrophic process has several advantages [[288] over phototrophic cultivation including-

- Elimination of the light requirement
- Good control of the cultivation process, and
- Low-cost for harvesting the biomass.

Both cell growth and biosynthesis of products are significantly influenced by medium nutrients and environmental factors in heterotrophic culture method. For the heterotrophic culture of microalgae, carbon sources are the most important element in the production of lipids. For example, although the green microalgae *C. protothecoides* can grow photoautotrophically or heterotrophically, the content of biomass as well as a lipid is much higher in case of heterotrophic growth of *C. protothecoides* using acetate, glucose, or other organic compounds as a carbon source [289].

Moreover, depending on the microalgal species, heterotrophic microalgae might utilize other carbon sources such as ethanol, glycerol, fructose etc. [290]. Liu et al. [291] reported that among several carbon sources glucose was chosen. Besides, cheaper carbon sources may be considered in order to lower the production cost of microalgal oils as biodiesel. For that reason, instead of glucose corn powder hydrolysate (CPH), molasses may be used as an organic carbon source in heterotrophic culture. It was reported that as CPH contained some beneficial components to Chlorella, CPH was considered as superior to glucose solution. In the consequence, *C. protothecoides* produced 55.2% crude lipids in the cells with a cell dry weight concentration of 15.5 g/L [281].

Heterotrophically grown microalgae usually accumulate more lipids than those cultivated photoautotrophically [292]. Some phototrophic microorganisms could also be grown on cheap organic substrates heterotrophically [287]. Evidence has been shown that the ''dark metabolism" of photosynthetic plants and microalgae is similar to that of non-photosynthetic organisms such as yeasts. Recently, an obligate photoautotrophic microalgae *Phaeodactylum tricornutum* was grown heterotrophically when a single gene (Glut 1) that encoded the glucose transporter protein (Glut 1) was introduced into this algae [293]. For further investigation of microalgae as biofuels by heterotrophic cultivation, such studies on the production of fatty acids and lipids could be useful.

11. Power Generation by Microbial Biofuel Cells

A biological fuel cell (abbreviated as *biofuel cell)* is the brood of two parent technologies: biotechnology and fuel cells. Like conventional fuel cells, biofuel cells comprise an anode and a cathode separated by a barrier that is selective for the passage of positively charged ions. However, they require the addition of fuel to generate power. Biofuel cells utilize enzymatic catalysts, either as they occur in microorganisms or as isolated proteins whereas conventional fuel cells usually use precious metals as catalysts.

There are several uses of biofuel cells. Firstly, they may be able to use fuels such as wastes containing paper, glucose or other biologically derived materials that cannot be handled by any existing transition metal based fuel cell. Secondly, biofuel cells are innately non-polluting and their enzymes operate at pH 7 and in low temperatures water. Thirdly, enzymes can be fabricated inexpensively by fermentation using genetic engineering. Fourthly, enzymes are able to contribute to the solution of problems such as the overpotential at the dioxygen cathode that have resisted solution by conventional fuel cell technology. In biological systems, oxidative metabolism using dioxygen is highly optimized. Finally, biofuel cells perhaps applicable to certain speculative but significant applications such as a fuel cell that could be implanted in humans and give power for other implanted prosthetic devices.

11.1. Microbial Biofuel Cells

In 1962 Davis and Yarbrough demonstrated one of the earliest examples of electrical energy derived from both microorganisms and enzymes [294]. The microorganisms used in the fuel cell were *Escherichia coli and Nocardia* and the enzyme used was glucose oxidase (GOD).

The authors started their study using a fuel cell configured with GOD, buffer, presumably containing bound flavin adenine dinucleotide (FAD), glucose, and a Pt-sheet electrode in the dinitrogen purged anodic compartment. The cathodic compartment, containing a buffer and a Pt-sheet electrode was saturated with dioxygen. GOD catalyzes the oxidation of p-D-glucose to GD-gluconolactone with the concurrent reduction of a FAD to $FADH_2$ [295]. $FADH_2$ is reoxidized by dioxygen under biological conditions, [295]. However, this reaction did not occur because the anodic chamber had been purged to remove dioxygen. Methylene blue (MB) was united to the anode compartment both to shuttle the reducing equivalents made available by the enzymatic reaction to the anode and reoxidize $FADH_2$ [295].

$$\beta\text{-D-glucose} + \text{GOD/FAD} \rightarrow \delta\text{-D-gluconolactone} + \text{GOD/FADH}_2 \quad (7)$$

$$\text{GOD/FADH}_2 + O_2 \rightarrow \text{GOD/FAD} + H_2O_2 \quad (8)$$

$$\text{GOD/FADH}_2 + \text{MB} \rightarrow \text{GOD/FAD} + \text{MBH} \quad (9)$$

$$\text{MBH} \rightarrow \text{MB} \quad (10)$$

The fuel cell yielded no electricity in the absence of MB. V_{cell} at open -circuit was 0.28 V and voltages between 0.075 to 0.09 V could be maintained under a load of 1 kΩ. The glucose-GOD-MB system under a dinitrogen atmosphere generated reduced MB that produced electricity. The authors reasoned that a facultative anaerobe would also produce a measurable current. AV_{ceu} of 0.625 V at open -circuit was obtained when *Escherichia coli* was added to the anode compartment in place of GOD. Under a 1 kΩ load, V_{cell} could be sustained at 0.52 V for over 1 h. The use of ethane as fuel was practiced using an ethane oxidizing *Nocardia*as catalyst. In this case, no current was produced even in the presence of MB. The increase of the metabolic activity was found when glucose was substituted for ethane. An open -circuit with a maximum current of 2 mA under a 1 kΩ load and voltage of 0.3 V were reported. This value of maximum current seems unlikely, given the upper limit of 0.3 mA inducted by Ohms law (V =IR.). A unique microbial biofuel using methane as fuel was demonstrated by van Hees [296]. To our knowledge, no other methane fed biofuel cell has been reported in the literature. The anodic compartment, containing1-naphthol-2-sulfonate indo-2, *Pseudomonas methanica,* a platinum anode and 6-dichlorophenol (NSIDCP) was purged of dioxygen and supplied with methane. The cathode compartment supplied with dioxygen in the form of air and houseda platinum cathode in a sterile medium. This dye did not function as an electrochemical mediator because; inclusion of NSIDCP in the anode compartment had no upshot on V_{cell} at open circuit. Moreover, this conclusion was supported by shutting the microorganisms in dye permeable dialysis tubing and observing that the value for V_{cell} at open-circuit was established only very slowly even in the presence of NSIDCP. This observation suggests that microbial contact with the anode was essential for electron transfer. The power density was little prolonged open-circuit conditions or high current drain was audited to be detrimental to the obligate aerobe used in this fuel cell. In summary, to functioning under moderate current drain the authors concluded that biofuel cells exploiting obligate aerobes are best suited.

Karube et al. and Suzuki et al. studied many microorganisms for their production of dihydrogen from glucose [297, 298]. Of the organisms studied, *Clostridium butyricurn* IF03847 produced dihydrogen at the highest rate. For this reason, *Clostridium butyricurn* IF03847 was used to supply dihydrogen to the anode compartment of a conventional fuel cell. Immobilization of the microorganism in a gel of agar-acetyl cellulose increased the rate of dihydrogen production and stabilized the dihydrogen producing system. For every mole of glucose consumed, 0.2 mol of formic acid and 0.6 mol of dihydrogen were produced. 40 ml (1.7 mmol) of dihydrogen per min, equivalent to near 5 A was produced from two kg of wet gel containing about 200 g of wet microorganism.

Four different fuel cell configurations were studied; the two most recent are described here. One configuration, a fuel cell stack of five fuel cells using Pd-blackened Ni-gauze cathodes and blackened Ni-gauze anodes were operated at room temperature. Dioxygen was passed through the cathode compartments while dihydrogen was passed from a 5 L reservoir containing about 2 kg of immobilized whole cells to the anode compartments of the fuel cell stack at a rate of 40 ml/min. The fuel cell system (dihydrogen producing microbial reservoir and five fuel cells) was maintained a current of 0.8 A at 2.2 V and operated for 7 days. A high temperature phosphoric acid fuel cell stack was used in the subsequent report, however; the poor current efficiency was not explained.

A second configuration used about 3 kg of immobilized microorganisms in a 200 L fermentation vessel containing 150 L of molasses. Dioxygen was passed into the cathode compartment at a rate of 1-1.5 L/min whereas dihydrogen and carbon dioxide, was passed into the anode compartment of a phosphoric acid fuel cell stack (two fuel cells in parallel operated at 200 °C) at a rate of 400-800 ml/min to produce biogas. A stable current of 10-14 A was maintained for four hours with a power output of 9-13 W while the open-circuit voltage was between 0.9-0.93 V. Power density, current density, and electrode size were not given. Only 1% of the energy consumed to operate the system (pumps and heaters) was generated by the microbial fuel cell and this was found by the authors by analyzing the electrical energy balance for the microbial/phosphoric acid fuel cell configuration. Delaney, et al., smeared fuel cell performance treating different microorganism mediator- substrate combinations [299]. Based on the difference between $E^\circ$of the different mediators used in the anode compartment and E° of the potassium ferricyanide used in the cathode compartment, V_{cell} at open-circuit was predicted to range from 0.9-1.2 V. However, V_{cell} was observed to be 0.2-0.4 V lower than predicted, indicating activation over potentials at the electrodes. The fuel cell that gave the best performance contained glucose, *Proteus vulgaris* and thionine (TH^+) in the anodic compartment. By discharging the fuel cell through a range of loads typical current-voltage curves were obtained. The coulombic yield of the fuel cell was measured by integration of the area under the current-time curve (Coulombic yield is the total number of coulombs passed through the fuel cell, as determined by integration of the current-time curve in units of amp-s = coulombs, relative to the theoretical number that might be obtained by complete conversion of the reactants to products). Based on the theoretical maximum coulombic equivalent for complete oxidation of glucose in accordance with equation 11, coulombic yields of 32-62% were obtained [295]. (The theoretical maximum coulombic equivalent of glucose is 24 mol electron/mol glucose × 96,487 C/mol electron = 2.3 x 106 C/mol glucose):

$$C_6H_{12}O_6 + 6\,H_2O \rightarrow 6CO_2 + 12\,H^+ + 24\,e^- \quad (11)$$

Thurston et al. argue that if fuels such as carbohydrates are to be used, microorganisms will be more economical catalysts than individual enzymes. This is because the multi-step oxidations employed with complex fuels and that would require multiple, expensive enzyme purifications [300]. 14C-labeled glucose was used as fuel to establish the distribution of products and *Proteus vulgaris* was chosen as the biocatalyst for glucose oxidation. By scintillation methods and a sodium hydroxide trap, the amount of CO_2 evolved from the oxidation of glucose was determined. Coulombic yields indicated near unit current efficiency (based on a 50% conversion of glucose) and conversion of glucose to CO_2 was 40-50%. The most significant loss in energy (30%) was observed in a side reaction involving the conversion of glucose to acetate. However; it is possible to avoid this side reaction by metabolic or genetic modification of the microorganism.

Bennetto et al. used a TH^+ mediated microbial biofuel cell containing *Proteus vulgaris* as biocatalyst and sucrose as fuel [301]. Equation 12 [295] gives the stoichiometry of complete oxidation of sucrose.

$$C_{12}H_{22}O_{11} + 13\,H_2O \rightarrow 12\,CO_2 + 48\,H^+ + 48\,e^- \qquad (12)$$

After the addition of various amounts of sucrose, the biofuel cell was discharged through a resistance of 100 Ω. The amount of CO_2 originated from the anode compartment detained near complete oxidation of sucrose. From both coulombic yield and CO_2 evolution it was determined the cell current peaked at 3.5 mA and the production of electricity from the oxidative conversion of sucrose was >95%. At higher concentrations of sucrose, due to the acid formation, the buffering capacity of the anodic medium was exceeded. It was found from preliminary tests that >70% of the carbohydrate was oxidized using molasses as fuel, where, neither the open-circuit voltage nor electrode size was given.

Vega and Fernandez studied mediation using several femc chelates, ferric ethylenediarninediacetic acid dipropionic acid (FeEDADPA), FeCyDTA, femc diethylene triamine penta acetic acid (FeDTPA), ferric triethylene tetramine hexa acetic acid (FeTTHA) and ferric ethylenediaminetetraacetic acid (FeEDTA), with microbial biofuel cells containing *Streptococcus lactis, Lactobacillus plantarum,* and *Erwinia dissolvens* [302]. *Erwinia dissolvens* grows on coffee wastes whereas *Streptococcus lactis* and *Lactobacillus plantarum* grow on dairy products. The objective of this study was to determine which microorganism/mediator combination would be the most proficient in a microbial biofuel cell. Fuel cells with coulombic yields between 80 and 90% ran for five consecutive days by using *Erwinia dissolvens* mediated with FeCyDTA, *Streptococcus lactis* and *Lactobacillus plantarum* did not fabricate any important electrical output with any of the ferric chelates. The main results from this study were: 1) biofuel cells using *E. dissolvens* as the biocatalyst gave higher coulombic yields than those using *E. Coli.*, 2) different microorganisms, using the same mediators and the same amount of fuel, gave different coulombic yields. The interaction of the microbial membrane with the mediator was regarded to be the source of these effects.

11.2. Miniature Microbial Biofuel Cells and Its State of the Art

One of the early efforts on miniature microbial biofuel cells reported a current density of 150 mA/m^2 based on the 10 μm diameter circular anodic electrode and a surface power output of 0.023 mW/m^2 and [303]. The miniature microbial biofuel cells were limited by coulombic efficiency due to their high internal resistances and comparatively low volumetric power density. Compared to macroscale microbial biofuel cells using the same microbes [304], miniature microbial biofuel cells generated lower volumetric power density with a similar volumetric current density, which is inadequate for the anticipated applications. The internal resistances of miniature microbial biofuel cells were around 40 fold higher than that in the macro scale microbial biofuel cells. With the same catholyte and anolyte, the ohmic loss was higher in the micropillar devices. However, they generated higher volumetric power density (32 A/m^3) than the serpentine-channel devices (0.5 A/m^3) [304]. The good microbe adaptivity of the micropillar electrodes decreased the anode resistance and the high surface area-to-volume ratio resulted in higher volumetric power output. Carbon-based anodes are widely used in macroscale microbial biofuel cells because they are known for easy adaptation of microorganism and high surface area-to-volume ratio. The recent investigations provide promising solutions for fabricating carbon-based anodes in miniature microbial biofuel cells by using carbon nanotubes (CNT) as electrodes [305]. The CNT based electrodes showed great improvement in the biocompatibility and electricity generation. Its highest power density was 42 mW/m^2 using *E. coli* as the microbial catalyst.

Various strategies were employed in the pursuit to enhance the miniature microbial biofuel cell performance, such as improving coupling of a microorganism to anode surface, increasing the anode surface area, decreasing proton diffusion resistance and developing electrochemically active microbes. In a nutshell, the enhancement strategies resulted in improved reaction kinetics, prolonged mass transport, and reduced ohmic resistance. Based on these developments, the ability to generate sufficient current and power from miniature devices was realized. This also breaks the conventional concept that small scale microbial biofuel cells would redact unsatisfactorily due to the limited quantity of substrates and microorganism. Hence, the volumetric power density and coulombic efficiency have been increased over 5 times since the development of the first miniature microbial biofuel cells in 2006.

The output potential from the miniature MFCs is still insufficient for powering conventional equipment. Although they are promising options for on-chip power sources, especially for medical implants, which only need several mill volts to operate. However, massive ohmic heating and electrolysis during the operation were observed due to higher current density. Therefore, the stability of the miniature MFCs, study in optimizing current density, microbe leakage, electrode design and device configuration for implantation rejection, analysis of the composition and distribution of internal resistances is necessary before further implementation in practical applications.

12. Microbial Biotechnology in the Monitoring of Environmental Pollution

The chemicals present in the environment can be monitored in two general ways. The first approach is based on chemical or physical analysis, termed as traditional and has some advantages such as its necessity for regulatory purposes and also to understand both the causes of pollution and potential remediation allowing with a highly accurate and sensitive determination of the exact composition of any sample. The methodologies do not have the capability to provide the data on the bioavailability of pollutants, their effects on the living systems and their synergistic behavior in mixtures. Also, this approach faces some disadvantages for its complexity and cost because it requires specialized laboratory arrangements.

The second approach, based on the use of living systems in different environmentally oriented bioassays is termed as complimentary which use numerous biological systems ranging from living organism assay including fish toxicity tests to others based on sub-cellular components or enzymes. All of them assay the effect of targeted chemicals instead of identifying them and this is done by using a special position of the test organisms which is held by a unicellular microorganism such as bacteria. The use of bacteria is a profitable option for pollution monitoring due to the large population size, rapid growth rates, low costs and facile maintenance and also they can be genetically modified to respond by a detectable signal to pre-specified changes in their environmental conditions. To monitor the chemicals in the environment, recently several reviews [306-310] have reported the different aspects of the use of genetically engineered microorganisms.

12.1. Biosensors

Development of new technologies and suitable methodologies for environmental monitoring are encouraged by the need for disposable systems or tools. In this respect, in recent years, biosensors possessed the greatest potential and that is why proposed as analytical tools for effective monitoring in many programs. By IUPAC, a biosensor is however defined as a self-contained integrated device which can provide specific quantitative or semi-quantitative analytical information by the help of biological recognition element/biochemical receptor retaining in direct spatial contact with a transduction element [311].

Although many of the biosensors cannot actually compete with the conventional analytical methods in terms of accuracy and reproducibility, they can be used by regularity authorities and in industries to provide enough information for routine testing and screening of samples. The possibility of probability, miniaturization, working on-site and the ability to measure pollutants in complex matrices with minimal sample preparation are some advantages that biosensors provide over conventional analytical techniques [312].

Enzymatic biosensors are useful in the determination of the toxicity of certain compounds in the cells by using the selective inhibition of specific enzymes. Different classes of compounds can be used to quantify the inhibition based on the decrease in activity of the immobilized enzyme in the presence of the target analyte. A wide range of enzymes based on their catalytic properties or substrate specificity which can be modified by genetic engineering and whole cells of living organisms, such as bacteria, fungi, yeasts, plants or animal cells, tissue slices based on their general metabolic status can be used as recognition components [313].

12.1.1. Type of Microbial Biosensors (Transducers)

Different types of microbial biosensors have been used recently for this certain purpose. They are as follows:

12.1.1.1. Optical Biosensor

It is a device where an optical transducer is used to change the optical properties such as adsorption, fluorescence, luminescence, refractive index etc. are directly proportional to the concentration of the analytes [314]. Biosensors based on fluorescence, bioluminescence, and colorimeter are most popular nowadays due to some important properties like small probe size, compactness, selectivity, sensitivity, flexibility, and resistance to the electrical nose [315].

12.1.1.2. Fluorescent Microbial Biosensor

These can emit fluorescent light which is directly proportional to the very dilute concentration of analytes [316, 317].

12.1.1.3. Bioluminescent Microbial Biosensor

Because of their ability to reflect toxicity, these have been widely used for detection of toxicity in environmental monitoring [318].

12.1.1.4. Colorimetric Microbial Biosensor

They utilize the changes in the color of the special compound to determine the concentration of the ascertain analytes. A colorimetric microbial biosensor for detection of methyl parathion was developed by immobilizing the *Sphingomonas* bacteria on the inner epidermis of onion bulb scale and a stable characteristic linear range was achieved [319].

12.1.1.5. MFC Biosensor

Microbial fuel cells (MFCs) are now using as transducers due to their ability to convert organic substrates into electricity through microbial catabolism [320] which usually are two chambered cell consists of an anodic and a cathodic chamber separated by a proton exchange membrane [321].

12.2. Different Types of Pollutant and Their Detection

Table 6. Biosensors for determination of organic and inorganic compounds and relevant parameters in the environment

Environmentally relevant compounds or parameters	Transducing element	Biorecognition element	Features	References
Toxicity	Optical (bioluminescence)	Recombinant bioluminescent Bacteria	Portable	[322]
Genotoxicity	Optical (bioluminescence)	Recombinant *Escherichia coli*	–	[323]
EDCs	Optical (fluorescence microscopy)	Recombinant yeast	–	[324]
BOD	Electrochemical (amperometric)	Multispecies culture	Minimum measurable BOD=0.088 mg/L O_2; (biosensor BOD/BOD_5) ratio=0.80	[325]

Environmentally relevant compounds or parameters	Transducing element	Biorecognition element	Features	References
Low BOD	Optical (optic fiber)	*Pseudomonas putida*	Minimum measurable BOD=0.5 mg/L O_2; comparison with BOD_5: R2=0.971	[326]
Chlamydia trachomatis (DNA)	Electrochemical (chronopotentiometric)	DNA (hybridisation)	Previous PCR amplification LOD: 0.2 mg/L	[327]
Salmonella enteriditis, Lysteria monocytogenes	Optical (SPR)	Antibodies	10^6 cell/mL	[328]
Salmonella Typhymurium	Acoustic	Antibodies	100 cells/mL	[329]
Escherichia Coli	Electrochemical (potentiometric)	Antibodies	10 cells/mL	[330]
		Organic compounds		
Isoproturon, diuron Simazine	Optical (fluorescence	*Chlorella vulgaris* (Algae cells)	0.025 µg/L 0.5 µg/L	[331]
Propanil	Optical (fluorescence)	Antibodies	0.6 ng/L	[332]
Carbaryl	Optical (SPR)	Antibodies	Self-assembled monolayers LOD: 1.38 µg/L	[333]
Organophosphorus Compounds	Electrochemical (amperometric)	Enzyme (organophosphorus hydrolase)	Modified carbon nanotube	[334]
Dichlorvos, parathion, anzinphos (organophosphorous pesticides)	Electrochemical, sonochemical	Enzyme (AChE)	Array of microelectrodes	[335]
Paraoxon and carbofuran (pesticides)	Electrochemical (amperometric)	Enzyme (AChE)	Discrimination between different AChEinhibitors by neuronal networks LOD: 0.2 µg/L	[336]
Pesticides and estrone	Optical (fluorescence)	Antibodies	Natural water samples	[337-340]
Hormones (estrone, progesterone, testosterone)	Optical (fluorescence)	Antibodies	LOD: sub-ng/L	[341-343]
PCBs	Optical (fluorescence)	Antibodies	LOD: 0.1 ng/L	[344]
Dioxine and dioxine-like Chemicals	Optical (luminescence) Recombinant	Recombinant hepatome cells	LOD: 10 pM	[345]
Phenols	Electrochemical (amperometric)	Enzymatic (cellobiose dehydrogenase and quinoproteindependent glucose dehydrogenase)	In-field measurements LOD: 0.8 µg/L	[346]

Table 6. (Continued)

Environmentally relevant compounds or parameters	Transducing element	Biorecognition element	Features	References
Chlorophenols	Optical chemiluminescence (fiber-optic)	Enzymatic (horseradish peroxidase)	LOD: 1.4–1975 µg/L	[347]
Bisphenol A	Optical (fluorescence)	Antibodies	Monitoring in a waterworks	[348]
Surfactants	Electrochemical (amperometric)	*Pseudomonas* and *Achromobacter* (plasmid for anionic surfactant degradation)	LOD: 0.25 mg/L (SDS)	[349, 350]
LAS	Electrochemical (amperometric) (O_2 consumption measurement)	*Trichosporoncutaneum* (LAS-degrading bacteria)	In-situ measurement LOD: 0.2 mg/L	[351]
Alkylphenols and their Ethoxylates	Electrochemical	Antibodies	LOD: µg/L range	[352]
Nonylphenol	Electrochemical (amperometric)	Antibodies	LOD: 10 µg/L range	[353]
Benzene and itsderivatives	Optical (luminescence)	Recombinant *Escherichia coli*	Air determination LOD: 0.5 mg/L LOD: 0.5 mg/L	[354]
Naphthalene and Phenanthrene	Optical (fluorescence)	DNA	Sol–gel array	[355]
Daunomicyn PCBs, Aflatoxin	Electrochemical (chronopotentiometry)	DNA	LOD: 0.3 mg/L LOD: 0.2 mg/L	[327]
Antibiotics	Optical (fluorescence)	Antibodies		[356]
Inorganic compounds				
Mercury arsenite	Optical (bioluminescence)	Pseudomonas fluorescens (with sensors plasmids)	0.003 µg/kg 0.7 µg/kg	[357]
Cadmium, copper, and Mercury	Electrochemical, optical and mass signal	Enzyme (AChE and urease)	Array biosensor for the measurement of pH, urea, heavy metals and ACh	[358]
Heavy metals	Optical (bioluminescence)	Recombinant *Escherichia coli*	Bioavailable fraction in soils	[359]
Inorganic phosphate	Electrochemical (amperometric)	Enzymatic (trienzymatic configuration)	LOD: 0.57 mg/L	[312]
Nitrate	Optical (fluorescence)	Recombinant *Escherichia coli*	Without interference of phosphate, chloride or nitrite	[360]

CONCLUSION

Modern technologies are now significantly focusing on the use of microorganism for maintaining a sustainable environment. In agriculture, the use of plant growth promoting microorganisms will soon replace the use of chemical fertilizers. These organisms have the ability to detoxify the soil from various pesticides residues. Thus, it can maintain soil microflora, soil nutritional status, physical and chemical properties of the soil and, which will ultimately help to enhance crop growth.

Microbial fuel cell or microbial electrolysis cell has emerged out as a reliable energy recovery and wastewater treatment method. Microorganisms are widely used in nitrification, denitrification, and anaerobic oxidation of ammonia. It can degrade chloro-organic compounds which are known as the most persistent environmental contaminant. It can remove heavy metals, dyes, surfactants, aromatics, heterocyclic compounds from wastewater.

There is vast biodiversity of microalgae and these are being used in the production of biodiesel, bio-oil, bio-syngas, and bio-hydrogen. Biofuel cell is currently under research to produce power from it. Low current density, handling problem of microbial catalysts and the proper understanding of biocatalytic redox reactions are the challenges to be improved for biofuel cell. Even conventional fossil fuel treatment use microbes for desulphurization.

Treatment of municipal wastes and wastewater at present is not possible without the use of microbes. Microbes multiples naturally and this is why this treatment is less costly. In the end, it can be averred that the benefits of the use of microbial catalysts in different technological processes are enormous. The role of microbial catalysts in the technological advancement of this time is inevitable for a sustainable environment.

REFERENCES

[1] Vallero, D. *Environmental biotechnology: a biosystems approach*, Academic Press, 2010.

[2] Parr, J. F., and Hornick, S. B. (1992). Agricultural use of organic amendments: A historical perspective. *Am J Alter Agr., 7*: 181-189.

[3] Onwurah, I. (2004). An integrated environmental biotechnology for enhanced bioremediation of crude oil contaminated agricultural land. *Bio-Research, 1*: 51-60.

[4] OECD. (2001). The Application of Biotechnology to Industrial Sustainability. *Organisation for Economic Co-operation and Development.*

[5] Parr, J., and Hornick, S. (1994). Assessment of the Third International Conference on Kyusei Nature Farming: Round Table Discussion by USDA Scientists, October 7, 1993. *Published by the Nature Farming Research and Development Foundation, Lompoc, California, USA.*

[6] Santos, V. B., Araújo, A. S., Leite, L. F., et al. (2012). Soil microbial biomass and organic matter fractions during transition from conventional to organic farming systems. *Geoderma, 170*: 227-231.

[7] Youssef, M., and Eissa, M. (2014). Biofertilizers and their role in management of plant parasitic nematodes. A review. *E3 J Biotechnol. Pharm Res., 5*: 1-6.

[8] Logan, B. E., Hamelers, B., Rozendal, R., et al. (2006). Microbial fuel cells: methodology and technology. *Env Sci Technol., 40*: 5181-5192

[9] Rabaey, K., Rodriguez, J., Blackall, L. L., et al. (2007). Microbial ecology meets electrochemistry: electricity-driven and driving communities. *The ISME journal, 1*: 9-18.

[10] Gorby, Y. A., Yanina, S., McLean, J. S., et al. (2006). Electrically conductive bacterial nanowires produced by Shewanella oneidensis strain MR-1 and other microorganisms. *Proceedings of the National Academy of Sciences, 103*: 11358-11363.

[11] Stams, A. J., De Bok, F. A., Plugge, C. M., et al. (2006). Exocellular electron transfer in anaerobic microbial communities. *Environ Microbiol., 8*: 371-382.

[12] Kim, B.-H., Kim, H.-J., Hyun, M.-S., et al. (1999). Direct electrode reaction of Fe (III)-reducing bacterium, Shewanella putrefaciens. *J Microbiol Biotechn., 9*: 127-131.

[13] Potter, M. C. (1911). Electrical effects accompanying the decomposition of organic compounds. *Proceedings of the Royal Society of London. Series B, Containing Papers of a Biological Character*: 260-276.

[14] Logan, B. E., and Regan, J. M. (2006). Electricity-producing bacterial communities in microbial fuel cells. *Trends Microbiol., 14*: 512-518.

[15] Liu, H., Hu, H., Chignell, J., et al. (2010). Microbial electrolysis: novel technology for hydrogen production from biomass. *Biofuels, 1*: 129-142.

[16] Cao, X., Huang, X., Liang, P., et al. (2009). A new method for water desalination using microbial desalination cells. *Environ Sci Technol., 43*: 7148-7152.

[17] Heijnen, J. J. (1999). Bioenergetics of microbial growth. *Encyclopedia of Bioprocess Technology*.

[18] Call, D., and Logan, B. E. (2008). Hydrogen production in a single chamber microbial electrolysis cell lacking a membrane. *Environ Sci Technol., 42*: 3401-3406.

[19] Rozendal, R. A., Hamelers, H. V., Rabaey, K., et al. (2008). Towards practical implementation of bioelectrochemical wastewater treatment. *Trends Biotechnol., 26*: 450-459.

[20] He, Z. (2013) Energy Issues in Bioelectrochemical Treatment of Water and Wastewater, In *224th ECS Meeting (October 27–November 1, 2013)*, Ecs.

[21] Cheng, K. Y., Ho, G., and Cord-Ruwisch, R. (2012). Energy-efficient treatment of organic wastewater streams using a rotatable bioelectrochemical contactor (RBEC). *Bioresource Technol., 126*: 431-436.

[22] Yetilmezsoy, K., and Sapci-Zengin, Z. (2009). Recovery of ammonium nitrogen from the effluent of UASB treating poultry manure wastewater by MAP precipitation as a slow release fertilizer. *J Hazard Mater., 166*: 260-269.

[23] Knight, R. L., Payne Jr, V. W., Borer, R. E., et al. (2000). Constructed wetlands for livestock wastewater management. *Ecol Eng., 15*: 41-55.

[24] De-Bashan, L. E., Hernandez, J.-P., Morey, T., et al. (2004). Microalgae growth-promoting bacteria as "helpers" for microalgae: a novel approach for removing ammonium and phosphorus from municipal wastewater. *Water Res., 38*: 466-474.

[25] Gupta, A., and Gupta, S. (2001). Simultaneous carbon and nitrogen removal from high strength domestic wastewater in an aerobic RBC biofilm. *Water Res., 35*: 1714-1722.

[26] Zeng, W., Zhang, Y., Li, L., et al. (2009). Control and optimization of nitrifying communities for nitritation from domestic wastewater at room temperatures. *Enzyme MicrobTtech., 45*: 226-232.

[27] Karthikeyan, O. P., and Joseph, K. (2006) Anammox" a novel process for nitrogen management in bioreactor landfills-a review, In *National seminar on Integrated municipal solid waste management", Periyar Maniammai College of Technology for Women, Thanjavur, India, April*, 27-28.

[28] Peyton, B. M., Mormile, M. R., and Petersen, J. N. (2001). Nitrate Reduction with Halomonas campisalis: Kinetics of Denitrification at pH 9 and 12.5% NaCl. *Water Res., 35*: 4237-4242.

[29] Park, J., Craggs, R., and Sukias, J. (2009). Removal of nitrate and phosphorus from hydroponic wastewater using a hybrid denitrification filter (HDF). *Bioresource Technol., 100*: 3175-3179.

[30] Munz, G., Gori, R., Cammilli, L., et al. (2008). Characterization of tannery wastewater and biomass in a membrane bioreactor using respirometric analysis. *Bioresource Technol., 99*: 8612-8618.

[31] Chimenos, J., Fernandez, A., Villalba, G., et al. (2003). Removal of ammonium and phosphates from wastewater resulting from the process of cochineal extraction using MgO-containing by-product. *Water Res., 37*: 1601-1607.

[32] Shen, J., He, R., Han, W., et al. (2009). Biological denitrification of high-nitrate wastewater in a modified anoxic/oxic-membrane bioreactor (A/O-MBR). *J Hazard Mater., 172*: 595-600.

[33] Carrera, J., Vicent, T., and Lafuente, J. (2004). Effect of influent COD/N ratio on biological nitrogen removal (BNR) from high-strength ammonium industrial wastewater. *Process Biochem., 39*: 2035-2041.

[34] Oguz, M. (2005). Investigation of nitrifying bacterial activities by monitoring nitrite oxidation, nitrate formation and carbon dioxide fixation during activated sludge treatment in the presence of metabolic inhibitors allylthiourea and azide. *Erciyes Üniversitesi Fen Bilimleri Enstitüsü Dergisi, 21*: 154-165.

[35] Watkinson, A., Murby, E., and Costanzo, S. (2007). Removal of antibiotics in conventional and advanced wastewater treatment: Implications for environmental discharge and wastewater recycling. *Water Res., 41*: 4164-4176.

[36] Sedlak, R. I. *Phosphorus and nitrogen removal from municipal wastewater: principles and practice*, CRC Press, 1991.

[37] Sabalowsky, A. R. (1999) *An investigation of the feasibility of nitrification and denitrification of a complex industrial wastewater with high seasonal temperatures*, Virginia Polytechnic Institute and State University.

[38] Pennington, P., and Ellis, R. (1993). Autotrophic and heterotrophic nitrification in acidic forest and native grassland soils. *Soil Biol Biochem., 25*: 1399-1408.

[39] Rittmann, B. E., and McCarty, P. L. *Environmental biotechnology: principles and applications*, Tata McGraw-Hill Education, 2012.

[40] Cyplik, P., Grajek, W., Marecik, R., et al. (2007). Effect of macro/micro nutrients and carbon source over the denitrification rate of Haloferax denitrificans archaeon. *Enzyme Microb Tech., 40*: 212-220.

[41] Lai, E., Senkpiel, S., Solley, D., et al. (2004). Nitrogen removal of high strength wastewater via nitritation/denitritation using a sequencing batch reactor. *Water Sci Technol., 50*: 27-33.

[42] Guo, H., Zhou, J., Su, J., et al. (2005). Integration of nitrification and denitrification in airlift bioreactor. *Biochem Eng J., 23*: 57-62.

[43] Gálvez, J., Gómez, M., Hontoria, E., et al. (2003). Influence of hydraulic loading and air flowrate on urban wastewater nitrogen removal with a submerged fixed-film reactor. *J Hazard Mater., 101*: 219-229.
[44] Chiu, Y.-C., and Chung, M.-S. (2003). Determination of optimal COD/nitrate ratio for biological denitrification. *Int Biodeter Biodegr., 51*: 43-49.
[45] Tal, Y., Watts, J. E., and Schreier, H. J. (2006). Anaerobic ammonium-oxidizing (anammox) bacteria and associated activity in fixed-film biofilters of a marine recirculating aquaculture system. *Appl Environ Microb., 72*: 2896-2904.
[46] Jetten, M. S., Wagner, M., Fuerst, J., et al. (2001). Microbiology and application of the anaerobic ammonium oxidation ('anammox') process. *Curr Opin Biotech., 12*: 283-288.
[47] Jetten, M. S., Strous, M., Pas-Schoonen, K. T., et al. (1998). The anaerobic oxidation of ammonium. *FEMS Microbiol Rev., 22*: 421-437.
[48] Gallert, C., and Winter, J. (1997). Mesophilic and thermophilic anaerobic digestion of source-sorted organic wastes: effect of ammonia on glucose degradation and methane production. *Appl Microbiol Biot., 48*: 405-410.
[49] Held, C., Wellacher, M., Robra, K.-H., et al. (2002). Two-stage anaerobic fermentation of organic waste in CSTR and UFAF-reactors. *Bioresource Technol., 81*: 19-24.
[50] Strous, M., Fuerst, J. A., Kramer, E. H., et al. (1999). Missing lithotroph identified as new planctomycete. *Nature, 400*: 446-449.
[51] Strous, M., Heijnen, J., Kuenen, J., et al. (1998). The sequencing batch reactor as a powerful tool for the study of slowly growing anaerobic ammonium-oxidizing microorganisms. *Appl Microbiol Biot., 50*: 589-596.
[52] Strous, M., Kuenen, J. G., and Jetten, M. S. (1999). Key physiology of anaerobic ammonium oxidation. *Appl Environ Microb., 65*: 3248-3250.
[53] Third, K., Sliekers, A. O., Kuenen, J., et al. (2001). The CANON system (completely autotrophic nitrogen-removal over nitrite) under ammonium limitation: interaction and competition between three groups of bacteria. *Syst Appl Microbio.l, 24*: 588-596.
[54] Burton, S. A., and Watson-Craik, I. A. (1998). Ammonia and nitrogen fluxes in landfill sites: applicability to sustainable landfilling. *Waste Manage Res., 16*: 41-53.
[55] Lema, J., Mendez, R., and Blazquez, R. (1988). Characteristics of landfill leachates and alternatives for their treatment: a review. *Water, Air, Soil Poll., 40*: 223-250.
[56] Barton, P. K., and Atwater, J. W. (2002). Nitrous oxide emissions and the anthropogenic nitrogen in wastewater and solid waste. *J Environ Eng., 128*: 137-150.
[57] Benitez, F., Beltran-Heredia, J., Real, F., et al. (1999). Enhancement of the ozonation of wine distillery wastewaters by an aerobic pretreatment. *Bioprocess Eng., 21*: 459-464.
[58] Vlissidis, A., and Zouboulis, A. (1993). Thermophilic anaerobic digestion of alcohol distillery wastewaters. *Bioresource Technol., 43*: 131-140.
[59] Coetzee, G., Malandra, L., Wolfaardt, G., et al. (2004). Dynamics of a microbial biofilm in a rotating biological contactor for the treatment of winery effluent. *Water SA, 30*: 407-412.
[60] Wilkie, A. C., Riedesel, K. J., and Owens, J. M. (2000). Stillage characterization and anaerobic treatment of ethanol stillage from conventional and cellulosic feedstocks. *Biomass Bioenerg., 19*: 63-102.
[61] Wolmarans, B., and De Villiers, G. H. (2004). Start-up of a UASB effluent treatment plant on distillery wastewater. *Water SA, 28*: 63-68.

[62] Dahiya, J., Singh, D., and Nigam, P. (2001). Decolourisation of molasses wastewater by cells of Pseudomonas fluorescens immobilised on porous cellulose carrier. *Bioresource Technol., 78*: 111-114.

[63] Kort, M. (1979). Colour in the sugar industry. *Science and Technology. Applied Science, London*: 97-130.

[64] Pazouki, M., Shayegan, J., and Afshari, A. (2008). Screening of microorganisms for decolorization of treated distillery wastewater. *Iran J Sci Technol., 32*: 53.

[65] Mohana, S., Desai, C., and Madamwar, D. (2007). Biodegradation and decolourization of anaerobically treated distillery spent wash by a novel bacterial consortium. *Bioresource Technol, 98*: 333-339.

[66] Kumar, S., and Viswanathan, L. (1991). Production of biomass, carbon dioxide, volatile acids, and their interrelationship with decrease in chemical oxygen demand, during distillery waste treatment by bacterial strains. *Enzyme Microb Tech., 13*: 179-187.

[67] Jain, N., Minocha, A., and Verma, C. (2002). Degradation of predigested distillery effluent by isolated bacterial strains. *Indian J Exp Biol., 40*: 101-105.

[68] Nakajima-Kambe, T., Shimomura, M., Nomura, N., et al. (1999). Decolorization of molasses wastewater by Bacillus sp. under thermophilic and anaerobic conditions. *J Biosci Bioeng., 87*: 119-121.

[69] Ohmomo, S., Daengsubha, W., Yoshikawa, H., et al. (1988). Screening of Anaerobic Bacteria with the Ability to Decolorize Molasses Melanoidin (Microbiology & Fermentation Industry). *Agr Biol Chem., 52*: 2429-2435.

[70] Sirianuntapiboon, S., Phothilangka, P., and Ohmomo, S. (2004). Decolorization of molasses wastewater by a strain No. BP103 of acetogenic bacteria. *Bioresource Technol., 92*: 31-39.

[71] Dahiya, J., Singh, D., and Nigam, P. (2001). Decolourisation of synthetic and spentwash melanoidins using the white-rot fungus Phanerochaete chrysosporium JAG-40. *Bioresource Technol., 78*: 95-98.

[72] Kumar, P., and Chandra, R. (2006). Decolourisation and detoxification of synthetic molasses melanoidins by individual and mixed cultures of Bacillus spp. *Bioresource Technol., 97*: 2096-2102.

[73] Fahy, V., FitzGibbon, F., McMullan, G., et al. (1997). Decolourisation of molasses spent wash by Phanerochaete chrysosporium. *Biotechnol Lett., 19*: 97-99.

[74] Kumar, V., Wati, L., Nigam, P., et al. (1998). Decolorization and biodegradation of anaerobically digested sugarcane molasses spent wash effluent from biomethanation plants by white-rot fungi. *Process Biochem., 33*: 83-88.

[75] Benito, G. G., Miranda, M. P., and de los Santos, D. R. (1997). Decolorization of wastewater from an alcoholic fermentation process with Trametes versicolor. *Bioresource Technol., 61*: 33-37.

[76] Jiménez, A. M., Borja, R., and Martın, A. (2003). Aerobic–anaerobic biodegradation of beet molasses alcoholic fermentation wastewater. *Process Biochem., 38*: 1275-1284.

[77] Patil, P., Kapadnis, B., and Dhamankar, V. (2003). Decolorisation of synthetic melanoidin and biogas effluent by immobilised fungal isolate of Aspergillus niger UM2. *Int Sugar J., 105*: 8, 10-13.

[78] Miyata, N., Iwahori, K., and Fujita, M. (1998). Manganese-independent and-dependent decolorization of melanoidin by extracellular hydrogen peroxide and peroxidases from Coriolus hirsutus pellets. *J Ferment Bioeng., 85*: 550-553.

[79] Raghukumar, C., Mohandass, C., Kamat, S., et al. (2004). Simultaneous detoxification and decolorization of molasses spent wash by the immobilized white-rot fungus Flavodon flavus isolated from a marine habitat. *Enzyme Microb Tech., 35*: 197-202.

[80] Kahraman, S., and Yeşilada, O. (2003). Decolorization and bioremediation of molasses wastewater by white-rot fungi in a semi-solid-state condition. *Folia microbiol., 48*: 525-528.

[81] Sirianuntapiboon, S., Somchai, P., Ohmomo, S., et al. (1988). Screening of Filamentous Fungi Having the Ability to Decolorize Molasses Pigments (Microbiology & Fermentation Industry). *Agr Biol Chem., 52*: 387-392.

[82] Angayarkanni, J., Palaniswamy, M., and Swaminathan, K. (2003). Biotreatment of distillery effluent using Aspergillus niveus. *B Environ Contam Tox., 70*: 0268-0277.

[83] Watanabe, Y., Sugi, R., TANAKA, Y., et al. (1982). Enzymatic decolorization of melanoidin by Coriolus sp. No. 20. *Agr Biol Chem., 46*: 1623-1630.

[84] Shayegan, J., Pazouki, M., and Afshari, A. (2005). Continuous decolorization of anaerobically digested distillery wastewater. *Process Biochem., 40*: 1323-1329.

[85] Rajendran, P., and Gunasekaran, P. (2006), 138-159, MJP Publishers.

[86] Robinson, T., McMullan, G., Mardhant, R., et al. (2001). *Bioresource Technol, 77*: 247255.

[87] O'Neill, C., Hawkes, F. R., Lourenco, N. D., et al. (1999). *J Chem Technol Biotechnol, 74*: 1009- 1018.

[88] Caselattoa, A. M. F., Coelhoa, D. F., Silveirab, E., et al. (2014). Determination of Strains from Tropical Culture Collection for Biological Treatment of Effluents from Textile Industry. *The Italian Association of Chemical Engineering 37*: 841-846.

[89] Chang, Chou, J.-S., Chien, C., et al. (2001). *Process Biochem, 36*: 757-763.

[90] Kuhad, R., Sood, N., Tripathi, K., et al. (2004). Developments in microbial methods for the treatment of dye effluents. *Adv Appl Microbiol., 56*: 185-213.

[91] Máximo, C., Amorim, M. T. P., and Costa-Ferreira, M. (2003). Biotransformation of industrial reactive azo dyes by Geotrichum sp. CCMI 1019. *Enzyme Microb Tech., 32*: 145-151.

[92] Martins, M. A. M., Queiroz, M. J., Silvestre, A. J., et al. (2002). Relationship of chemical structures of textile dyes on the pre-adaptation medium and the potentialities of their biodegradation by Phanerochaete chrysosporium.*Res Microbiol., 153*: 361-368.

[93] Shin, K.-S. (2004). The role of enzymes produced by white-rot fungus Irpex lacteus in the decolorization of the textile industry effluent. *J Microbiol-Seoul., 42*: 37-41.

[94] Selvam, K., and Priya, M. S. (2012). Biological treatment of Azo dyes and textile industry effluent by newly isolated White rot fungi Schizophyllum commune and Lenzites eximia. *International J Environ Sci., 2*: 1926-1935.

[95] Banat, I. M., Mcmullan, G., Mechan, C., et al. (1999), 1-6, In *Proceedings of the Industrial Waste Technical Conference Indian Policy.*

[96] Yu, Z., and Wen, X. (2005). Screening and identification of yeasts for decolorizing synthetic dyes in industrial wastewater. *Int Biodeter Biodegr., 56*: 109-114.

[97] Jadhav, J., and Govindwar, S. (2006). Biotransformation of malachite green by Saccharomyces cerevisiae MTCC 463. *Yeast, 23*: 315-323.

[98] Jinqi, L., and Haitian, L. (1992). *Environ Pollut, 75*: 279-283.

[99] Jinqi, L., and Houtian, L. (1992). Degradation of azo dyes by algae. *Environ Pollut., 75*: 273-278.

[100] Aksu, Z. (2005). Application of biosorption for the removal of organic pollutants: a review. *Process Biochem., 40*: 997-1026.

[101] Özer, A., Akkaya, G., and Turabik, M. (2006). The removal of Acid Red 274 from wastewater: Combined biosorption and biocoagulation with Spirogyra rhizopus. *Dyes Pigments., 71*: 83-89.

[102] Daneshvar, N., Ayazloo, M., Khataee, A., et al. (2007). Biological decolorization of dye solution containing Malachite Green by microalgae Cosmarium sp. *Bioresource Technol., 98*: 1176-1182.

[103] Ertugrul, S., Bakıra, M., and Donmez, G. (2008). *Ecol Eng, 32*: 224-244.

[104] An, S. Y., Min, S. K., Cha, I. H., et al. (2004). *Biotechnol, 24*: 1037-1040.

[105] Baldrian, P. (2006). *FEMS Microbiol Rev, 30*: 215-292.

[106] Leena, R., and Selva Raj, D. (2008). Bio-decolorization of Textile Effluent Containing Reactive Black-B by Effluent-adapted and Non-adapted bacteria. *Afr J Biotechnol., 7*: 3309-3313.

[107] Khelifi, E., Ayed, L., Bouallagui, H., et al. (2009). *Hazard Mater. 163*: 1056-1062.

[108] Togo, C. A., Mutambanengwe, C. C. Z., and Whiteley, C. G. (2008). Decolourisation and degradation of textile dyes using a sulphate reducing bacteria (SRB)–biodigester microflora co-culture. *Afr J Biotechnol., 7*: 114-121.

[109] Elisangela, F., Andrea, Z., Fabio, D. G., et al. (2009). Biodegradation of textile azo dyes by a facultative Staphylococcus arlettae strain VN-11 using a sequential microaerophilic/aerobic process. *Int Biodeter Biodegr., 63*: 280-288.

[110] Hsueh, C.-C., and Chen, B.-Y. (2008). Exploring effects of chemical structure on azo dye decolorization characteristics by Pseudomonas luteola. *J Hazard Mater., 154*: 703-710.

[111] Pandey, A., Singh, P., and Iyengar, L. (2007). Bacterial decolourisation and degradation of azo dyes. *Int. Biodeterior. Biodegrad., 59*: 73-84.

[112] Vijaykumar, M., Vaishampayan, P. A., Shouche, Y. S., et al. (2007). Decolourization of naphthalene-containing sulfonated azo dyes by Kerstersia sp. strain VKY1. *Enzyme Microb Tech., 40*: 204-211.

[113] Tan, N. C., Van Leeuwen, A., Van Voorthuizen, E. M., et al. (2005). Fate and biodegradability of sulfonated aromatic amines. *Biodegradation, 16*: 527-537.

[114] Blumel, S., Knackmuss, H.J., et al. (2002). Molecular cloning and characterization of the gene coding for the aerobic azoreductase from Xenoplylus azovorans KF46F. *Appl Environ Microb., 68*: 3948–3955.

[115] Franciscon, E., Grossman, M. J., Paschoal, J. A. R., et al. (2012). *Decolorization and biodegradation of reactive sulfonated azo dyes by a newly isolated Brevibacterium sp. strain VN-15*. SpringerPlus, *1*: 37.

[116] Shah, M. Exploited application of Pseudomonas stutzeri ETL-4 in microbial degradation of Congo red

[117] Shah, M. P. (2014). Exploring the Strength of Pseudomonas Putida ETL-7 in Microbial Degradation and Decolorization of Remazol Black-B. *International Journal, 2*: 12-17.

[118] Shah, M., Patel, K., Nair, S., et al. (2013). Microbial Degradation of Reactive Red by Pseudomonas spp. *MPS-2. J Bioremed Biodeg, 4*: 2.

[119] Bheemaraddi, M. C., Patil, S., Shivannavar, C. T., et al. (2014). Isolation and Characterization of Paracoccus sp. GSM2 Capable of Degrading Textile Azo Dye Reactive Violet 5. *The Scientific World Journal, 2014*.

[120] Zhou, W., and Zimmermann, W. (1993). Decolorization of industrial effluents containing reactive dyes by actinomycetes. *FEMS Microbiol Lett., 107*: 157-161.

[121] Molina-Guijarro, J. M., Pérez, J., Muñoz-Dorado, J., et al. (2010). Detoxification of azo dyes by a novel pH-versatile, salt-resistant laccase from Streptomyces ipomoea. *Int Microbiol., 12*: 13-21.

[122] Saratale R. G., S. G. D., Parshetti G. K., Chang J. S., Govindwar S. P. (2009). Outlook of bacterial decolorization and degradation of azo dyes: a review. *Int J Environ Res Pu.,*

[123] Rao, M., Scelza, R., Scotti, R., et al. (2010). Role of enzymes in the remediation of polluted environments. *J Soil Sci Plant Nut,. 10*: 333-353.

[124] Dec, J., and Bollag, J. M. (1994). Use of plant material for the decontamination of water polluted with phenols. *Biotechnol Bioeng., 44*: 1132-1139.

[125] Mayer, A. M., and Harel, E. (1979). Polyphenol oxidases in plants. *Phytochemistry, 18*: 193-215.

[126] Alberti, B. N., and Klibanov, A. M. (1981) Enzymatic removal of dissolved aromatics from industrial aqueous effluents, In *Biotechnol. Bioeng. Symp.;(United States)*, Massachusetts Inst. of Tech., Cambridge.

[127] Gianfreda, L., Iamarino, G., Scelza, R., et al. (2006). Oxidative catalysts for the transformation of phenolic pollutants: a brief review. *Biocatal Biotransfor., 24*: 177-187.

[128] Rodríguez Couto, S., and Toca Herrera, J. L. (2006). Industrial and biotechnological applications of laccases: a review. *Biotechnol Adv., 24*: 500-513.

[129] Bollag, J. M. (1992). Decontaminating soil with enzymes. *Environ Sci Technol., 26*: 1876-1881.

[130] Dec, J., and Bollag, J.-M. (1990). Detoxification of substituted phenols by oxidoreductive enzymes through polymerization reactions. *Arch Environ Con Tox., 19*: 543-550.

[131] Arora, P. K., Kumar, M., Chauhan, A., et al. (2009). OxDBase: a database of oxygenases involved in biodegradation. *BMC Research Notes, 2*: 67.

[132] Karigar, C. S., and Rao, S. S. (2011). Role of microbial enzymes in the bioremediation of pollutants: a review. *Enzyme research, 2011.*

[133] Gianfreda, L., Xu, F., and Bollag, J.-M. (1999). Laccases: a useful group of oxidoreductive enzymes. *Bioremediat J., 3*: 1-26.

[134] Fox, B. G., Borneman, J. G., Wackett, L. P., et al. (1990). Haloalkene oxidation by the soluble methane monooxygenase from Methylosinus trichosporium OB3b: mechanistic and environmental implications. *Biochemistry, 29*: 6419-6427.

[135] Grosse, S., Laramee, L., Wendlandt, K.-D., et al. (1999). Enzymology and Protein Engineering-Purification and Characterization of the Soluble Methane Monooxygenase of the Type II Methanotrophic Bacterium Methylocystis sp Strain WI 14. *Appl Environ Microb., 65*: 3929-3935.

[136] Arora, P., Srivastava, A., and Singh, V. (2010). Application of monooxygenases in dehalogenation, desulphurization, denitrification and hydroxylation of aromatic compounds. *J Bioremed Biodegrad, 1*: 112.

[137] Ullah, M. A., Bedford, C. T., and Evans, C. S. (2000). Reactions of pentachlorophenol with laccase from Coriolus versicolor. *Appl Microbiol Biot., 53*: 230-234.

[138] Dedeyan, B., Klonowska, A., Tagger, S., et al. (2000). Biochemical and molecular characterization of a laccase from Marasmius quercophilus. *Appl Environ Microb., 66*: 925-929.

[139] Piontek, K., Smith, A., and Blodig, W. (2001). Lignin peroxidase structure and function. *Biochem Soc T., 29*: 111-116.

[140] Wong, D. W. (2009). Structure and action mechanism of ligninolytic enzymes. *Appl Biochem Biotech., 157*: 174-209.

[141] Ruiz-Dueñas, F. J., Morales, M., Pérez-Boada, M., et al. (2007). Manganese oxidation site in Pleurotus eryngii versatile peroxidase: a site-directed mutagenesis, kinetic, and crystallographic study. *Biochemistry, 46*: 66-77.

[142] Sánchez-Porro, C., Martin, S., Mellado, E., et al. (2003). Diversity of moderately halophilic bacteria producing extracellular hydrolytic enzymes. *J Appl Microbiol., 94*: 295-300.

[143] Schmidt, O. (2006). *Wood and tree fungi*. Springer-Verlag Berlin Heidelberg.

[144] MP, P., and Manjunath, K. (2011). Comparative study on biodegradation of lipid-rich wastewater using lipase producing bacterial species. *Indian J Biotechnol., 10*: 121-124.

[145] Sun, Y., and Cheng, J. (2002). Hydrolysis of lignocellulosic materials for ethanol production: a review. *Bioresource Technol., 83*: 1-11.

[146] Rashed, I. G. A.-A., Akunna, J., El-Halwany, M. M., et al. (2010). Improvement in the efficiency of hydrolysis of anaerobic digestion in sewage sludge by the use of enzymes. *Desalin Water Treat., 21*: 280-285.

[147] GUO, X.-s., LIU, J.-x., WEI, Y.-s., et al. (2007). Sludge reduction with Tubificidae and the impact on the performance of the wastewater treatment process. *J Environ Sci., 19*: 257-263.

[148] Ayol, A. (2005). Enzymatic treatment effects on dewaterability of anaerobically digested biosolids-I: performance evaluations. *Process Biochem., 40*: 2427-2434.

[149] Roman, H., Burgess, J., and Pletschke, B. (2006). Enzyme treatment to decrease solids and improve digestion of primary sewage sludge. *Afr J Biotechnol., 5*.

[150] Watson, S., Akhurst, T., Whiteley, C., et al. (2004). Primary sludge floc degradation is accelerated under biosulphidogenic conditions: Enzymological aspects. *Enzyme Microb Tech., 34*: 595-602.

[151] Dharmsthiti, S., and Kuhasuntisuk, B. (1998). Lipase from Pseudomonas aeruginosa LP602: biochemical properties and application for wastewater treatment. *J Ind Microbiol Biotechnol., 21: 75-80.*

[152] Parmar, N., Singh, A., and Ward, O. (2001). Enzyme treatment to reduce solids and improve settling of sewage sludge. *J Ind Microbiol Biotechnol., 26*: 383-386.

[153] Domingues, F., Queiroz, J., Cabral, J., et al. (2000). The influence of culture conditions on mycelial structure and cellulase production by Trichoderma reesei Rut C-30. *Enzyme Microb Tech., 26*: 394-401.

[154] Hageskal, G., Lima, N., and Skaar, I. (2009). The study of fungi in drinking water. *Mycol Res., 113*: 165-172.

[155] Kezhang, Z. (1995) *Ethanol and Distillation liquor Technology*, 294-295, China Light Industry Press Beijing, Chaina.

[156] Moraïs, S., Heyman, A., Barak, Y., et al. (2010). Enhanced cellulose degradation by nano-complexed enzymes: Synergism between a scaffold-linked exoglucanase and a free endoglucanase. *J Biotechnol., 147*: 205-211.

[157] Naika, G. S., and Tiku, P. K. (2010). Characterization of functional intermediates of endoglucanase from Aspergillus aculeatus during urea and guanidine hydrochloride unfolding. *Carbohyd Res., 345*: 1627-1631.

[158] Warren, R. (1996). Microbial hydrolysis of polysaccharides. *Annual Reviews in Microbiology, 50*: 183-212.

[159] Xue, Y., Jin, L., Liu, Z., et al. (2008). Purification and characterization of β-glucosidase from Reticulitermes flaviceps and its inhibition by valienamine and validamine. *Afr J Biotechnol., 7*.

[160] Parr, J., Papendick, R., Hornick, S., et al. (1992). Soil quality: attributes and relationship to alternative and sustainable agriculture. *Am J Alternative Agr., 7*: 5-11.

[161] Reganold, J. P., Papendick, R. I., and Parr., J. F. (1990). Sustainable Agriculture. *Sci Am.,262*: 112-120.

[162] Sciences., N. A. o. (1989) Alternative Agriculture., In *Committee on the Role of Alternative Agriculture Farming Methods in Modern Production Agriculture.*, 448, National Research Council, Board on Agriculture., National Academy Press, Washington, D.C., USA.

[163] Raja, N. (2013). Biopesticides and biofertilizers: ecofriendly sources for sustainable agriculture. *J Biofertil Biopestici, 4*: e112.

[164] Araújo, A., Santos, V., and Monteiro, R. (2008). Responses of soil microbial biomass and activity for practices of organic and conventional farming systems in Piauí state, Brazil. *Eur J Soil Biol., 44*: 225-230.

[165] Higa, T., and Wididana, G. (1991) Changes in the soil microflora induced by effective microorganisms, In *Proceedings of the First International Conference on Kyusei Nature Farming. US Department of Agriculture, Washington, DC, USA*, 153-161.

[166] Higa, T., and Parr, J. F. *Beneficial and effective microorganisms for a sustainable agriculture and environment*, Vol. 1, International Nature Farming Research Center Atami,, Japan, 1994.

[167] Chew, Tin, and Lee. (2008). *Application Of Beneficial Microorganisms.*

[168] Artursson, V., Finlay, R. D., and Jansson, J. K. (2006). Interactions between arbuscular mycorrhizal fungi and bacteria and their potential for stimulating plant growth. *Environ Microbiol., 8*: 1-10.

[169] Sinha, R. K., Valani, D., Chauhan, K., et al. (2010). Embarking on a second green revolution for sustainable agriculture by vermiculture biotechnology using earthworms: Reviving the dreams of Sir Charles Darwin. *Journal of Agricultural Biotechnology and Sustainable Development, 2*: 113-128.

[170] Singh, J. S., Pandey, V. C., and Singh, D. (2011). Efficient soil microorganisms: a new dimension for sustainable agriculture and environmental development. *Agriculture, ecosystems & environment, 140*: 339-353.

[171] Adesemoye, A. O., and Kloepper, J. W. (2009). Plant–microbes interactions in enhanced fertilizer-use efficiency. *Appl Microbiol Biot., 85*: 1-12.

[172] Megali, L., Glauser, G., and Rasmann, S. (2014). Fertilization with beneficial microorganisms decreases tomato defenses against insect pests. *Agron Sustain Dev., 34*: 649-656.

[173] Sahoo, R. K., Ansari, M. W., Dangar, T. K., et al. (2014). Phenotypic and molecular characterisation of efficient nitrogen-fixing Azotobacter strains from rice fields for crop improvement. *Protoplasma, 251*: 511-523.

[174] Ahemad, M., and Kibret, M. (2014). Mechanisms and applications of plant growth promoting rhizobacteria: Current perspective. *Journal of King Saud University-Science, 26*: 1-20.

[175] Bashan, Y., and Levanony, H. (1990). Current status of Azospirillum inoculation technology: Azospirillum as a challenge for agriculture. *Can J Microbiol., 36*: 591-608.
[176] James, E. K., and Olivares, F. L. (1998). Infection and colonization of sugar cane and other graminaceous plants by endophytic diazotrophs. *Crc Rev Plant Sci., 17*: 77-119.
[177] Richardson, A. E. (2001). Prospects for using soil microorganisms to improve the acquisition of phosphorus by plants. *Funct Plant Biol., 28*: 897-906.
[178] Rodríguez, H., and Fraga, R. (1999). Phosphate solubilizing bacteria and their role in plant growth promotion. *Biotechnol Adv., 17*: 319-339.
[179] Bar-Yosef, B. (1991). Root excretions and their environmental effects: influence on availability of phosphorus. *Eds. Waisel Y, Eshel A y Kafkafi U. Plant roots: The hidden half. Marcel Dekker, New York*: 581-605.
[180] Rodriguez, H., Gonzalez, T., Goire, I., et al. (2004). Gluconic acid production and phosphate solubilization by the plant growth-promoting bacterium Azospirillum spp. *Naturwissenschaften, 91*: 552-555.
[181] G. C. Tao, S. J. T., M. Y. Cai, G. H. Xie. (2008). Phosphatesolubilizing and -mineralizing abilities of bacteria isolated from soils1 1 project supported by the Scienti. *Pedosphere, 18*: 515-523.
[182] Guerinot, M. L., and Yi, Y. (1994). Iron: nutritious, noxious, and not readily available. *Plant Physiol., 104*: 815.
[183] Loper, J. E., and Buyer, J. S. (1991). Siderophores in microbial interactions on plant surfaces. *Mol. Plant-Microbe Interact, 4*: 5-13.
[184] Hider, R. C., and Kong, X. (2010). Chemistry and biology of siderophores. *Nat Prod Rep., 27*: 637-657.
[185] Neilands, J. (1981). Iron absorption and transport in microorganisms. *Annu Rev Nutr., 1*: 27-46.
[186] Jahanian, A., Chaichi, M. R., Rezaei, K., et al. (2012). The effect of plant growth promoting rhizobacteria (pgpr) on germination and primary growth of artichoke (Cynara scolymus). *Int. J. Agric. Crop Sci, 4*: 923-929.
[187] Ma, Y., Rajkumar, M., Luo, Y., et al. (2011). Inoculation of endophytic bacteria on host and non-host plants—effects on plant growth and Ni uptake. *J Hazard Mater., 195*: 230-237.
[188] Ma, Y., Rajkumar, M., Vicente, J., et al. (2010). Inoculation of Ni-resistant plant growth promoting bacterium Psychrobacter sp. strain SRS8 for the improvement of nickel phytoextraction by energy crops. *Int J Phytoremediat., 13*: 126-139.
[189] Zahir, Z., Shah, M. K., Naveed, M., et al. (2010). Substrate-dependent auxin production by Rhizobium phaseoli improves the growth and yield of Vigna radiata L. under salt stress conditions. *J Microbiol Biotechnol, 20*: 1288-1294.
[190] Wani, P. A., and Khan, M. S. (2010). Bacillus species enhance growth parameters of chickpea (Cicer arietinum L.) in chromium stressed soils. *Food Chem Toxicol., 48*: 3262-3267.
[191] Dary, M., Chamber-Pérez, M., Palomares, A., et al. (2010). "In situ" phytostabilisation of heavy metal polluted soils using Lupinus luteus inoculated with metal resistant plant-growth promoting rhizobacteria. *J Hazard Mater., 177*: 323-330.
[192] Phi, Q.-T., Park, Y.-M., Seul, K.-J., et al. (2010). Assessment of root-associated Paenibacillus polymyxa groups on growth promotion and induced systemic resistance in pepper. *J. Microbiol. Biotechnol, 20*: 1605-1613.

[193] Gholami, A., Shahsavani, S., and Nezarat, S. (2009). The effect of plant growth promoting rhizobacteria (PGPR) on germination, seedling growth and yield of maize. *Int J Biol Life Sci, 1*: 35-40.

[194] Rodrigues, E. P., Rodrigues, L. S., de Oliveira, A. L. M., et al. (2008). Azospirillum amazonense inoculation: effects on growth, yield and N2 fixation of rice (Oryza sativa L.). *Plant Soil., 302*: 249-261.

[195] Lawongsa, P., Boonkerd, N., Wongkaew, S., et al. (2008). Molecular and phenotypic characterization of potential plant growth-promoting Pseudomonas from rice and maize rhizospheres. *World Journal of Microbiology and Biotechnology, 24*: 1877-1884.

[196] Ganesan, V. (2008). Rhizoremediation of cadmium soil using a cadmium-resistant plant growth-promoting rhizopseudomonad. *Curr Microbiol., 56*: 403-407.

[197] Sinha, S., and Mukherjee, S. K. (2008). Cadmium–Induced Siderophore Production by a High Cd-Resistant Bacterial Strain Relieved Cd Toxicity in Plants Through Root Colonization. *Curr Microbiol., 56*: 55-60.

[198] Remans, R., Beebe, S., Blair, M., et al. (2008). Physiological and genetic analysis of root responsiveness to auxin-producing plant growth-promoting bacteria in common bean (Phaseolus vulgaris L.). *Plant and soil, 302*: 149-161.

[199] Beneduzi, A., Peres, D., Vargas, L. K., et al. (2008). Evaluation of genetic diversity and plant growth promoting activities of nitrogen-fixing bacilli isolated from rice fields in South Brazil. *Appl Soil Ecol., 39*: 311-320.

[200] Anjum, M. A., Sajjad, M. R., Akhtar, N., et al. (2007). Response of cotton to plant growth promoting rhizobacteria (PGPR) inoculation under different levels of nitrogen. *Journal of Agricultural Research (Pakistan)*.

[201] Wani, P., Khan, M., and Zaidi, A. (2007). Co-inoculation of nitrogen-fixing and phosphate-solubilizing bacteria to promote growth, yield and nutrient uptake in chickpea. *Acta Agronomica Hungarica, 55*: 315-323.

[202] Sheng, X. F., Jiang, C. Y., and He, L. Y. (2008). Characterization of plant growth-promoting Bacillus edaphicus NBT and its effect on lead uptake by Indian mustard in a lead-amended soil. *Can J Microbiol., 54*: 417-422.

[203] Faisal, M., and Hasnain, S. (2006). Growth stimulatory effect of Ochrobactrum intermedium and Bacillus cereus on Vigna radiata plants. *Lett Appl Microbiol., 43*: 461-466.

[204] Valverde, A., Burgos, A., Fiscella, T., et al. (2006). Differential effects of coinoculations with Pseudomonas jessenii PS06 (a phosphate-solubilizing bacterium) and Mesorhizobium ciceri C-2/2 strains on the growth and seed yield of chickpea under greenhouse and field conditions. *Plant and soil, 287*: 43-50.

[205] Gupta, A., Rai, V., Bagdwal, N., et al. (2005). In situ characterization of mercury-resistant growth-promoting fluorescent pseudomonads. *Microbiol Res., 160*: 385-388.

[206] Khalid, A., Arshad, M., and Zahir, Z. (2004). Screening plant growth-promoting rhizobacteria for improving growth and yield of wheat. *J Appl Microbiol., 96*: 473-480.

[207] Urrutia, M. (1997). General bacterial sorption processes. *Biosorbents for metal ions*: 39-66.

[208] Volesky, B. (2007). Biosorption and me. *Water research, 41*: 4017-4029.

[209] Mendil, D., Tuzen, M., and Soylak, M. (2008). A biosorption system for metal ions on Penicillium italicum loaded on Sepabeads SP 70 prior to flame atomic absorption spectrometric determinations. *J Hazard Mater., 152*: 1171-1178.

[210] Svecova, L., Spanelova, M., Kubal, M., et al. (2006). Cadmium, lead and mercury biosorption on waste fungal biomass issued from fermentation industry. I. Equilibrium studies. *Sep Purif Technol., 52*: 142-153.
[211] Rafiu Awofolu, O., Okechukwu Okonkwo, J., Badenhorst, J., et al. (2006). A new approach to chemical modification protocols of Aspergillus niger and sorption of lead ion by fungal species. *Electronic J Biotechnol., 9*: 0-0.
[212] Çabuk, A., Ilhan, S., Filik, C., et al. (2005). Pb 2+ Biosorption by Pretreated Fungal Biomass. *Turk J Biol., 29.*
[213] Say, R., Yilmaz, N., and Denizli, A. (2003a). Biosorption of cadmium, lead, mercury, and arsenic ions by the fungus Penicillium purpurogenum. *Sep Sci Technol 38*: 2039–2053.
[214] Say, R., Yilmaz, N., and Denizli, A. (2003). Removal of heavy metal ions using the fungus Penicillium canescens. *Adsorpt Sci Technol., 21*: 643-650.
[215] Vieira, R. H., and Volesky, B. (2010). Biosorption: a solution to pollution? *Int Microbiol., 3*: 17-24.
[216] Kapoor, A., and Viraraghavan, T. (1995). Fungal biosorption—an alternative treatment option for heavy metal bearing wastewaters: a review. *Bioresource Technol., 53*: 195-206.
[217] Filipović-Kovačević, Ž., Sipos, L., and Briški, F. (2000). Biosorption of chromium, copper, nickel and zinc ions onto fungal pellets of Aspergillus niger 405 from aqueous solutions. *Food Technol Biotech., 38.*
[218] Dursun, A., Uslu, G., Tepe, O., et al. (2003). A comparative investigation on the bioaccumulation of heavy metal ions by growing Rhizopus arrhizus and Aspergillus niger. *Biochem Eng J., 15*: 87-92.
[219] Kogej, A., and Pavko, A. (2001). Comparison of Rhizopus nigricans in a pelleted growth form with some other types of waste microbial biomass as biosorbents for metal ions. *World Journal of Microbiology and Biotechnology, 17*: 677-685.
[220] Podgorskiĭ, V., Kasatkina, T., and Lozovaia, O. (2003). [Yeasts--biosorbents of heavy metals]. *Mikrobiolohichnyi zhurnal* (Kiev, Ukraine: 1993)*, 66*: 91-103.
[221] Volesky, B. *Biosorption of heavy metals*, CRC press, 1990.
[222] Volesky, B. (1990). Biosorption by fungal biomass. *Biosorption of heavy metals*: 139-172.
[223] Wang, J., and Chen, C. (2009). Biosorbents for heavy metals removal and their future. *Biotechnol Adv., 27*: 195-226.
[224] Kuyucak, N., and Volesky, B. (1988). Biosorbents for recovery of metals from industrial solutions. *Biotechnol Lett., 10*: 137-142.
[225] Ercole, C., Veglio, F., Toro, L., et al. (1995). Immobilization of microbial cells for metal adsorption and desorption. *Minerals Bioprocessing II*: 181-188.
[226] Nanda, M., Sharma, D., and Kumar, A. (2011). Removal of heavy metals from industrial effluent using bacteria. *International J Environ Sci., 2*: 765-780.
[227] Bruins, M. R., Kapil, S., and Oehme, F. W. (2000). Microbial resistance to metals in the environment. *Ecotox Environ Safe., 45*: 198-207.
[228] Nies, D. H. (1999). Microbial heavy-metal resistance. *Appl Microbiol Biot., 51*: 730-750.
[229] Ji, G., and Silver, S. (1995). Bacterial resistance mechanisms for heavy metals of environmental concern. *Journal of Industrial Microbiology, 14*: 61-75.

[230] Kachur, A. V., Koch, C. J., and Biaglow, J. E. (1998). Mechanism of copper-catalyzed oxidation of glutathione. *Free radical Res., 28*: 259-269.

[231] Carine, F., Enrique, A.-G., and Stéven, C. (2009). Metal effects on phenol oxidase activities of soils. *Ecotox Environ Safe., 72*: 108-114.

[232] Olaniran, A. O., Balgobind, A., and Pillay, B. (2013). Bioavailability of heavy metals in soil: impact on microbial biodegradation of organic compounds and possible improvement strategies. *Int J Mol Sci., 14*: 10197-10228.

[233] Akpor, O., and Muchie, M. (2010). Remediation of heavy metals in drinking water and wastewater treatment systems: Processes and applications. *Int J Phys Sci., 5*: 1807-1817.

[234] Finnerty, W. R. (1992). Fossil resource biotechnology: challenges and prospects. *Curr Opin Biotech., 3*: 277-282.

[235] Kilbane, I., and John, J. (1990). Sulfur-specific microbial metabolism of organic compounds. *Resources, Conservation and Recycling, 3*: 69-79.

[236] Monticello, D. (1993). Biocatalytic desulfurization of petroleum and middle distillates. *Environ Prog., 12*: 1-4.

[237] Monticello, D., Bakker, D., and Finnerty, W. (1985). Plasmid-mediated degradation of dibenzothiophene by Pseudomonas species. *Appl Environ Microb., 49*: 756-760.

[238] Kropp, P., and Gerber, W.-D. (1998). Prediction of migraine attacks using a slow cortical potential, the contingent negative variation. *Neuroscience letters, 257*: 73-76.

[239] Kertesz, L. (2001). Building a scientific foundation for prevention. *Healthplan, 42*: 44-47.

[240] Kawatra, S. K., and Eisele, T. C. *Coal desulfurization: high-efficiency preparation methods*, Taylor & Francis, 2001.

[241] Shennan, J. (1997) Microbial attack on sulfur-containing hydrocarbons: implications for the biodesulfurization of oils and coals, In *Fuel and Energy Abstracts*, Elsevier Science.

[242] McFarland, B. L., Boron, D. J., Deever, W., et al. (1998). Biocatalytic sulfur removal from fuels: applicability for producing low sulfur gasoline. *Crit Rev Microbiol. 24*: 99-147.

[243] Marcelis, C. (2002). *Anaerobic biodesulfurization of thiophenes.* Thesisi Wageningen University, ISBN: 90-5808.

[244] Shafi, R., and Hutchings, G. J. (2000). Hydrodesulfurization of hindered dibenzothiophenes: an overview. *Catal Today., 59*: 423-442.

[245] Guerinik, K., and Al-Mutawah, Q. (2003). Isolation and characterization of oil-desulphurizing bacteria. *World Journal of Microbiology and Biotechnology, 19*: 941-945.

[246] Li, W., Zhang, Y., Wang, M. D., et al. (2005). Biodesulfurization of dibenzothiophene and other organic sulfur compounds by a newly isolated Microbacterium strain ZD-M2. *FEMS microbiology letters, 247*: 45-50.

[247] Ansari, F. (2008) The Use of Magnetic Nanoparticles to Enhance Biodesulfurization, (University, C., Ed.).

[248] Mohebali, G., and Ball, A. S. (2008). Biocatalytic desulfurization (BDS) of petrodiesel fuels. *Microbiology, 154*: 2169-2183.

[249] Kirimura, K., Furuya, T., Sato, R., et al. (2002). Biodesulfurization of naphthothiophene and benzothiophene through selective cleavage of carbon-sulfur bonds by Rhodococcus sp. strain WU-K2R. *Appl Environ Microb., 68*: 3867-3872.

[250] Borgne, S. L., and Quintero, R. (2003). Biotechnological processes for the refining of petroleum. *Fuel Process Technol., 81*: 155-169.

[251] Izumi, Y., Ohshiro, T., Ogino, H., et al. (1994). Selective desulfurization of dibenzothiophene by Rhodococcus erythropolis D-1. *Appl Environ Microb., 60*: 223-226.

[252] Konishi, J., Ishii, Y., Onaka, T., et al. (1997). Thermophilic carbon-sulfur-bond-targeted biodesulfurization. *Appl Environ Microb., 63*: 3164-3169.

[253] Kargi, F., and Robinson, J. M. (1982). Removal of sulfur compounds from coal by the thermophilic organism Sulfolobus acidocaldarius. *Appl Environ Microb., 44*: 878-883.

[254] Xiaojuan, T., Lingtian, T., Li'e, P., et al. (2008). Research on identification and screen of microbial desulfurization strains for petroleum. *Earth Science Frontiers, 15*: 192-198.

[255] Kilbane, J. J. (1989). Desulfurization of coal: the microbial solution. *Trends Biotechnol., 7*: 97-101.

[256] Ohshiro, T., Hirata, T., and Izumi, Y. (1996). Desulfurization of dibenzothiophene derivatives by whole cells of Rhodococcus erythropolis H-2. *FEMS Microbiol Lett., 142*: 65-70.

[257] Omori, T., Monna, L., Saiki, Y., et al. (1992). Desulfurization of dibenzothiophene by Corynebacterium sp. strain SY1. *Appl Environ Microb., 58*: 911-915.

[258] Constanti, M., Giralt, J., and Bordons, A. (1994). Desulphurization of dibenzothiophene by bacteria. *World Journal of Microbiology and Biotechnology, 10*: 510-516.

[259] Gupta, N., Roychoudhury, P., and Deb, J. (2005). Biotechnology of desulfurization of diesel: prospects and challenges. *Appl Microbiol Biot., 66*: 356-366.

[260] Bressler, D. C., Norman, J. A., and Fedorak, P. M. (1997). Ring cleavage of sulfur heterocycles: how does it happen? *Biodegradation, 8*: 297-311.

[261] Kim, T. S., Kim, H. Y., and Kim, B. H. (1990). Petroleum desulfurization by<i>Desulfovibrio desulfuricans</i> M6 using electrochemically supplied reducing equivalent. *Biotechnol Lett., 12*: 757-760.

[262] Sohrabi, M., and Jamshidi, A. M. (2008) A Kinetic Study On Biodesulfurization of Dibenzothiophene In a Down Flow Jet Loop Bioreactor With Coaxial Draft Tube, In *19th World Petroleum Congress*, World Petroleum Congress.

[263] Ohshiro, T., and Izumi, Y. (1999). Microbial desulfurization of organic sulfur compounds in petroleum. *Bioscience, Biotechnology, and Biochemistry, 63*: 1-9.

[264] McFarland, B. L. (1999). Biodesulfurization. *Curr Opin Microbiol. 2*: 257-264.

[265] Labana, S., Pandey, G., and Jain, R. (2005). Desulphurization of dibenzothiophene and diesel oils by bacteria. *Lett Appl Microbiol., 40*: 159-163.

[266] Aggelis, G., and Sourdis, J. (1997). Prediction of lipid accumulation-degradation in oleaginous micro-organisms growing on vegetable oils. *Antonie van Leeuwenhoek, 72*: 159-165.

[267] Metzger, P., and Largeau, C. (2005). Botryococcus braunii: a rich source for hydrocarbons and related ether lipids. *Appl Microbiol Biot., 66*: 486-496.

[268] Spolaore, P., Joannis-Cassan, C., Duran, E., et al. (2006). Commercial applications of microalgae. *J Biosci Bioeng., 101*: 87-96.

[269] Gavrilescu, M., and Chisti, Y. (2005). Biotechnology—a sustainable alternative for chemical industry. *Biotechnol Adv., 23*: 471-499.

[270] Kapdan, I. K., and Kargi, F. (2006). Bio-hydrogen production from waste materials. *Enzyme Microb Tech., 38*: 569-582.

[271] Guil-Guerrero, J., Navarro-Juárez, R., López-Martınez, J., et al. (2004). Functional properties of the biomass of three microalgal species. *J Food Eng., 65*: 511-517.
[272] Meng, X., Yang, J., Xu, X., et al. (2009). Biodiesel production from oleaginous microorganisms. *Renew Energ., 34*: 1-5.
[273] Alvarez, H., and Steinbüchel, A. (2002). Triacylglycerols in prokaryotic microorganisms. *Appl Microbiol Biot., 60*: 367-376.
[274] Ratledge, C. (1993). Single cell oils—have they a biotechnological future? *Trends biotechnol., 11*: 278-284.
[275] Ratledge, C. (1982). Microbial oils and fats: an assessment of their commercial potential [Algae, yeasts, fungi]. *Progress in industrial microbiology.*
[276] Kavadia A, K. M., Chevalot I, Blanchard F, Marc I, Aggelis G. . (2001). Lipid and gamma-linolenic acid accumulation in strains of Zygomycetes growing on glucose. *J Am Oil Chem Soc 78*: 341-346.
[277] EVANS, C. T., and RATLEDGE, C. (1984). Influence of nitrogen metabolism on lipid accumulation by Rhodosporidium toruloides CBS 14. *J Gen Microbiol., 130*: 1705-1710.
[278] Li, Y., Zhao, Z. K., and Bai, F. (2007). High-density cultivation of oleaginous yeast<Rhodosporidium toruloides Y4 in fed-batch culture. *Enzyme Microb Tech., 41*: 312-317.
[279] Takeno, S., Sakuradani, E., Tomi, A., et al. (2005). Transformation of oil-producing fungus,Mortierella alpina1S-4, using Zeocin, and application to arachidonic acid production. *J Biosci Bioeng., 100*: 617-622.
[280] Chisti, Y. (2007). Biodiesel from microalgae. *Biotechnol Adv., 25*: 294-306.
[281] Xu, H., Miao, X., and Wu, Q. (2006). High quality biodiesel production from a microalga Chlorella protothecoides by heterotrophic growth in fermenters. *J Biotechnol., 126*: 499-507.
[282] Huang, G., Chen, F., Wei, D., et al. (2010). Biodiesel production by microalgal biotechnology. *Appl Energ, 87*: 38-46.
[283] LM., B. (1996). Uptake of carbon dioxide from fuel gas by microalgae. *Energy Convers Manage, 37*: 1363-1367.
[284] Yun, Y. S., Lee, S. B., Park, J. M., et al. (1997). Carbon dioxide fixation by algal cultivation using wastewater nutrients. *J Chem Technol Biot., 69*: 451-455.
[285] Wen, Z.-Y., and Chen, F. (2003). Heterotrophic production of eicosapentaenoic acid by microalgae. *Biotechnol Adv., 21*: 273-294.
[286] Myers, J., and Burr, G. (1940). Studies on photosynthesis: some effects of light of high intensity on Chlorella. *The Journal of general physiology, 24*: 45.
[287] Chen, F. (1996). High cell density culture of microalgae in heterotrophic growth. *Trends Biotechnol., 14*: 421-426.
[288] Chen, F., and Johns, M. R. (1991). Effect of C/N ratio and aeration on the fatty acid composition of heterotrophicChlorella sorokiniana. *J Appl Phycol., 3*: 203-209.
[289] Wu, Q., Yin, S., Sheng, G., et al. (1994). New discoveries in study on hydrocarbons from thermal degradation of heterotrophically yellowing algae. *Science in China. Series B, Chemistry, life sciences & earth sciences, 37*: 326-335.
[290] Yokochi, T., Honda, D., Higashihara, T., et al. (1998). Optimization of docosahexaenoic acid production by Schizochytrium limacinum SR21. *Appl Microbiol Biot., 49*: 72-76.
[291] Liu SM, C. F., Liang SZ. . (1999). Researches on the heterotrophic culture of Chlorella vulgaris. *J South China Univ Technol 27*: 111-115.

[292] Miao, X., and Wu, Q. (2006). Biodiesel production from heterotrophic microalgal oil. *Bioresource Technol., 97*: 841-846.
[293] Zaslavskaia, L., Lippmeier, J., Shih, C., et al. (2001). Trophic conversion of an obligate photoautotrophic organism through metabolic engineering. *Science, 292*: 2073-2075.
[294] Davis, J. B., and Yarbrough, H. F. (1962). Preliminary experiments on a microbial fuel cell. *Science, 137*: 615-616.
[295] Himmel, M. E., Baker, J. O., and Overend, R. P. *Enzymatic conversion of biomass for fuels production*, American Chemical Society, 1994.
[296] van Hees, W. (1965). A bacterial methane fuel cell. *Journal of the Electrochemical Society, 112*: 258-262.
[297] Karube, I., Suzuki, S., Matsunaga, T., et al. (1981). Biochemical energy conversion by immobilized whole cells. *Ann Ny Acad Sci., 369*: 91-98.
[298] Suzuki, S., Karube, I., Matsuoka, H., et al. (1983). Biochemical energy conversion by immobilized whole cells. *Ann Ny Acad Sci., 413*: 133-143.
[299] Delaney, G. M. B., H. P.; Mason, J. R.; Roller, S. D.; Stirling, J. L.; Thurston, C. F. J. . (1984). *Chem. Technol. Biotechnol., 34B*: 13.
[300] Thurston, C. F. B., H. P.; Delaney, G. M.; Mason, J. R.; Roller, S. D.; Stirling, J. L. J. . (1985). *Gen. Microbiol., 131*: 1393.
[301] Bennetto, H. P. D., G. M.; Mason, Roller, S. D.; Stirling, J. L.; Thurston, C. F. . (1985). *Biotech. LRn., 7*: 699.
[302] Vega, C. A. F., I. (1987). *Bioelectrochem. Bioenerg., 17*: 217.
[303] Chiao, M., Lam, K. B., and Lin, L. (2006). Micromachined microbial and photosynthetic fuel cells. *J Micromech Microeng., 16*: 2547.
[304] Ringeisen, B. R. H., E.; Wu, P.K.; Pietron, J.; Ray, R.; Little, B.; Biffinger, J.C. & Jones-Meehan, J.M. . (2006). High power density from a miniature microbial fuel cell using Shewanella oneidensis DSP10. *Environ Sci Technol., 40*: 2629– 2634.
[305] Qiao, Y., Li, C. M., Bao, S.-J., et al. (2007). Carbon nanotube/polyaniline composite as anode material for microbial fuel cells. *J Power Sources., 170*: 79-84.
[306] Belkin, S. (2003). Microbial whole-cell sensing systems of environmental pollutants. *Curr Opin Microbiol., 6*: 206-212.
[307] Cheng Vollmer, A., and Van Dyk, T. K. (2004). Stress responsive bacteria: biosensors as environmental monitors. *Adv Microb Physiol., 49*: 131-174.
[308] Daunert, S., Barrett, G., Feliciano, J. S., et al. (2000). Genetically engineered whole-cell sensing systems: coupling biological recognition with reporter genes. *Chemical Reviews, 100*: 2705-2738.
[309] Gu, M. B., Mitchell, R. J., and Kim, B. C. (2004) Whole-cell-based biosensors for environmental biomonitoring and application, In *Biomanufacturing*, 269-305, Springer.
[310] Hansen, L. H., and Sørensen, S. J. (2001). The use of whole-cell biosensors to detect and quantify compounds or conditions affecting biological systems. *Microb Ecol., 42*: 483-494.
[311] Thévenot, D. R., Toth, K., Durst, R. A., et al. (2001). Electrochemical biosensors: recommended definitions and classification. *Biosens Bioelectron., 16*: 121-131.
[312] Parellada, J., Narvaez, A., Lopez, M., et al. (1998). Amperometric immunosensors and enzyme electrodes for environmental applications. *Anal Chim Acta., 362*: 47-57.
[313] Wang, J., Rivas, G., Cai, X., et al. (1997). Sequence-specific electrochemical biosensing of M. tuberculosis DNA. *Anal Chim Acta., 337*: 41-48.

[314] Velasco-Garcia, M. (2009) Optical biosensors for probing at the cellular level: A review of recent progress and future prospects, In *Seminars in cell & developmental biology*, 27-33, Elsevier.

[315] Arora, P., Sindhu, A., Dilbaghi, N., et al. (2011). Biosensors as innovative tools for the detection of food borne pathogens. *Biosens Bioelectron., 28*: 1-12.

[316] D'souza, S. (2001). Microbial biosensors. *Biosens Bioelectron., 16*: 337-353.

[317] Lei, Y., Chen, W., and Mulchandani, A. (2006). Microbial biosensors. *Anal Chim Acta., 568*: 200-210.

[318] Steinberg, S. M., Poziomek, E. J., Engelmann, W. H., et al. (1995). A review of environmental applications of bioluminescence measurements. *Chemosphere, 30*: 2155-2197.

[319] Kumar, J., and D'Souza, S. (2011). Immobilization of microbial cells on inner epidermis of onion bulb scale for biosensor application. *Biosens Bioelectron., 26*: 4399-4404.

[320] Choi, S., and Chae, J. (2012). An array of microliter-sized microbial fuel cells generating 100μW of power. *Sens Actuators A: Physical, 177*: 10-15.

[321] Du, Z., Li, H., and Gu, T. (2007). A state of the art review on microbial fuel cells: a promising technology for wastewater treatment and bioenergy. *Biotechnol Adv., 25*: 464-482.

[322] Choi, S. H., and Gu, M. B. (2002). A portable toxicity biosensor using freeze-dried recombinant bioluminescent bacteria. *Biosens Bioelectron., 17*: 433-440.

[323] Kostrzynska, M., Leung, K. T., Lee, H., et al. (2002). Green fluorescent protein-based biosensor for detecting SOS-inducing activity of genotoxic compounds. *J Microbiol Meth., 48*: 43-51.

[324] Wozei, E., Hermanowicz, S., and Holman, H. (2006). Developing a biosensor for estrogens in water samples: Study of the real-time response of live cells of the estrogen-sensitive yeast strain RMY/ER-ERE using fluorescence microscopy. *Biosens Bioelectron., 21*: 1654-1658.

[325] Tan, T., and Wu, C. (1999). BOD sensors using multi-species living or thermally killed cells of a BODSEED microbial culture. *Sens Actuators B: Chemical, 54*: 252-260.

[326] Chee, G.-J., Nomura, Y., Ikebukuro, K., et al. (2000). Optical fiber biosensor for the determination of low biochemical oxygen demand. *Biosens Bioelectron., 15*: 371-376.

[327] Marrazza, G., Chianella, I., and Mascini, M. (1999). Disposable DNA electrochemical biosensors for environmental monitoring. *Anal Chim Acta., 387*: 297-307.

[328] Koubova, V., Brynda, E., Karasova, L., et al. (2001). Detection of foodborne pathogens using surface plasmon resonance biosensors. *Sens Actuators B: Chemical, 74*: 100-105.

[329] Pathirana, S., Barbaree, J., Chin, B., et al. (2000). Rapid and sensitive biosensor for< i> Salmonella. *Biosens Bioelectron., 15*: 135-141.

[330] Ercole, C., Gallo, M. D., Pantalone, M., et al. (2002). A biosensor for Escherichia coli based on a potentiometric alternating biosensing (PAB) transducer. *Sens Actuators B: Chemical, 83*: 48-52.

[331] Védrine, C., Leclerc, J.-C., Durrieu, C., et al. (2003). Optical whole-cell biosensor using Chlorella vulgaris designed for monitoring herbicides. *Biosens Bioelectron., 18*: 457-463.

[332] Tschmelak, J., Proll, G., and Gauglitz, G. (2004). Ultra-sensitive fully automated immunoassay for detection of propanil in aqueous samples: steps of progress toward sub-nanogram per liter detection. *Anal Bioanal Chem., 379*: 1004-1012.

[333] Mauriz, E., Calle, A., Abad, A., et al. (2006). Determination of carbaryl in natural water samples by a surface plasmon resonance flow-through immunosensor. *Biosens Bioelectron., 21*: 2129-2136.

[334] Deo, R. P., Wang, J., Block, I., et al. (2005). Determination of organophosphate pesticides at a carbon nanotube/organophosphorus hydrolase electrochemical biosensor. *Anal Chim Acta., 530*: 185-189.

[335] Law, K. A., and J Higson, S. P. (2005). Sonochemically fabricated acetylcholinesterase micro-electrode arrays within a flow injection analyser for the determination of organophosphate pesticides. *Biosens Bioelectron., 20*: 1914-1924.

[336] Bachmann, T. T., and Schmid, R. D. (1999). A disposable multielectrode biosensor for rapid simultaneous detection of the insecticides paraoxon and carbofuran at high resolution. *Anal Chim Acta., 401*: 95-103.

[337] Mallat, E., Barzen, C., Abuknesha, R., et al. (2001). Fast determination of paraquat residues in water by an optical immunosensor and validation using capillary electrophoresis-ultraviolet detection. *Anal Chim Acta., 427*: 165-171.

[338] Mallat, E., Barzen, C., Abuknesha, R., et al. (2001). Part per trillion level determination of isoproturon in certified and estuarine water samples with a direct optical immunosensor. *Anal Chim Acta., 426*: 209-216.

[339] Mallat, E., Barzen, C., Klotz, A., et al. (1999). River analyzer for chlorotriazines with a direct optical immunosensor. *Environ Sci Technol., 33*: 965-971.

[340] Rodriguez-Mozaz, S., Reder, S., Lopez de Alda, M., et al. (2004). Simultaneous multi-analyte determination of estrone, isoproturon and atrazine in natural waters by the RIver ANAlyser (RIANA), an optical immunosensor. *Biosens Bioelectron., 19*: 633-640.

[341] Hua, P., Hole, J., Wilkinson, J., et al. (2005). Integrated optical fluorescence multisensor for water pollution. *Optics Express, 13*: 1124-1130.

[342] Tschmelak, J., Kumpf, M., Käppel, N., et al. (2006). Total internal reflectance fluorescence (TIRF) biosensor for environmental monitoring of testosterone with commercially available immunochemistry: Antibody characterization, assay development and real sample measurements. *Talanta, 69*: 343-350.

[343] Tschmelak, J., Proll, G., and Gauglitz, G. (2004). Sub-nanogram per litre detection of the emerging contaminant progesterone with a fully automated immunosensor based on evanescent field techniques. *Anal Chim Acta., 519*: 143-146.

[344] Endo T, O. A., Matsubara Y, Nishi K, Kobayashi M,, and Yamamura S, M. Y., Takamura Y, Mizukami H, Tamiya E. (2005). *Anal Chim Acta, 531*: 7-13.

[345] Kasai, A., Hiramatsu, N., Meng, Y., et al. (2005). Fast-track DRESSA: a bioassay for fast, sensitive, and selective detection of halogenated and polycyclic aromatic hydrocarbons. *Anal Biochem., 337*: 84-88.

[346] Nistor, C., Rose, A., Farré, M., et al. (2002). In-field monitoring of cleaning efficiency in waste water treatment plants using two phenol-sensitive biosensors. *Anal Chim Acta., 456*: 3-17.

[347] Degiuli A., and L.J., B. (2000). Flow injection chemiluminiscence: detection of chlorophenols with a fibre optic biosensor. *J. Med. Biochem., 4*: 32-42.

[348] Rodriguez-Mozaz, S., De Alda, M. L., and Barceló, D. (2005). Analysis of bisphenol A in natural waters by means of an optical immunosensor. *Water Res., 39*: 5071-5079.

[349] Reshetilov, A. N., Semenchuk, I. N., Iliasov, P. V., et al. (1997). The amperometric biosensor for detection of sodium dodecyl sulfate. *Anal Chim Acta., 347*: 19-26.

[350] Taranova, L., Semenchuk, I., Manolov, T., et al. (2002). Bacteria-degraders as the base of an amperometric biosensor for detection of anionic surfactants. *Biosens Bioelectron., 17*: 635-640.

[351] Nomura, Y., Ikebukuro, K., Yokoyama, K., et al. (1998). Application of a linear alkylbenzene sulfonate biosensor to river water monitoring. *Biosens Bioelectron., 13*: 1047-1053.

[352] Rose, A., Nistor, C., Emnéus, J., et al. (2002). GDH biosensor based off-line capillary immunoassay for alkylphenols and their ethoxylates. *Biosens Bioelectron., 17*: 1033-1043.

[353] Evtugyn, G., Eremin, S., Shaljamova, R., et al. (2006). Amperometric immunosensor for nonylphenol determination based on peroxidase indicating reaction. *Biosens Bioelectron., 22*: 56-62.

[354] Berno, E., Pereira Marcondes, D., Ricci Gamalero, S., et al. (2004). Recombinant Escherichia coli for the biomonitoring of benzene and its derivatives in the air. *Ecotox Environ Safe., 57*: 118-122.

[355] Doong, R.-a., Shih, H.-m., and Lee, S.-h. (2005). Sol–gel-derived array DNA biosensor for the detection of polycyclic aromatic hydrocarbons in water and biological samples. *Sens Actuators B: Chemical, 111*: 323-330.

[356] Tschmelak, J., Proll, G., and Gauglitz, G. (2005). Optical biosensor for pharmaceuticals, antibiotics, hormones, endocrine disrupting chemicals and pesticides in water: Assay optimization process for estrone as example. *Talanta, 65*: 313-323.

[357] Petänen, T., and Romantschuk, M. (2002). Use of bioluminescent bacterial sensors as an alternative method for measuring heavy metals in soil extracts. *Anal Chim Acta., 456*: 55-61.

[358] Tsai, H.-c., and Doong, R.-a. (2005). Simultaneous determination of pH, urea, acetylcholine and heavy metals using array-based enzymatic optical biosensor. *Biosens Bioelectron., 20*: 1796-1804.

[359] Liao, V. H.-C., Chien, M.-T., Tseng, Y.-Y., et al. (2006). Assessment of heavy metal bioavailability in contaminated sediments and soils using green fluorescent protein-based bacterial biosensors. *Environ Pollut., 142*: 17-23.

[360] Taylor, C. J., Bain, L. A., Richardson, D. J., et al. (2004). Construction of a whole-cell gene reporter for the fluorescent bioassay of nitrate. *Anal Biochem., 328*: 60-66.

In: Microbial Catalysts. Volume 1
Editors: S. M. Abdel-Aziz, N. Garg, A. Aeron et al.
ISBN: 978-1-53614-679-0

Chapter 8

MICROBIAL ENZYMES: THE TYPES AND ROLES IN THE BIO-CONTROL OF FUNGAL PHYTOPATHOGENS

***Dweipayan Goswami*[1], *Pinakin Dhandhukia*[2] and *Janki N. Thakker*[3,*]**

[1]Department of Biochemistry and Biotechnology,
St. Xavier's College (Autonomous), Ahmedabad, India
[2]Ashok and Rita Patel Institute of Integrated Study and Research in Biotechnology and Allied Sciences, New Vidyangar, Anand, Gujarat, India
[3]Department of Biotechnology, P. D. Patel Institute of Applied Sciences,
Charotar University of Science and Technology,
Changa, Anand, Gujarat, India

ABSTRACT

Fungal pathogens are the major culprit causing infections in plants. An estimated, two-third of the total diseased plants are infected by fungi. Aggregate of 125 million tons of the main five sustenance crops in particular rice, wheat, maize, potatoes and soybeans are being spoiled by parasitic diseases all around every year. Diseases caused by phytopathogenic fungi constitute of rusts, smuts, needle casts, leaf curls, mildew, sooty molds, flower spots, cankers, blights, scabs (in root, stem and fruit), wood rots, wilts (in leaf, shoot) and bud galls. Parasitic contamination by fungi in rice, wheat and maize itself represents the loss of $60 billion every year. Fungal diseases are controlled through the use of fungicides, however new races of fungi that are resistant to various fungicides are evolving that are resistant to various fungicides. Moreover, the concerns regarding the public safety has revoked the use of toxic chemical fungicides use leading to the development of other alternatives for suppressing fungal phytopathogens. Thus, plant beneficial microbes including both bacteria and other non-pathogenic fungal strains that can antagonize growth of phytopathogens on plant tissues are exploited. Such microbes which can suppress fungal phytopathogens are known as bio-control agents. These agents produce fungal cell wall degrading enzymes, β-1,3-glucanase, chitinase, cellulase, and

[*] Corresponding Author's Email: jankithakker.bt@ecchanga.ac.in

protease that primarily acts to impede the invasion of pathogens. Among bacterial strains, *Bacillus licheniformis, B. subtilis, B. cereus, B. circulans, B. cepacia* and *B. thuringiensis, Paenibacillus* spp., *Streptomyces* spp. *Serratia marcescens, Enterobacter agglomerans, Pseudomonas aeruginosa,* and *P. fluorescens* are some of the efficient chitinase and glucanase producers. Among fungal bio-control agents, *Trichoderma viride* and *Trichoderma harzianum* are most widely studied for production of chitinase and glucanase. All these microbes are reported to successfully suppress the growth of *Rhizoctonia solani, Sclerotium rolfsii, Fusarium oxysporum, Pythium ultimum, Aspergillus flavus and Aspergillus niger* which are the most compelling fungal pathogens. The following discourse in the form of chapter focuses on role of microbial enzymes that act against fungal phytopathogens.

Keywords: chitinases, glucanases, cellulases, mycoparasitism

1. INTRODUCTION

Infections in plant are common and greater part of plant illness occur by microbes. Out of all the microbial administrators causing ailment in plants, fungi represent the largest group. Organisms which that are evolved for causing malady in plants are known as phytopathogens. [1] (Dufresne and Osbourn 2001). Fungi are unicellular eukaryotic organisms devoid of chlorophyll and possess the ability to utilize diverse of organic matter for its metabolism. The majority of these fungi belong to the Ascomycetes, Basidiomycetes and Oomycetes [2]. More than 100,000 fungal species are saprophytes. However, over 20,000 species of fungi are parasites and cause disease in crops and plants (USEPA 2005). Fungal plant diseases, in past have caused loss of crops which had resulted in hunger and starvation. In less developed countries, where there is lack of availability to disease-control methods, annual losses ranging from 30 to 50 percent have been reported for major crops [3]. Diverting the focus to the history in Europe, where devastating outbreak of late blight of potato (*Phytophthora infestans*) in the year 1845 brought about the Irish famine caused starvation, death, and mass migration of the Irish population [4]. Statistics reflected that, from the total population of eight million, approximately one million (about 12.5 percent) died of starvation and 1.5 million (almost 19 percent) moved to United States, as refugees from the destructive blight. This was the portrayal of the epidemic late blight of potato that had posed a tremendous influence on the economic, political, and cultural development in Europe and the United States. Another incidence where late blight damage to the potato crop was in Germany during World War-I and starvation caused by this event may have helped to end the war [4]. Losses incurred from plant diseases also impose significant economic impact, resulting reduction in income for crop producers and distributors and subsequently higher prices for consumers. Till date infections caused by fungal phytopathogens are in high numbers, where annually 125 million tons of the top five food crops namely rice, wheat, maize, potatoes and soybeans are spoilt [5]. It is estimated that fungal diseases reduce gross plant yields up to 20% in various countries where loss in more prominent in less developed countries and vice-versa. Thus, persistent developments in the science of plant pathology are mandatory to improve disease control by developing new and advanced eco-friendly methods to control fungal diseases in plants [4, 5].

Mechanism of fungal infection can be described as the plant growing in unfavorable environment, defined by improper humidity, temperature and nutrition can easily be infected byfungal spores which are generally living as saprophytes which later causes local or extensive necrosis plants. Thus, for infection to occur it is necessary for the dormant spores of fungus to be in the vicinity of suitable host, as most of the phytopathogenic fungi are host specific. Followed by necrosis, it inhibits normal growth (hypotrophy) or induce excessive abnormal growth (hypertrophy or hyperplasia) in a limited portion of or throughout an entire plant [6]. Symptoms associated with necrosis include leaf spots, blight, scab, rots, damping-off, anthracnose, dieback, and canker. Symptoms associated with hyperplasia include club root, galls, warts, and leaf curls. In some instances, the fungus infecting the plant may produce growth or structures on the plant, stems, or leaves such as masses of mycelium or aggregates of spores with a characteristic appearance. These developments are referred to as the signs of infection, in contrast to symptoms, which refer specifically to the plant or plant tissue [1, 3, 4]. Chemical fungicides are routinely used to treat such fungal infection of plants. Nevertheless, new races of fungi often evolve that are resistant to various fungicides. To quote few examples, sugar beet leaf blotch remains resistant to azoles; *Botrytis cinerea* is resistant to both azoles and dicarboximide fungicides; Black Sigatoka which is an economically important pathogen of banana, is resistant to the QoI fungicides, etc. Further, the concerns regarding the public safety has revoked their use leading to development of other alternatives to suppress fungal phytopathogens [7, 8].

Advances in the approaches to improve fungal disease control has obtained an alternative for the use of chemical fungicides in the form of microbial fungicides. Microbial fungicides represent a group of phytofriendly microbes which inhibit growth of fungal phytopathogen ultimately leading to the reduction in fungal diseases [9-11]. These microbial fungicides can be both, bacteria and fungi as they antagonize the growth of phytopathogenic fungi by competition for survival, by producing antifungal agents, or by directly killing fungal phytopathogen by the production of fungal cell wall degrading enzymes. Here our focus will narrow down in understanding the enzymes and their roles in suppression of fungal phytpathogens.

2. FUNGAL DISEASES IN PLANTS

Fungi are eukaryotic organisms that are devoid of chlorophyll and they depend on other sources for food, as they do not possess ability to produce own food [12]. They obtain nutrients by absorption through hyphae which are tiny thread-like filaments which has an ability to branch in various directions all over a substrate. A collection of hyphae is referred to as mycelium where bunch of mycelium is called as mycelia. The hyphae possess protoplasm that contains nuclei and other organelles. Mycelia are the key diagnostic sign associated with diseases caused by fungi as they are macroscopic and can be easily visualized by naked eye, the same way where myceliums growing on old bread or rotten fruit or vegetables are visibly characterized as molds or mildew [13].

Table 1. The list of few fungal causative agents infecting some of the economically important crops

Disease	Causative agent	Hosts	Symptoms and signs	References
Late blight of potato	Phytophthora infestans	Potato	Water-soaked dark green to black or purplish lesions with pale green margins on lower leaves, white mildew at edge of lesions, responsible for Irish famine	Vleeshouwers et al. [19]
Vascular wilt	*Verticillum alboatrum*	Groundnut	Wilting of leaflets and petioles, Leaflets are curled and	Pal et al. [20]
Chestnut blight	Cryphonectria parasitica	Chestnut Tree	Yellowish to reddish brown patches appear on bark; lesions spread quickly and girdle twigs or limbs, which die	Robin et al. [21]
Seed, seedling and stem rot	Rhizopus arrhizus and Sclerotium rolfsii	Groundnut	Sudden wilting of lateral branches that are in contact with soil and white coating of fungus mycelium on infected plant appears	Le et al. [22]
Dutch elm disease	Ceratocystis ulmi	Elm Tree	Leaves wilt, turn dull green to yellow or brown, and drop off; branches die (the causative fungus is believed to have entered Europe from Asia during World War I and was later transported to the United States (1930))	Forgetta et al. [23]
Black stem rust of wheat	Puccinia graminis	Wheat; Many Grasses	On wheat, rust-coloured pustules with spores, chlorosis of surrounding tissue, followed by development of black teliospores; on barberry, chlorosis and hypertrophy of infected tissue, orange spore masses (in 1935 it destroyed about 60 percent of the total hard red spring wheat crop in Minnesota and South Dakota)	Jin [24]
Coffee rust	Hemileia vastatrix	Coffee	Orange-yellow powdery spots on lower side of leaves; centres turn brown and leaves fall, most destructive disease of coffee	Carvalho et al. [25]
White-pine blister rust	Cronartium ribicola	White Pine Tree	Small, discoloured, spindle-shaped cankers surrounded by narrow band of yellow-orange bark; blisters exude secretion followed by bright orange pustules (one of the most important forest diseases in the United States)	Wilson et al. [26]
Corn smut	Ustilago maydis	Corn	Minute galls form on young corn seedlings; on older plants, large galls are produced on the silk of ears and on tassels, leaves, and stalks	Foglio et al. [27]
Loose smut	Ustilago nuda, Ustilago *tritici*	Barley, Oats, Wheat	Infected heads are covered with masses of olive-green spores	Wunderle et al. [28]
Downy mildew	*Pseudoperospora cubensis* and many species of the family Peronosporaceae,	Many Types of Plants: Grapes, cucumber,Vegetables, And Others	Yellow irregular spots appear on upper leaf surface; downy fungus growth appears on underside; leaves die	Nostar et al. [29]
Powdery mildew	many species of the family Erysiphaceae	Many Types of Plants: Grasses, Vegetables, Shrubs, And Trees	Spots of powdery mildew growth that enlarge to cover leaves or other plant organs	Wicker et al. [30]

Disease	Causative agent	Hosts	Symptoms and signs	References
Apple scab	Venturia inaequalis	Apple	Small olive-coloured areas appear on young leaves, later turn black, and may coalesce; black circular spots appear on fruit	Xu et al. [31]
Black spot of rose	Diplocarpon rosae	Rose	Large circular black lesions on leaves; leaves turn yellow and fall off	Kumar et al. [32]
Anthracnose of grape	Elsinae ampelina	Grape	(As above)	Carisse and Morissette-Thomas [33]
Nectria canker	Nectria galligena	Apple And Pear And Many Hardwood Forest Trees	Initially small circular brown areas that enlarge and become depressed with raised edges; callus tissue produced around canker	Weber et al. [34]
Black knot disease of cherry	*Dibotryon morbosum*	Chery	Small black knotty swellings on twigs and branches	Zhang et al. [35]
Brown rot	Monilinia fructicola	Stone Fruits	Brown spots on blossoms; twigs develop small sunken brown cankers; fruit develops brown spots that spread rapidly	Casals et al. [36]
Soft rot	Rhizopus species	Flowers, Fruits, And Vegetables With Fleshy Organs	Tissues become soft with water-soaked appearance that often spreads rapidly, followed by development of fuzzy gray mycelium and black spores	Zhang et al. [37]
Fusarium wilt of tomato	Fusarium oxysporum f.sp. *lycopersici*	Tomatoes	Leaves are bent down, growth is stunted, plant dies; dark streaks appear in vascular tissue (one of the most destructive diseases of tomato)	Borrero et al. [38]
Fusarium wilt of banana	Fusarium oxysporum f.sp. cubense	Banana Plant	Plant causing the secretion of a gel. This is followed by the formation of tylose in the vascular vessels, followed by infection in rhizome, lastly xylem turn a reddish-brown colour and leaves turns yellow.	Thakker et al. [10]
Other fusarium wilts	*Fusarium oxysporum f.sp. ciceris*, *Fusarium* oxysporum f.sp. albedinis, *Fusarium oxysporum f.sp. asparagi*, *Fusarium oxysporum f.sp. coffea*, *Fusarium oxysporum f. sp. melonis*	Chick Pea, Dates, Asparagus, Coffee, Melon	Species of fusarium are specific to host and are capable to produce wilts. Various fusarium phytopathgens are capable to cause wilts in coffee, tomatoes, chickpea, dates, sweet potatoes and several other plats. The only thing to notice here is each of fungal specie of fusarium is highly host specific	Ma et al. [39]
Wilts of vegetables, flowers, and some trees	Verticillium species	Cotton, Potato, Tomato, Alfalfa, Shade Trees, And Others	Similar to fusarium wilts; develops primarily in seedlings that die shortly after infection; older plants also are attacked	Lu et al. [40]
Groundnut rust	Puccinia arachidis	Groundnut Plant Leaf	The disease attacks all aerial parts of the plant. Small brown to chestnut dusty pustules (uredosori) appear on the lower surface of leaves. The rust pustules may be seen on petioles and stem. The severe infection leads to production of small and shriveled seeds.	Mondal et al. [41]

The large population of fungal phytopathogens belongs to the class of Ascomycetes, Basidiomycetes, Oomycetes and Chytridiomycota. Though, fungi belonging to phylum Zygomycota are known to decay stored food as they are obligate saprophytes [14]. (1) Ascomycetes, fungi belonging to this class are 'spore shooters'. They are the fungi which produce microscopic spores inside special, elongated cells or sacs, known as 'asci', which gave the group its name. Some widespread fungi belonging to this group are, *Fusarium* spp. (causal agents of Fusarium wilt disease) [9, 10], *Thielaviopsis* spp. (causative agents of: canker rot, black root rot, Thielaviopsis root rot), *Verticillium* spp., *Magnaporthe grisea* (causative agent of blast of rice and gray leaf spot in turfgrasses), *Sclerotinia sclerotiorum,* the one that causes white mold [15, 16]. (2) Basidiomycetes, fungi belonging to this class include those members that produce basidia and basidiospores on or in a basidiocarp. Basidia are typically somewhat club-shaped, where mushroom being the most common example of fungus belonging to this class. Some wellknown fungal phytopathogens belonging to this group include, *Ustilago* spp., *Rhizoctonia* spp., *Phakospora pachyrhizi* (causal agent of soybean rust), *Puccinia* spp. (causative agents of severe rusts of virtually all cereal grains and cultivated grasses), *Armillaria* spp. (the so-called honey fungus species, which are virulent pathogens of trees and produce edible mushrooms) [16, 17]. (3) Oomycetes; these are not true fungi but are fungus-like organisms. They include some of the most destructive plant pathogens including the genus *Phytophthora*, which includes the causative agents of potato late blight and sudden oak death. Few species of oomycetes are also responsible for root rot. Despite not being closely related to the fungi, the oomycetes have developed very similar infection strategies. Oomycetes can use effector proteins to turn off a plant's defenses in its infection process. Plant pathologists commonly group them with fungal pathogens. Some common fungal phytopathogen belonging to this group are *Pythium* spp., *Phytophthora* spp.; also the pathogenic agent of the Great Irish Famine belongs to this group (1845–1849) [4, 18]. (4) Zygomycota are characterized by the production of characteristic Zygophores which, are chemotropic aerial hyphae. Fungus belonging to the family of Zygomycota are not known to act as phytopathogens, as they are terrestrial in habitat, living in soil or on decaying plant or animal material and thus, they are considered as obligate saprophytes. *Rhizopus* and *Mucor* are the most well-known fungus belong to this class known to decay stored food, however they rarely act as the pathogen for living plants [5-9]. (5) Chytridiomycota are characterized by the production of zoospores with one posteriorly directed, whiplash flagellum. Chytridiomycota are also known as Chytrids. *Synchytrium endobioticum* is one of the most studied phytopathogen belonging to this phylum, it causes black wart disease of potato. While, other fungi of this phylum are saprotrophs and uses substrates such as cellulose, chitin, and keratin as a primary carbon source. Thus, it is evident that large population of fungi can infect several plants (Table 1).

Consequently, it is evident that several fungal phytopathogens that cause diseases in crops should be controlled to avoid the productivity losses incurred by their infection.

3. Composition of Fungal Cell Wall

A variety of microorganisms' exhibit hyper-parasitic activity, antagonizing pathogens by excreting cell wall hydrolases by the production of hydrolytic enzymes. The enzymes that cause hydrolysis of the fungal cell wall are Chitinases, Glucanases, Amylases and Proteases [42]. To understand how these enzymes, exhibit their role in fungal cell degradation, it is necessary to understand the composition of fungal cell wall.

Fungal cell wall is highly rigid, and its major component consists polymer of polysaccharides. Briefly, the walls contain fibrillar materials bound together by sugars, proteins, lipids and a variety of polysaccharides. The outer-most layer of fungus, external to the plasma membrane provides the mechanical support for its survival. Fungal cell wall is composed of fibres which are contained in a complex gel-like matrix. This part of cell wall fibers possess three main components; (i) Chitin, (ii) β -Glucans and (iii) proteins whereas, gel-like polymer matrix is made of (i) glycoprotein and (ii) α-Glucans [10, 11]. Cell wall constituents found in different divisions of fungi are described in Table 2.

Chitin is a polymer of β-(1,4)-linked-N-Acetylglucosamine. In Ascomycota and Basidiomycota, chitin is composed of unbranched chains of β-(1,4)-linked-N-Acetyl-glucosamine. Monomer units of N-Acetylglucosamine form covalent β-1,4 linkages (like the linkages between glucose units forming cellulose). Therefore, chitin may be described as cellulose with one hydroxyl group on each monomer replaced with an 'acetyl amine' group. This allows increased hydrogen bonding between adjacent polymers, giving the chitin-polymer matrix enhancd strength. Whereas, in Zygomycota chitosan as a major cell wall component. Chitosan is basically deacetylated chitin. Both chitin and chitosan are synthesized and extruded at the plasma membrane [1, 4, 9-11]. Structure of N-Acetylglucosamine and its polymer chitin possessing β-1,4 linkage between its repeating unit is described in Figure 1.

Glucans are polymers of glucose that are linked with β or α linkages, their function is to cross-link chitin or chitosan polymers that enhances the rigidity and compactness of fungal cell wall. β-glucans are glucose molecules linked by β-(1,3)- or β-(1,6)- bonds and provide rigidity to the cell wall while α-glucans are defined by α-(1,3)- and/or α-(1,4) bonds and function as part of the matrix [10-12]. Generally, α-Glucans are not the part of cell wall fibers, but they are present in gel like matrix where these cell wall fibers are contained within.

Proteins constitute a small fraction of wall material, rarely more than 20%, and often as glycoprotein. Not all proteins have a structural role. Other proteins which are enzymes necessary for cell wall synthesis, in addition to structural proteins, transporter proteins are also present in the cell wall. Most of the structural proteins found in the cell wall are glycosylated and contain mannose, thus these proteins are addressed as mannoproteins or mannans [3, 7, 9].

Table 2. Common wall constituents found in each division of fungi [4, 18, 43].

Division	Fibrous	Gel-like Polymer
Basidiomycota	Chitin β -(1,3), β-(1,6) Glucan	Xylomannoproteins α (1,3) Glucan
Ascomycota	chitin β -(1,3), β-(1,6) Glucan	Galactomannoproteins α (1,3) Glucan
Zygomycota	Chitin Chitosan	Polyglucuronic acid Glucuronomannoproteins Polyphosphate
Chytridiomycota	Chitin Glucan	Glucan
Oomycetes	Cellulose β -(1,3), β-(1,6) Glucan	Glucan

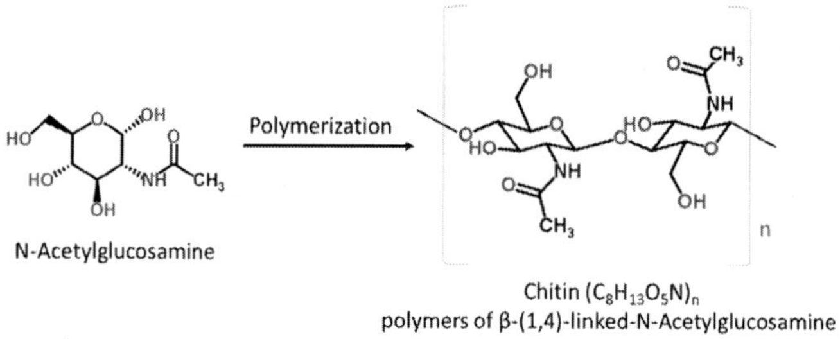

Figure 1. The 2D chemical structure of N-Acetylglucosamine (on the left) and its polymer chitin (on the right) where the repeating units are joined by covalent β-1,4 linkage.

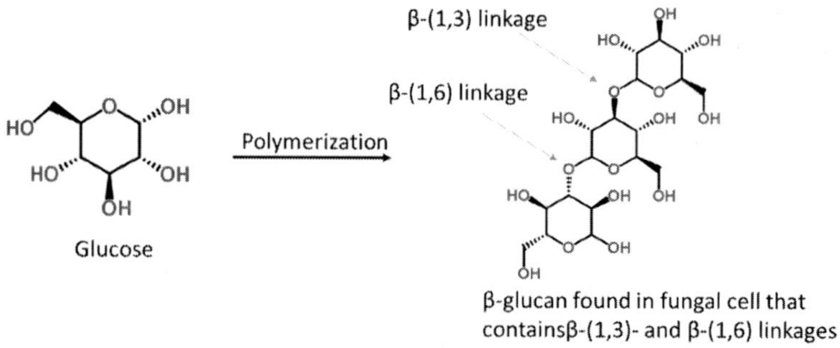

Figure 2. The 2D chemical structure of Glucose (on the left) and its Glucan polymer (on the right) usually found in the fungal cell wall. This Glucan polymer possess β-(1,3)- or β-(1,6) and thus broadly known as β-Glucans.

4. Role of Microbial Enzymes in Bio-Control of Fungal Phytopathogens

Fungal cell wall is composed of Chitin and β-Glucans which provides mechanical strength to the cell wall, so any enzyme which can depolymerize these carbohydrate polymers can aid to degrade fungal wall that can ultimately leadto antagonize fungal growth.

4.1. Chitinases

Chitinases belong to the group of glycoside hydrolases that can hydrolyze the β-1,4-glycosidic bonds between the N-acetyl-D-glucosamine residues of chitin [44, 45]. The complete enzymatic hydrolysis of chitin to free N-acetylglucosamine is performed by a chitinolytic system composed of a group of heterogeneous enzymes which synergistically catalyze the hydrolytic depolymerization of chitin. Presence of Chitinases have been found in microorganisms, plants, and animals [45, 46]. Although all chitinases have a role on the hydrolysis of β-1,4-glycosidic linkages, yet they are highly specific in their ability to utilize the type of chitin, as chitin polymers from different sources have different structures and complexities. Thus, chitinases from various sources are classified in to different families based on their specificities to utilize substrate [47, 48].

Wide variety of chitinolytic enzymes are produced by different organisms that exhibit different substrate specificities. In bacteria, chitinases play roles in nutrition, parasitism, and recycling chitin in nature, whereas in fungi, protozoa, and invertebrates, they are also involved in morphogenesis. Chitinases are involved in the defense mechanism of higher plants, vertebrates, and human against fungal pathogens [22, 46, 48, 49, 50].

Broadly, chitinases are classified as, endo-chitinases (EC 3.2.1.14) and exo-chitinases. The endo-chitinases randomly cleaves glycosidic linkages at the interior of the chitin chain, producing low molecular oligomers of GlcNAc, such as chitotetraose and chitotriose, eventually giving diacetylchitobiose as predominant products [22, 48]. Exo-chitinases can be further categorized as Exo-chitinases A and Exo-chitinases B. Exo-chitinases A reacts at reducing end and catalyze the release of diacetylchitobiose from the reducing end of chitin chain. Exo-chitinases B are further classified into two subcategories: chitobiosidases and β-(1,4)-N-acetyl-glucosaminidases. Chitobiosidases (EC 3.2.1.29) react at nonreducing end of chitin chain and catalyze the release of diacetylchitobiose (which are the sole products) where no monosaccharides or oligosaccharides are formed. β-(1,4)-N-acetyl-glucosaminidases (EC 3.2.1.30) or chitobiases split diacetylchitobiose and higher chitin polymers, including chitotriose and chitotetraose, into GlcNAc monomers in an exo-type fashion (Li 2006; Neeraja et al. 2010). Such enzymes have broad specificity and it are also called β-N-acetylhexosaminidases (EC 3.2.1.52) [22, 48, 49, 50]. Another enzyme chitin deacetylase causes deacetylation of chitin to chitosan, and its product is acted upon by chitosanase which yields glucosamine as the end product [48-0] (Figure 3).

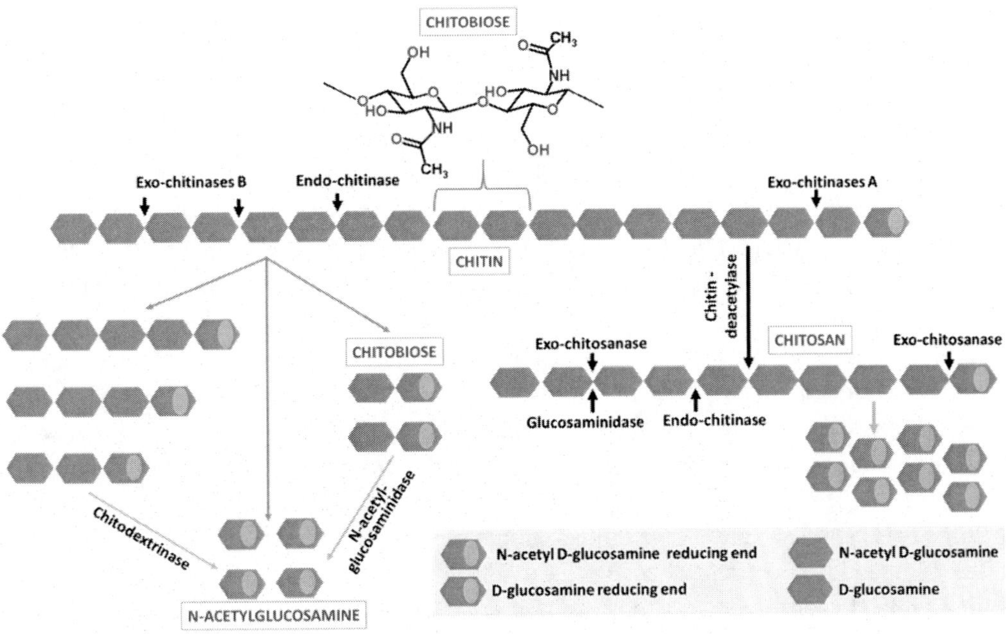

Figure 3. Representation enzymes involved in chitin depolymerization. Chitin polymer acted upon by endochitinase, exochitinase A from the reducing end; exochitinase B from nonreducing end to produce oligomers/chitobiose; chitin can be deacetylated by chitin deacetylase resulting in chitosan, which can be cleaved by endo-/exo-enzymes. Chitin oligomers and chitobiose can be further hydrolyzed by chitodextrinase and *N*-acetylglucosaminidase, respectively, to generate *N*-acetylglucosamine (as described Neeraja et al. [50] with modifications).

4.1.1. Family of Chitinases

Chitinolytic organisms produce multiple isomeric forms of chitinases that is because enzymes are highly specific to the different structures and complexities of chitin. This isomeric forms of chitinases result from post-translational processing of a single-gene product or the product of multiple genes (Gortari and Hours 2008). The heterogeneity of chitinases was attributed to post-translational modifications such as differential glycosylation and/or proteolysis [49]. Chitinases and N-acetylhexosaminidases has been grouped into three families namely, Family-18, Family-19, and Family-20, this grouping is based on the analysis of amino acid sequences in the catalytic domains [28, 48, 49]. Family-18 contains chitinases primarily from fungal sources, but includes some enzymes from bacteria, viruses, animals, insects, and plant sources, and hence the family is diverse in evolutionary terms. Family-19 consists of plant chitinases (Class I, II, and IV) and some of bacterial origin such as the 'C chitinase' from *Streptomyces griseus*. Family-20 includes N-acetylglucosamidases (EC 3.2.1.30) from *Vibrio harveyi* and N-acetylhexosaminidases (EC 3.2.1.52) from human and *Dictyostelium discoideum* [44, 45, 49, 51]. Chitinases consist of discrete domains which can be arranged in different orders in different proteins. Domains which are involved in catalyzing the hydrolysis of the glycosidic bonds (the catalytic domain) and the domains involved in the binding of the enzyme to the substrate (carbohydrate binding-domain) are the most important domains. Chitinases belonging to Family-18 and Family-19 of glycohydrolases differ in their amino acid sequences of catalytic domain, carbohydrate binding domains and even in their 3D structures. Further, chitinases belonging to these two families possess several differences in their biochemical properties as well [49-51].

4.1.1.1. Family of Bacterial Chitinases

To our interest, bacteria produce multiple chitinases to degrade the variety of chitin polymers present in nature. As described earlier, chitinolytic enzymes are grouped into families 18, 19, and 20 of glycosyl hydrolases based on amino acid sequences [44, 54]. Family-18 contains chitinases from bacteria, fungi, viruses, animals, and some plants thus this family is diverse. Family-19 include chitinases from plants (classes I, II, and IV) and chitinases from some *Streptomyces* [52]. Family-20 includes the β-*N*-acetylhexosaminidases from bacteria, *Streptomycetes*, and humans. Watanabe et al. [53] classified bacterial chitinases into three major subfamilies, A, B, and C, based on the amino acid sequence of individual catalytic domains. *Bacillus circulans* produces six different chitinases of which chitinase A1 is most efficient chitin degrading enzyme. Whereas, *Bacillus licheniformis* produces five chitinases. *S. marcescens* strain-NIMA possess chitinase activity 43 times greater than those produced by other strains, as it produces three efficient chitinase enzyme: an endochitinase (Chi60), an exochitinase (Chi50), and an *N*-acetylglucosaminidase [54]. *Pseudomonas aeruginosa* K-187A produce bifunctional enzyme that act as chitinase and as lysozyme [55, 56]. The chitinase (tk-ChiA) from the hyperthermophilic archea, *Thermococcus kodakaraensis* KOD1, contains two catalytic domains and three chitin-binding domains, and studies performed on the structure and function of this chitinase showed of a supportive role of two catalytic domains in the degradation of single chitin chain [57].

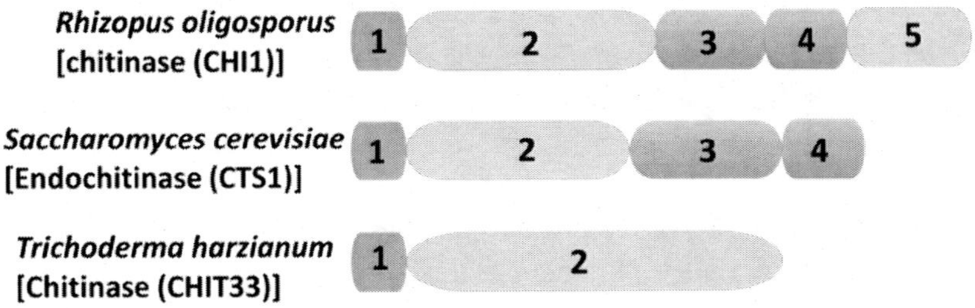

Figure 4. The structure of fungal chitinase of *Saccharomyces cerevisiae* endochitinase (CTS1), *Rhizopus oligosporus* chitinase (CHI1), *Trichoderma harzianum* chitinase (CHIT33); where, 1: signal peptide region, 2: catalytic domain, 3: serine/theronine-rich region, 4: chitin-binding domain, 5: C-terminal extension region [59].

4.1.1.2. Family of Fungal Chitinases

Chitinases from fungi belong to Family-18 of the glycohydrolase superfamily [44, 45]. Multi-domain structure is a typicality of the Family-18 chitinases. Family-18 chitinases have following five domains and regions: (1) N-terminal signal peptide region, (2) catalytic domain, (3) serine/theronine-rich region, (4) chitin-binding domain, and (5) C-terminal extension region. In most of the cases fungal chitinases lack the last three domains, as they do not need to be necessary for chitinase activity. Amino-acid sequence of endochitinase (CTS1) from *Saccharomyces cerevisiae*, shows four distinct domains: (1) a signal sequence, (2) a catalytic domain, (3) a serine/threonine rich region and (4) chitin-binding domain [58]. Whereas, two chitinases (chitinase-I and chitinase-II) of *Rhizopus oligosporus* have five distinct domains: (1) a signal sequence, (2) a catalytic domain, (3) a serine/theronine-rich region, (4) chitin-binding domain, and (5) an additional C-terminal domain. Chitinase of *Trichoderma harzianum* has 33 kDa mass and it possess a putative signal peptide region and a catalytic domain, but lacks a serine/theronine-rich region, a chitin-binding region and a C-terminal region [59] (Figure 6).

Till date crystal structure of only two chitinases (CiX1 and ChiB1) of fungal origin are available. The three-dimensional model of a chitinase 'CiX1' from fungal isolate *Coccidioides immitis*, and chitinase 'ChiB1' from fungus *Aspergillus fumigatus* has been reported [60, 61]. For, CiX1: Its active site is composed of conserved amino acid residues (Trp47,131,315,378, Tyr239,293, Arg52,295). The amino acid residue Glu171 is catalytic amino acid, as site-directed mutagenesis of Glu171 eradicates complete enzyme activity. Model of enzyme structure reveals that it is an eight-stranded β/α-barrel. Briefly, the barrel's core consists of eight parallel β-strands, which is surrounded by eight α-helices connected to the barrel. Groove appear on the enzyme's surface which is believed to be formed due to this unique barrel structure-. This groove binds to the sugar units of chitin, possibly $(Glc\text{-}NAc)_6$ moieties are considered as the active center, which are subsequently cleaved [62].

4.1.2. Bacterial Chitinases as Microbial Fungicides

Bacterial strains capable of producing chitinases are -exploited to treat fungal infections in plants. Use of antifungal chitin-degrading enzymes, directly in formulations, has opened a new era of biological control of plant fungal diseases that does not depend on the physical presence of the biocontrol bacteria [50]. Over the period of few decades, chitinolytic enzymes have gained importance due to their overwhelming potentials to treat fungal infections in plants [51]. Highly capable strains in producing chitinases belongs to genus of *Bacillus, Pseudomonas,*

Paenibacillus, Serratia and *Enterobacter* and are exploited for commercialization. From the strains of *Bacillus*, *Bacillus amyloliquefaciens* V656 capable of producing chitinases has been reported to inhibit the growth of *Fusarium oxysporum*, here the strain itself and its purified chitinases showed inhibitory effect on *F. oxysporum* [55, 56, 63-66]. *Bacillus thuringiensis* is also a chitinase producer as reported by Reyes-Ramírez et al. [67] and has contributed in the biocontrol of *Sclerotium rolfsii* and other phytopathogenic fungi in soybean seeds. *Bacillus subtilis* is another bacillus strain well-known for its ability to produce chitinases, where Chang et al. [68] showed the purified chitinases from *Bacillus subtilis* inhibits several phytopathogenic fungi. As per the report by Podile and Prakash [69], *Bacillus subtilis* is one of the most popular bacterial biocontrol agents possessing chitinolytic activity and to prove this, they showed the growth suppression of a causative agent of groundnut rust, *Puccinia arachidis,* by *Bacillus subtilis* AF1. Several biocontrol products as commercial biofungicides, containing *Bacillus subtilis* as a main ingredient are available globally (Companion, Epic, Kodiak, Rhizo-Plus, etc.) [50]. Someya et al. [70] reported the biocontrol of *B. cinerea, Rhizoctonia solani,* and *F. oxysporum* by the extracted chitinases from *Serratia marcescens.* Further, El-Tarabily et al. [72] reported the control of *Sclerotinia minor* which is the causal agent of basal drop diseases of lettuce by the chitinases from *S. marcescens.* Zhang and Yuen [73] reported the control of leaf spot on *Festuca arundinacea* caused by *Bipolaris sorokiniana* by *S. maltophilia.* Chitinase from another strain, that is from *Serratia proteamaculans* is reported to suppress fungal spore germination and hyphal growth of pathogenic *F. oxysporum* and *Aspergillus niger* [74]. Among the strains of *Pseudomonas*, *Pseudomonas aureofaciences* AB254 is reported to suppress phytopathogens of sweet corn belonging to genus of *Pythium spp.* [75]. *P. fluorescens* VO61 is reported to suppress damping-off and root rot disease caused by *Pythium ultimum* in *Lotus corniculatus* [76]. Chitinolytic *P. fluorescens* WCS358 is reported to suppress *F. oxysporum f.* sp. *lini* which causes fusarium wilt in Flax plant [77]. For commercial production of chitinases and chitinase expressing biocontrol strains, gram-negative bacteria such as *Pseudomonas* sp., *Erwinia* sp., and *Agrobacterium* sp. are difficult to commercialize due to short shelf life as, they fail to produce endospores that are resistant to heat, radiation, organic solvents, and other adverse environments. Despite of several gram-negative strains strains capable of producing chitinases, biocontrol formulations containing gram-positive *Bacillus* species are dominated in the commercial market.

The bacterial strains possessing chitinolytic activity are applied to diseased plant along with the inducer molecule (chitin) which enhances the overall volume of production of chitinases. This is an efficient approach to enhance the bio-control ability of chitinolytic bacterial strains. Repots suggests that the bacterial formulation with 1% colloidal chitin (induces chitinase production in bacterial strains) can remarkably increase disease control efficiency [78]. Bioformulation of *B. subtilis* AF1 containing colloidal chitin improved the biocontrol activity against *Aspergillus niger* and *Fusarium udum* phytopathogen of groundnut and pigeon pea respectively [79, 80]. Chang et al. [68] showed *Bacillus cereus*, grown on shellfish chitin waste, significantly subdued the growth of *Fusarium oxysporum, Fusarium solani* and *Pythium ultimum* (phytopathogens of Chinese cabbage). Talc-based formulation (with and without chitin) containing the mixture of two *Pseudomonas fluorescens* strains (PF1 and FP7) were tested against sheath blight and leaf folder in rice. The application of talc formulation through seed, root, soil, and foliar spray significantly reduced the sheath blight and leaf-folder incidence both under greenhouse and field conditions and it was observed that formulation containing chitin was more effective than the formulation not amended with chitin

[81]. Zeolite-based chitosan amended formulation containing chitinolytic strain of *Paenibacillus* sp. 300 and *Streptomyces* sp. 385, suppressed *Fusarium oxysporum* f. sp. *Cucumerinum,* a causative agent of *Fusarium* wilt of cucumber [82, 83]. Other bacterial strains viz. *Pseudomonas fluorescens, Pseudomonas aeruginosa, Bacillus licheniformis,* etc. which are categorized as plant growth promoting bacteria (PGPR) also show bio-control capability against phytopathogenic fungi due to their ability to produce chitinases [63-66].

4.1.3. Fungal Chitinases as Microbial Fungicides

Fungal strains are known to inhibit the phytopathogens, this mechanism exhibited by fungal biocontrol strains is known as mycoparasitism, which is mainly accredited by fungal cell wall degrading enzymes. Chitinases are among the most important hydrolytic enzymes produced by fungal strains [84]. As mentioned earlier, chitin is an essential structural component found in fungal cell wall and so it is considered a target molecule by fungicides [85].

Mycoparasitic *Fusarium chlamydosporum* is reported to suppress the groundnut rust caused by the phytopathogen *Puccinia arachidis,* mainly due to its ability to produce an endo-chitinase which is responsible for the inhibition uredospores germination of the rust fungus. This indicates the significant contribution of chitinase in the biocontrol of groundnut rust [89]. Large number of strains belonging to *Fusarium* genus are classified as phytopathogesn, however, there are several other strains of *Fusarium* which are non-pathogenic to plants. Further, there are several reports suggesting that such non-pathogenic strains of *Fusarium* show mycoparasitism attributed by the production of chitinases, against other strains of *Fusarium* which causes Fusarium-wilt in plants. Non-pathogenic strains of *Fusarium* can suppress *Fusarium oxysporum* f.sp. *ciceris*, *Fusarium* oxysporum f.sp. *lycopersici*, *Fusarium oxysporum* f.sp. *asparagi, Fusarium oxysporum* f.sp. *Melonis* [90]. *Myrothecium verrucaria* is another bio-control agent known to produce high levels of extracellular lytic enzymes viz. chitinase and β-1,3-glucanase [91, 92]. The crude extract of this biocontrol agent containing chitinase and β-1,3-glucanase in the form of crude extract suppressed the infection caused by *Sclerotium rolfsii* in the roots of groundnut. Similar findings are also reported by Govindsamy et al. [93], where chitinase was purified from crude fungal extract of *Myrothecium verrucaria* showed to control groundnut rust that is caused by *Puccinia arachidis.* However, out of all the genus of fungi, strains belonging to the genus of *Trichoderma* are most widely studied for their bio-control ability, where these strains possess ability to produce lytic enzymes, mainly chitinases and glucanases [84]. Chitinolytic enzymes of *Trichoderma harzianum,* a most studied mycoparasitic fungus, were found to be inhibitory to the wider range of deleterious fungi than similar enzymes from other sources [94, 95]. No other chitinases have been reported with such a broad range of activity, synergy between different enzymes, or synergy with several other biological and chemical control agents [51]. Way back in 1980, it was reported that *Trichoderma harzianum* can efficiently suppress *Sclerotium rolfsii* and *Rhizoctonia solani* which infects *Phaseolus vulgaris*. The mechanism proposed was that, the strain of *T. harzianum* could easily lyse the mycelium of these phytopathogenic fungi, which was attributed by the production of cell wall degrading enzymes. Later, *T. harzianum* was reported to control several phytopathogenic fungi and now it is being considered as one of the most effective fungal bio-control agent [96, 97]. Other efficient strains of *Trichoderma* capable of producing chitinases as an effective biocontrol agent include *T. virens, T. koningii, T. lignorum, T. viride* [96]. Metcalf and Wilson [98] described *T. koningii* (Tr5) can suppress the infection of *Sclerotium cepivorum* in the onion roots at ease. Here, the observation educed that the Hyphae of the biocontrol agent penetrates into infected epidermal and cortical tissue of the root destroys the hyphae of the pathogen, with no damage to uninfected plant tissue. The authors claimed production of endo- and exo-chitinases by *T. koningii* to play role in the lysis of Hyphae. To confirm the role of chitinase in the bio-control activity of *Trichoderma* spp., Woo et al. [99] disrupted chitinase (ech42) gene *T. harzianum* (P1) to block the production of chitinase enzyme, which cause reduction in its bio-control activity against *Botrytis cinerea* on bean leaves. Thus, it was evident that chitinase enzyme play a major role in the bio-control activity as shown by fungal-fungicides. To confirm the role of chitinolytic enzymes in bio-control was further supported by the work of Lorito and colleagues [94, 95], as they transferred the gene encoding endochitinase from *T. harzianum* (P1) into tobacco and potato which demonstrated a high level and broad spectrum of resistance against several plant pathogens. Later, Bolar et al. [100] developed transgenic apple plant which contained endo and exo-chitinases form *Trichoderma atroviride* (P1) that showed enhanced resistance to apple scab incited by *Venturia inaequalis*. Recently, Chen et al. [101] developed transgenic *Nicotiana benthamiana* possessing sporamin gene from sweet potato and chitinase gene from fungus *Paecilomyces javanicus*. This transgenic plant showed efficient suppression of leaf spot diseases causing pathogen *Alternaria alternata* and soft rot disease causing pathogen *Pectobacterium carotovorum* subsp. *carotovorum*. Further, this transgenic crop also showed suppression of insects, *Spodoptera litura* and *Spodoptera exigua*. Thus, by using the genes coding for various types of chitinases form fungal bio-control agents to develop disease resistant transgenic plants has opened a new horizon to exploit microbial enzymes.

4.2. Glucanases

Glucan is a polymer of glucose, and the enzymes that are capable of hydrolyzing glucosidic bond are known as Glucanases. Glucose subunits can polymerize in several ways and produce variety of polymers. Broadly, a molecule of glucose can from α- or β-glucosidic linkage with neighboring glucose subunit in the polymer. Polymers of glucose are categorized based on the type of glucosidic linkage between two adjacent glucose subunits. Similarly, Glucanases are broadly classified as α- or β-Glucanases. α-Glucanases are the enzymes that can cleave only α-glucosidic linkages. These enzymes are further categorized as (1) α-1,4-glucanase, an enzyme that breaks down α-1,4-glucans and (2) α-1,6-glucanase, an enzyme that breaks down α-1,6-glucans. Similarly, β-Glucanases are the enzymes that can cleave only β-glucosidic linkages. These enzymes are further categorized as (1) β-1,3-glucanase, an enzyme that breaks down β-1,3-glucans such as callose or curdlan (2) β-1,6 glucanase, an enzyme that breaks down β-1,6-glucans (3) Cellulase, an enzyme that perform the hydrolysis of 1,4-beta-D-glucosidic linkages of cellulose, lichenin and cereal β-D-glucans. Both α- and β-Glucanases enzymes can further be classified as endo or exo glucanases based on the location where they cleave glucosidic linkages within the glycan polymer.

To our interest, β-Glucanases which can cleave the β-1,3 and β-1,6 linkages of the glucan found in the fungal cell wall are of use and can be exploited to weaken the fungal cell wall of phytopathogenic fungi.

4.2.1. β-Glucanases from Microbes and Their Action as Fungicides

Microbes, both bacteria and fungi are capable of producing several types of β-Glucanases. However, the most efficient enzyme they produce is 'Laminarinases' which is also addressed as'endo-β-1,3-glucanases EC3.2.1.39 and endo-β-1,3-1,4-glucanases'. It is known that structure and composition of laminarin (glucan found in brown algae) is highly similar to that of β-glucan found in the fungal cell wall. Thus, any bacterial or fungal strain, showing Laminarinase activity is most likely to cleave fungal cell wall. Another important enzyme produced by bacteria include exo-β-1,3-glucosidase (glucan 1,3-β-glucosidase EC3.2.1.58), lichenase (1,3-1,4-β-D-glucan 4-glucanohydrolase EC 3.2.1.73) and β-glucosidases (EC 3.2.1.21) that can cleave β-glucosidic bonds in glucans other than cellulose [102, 103]. Lichenase is an enzyme that has been significant because of its β-glucanase activity. Most of this type are bacterial in origin, with the *Bacillus* species being the most well-known. All β-glucanases are highly similar in action and if any of these β-glucanase activity is detected in bacterium it is likely that the strain might also produce other β-glucanases [102].

There are several evidences suggesting the role of *Bacillus* spp. -in bio-control of fungal phytopathogens. To quote few, Singh et al. [82, 83] reported the suppression of root rot fungus *Macrophomina phaseolina,* a phytopathogens of *Pinus roxburghii* by *Bacillus subtilis* BN1 which showed dual capability to produce β-Glucanase and act as Plant Growth Promoting Rhizobacteria (PGPR). Another strain, *Bacillus sublitis* AF1 was thoroughly studied by Manjula and Podile [80] for its biocontrol properties. AF1 showed dual nature to produce chitinase and β-Glucanase where authors showed the strain to control groundnut rust and soft rot of lemons. Alternative strain, NSRS 89-24 of *Bacillus subtilis* is reported to suppress *Pyricularia grisea* and *Rhizoctonia solani* which are the causative agents of rice blast and sheath blight respectively [104]. Bargabus et al. [105] reported *Bacillus mycoides* capable to produce β-Glucanases could effectively suppresses the growth of *Cercospora beticola* which causes leaf spot disease in sugar beet. More recently, Abraham et al. [106] reported β-1,3-glucanase producing *Bacillus amyloliquefaciens* (a bacterial endophyte of *Hevea brasiliensis*) could effectively antagonize phytopathogenic *Phytophthora meadii.* Martínez-Absalón et al. [8] showed *Bacillus thuringiensis* UM96 capable to produce β-Glucanase could suppress gray mold phytopathogen *Botrytis cinerea.*

Other than the strains of *Bacillus*, *Pseudomonas* spp. are also widely studied for their ability to produce β-Glucanases that attributes to their bio-control ability. Back in the year 1993, β-1,3-glucanase producing *Pseudomonas cepacia* was reported by Fridlender et al. [107] that showed to control soilborne plant pathogens, *Rhizoctonia solani*, *Sclerotium rolfsii* and *Pythium ultimum,* where disease incidence was reduced by 85, 48 and 71%, respectively. β-1,3-glucanase produced by *Pseudomonas stutzeri* YPL-1 was identified as one of the key enzymes in decomposition of fungal cell walls as reported way back in 1995, by Lim and Kim [108] where the enzyme inhibited 53% mycelial growth of the fungus *Fusarium solani.* The optimum pH and temperature for β-1,3 glucanase to operate was found to be pH 5.5 and 40°C, respectively. β-1,3- and β-1,4-glucanase produced by *Pseudomonas* sp. PE2 was found to be an essential which hydrolyzed *Pythium porphyrae* cell walls [109]. *Pseudomonas fluorescens* was reported to produce β-1,3 and β-1,4-glucanase by Diby et al. [110] where the enzyme was responsible for the lysis of *Phytophthora capsici* mycelium. Later, Arora et al. [111] in 2007 showed *Pseudomonas fluorescens* strains GRC3 and GRC4 -efficiently produce β-1,3-glucanase played crucial role in the inhibition of phytopathogenic fungi *Rhizoctonia solani.* In 2010, *Pseudomonas fluorescens* strain producing enzymes such as β-1,3- and β-1,6-glucanases was used by Dev and Dawande [112] in the biocontrol of soil borne fungal plant pathogen, *Rhizoctonia solani*. Most recently, Yasmin et al. [113] reported *Pseudomonas aeruginosa* Z5 isolated from cotton rhizosphere was capable to produce β-1,3-glucanase and showed bio-control of *Fusarium oxysporum*, *Fusarium moniliforme*, *Fusarium solani* and *Rhizoctonia solani* significantly with percent inhibition ranging from 25% to 91.5%. On the whole, it can be deduced that strains of *Bacillus* and *Pseudomonas* have been exploited as bio-control agents against phytopathogens, as they hold ability to produce fungal cell wall degrading β-glucanases.

4.2.2. β-Glucanases from Microbes and Their Action as Fungicides

Production of β-Glucanases is quite common and it's been studied over many years. The primary reason to produce these enzymes by fungi include roles in the fungal cell maintainance, including cell wall extension, hyphal branching, sporulation, budding, and autolysis [114]. Fungal species belonging to *Trichoderma* and *Aspergillus* are most widely studied for β-Glucanases production for suppression of fungal phytopathodens [114]. Grishutin et al. [103] in 2006 reported a lichenase–like endo-β-1,4-glucanase, isolated from *Aspergillus japonicas*. Interestingly, this enzyme resembles to that of bacterial lichenase, with a minor difference in that this enzyme is that it possess low activity toward Carboxymethyl cellulose (modified β-1,4-glucan).

Several strains of fungi produce β-Glucanases however, for using fungal strains as bio-control agents, *Trichoderma* spp. are most widely exploited. To reflect the role of β-Glucanases in the bio-control capabilities of *Trichoderma* spp., Migheli et al. [115] overexpressed β-1,4-endoglucanase gene '*egl1*' in *Trichoderma longibrachiatum,* where it was observed that such overexpression enhanced the bio-control ability of *Trichoderma longibrachiatum* against *Pythium ultimum* (a phytopathogen of cucumber). Thus, this concluded that β-Glucanases produced by the strains of *Trichoderma* aids in fungal cell wall lysis, the similar way where fungal cell wall lysis is aided by chitinases. Other reports for the use of strains of *Trichoderma* for the control of fungal phytopathogens are quoted as follows; Ganesan and Sekar [116] examined six strains of *Trichoderma* namely, *T. harzianum* (ITCC-4572), *T. koningii* (MTCC-2385), *T. viride* (MTCC-800), *T. pseudokoningii* (MTCC-3011), *T. hamatum* (MTCC-2577), *T. reesii* (MTCC-798) to check their bio-control ability against *Rhizoctonia solani* (causative agent of web blight disease in Groundnut), where authors showed that *T. koningii* showed maximum β-Glucanase activity with 60.40 IU/ml yet all the strains of *Trichoderma* could effectively lyse fungal cell wall of *R. solani*. Djonović et al. [117] reported that *Trichoderma virens* could produce an enzyme 'Tvbgn3' which is a type of β-1,6-glucanase that can effectively suppress *Pythium ultimum*. Thus, it can be deduced that production of β-Glucanases aspects the bio-control ability showed by fungal strains.

5. Commercially Available Formulations of Bio-Control Agents Capable of Producing Fungal Cell Wall Degrading Enzymes

Bacterial and fungal strains are commercially available in the form of formulated products which are capable to produce chitinases and glucanases to feature their role to inhibit phytopathogenic fungi [46]. Fungal bio-control agents are usually prepared as a powder formulation, granular powder and fluid-bed granules using dextrin as binder. Alginate gels are used to prepare bacterial and fungal formulations [46, 118]. Bacterial bio-control agents are formulated in variety of ways and are available in the market. Formulation of the sporulating, gram positive bacteria are resistant to desiccation and thus, provide longer shelf life. Gram-positive microorganism possesses heat resistant spores that are exploited to formulate stable and dry powder products [46, 119]. Alternative to solid powdered formulation is the suspension of organisms in oil, where the purpose is to exclude oxygen which prevents respiration [120]. Addition of silica gel to oil formulation enhances the shelf life as it is reported to mutate conidia for fungal strains [120].

Few well known bio-control formulation commercially available are: Bio-formulation of *Trichoderma spp.* by Binab (Sweden) and it is effective against pathogenic fungi that cause wilt, root rot, and internal decay of wood products and decay in tree wounds. Bio-formulation of *Fusarium oxysporum* is commercialized by Biofox (www.biofox.com) and it is effective against *Fusarium moniliforme. Bacterial bio-formualtion of Pseudomonas aureofaciens* commercialized by Ecosoil (www.ecosoil.com) is effective against Dollar spot, Anthracnose, *Pythium aphanidermatum*, Michrochium patch (pink snow mold). *Streptomyces griseoviridis* strain K61 is commercially formulated by AgBio (http://www.agbio-inc.com) which is known to inhibit *Fusarium* spp., *Alternaria brassicola, Phomopsis* spp., *Botrytis* spp., *Pythium* spp., and *Phytophthora* spp. that cause seed, root, and stem rot, and wilt disease of ornamental and vegetable crops. A bio-control formulation of *Bacillus licheniformis* SB3086 is produced by Novozymes (http://www.novozymes.com) which is effective against Dollar spot disease of plants. Commercial bioformulation of *Coniothyrium minitans* is produced by BIOVED, Ltd., Hungary, is effective in suppressing *Sclerotinia sclerotiorum* and *Sclerotinia minor* which are phytopathogens infecting cucumber, lettuce, capsicum, tomato, and ornamental flowers.

Despite of several Gram-negative bacteria known to possess efficient bio-control ability, they are difficult to formulate as they do not produce endospores, their formulations have short shelf life and bacteria gets easily killed when the formulations are desiccated. The problems faced by bio-control developers is that crops are grown under a multiplicity of climatic and environmental conditions which include temperature, rainfall, soil type, crop variety which change from farm to farm or even within one field and such variations causes disparity in the potentiality of biocontrol agent. Also, biotic condition and type of pathogens need to be considered when attempting to develop new potential biocontrol agent for particular area [46, 121].

Conclusion and Future Perspectives

Worldwide growth in population, increase in the demand of global food production and environmental damage causing problems in agriculture yield are major concerns to the word. These problems may soon cause insufficiency to feed all the world's people. Current population of world 7 billion people and the population is expected to reach 10 billion by next 50 years. The agriculture production should be increased by at least 35% to meet the demand of food in future. Agricultural strategies to feed all these individuals is an important challenge in 21st century. For the same it is essential that agricultural productivity has to significantly increased within the next few decades. Out of several constrains faced by the farmer, constrain caused by the fungal pathogens by infecting and spoiling crops is the most alarming one in agriculture today. Annually crop spoilt by the fungal diseases is at least 30 percent of total production. If this spoilage by the fungal pathogens is controlled, the demand of food supply to the population of the world can be fulfilled. Use of chemical fungicides has created concerns regarding- public safety and thus, has revoked their use leading to development of other alternatives to suppress fungal phytopathogens. Alternative to chemical fungicides are the microbial fungicides possessing the trait to suppress pathogenesis of fungal plant pathogens by lysing their cell wall which is aided by the cell wall degrading enzymes. The progress in the development of bio-control agents from microbes that are capable to produce chitinase and glucanases is rapid, yet several efforts must be made to produce highly effective bio-control agents. The problems that are to be looked upon are, developing bio-control agents which can be effective for several pathogens, as there are thousands of phytopathogens are identified, so there cannot be one bio-control agent for one pathogen. Strains used for developing bio-control agents should be able to produce fungal cell wall degrading enzymes under several climatic conditions and should have broad spectrum of action. The enzyme produced should be effective at all the temperatures where plant can grow as plants for which these bio-control agents are applied will vary as different plant has different optimum temperature for its growth. The future researches should aim to discover novel fungal cell wall degrading enzymes. Biological functions of chitinase and glucanases in the antagonism of fungal pathogens must be studied in further depth, more knowledge regarding the microbial genes of chitinases and glucanases is required to develop transgenic crops with the capability to express microbial chitinases and glucanases that can enhance their immunity against several phytopathogens. Such cumulative knowledge will be helpful in protein engineering of microbial chitinases and glucanases that can help us to build disease free agriculture.

ACKNOWLEDGMENTS

Authors are thankful to Council of Scientific & Industrial Research (CSIR) for providing Senior Research Fellowship (SRF) and Charotar University of Science and Technology (CHARUSAT) for providing necessary facilities.

REFERENCES

[1] Dufresne, M. and Osbourn, A. (2001). Definition of tissue-specific and general requirements for plant infection in a phytopathogenic fungus. *Mol Plant-Microbe Interactions 14* (3): 300-307.

[2] Inglis, G. D. and Kawchuk, L. M. (2002). Comparative degradation of oomycete, ascomycete, and basidiomycete cell walls by mycoparasitic and biocontrol fungi. *Can J Microbiol., 48*: 60-70.

[3] Fisher, M., Henk, D., Briggs, J., et al. (2012). Emerging fungal threats to animal, plant and ecosystem health. *Nature 484*:186-194.

[4] Geber, J. and Murphy, E. (2012). Scurvy in the Great Irish Famine: Evidence of vitamin C deficiency from a mid-19th century skeletal population. *Amr J Phys Anthropol, 148*: 512-524.

[5] Chen, G., Nowocin, K. and Marathe, N. (2017). Toward Reducing Crop Spoilage and Increasing Small Farmer Profits in India: A Simultaneous Hardware and Software Solution. In *Proceedings of the Ninth International Conference on Information and Communication Technologies and Development*, p. 37. ACM.

[6] Mendgen, K., Hahn, M. and Deising, H. (1996). Morphogenesis and mechanisms of penetration by plant pathogenic fungi. *Ann Rev Phytopathol.*, 34:367-386.

[7] Marin, D., Romero, R., Guzmán, M. and Sutton, T. (2003). Black Sigatoka: an increasing threat to banana cultivation. *Plant Disease*, *87*: 208-222.

[8] Martínez-Absalón, S., Rojas-Solís, D., Hernández-León, R., et al. (2014). Potential use and mode of action of the new strain Bacillus thuringiensis UM96 for the biological control of the grey mould phytopathogen *Botrytis cinerea*. *Biocont Sci Technol.*, *24*: 1349-1362.

[9] Thakker, J. N., Patel, N. and Kothari, I. (2007). *Fusarium oxysporum* derived elicitor-induced changes in enzymes of banana leaves against wilt disease. *J Mycol Plant Pathol.*, *37*: 510-513.

[10] Thakker, J. N., Patel, P. and Dhandhukia, P. (2011). Induction of defence-related enzymes in susceptible variety of banana: Role of Fusarium-derived elicitors. *Arch Phytopathol Plant Prot.*, *44*: 1976-1984.

[11] Sang-Gyu, K., Ye-Lim, J., Hye-Young, K., et al. (2010). Comparison of microbial fungicides in antagonistic activities related to the biological control of phytophthora blight in chili pepper caused by *Phytophthora capsici*. *Plant Pathol J., 26*: 340-345.

[12] Minou N., Jason E., Chu, M., et al. (2010). De novo assembly of a 40 Mb eukaryotic genome from short sequence reads: *Sordaria macrospora*, a model organism for fungal morphogenesis. *PLoS genetics*, *6*: e1000891-e1000912.

[13] Horst, R. K. (2013). Rusts. In *Westcott's Plant Disease Handbook*. Springer Netherlands, pp. 341-362.

[14] Pain, A. and Hertz-Fowler, C. (2010). Plant killers on the spot. *Nat Rev Microbiol.*, *8*: 615. Doi: 10.1038/nrmicro2433.

[15] Almario, J., Moënne-Loccoz Y. and Daniel M. (2013). Monitoring of the relation between 2, 4-diacetylphloroglucinol-producing *Pseudomonas* and *Thielaviopsis basicola* populations by real-time PCR in tobacco black root-rot suppressive and conducive soils. *Soil Biol Biochem.*, *57*: 144-155.

[16] Arsenault-Labrecque, G., James G. and Richard R. (2012). Effect of silicon absorption on soybean resistance to *Phakopsora pachyrhizi* in different cultivars. *Plant Dis.*, *96*: 37-42.

[17] Radhajeyalakshmi, R., Velazhahan, R. and Prakasam, V. (2012). *In vitro* evaluation of solvent extracted compounds from edible macromycetes against phytopathogenic fungi. *Arch Phytopathol Plant Prot.*, 45: 93-300.

[18] Schena, L., Li D., Sanzani, S., et al. (2013). Development of quantitative PCR detection methods for phytopathogenic fungi and oomycetes. *J Plant Pathol.*, *2013*:7-24.

[19] Vleeshouwers, V., Raffaele, S., Vossen, J., et al. (2011). Understanding and exploiting late blight resistance in the age of effectors. *Ann Rev Phytopathol.*, *49*: 507-531.

[20] Pal, K., Dey, R. and Tilak, K. (2014). Fungal Diseases of Groundnut: Control and Future Challenges. In: *Future Challenges in Crop Protection Against Fungal Pathogens*, pp. 1-29. Springer, New York, NY.

[21] Robin, C., Stefan, S. and Daniel R. (2010). Dominance of natural over released biological control agents of the chestnut blight fungus Cryphonectria parasitica in south-eastern France is associated with fitness-related traits. *Biolog Cont., 53*: 55-61.

[22] Le, C., Kruijt, M. and Raaijmakers, J. (2012). Involvement of phenazines and lipopeptides in interactions between *Pseudomonas* species and *Sclerotium rolfsii*, causal agent of stem rot disease on groundnut. *J Appl Microbiol., 112*: 390-403.

[23] Forgetta, V., Gary, L., Dias, J., et al. (2013). Sequencing of the Dutch elm disease fungus genome using the Roche/454 GS-FLX Titanium System in a comparison of multiple genomics core facilities. *J Biomol Techniq., 24*: 39-49.

[24] Jin, Y. (2011). Role of *Berberis* spp. as alternate hosts in generating new races of *Puccinia graminis* and *P. striiformis*. *Euphytica, 179*: 105-108.

[25] Carlos R., Ronaldo C. and Guilherme M. (2011). Cryptosexuality and the genetic diversity paradox in coffee rust, Hemileia vastatrix. *Plos One* 6: e26387-e26393.

[26] Wilson, A., Beckerman, J. and Aime, M. (2014). First Report of the White Pine Blister Rust Fungus, *Cronartium ribicola*, on *Ribes odoratum* in Indiana. *Plant Dis., 98*: 277-277.

[27] Foglio, E., Sykes, J., Espiritu-Fuller, M., et al. (2013). Brain abscess caused by" Corn Smut" Ustilago species in an immunocompetent child. *J Ped Infect Dis., 8*: 143-148.

[28] Jan W., Andreas, L., Schaffrath, U., et al. (2012). Assessment of the loose smut fungi (*Ustilago nuda* and *U. tritici*) in tissues of barley and wheat by fluorescence microscopy and real-time PCR. *Eur J Plant Pathol., 133*: 865-875.

[29] Ozge, N., Filiz, O., Melike, B., et al. (2013). Combined effects of salt stress and cucurbit downy mildew (*Pseudoperospora cubensis* Berk. and Curt. Rostov.) infection on growth, physiological traits and antioxidant activity in cucumber (*Cucumis sativus* L.) seedlings. *Physiolog Mol Plant Pathol., 83*: 84-92.

[30] Thomas W., Simone, O., Francis, P., et al. (2013). The wheat powdery mildew genome shows the unique evolution of an obligate biotroph. *Nature Genetics, 45*: 1092-1096.

[31] Xu, Xiangming, Harvey, Nick, Roberts, Anthony and Barbara, Dez. 2013. "Population variation of apple scab (*Venturia inaequalis*) within mixed orchards in the UK. *Eur J Plant Pathol., 135*: 97-104.

[32] Kumar, S., Shakywar, R., Tomar, K., et al. (2013). Varietal reaction of rose against black spot caused by *Diplocarpon rosae* Wolf. in Arunachal Pradesh. *HortFlora Res Spect 2*: 56-59.

[33] Odile C. and Morissette-Thomas, V. (2013). Epidemiology of grape anthracnose: Factors associated with defoliation of grape leaves infected by *Elsinoë ampelina*. *Plant Dis., 97*: 222-230.

[34] Weber, R. W. (2014). Biology and control of the apple canker fungus Neonectria ditissima (syn. N. galligena) from a Northwestern European perspective. *Erwerbs-Obstbau, 56*: 95-107.

[35] Zhang, Y., Pedro W., Schoch, C., et al. (2011). A molecular, morphological and ecological re-appraisal of Venturiales—a new order of Dothideomycetes. *Fungal Diversity, 51*: 249-277.

[107] Fridlender, M., Inbar, J. and Chet, I. (1993). Biological control of soilborne plant pathogens by a β-1, 3 glucanase-producing *Pseudomonas cepacia*. *Soil Biol Biochem.*, *25*: 1211-1221.

[108] Ho-Seong, L. and Sang-Dal, K. (1995). The Role and Characterization of β-1, 3-Glucanase in Biocontrol of *Fusarium solani* by *Pseudomonas stutzeri* YPL-1. *J Microbiol.*, *33*: 295-301.

[109] Kitamura, E. and Kamei, Y. (2006). Molecular cloning of the gene encoding β-1, 3 (4)-glucanase A from a marine bacterium, *Pseudomonas* sp. PE2, an essential enzyme for the degradation of *Pythium porphyrae* cell walls. *Appl Microbiol Biotechnol.*, *71*: 630-637.

[110] Diby, P., Saju, K., Jisha, P., et al. (2005). Mycolytic enzymes produced by *Pseudomonas fluorescens* and *Trichoderma* spp. against *Phytophthora capsici*, the foot rot pathogen of black pepper (Piper nigrum L.). *Annals Microbiol.*, *55*: 129-133.

[111] Kumar, A., Kim, M., Kang, Chul, S., et al. (2007). Role of chitinase and β-1, 3-glucanase activities produced by a fluorescent pseudomonad and in vitro inhibition of *Phytophthora capsici* and *Rhizoctonia solani*. *Can J Microbiol.*, *53*: 207-212.

[112] Dev, N. and Dawande, A. (2010). Biocontrol of soil borne plant pathogen Rhiozoctonia solani using *Trichoderma* spp. and *Pseudomonas fluorescens*. *Asiatic J Biotechnol Res.*, *1*: 39-44.

[113] Hafeez, Y., Fauzia Y. and Ghulam, R. (2014). Evaluation of *Pseudomonas aeruginosa* Z5 for biocontrol of cotton seedling disease caused by *Fusarium oxysporum*. *Biocont Sci Technol.*, *24*: 1227-1242.

[114] Sinsabaugh, R. (2005). Fungal enzymes at the community scale. *Mycology Series*, *23*: 349-360.

[115] Migheli, Q., Luis G., Dealessi, L., et al. (1998). Transformants of *Trichoderma longibrachiatum* overexpressing the β-1, 4-endoglucanase gene egl1 show enhanced biocontrol of *Pythium ultimum* on cucumber. *Phytopathol.*, *88*: 673-677.

[116] Ganesan, S. and Sekar, R. (2011). Screening of biocontrol agents against Rhizoctonia solani causing web blight disease of groundnut (*Arachis hypogaea* L.). In: *Pesticides in the Modern World-Pests Control and Pesticides Exposure and Toxicity Assessment*. InTech Publisher.

[117] Slavica D., Pozo, M. and Charles K. (2006). Tvbgn3, a β-1, 6-glucanase from the biocontrol fungus *Trichoderma virens*, is involved in mycoparasitism and control of *Pythium ultimum*. *Appl Env Microbiol.*, *72*: 7661-7670.

[118] Desai, S., Reddy, M. and Kloepper, J. (2002). Comprehensive testing of biocontrol agents. In: *Biological control of crop diseases*. Pp. 387-420.

[119] Hornby, D. (1990). *Biological control of soilborne plant pathogens*. No. 632.96. CAB International UK.

[120] Honeycutt, E. and Benson, D. (2001). Formulation of binucleate *Rhizoctonia* spp. and biocontrol of *Rhizoctonia solani* on impatiens. *Plant Dis.*, *85*: 1241-1248.

[121] Duo-Chuan, L. (2006). Review of fungal chitinases. *Mycopathologia*, *161*: 345-360.

SECTION D. FOOD MICROBIOLOGY

In: Microbial Catalysts. Volume 1
Editors: S. M. Abdel-Aziz, N. Garg, A. Aeron et al. ISBN: 978-1-53614-679-0

Chapter 9

ENZYMES AND THEIR VITAL ROLE IN THE FOOD INDUSTRY

Shadia M. Abdel-Aziz[1,*], Mohamed El-Awady[2], Neelam Garg[3] and Abhinav Aeron,[4,5]

[1]Department of Microbial Chemistry, National Research Centre, Dokki, Giza, Egypt
[2]Department of Microbial Biotechnology, National Research Centre, Dokki, Giza, Egypt
[3]Department of Microbiology, Kurukshetra University, Haryana, India
[4] School of Basic Science and Research, Sharda University, Greater Noida, Uttar Pradesh, India
[5]Division of Biotechnology, CEBS, Chonbuk National University, Iksan, South Korea

ABSTRACT

Enzymes, substances act as catalysts in living organisms, regulate the rate at which chemical reactions proceed without being altered in the process. The catalytic properties of enzymes have long been exploited in many industrial sectors to carry out efficient and cost-effective bioconversions for biotechnological applications. In enzymes biotechnology, microbial enzymes are utilizing as biocatalysts for production of valuable products. Enzymes are used in the environmental processes or for analytical and diagnostic purposes. Safety regulations of foods associated with enzymes focus on allergenic properties and differ among countries. Food safety regulations must be ensured especially with the availability of enzymes with new and unusual properties. Application of nanotechnology and nanoparticles in food has gained great importance in recent years. Nanotechnology has the potential to improve the nutritional value of food and generate new safe-food packaging to extend shelf-life. However, this new technology also raises concern about risk assessment. Nanotechnology risks include environmental risks due to the release of nanoparticles into the environment and safety risks from nanoparticles for workers and consumers.

[*] Corresponding Author's Email: abdelaziz.sm@gmail.com.

Keywords: microbial enzymes, food industry, dairy, meat, bakery, beverages

1. Introduction

The catalytic properties of enzymes have long been exploited in many industrial sectors such as foods, cosmetics, agriculture, chemistry, energy, and pharmaceutical fields to carry out efficient and cost-effective bioconversions. Microbial enzymes play a significant role in different food processes due to their easiness and availability. Enzymes are mainly used by the food industry to optimize the production process, improve efficiency, quality, shelf life, and most importantly to achieve desired organoleptic characteristics of the final product [1]. In enzyme biotechnology, enzymes are utilizing as biocatalysts for food products (e.g., bread, cheese, beer, vinegar), fine chemicals (e.g., amino acids, vitamins), pharmaceuticals, environmental processes, or for analytical and diagnostic purposes [2].

A variety of enzymes from microorganisms catalyze fundamental chemical reactions with high substrate specificity and overproduction. Catalytic enzymes possess many favorable properties such as cost-effectiveness, stability under mild pH and temperature conditions, low toxicity levels, and ease of termination of activity. As efficient natural biocatalysts, enzymes provide great opportunities to carry out important chemical reactions in several industrial settings [3]. Biocatalysis is becoming a favorable alternative to chemical processes and a vital part of green technology.

Worldwide Food Enzymes Market is projected to surpass USD 3.6 billion by 2024; according to a new research report by Global Market Insights, Inc. Increasing consumer awareness regarding nutritional benefits offered by enzymes has resulted in increased product demand, thus driving the food enzymes market penetration [4]. Numerous food processing units are extensively using these additives to attain environment-friendly processes, as these extracts are used as biocatalysts to increase the production rate. Microorganisms are important sources of enzymes and are preferred over sources of plant and animal origin. As a result, more than 50% of the industrial enzymes are obtained from bacteria. Extremophilic bacteria are important groups of extremophilic microorganisms that have been studied during the last years. They are considered a source of enzymes due to the great diversity and can survive under extreme conditions [5]. Many enzymes produced by these microorganisms are of great importance and have found applications in several industries. Due to their activity and stability under extreme conditions, these enzymes offer new alternatives for current biotechnological and industrial applications.

Four features are to be remembered about enzymes: 1) specificity; 2) proteinaceous nature; 3) efficiency; and 4) catalytic properties; as one enzyme can react again over a million of times without being consumed. Microbial enzymes are of great importance in the development of industrial bioprocesses. Current applications are focused on many different markets including pulp and paper, leather, detergents and textiles, pharmaceuticals, chemical, food and beverages as well as biofuels, animal feed, and personal care, among others. Today there is a need for new improved and more versatile enzymes in order to develop more novel, sustainable, and economically competitive production processes [6]. Microbial diversity and modern molecular techniques such as metagenomics are being used to discover new microbial enzymes whose catalytic properties can be improved/modified by different strategies [6, 7, 8]. This chapter deals with enzymes as biocatalysts and their use in food industry with a special reference to the food safety regulations.

2. A Brief History

Microbiology has had a long rich history, initially centered in the causes of infectious diseases but now including practical applications of science. Microorganisms have been identified and exploited for many centuries. There is a great history beyond the fermentation processes by microorganisms which resulted in production of fermented foods and beverages [9]. In early history of microbiology, Historians are unsure who made the first observations of microorganisms, but the microscope was available during the mid-1600s, and an English scientist, *Robert Hooke*, made key observations. In the 1670s and the decades thereafter, a Dutch merchant named *Anton van Leeuwenhoek* made careful observations of microscopic organisms, which he called "animalcules". Until his death in 1723, van Leeuwenhoek revealed the microscopic world to scientists and is regarded as one of the first to provide accurate descriptions of protozoa, fungi, and bacteria [9]. In the mid-nineteenth century, Louis Pasteur understood the role of bacteria in fermented food, wine, alcohols, and beverages [10]. He identified many microbial processes and discovered the first principals for fermentation which revealed that microorganisms require specific substrates to produce primary and secondary metabolites as well as end products [11, 12]. In the new millennium, extensive application of bioprocesses has created a great field of biotechnology. Subsequently, many other products such as cheese, milk, yoghurt, and other dairy products as well as fuels and fine chemical industries are reported.

Application of biological sciences in industrial processes is known as "*Bioprocessing*". Nowadays most biological and pharmaceutical products are produced in well-defined industrial bioprocesses. There are hundreds of microbial and fungal products purely available in the biotechnology market. Growth of cells on a large scale is called: *Industrial Fermentation*, which is normally performed in a bioreactor under controlling aeration, pH, and temperature conditions [13]. Fermentation includes any process mediated by or involving microorganism in which a product of economic value is obtained. Biotechnology arose from the field of zymotechnology which began as a search for a better understanding of industrial fermentation, particularly beer [14]. Microorganisms are employed since several decades for large-scale production of a variety of biochemicals as well as in processing of foods and feeds. Benefits provided by microorganisms may include: *a*) alive or dead microbial cells; *b*) processed microbial biomass; *c*) components of microbial cells; *d*) intracellular or extracellular enzymes; or *e*) chemicals produced by strains utilizing constituents of the substrates provided [14]. Over the years, novel enzymes with actual potential for biotechnological applications have been produced from microorganisms living in extreme conditions [3, 5]. More stable enzyme variants are developing from such extremophiles.

3. Enzymes in General

Enzymes are used in industrial applications when extremely specific catalysts are required. However, enzymes in general lack the stability in organic solvents and at high temperatures. As a consequence, protein engineering involves attempts to create new enzymes with novel properties. As with all catalysts, enzymes are not consumed by the reactions they catalyze, nor do they alter the equilibrium of these reactions [23]. However, enzymes do differ from most other catalysts in that they are highly specific for their substrates. Specificity of enzymes to their substrates is often referred to as "lock and key" model [24]. In enzymatic reactions, substrates at the beginning of a process are converted into different molecules, called products. Like all catalysts, enzymes increase the rate of a reaction by which products are formed faster [24]. Most enzyme reaction rates are millions of times faster than those of comparable un-catalyzed reactions. Enzymes such as amylases and proteases break down large molecules (starch or proteins, respectively) into smaller one. Starch molecules, for example, are too large to be absorbed from the intestine, but enzymes hydrolyze starch chains into smaller molecules such as maltose and glucose, which can then be easily absorbed. In ruminants which have herbivorous diets, microorganisms in the gut produce cellulase to break down the cellulose cell walls of plant fiber [25]. Without enzymes, metabolism would neither progress through the same steps nor be fast enough to serve the needs of the cell.

3.1. Source of Enzymes

Enzymes can be obtained from animals, plants, and microorganisms. However, the nature of a source dictates the availability, cost of source material, ease of recovery and many other factors. Microbial enzymes, in principle, could be produced on an industrial scale, because of: *a*) abundant supplied and ability to control all phases of the enzyme production; *b*) commercial enzyme preparation is well standardized; *c*) the enzyme market has a lot of competing suppliers; *d*) microbial strains can often produce several different enzymes; *e*) Huge amounts of microbial enzymes can be produced in a short period; and *f*) the microbial strain can be genetically manipulated, through the DNA recombinant technique, to improve its enzyme production capability [26].

3.2. Production of Enzymes

As all commercially valuable enzymes are proteins, so they can be produced through processes using quite similar unit operations, i.e., extraction from an adequate source (animal, vegetal or microbial), filtration, centrifugation, precipitation, purification, drying, stabilization, standardization and packaging. In biotechnology, all unit operations comprised between filtration and packaging are generally called downstream processing [26]. Enzymes of animal and plant origin are produced through the maceration of tissues, organs, leaves and fruits, followed by the extraction with water or organic solvent. Differently, the microbial enzymes, obtained from either procariotic (bacteria) or eucariotic (yeasts and fungi, mainly) are produced by cell culture in a liquid or semi-solid medium, carried out in a special reactor called fermenter, in a process known as fermentation [26]. Fermentation is defined as the chemical transformation of organic compounds with the aid of enzymes. The overall biochemical reaction mechanisms to extract energy and form products under anaerobic conditions are called fermentation processes. Many microorganisms synthesize and excrete large quantities of enzymes into the surrounding medium. Enzymes are proteins that denatured by heat and extracted or precipitated by chemical solvents like ethanol and by inorganic salts like ammonium sulphate [27]. It is industrially applicable and economically feasible to produce, concentrate, extract and purify enzymes from bacteria and cultures of molds.

3.3. Specificity of Enzymes

Enzymes are usually very specific to the substrates involved in the reactions and are responsible for thousands of metabolic processes that sustain life. An enzyme catalyzes a specific reaction inside or outside a living cell and greatly accelerating the rate and enhancing the specificity of metabolic chemical reactions, from the digestion of food to the synthesis of DNA [28]. The action of enzymes depends on their ability to bind the substrate at a domain of the enzyme molecule called the active site. Enzymes are very specific because both the enzyme and substrate possess specific complementary geometric shapes (lock and key) model that fit exactly into one another [29]. Complementary shape, charge, and hydrophilic/ hydrophobic characteristics of enzymes and substrates are responsible for this specificity. Many enzymes are well known to be specific to more than one substrate.

3.4. Activity and Mechanisms of Action

Enzyme activity can be affected by other molecules, temperature, pH, enzyme concentration, substrate concentration, and presence of any inhibitors or activators. Many enzymes have very specific requirements of pH and temperature. However, new awareness, knowledge, and biotechnology can produce enzymes from microorganisms which can survive and function in extreme conditions. Enzymes can act in several ways, all of which lower $\Delta G^{\ddagger}$ (Gibbs energy) [30]:

1. Lowering the activation energy (which is defined as the minimum energy required to start a chemical reaction) by creating an environment in which the transition state is stabilized.
2. Lowering the energy of the transition state, but without distorting the substrate, by creating an environment with the opposite charge distribution to that of the transition state.
3. Providing an alternative pathway. For example, temporarily reacting with the substrate to form an intermediate Enzyme-Substrate complex, which would be impossible in the absence of the enzyme.
4. Reducing the reaction entropy change by bringing substrates together in the correct orientation to react. Considering $\Delta H^{\ddagger}$ alone overlooks this effect.
5. Increases in temperatures speed up reactions, thus, temperature increases help the enzyme function and develop the end product even faster. However, if heated too

much, the enzyme's shape deteriorates and the enzyme becomes denatured. Some enzymes like thermolabile enzymes work best at low temperatures. The function of a protein is dependent on its structure so that if the structure is denatured, the function will so disrupt [30].

3.5. Inhibition of Enzymes

Enzyme reaction rates can be decreased by various types of enzyme inhibitors due to the competitive inhibition action. *Inhibitor* of an enzyme is a molecule that binds to the enzyme and decreases its activity. Since blocking an enzyme's activity can kill a pathogen or correct a metabolic imbalance, many drugs are considered enzyme inhibitors. Not all molecules that bind to enzymes are inhibitors; *enzyme activators* bind to enzymes and increase their enzymatic activity. Inhibition of enzymes can be divided to competitive inhibition and noncompetitive inhibition. In *Competitive inhibition*, the inhibitor and substrate can't bind to the active site at the same time [31]. This usually results from the inhibitor having an affinity for the active site of an enzyme. Competitive inhibitors bind reversibly to the enzyme, preventing the binding of substrate. On the other hand, binding of substrate prevents binding of the inhibitor. Substrate and inhibitor compete for the enzyme. This type of inhibition can be overcome by sufficiently high concentrations of substrate. Often competitive inhibitors strongly resemble the real substrate of the enzyme. In *Noncompetitive inhibition*, the inhibitor can't bind to the free enzyme, only binds to the enzyme-substrate complex (ES-complex). The enzyme/inhibitor/substrate-complex formed is enzymatically inactive [31]. This type of inhibition is rare and causes V_{max} and K_m to decrease. Maximum velocity decreases as a result of removing the activated complex, and K_m decreases due to the better binding efficiency and effective elimination of the E-S complex, thus decreasing the K_m which indicates a higher binding affinity [31]. Since inhibitors modulate the function of enzymes they are often used as drugs. A common example of an inhibitor that is used as a drug is aspirin, which inhibits the COX-1 and COX-2 enzymes that produce the inflammation messenger prostaglandin, thus suppressing pain and inflammation. However, other enzyme inhibitors are poisons. For example, the poison cyanide is an irreversible enzyme inhibitor that combines with the copper and iron in the active site of the enzyme cytochrome c oxidase and blocks cellular respiration [32].

The binding of an inhibitor can stop a substrate from entering the enzyme's active site and/or hinder the enzyme from catalyzing its reaction. Inhibitor binding is either reversible or irreversible. *Irreversible inhibitors* usually react with the enzyme and change it chemically (e.g., via covalent bond formation). These inhibitors modify key amino acid residues needed for enzymatic activity. *Reversible inhibitors*, in contrast, bind non-covalently and different types of inhibition are produced depending on whether these inhibitors bind to the enzyme, the enzyme-substrate complex, or both [33]. Some enzymes do not need any additional components to show full activity, while others, however, are not reactive on their own and often possess cofactors [33]. Common cofactors include metals such as Zn and Fe, and organic molecules such as vitamins.

3.6. Kinetic of Enzymes

Enzyme kinetics is the investigation of how enzymes bind substrates and turn them into products. The enzyme reaction is achieved in two stages. *First*, the substrate binds reversibly to the enzyme, forming the ES-complex, that sometimes called the Michaelis complex. *Second*: the enzyme then catalyzes the chemical step in the reaction and releases the product [34]. Enzymes can catalyze up to several million reactions in seconds, and an enzyme rates depend on the solution conditions and substrate concentration [35]. Conditions that denature the protein abolish enzyme activity, such as high temperatures, extremes of pH or high salt concentrations. To find the maximum speed of an enzymatic reaction, the substrate concentration is increased until a constant rate of product formation is occurred [36]. Saturation happens because, as substrate concentration increases, more and more of the free enzyme is converted into the enzyme-substrate-bound form. At the maximum reaction rate (V_{max}) of an enzyme, all the enzyme active sites are bound to substrate, and the amount of ES-complex is the same as the total amount of enzyme. However, V_{max} is only one kinetic constant of enzymes. The amount of substrate needed to achieve a given rate of reaction is also important. This is given by the Michaelis-Menten constant (K_m), which means the substrate concentration required for an enzyme to reach one-half its maximum reaction rate; generally, each enzyme has a characteristic K_m for a given substrate [34].

4. Enzymatic Biocatalysis

The enzyme-catalyzed processes are gradually replacing chemical processes in many areas of industry. Enzymes have all the properties of true catalysts. In an enzyme process, a chemical reaction occurs at a much higher rate but the enzyme is not consumed by the reaction. The ability to perform very specific chemical transformations (bio-transformations) has made enzymes increasingly preferable in industries, where less specific chemical processes produce undesirable by-products [6]. Purity and predictability are of particular importance in food manufacture to avoid by-products that may affect flavor. Many industrial processes are designed to be run at elevated temperatures where viscosity is decreased, chemical reactions are faster, and problems with contamination are reduced.

4.1. Biocatalysis

Biocatalysis, a major part of biotechnology, can be defined as the use of biological molecules (biocatalysts) to catalyze and accelerate specific chemical reactions. Biocatalysis involves the application of whole microbial cells, cell extracts, purified enzymes, immobilized cells, or immobilized enzymes as catalysts for biological processes. The rapid development in biocatalysis is due to the recent advances in large-scale genome sequencing, protein expression, metabolic engineering, and structural biology [7, 8, 15]. Biocatalysis is found in industry most commonly in food production and processing [16]. Familiar examples are cheese, yoghurt, beer and wine production. Many food additives and sugars are also produced by enzymes.

4.2. Biocatalysts

Biocatalysts are biological molecules having catalytic properties with the ability to perform specific chemical transformations. Biocatalysts have widely been applied for a long time in various sectors include: *a*) brewing and food industries; *b*) pharmaceuticals and fine chemicals industry; and *c*) industrial bioprocesses in food, feed, and chemical industries [17, 18, 19]. Biocatalysts have now become the ideal choice for bioprocess in industry which is always reliable and scalable processes with minimal waste generation [16, 19]. Biocatalysts have all the properties of a true catalyst and carry out reactions in free or immobilized form, whole cells, or cellular extracts [20].

4.3. Enzymes

Enzymes are biocatalysts responsible for supporting almost all of the chemical reactions that maintain general homeostasis due to their role in maintaining life processes. Enzymes as biocatalysts are easy to be controlled and operated in the mild range of reaction conditions [21]. All enzymes are complex protein molecules composed of hundreds of amino acids and may or may not possess a non-protein prosthetic group [22]. Enzymes play a key role in almost all biological processes, accelerating a variety of metabolic reactions as well as controlling the energy for genetic information pathways. Compared to chemicals, enzymes are important because they are tools of nature. Enzymes as biocatalysts represent advantages and disadvantages features. Enzymes show advantages such as: *a*) acceleration of specific biochemical reactions and biological processes to produce a useful product or effect; *b*) specificity in their action and work at mild conditions; *c*) responsible for a number of reactions and biological activities in plant, animal, human beings and microorganisms; *d*) non-toxic when correctly used, consume low energy, and leave less by-products [2]; and *e*) widely used in modern industrial processes due to their low cost, large productivity and vast availability [16]. On other side, un-stability at high temperatures, un-stability at extreme pH-values, or in aggressive solvents are major disadvantages of enzymes. Furthermore, some enzymes are still very expensive, require expensive substrates, inhibited by some metal ions, and can be hydrolyzed by peptidases [2]. In addition, when inhaled or ingested, as all foreign proteins, enzymes are becoming potential allergens.

5. Enzymes Applied in Food Industry

Food industry is looking for a new creative product with certain attributes such as safety, sustainability, cost-effectively, and a more natural product. Ingredients and additives used in the food industry should: *a*) reduce waste; *b*) enhance nutrition; *c*) improve functionality; and *d*) improve characteristics in final product such as texture, shelf life, color, flavor, and odor [1]. Enzymes are being used for monitoring food processes, increasing bioavailability of minerals, and favor processes related to the dairy products as well as meat, baking, fermented beverages and juice [37].

Enzymes used in food industry can be derived from three sources; animal, plant, and microorganisms [38]. Enzymes derived from animal tissues and commonly used in food preparation include: *a*) rennet: an extract of fourth stomach of calf and it is rich in rennin E and pepsin E, both of them are proteases that cause milk to curdle into cheese; *b*) lipase, used to impart buttery flavor to oils by degrading some of lipids and to accelerate cheese ripening; c) pancreatin, rich in a plethora of protease E and used to modify protein to be more easily digested. Enzymes derived from plant sources include three major proteases which used commercially: *a*) papain is derived from papaya plant; *b*) bromelain is obtained from the pineapple plant and are commonly used as meat tenderizers; *c*) Ficin is obtained from the fig papain and it is more limited in use due to its more dangerous proteolytic activity; *d*) Barley amylase is used to make maltose syrup.

Many chemical transformation processes used in various industries have negative effects concerning the environmental point of view. Nonspecific reactions may result in poor product yields. The need for high temperature, acidity, alkalinity, and high pressure to drive reactions lead to high energy costs and may require large volumes of cooling water for downstream processes [39]. Therefore, application of enzymes in industry have many advantages include: *a*) industrial enzymes originate from biological systems which contribute to sustainable development using renewable resources; *b*) small amounts of enzymes can carry out the chemical reactions even on an industrial scale, at either solid or liquid enzyme preparations; *c*) enzymes reduce the impact of manufacturing on the environment by reducing consumption of chemicals, water, energy, and the subsequent generation of waste [39]. All enzymes used as food ingredients or food additives should generally be recognized as safe "GRAS", i.e., an acronym for the phrase Generally Recognized As Safe. According to the Federal Food, Drug, and Cosmetic: any substance that is intentionally added to food is a food additive that is subject to premarket review and approval by the Federal Food & Drug Administration (FDA, US). Being catalysts, enzymes start working and be in contact with their substrate as soon as conditions are suitable (water content, pH, temperature) [39]. Examples for some of microbial enzymes applied in the food industry are represented in Table 1.

Microbial enzymes are obtained by growth of microorganisms on a nutrient medium where a variety of enzymes are produced due to their natural metabolic function in a process commonly referred to as fermentation. Microbial enzymes can be purified and concentrated for use in industry or other applications. Application of enzymes as food ingredients or food additives in the food industry are discussed below.

5.1. Enzymes Used as Food Ingredients

5.1.1. Invertase in Confectionery

An example for enzymes used as food ingredients includes invertase. After enzymatic reaction in presence of the enzyme, invertase continues performing a technological function in the final food. The substrate, sucrose, is continuously reformed during storage and, therefore, not depleted [40]. This enzyme can be used at a very low dosage to hydrolyze the substrate sucrose into glucose and fructose.

5.1.2. Glucose Oxidase in Bottled Beverages and Bread

Another example for enzymes used as food ingredients is glucose-oxidase. This enzyme can be used to remove oxygen from bottled drinks (e.g., citrus based soft drinks) in order to reduce browning due to the oxidation. The enzyme is added just before bottling, canning, or packaging of the beverage and removes oxygen from the liquid and headspace. When oxygen in the headspace is depleted, the enzyme does not perform a technological function. However,

Table 1. Food Enzymes and source of microorganisms used in food industry [38]

Enzyme	Source	Field
α-Acetolactate-decarboxylase	*Bacillus subtilis* ToC46	Brewers' Mash, Distillers' Mash
Aminopeptidase	*Lactococcus lactis*	Cheddar cheese, Dairy based flavoring preparations, milk and vegetable protein.
Amylase	*Aspergillus niger var., A. oryzae var., B.subtilis var., Rhizopus oryzae var.,*	Beer, Malt liquor, Bread, Flour, Whole wheat flour, Chocolate syrups, Malt-flavored dried breakfast cereals, Plant-based beverages, Starch used in the production of dextrin, maltose, dextrose, glucose (glucose syrup) or glucose solids (dried glucose syrup).
Amylase	*B. licheniformis* MDT06-228	Infant cereal products.
Amylase	*B. licheniformis* JS1252	Bread, Flour, Whole wheat flour.
Asparaginase	*A. niger* AGN7-41, *A. niger* ASP72	Bread; Flour; Whole wheat flour, Unstandardize.d foods.
Glucoamylase	*A. niger*	Juice of Apple, Banana juice; Grape juice; Grapefruit juice; Lemon juice; Lime juice or lime fruit juice; Orange juice; Pear juice; Pineapple juice, Infant cereal products, Infant cereal products.
Glucose oxidase	*A. niger var., A. oryzae* Mtl-72	Soft drinks, Liquid egg-white, liquid whole egg or liquid yolk, Bread; Flour; Whole wheat flour, Unstandardized bakery products.
Glucose-Isomerase	*B. coagulans* var., *Streptomyces olivaceus* var., *S. murinus* DSM 3252, *S. rubiginosus* SYC 5406	Glucose (glucose syrup) to be partially or completely isomerized to fructose.
Invertase	*Aspergillus japonicas, Saccharomyces* sp.	Sucrose used in the production of fructooligosaccharides, liquid-centred confectionery, unstandardized bakery products.
Protease	*A. oryzae* var., *A. niger* var., *B. subtilis* var.	Bread, Flour, Whole wheat flour, Dairy-based flavouring preparations, Meat tenderizing preparations, Plant-based beverages, Unstandardized bakery products.

as soon as the bottle, can, or packing is opened by the consumer and beverage comes in contact with oxygen again, the enzyme resumes its catalytic activity (provided that the beverage still contains some glucose) due to the presence of oxygen substrate [40].

Chemical oxidants such as bromates, azodicarbonamide, and ascorbic acid have been widely used to strengthen the gluten when making bread. As an alternative, oxidases such as glucose oxidase can partially replace the use of these chemical oxidants and achieve better bread quality. Glucose oxidase can be used not only to replace bromate but also to give a larger bread volume. Glucose oxidase cross-links gluten to make weak dough stronger, drier, and more elastic [39]. Better bread quality depends on the enzymes used and their own substrate specificity. For example, in wheat flour dough, lipases work on the lipids, xylanase works on pentosans, and amylases work on the starch. Because the interaction of these substrates in dough and bread is rather complex, the use of enzyme combinations can have synergistic effects that are not seen when only one enzyme is used. Quite often an overdose of enzymes will have a detrimental effect on either the dough or the bread. An overdose of fungal alpha-amylase or hemicellulase/xylanase, for instance, may result in a dough that is too sticky to be handled [39]. It is beneficial therefore to use a combination of lower dosages of alpha-amylase and xylanase with low dosages of lipase or glucose oxidase to achieve optimal dough consistency, stability, and bread quality.

5.2. Enzymes Used as Food Processing Aids

The increasing use of enzymes as biocatalysts in processing of food has been primarily aimed at obtaining easier processes, efficient raw material utilization, health trends, and a consistent product quality. A processing aid is any substance intentionally used in the processing of raw materials, foods, or their ingredients to fulfil a certain technological purpose during treatment or processing. For enzymes used in food processing particularly, it should be clarified for costumer if the microbial source is an engineered strain. Enzymes produced by means of engineered microorganism may subject to completely different testing for approval [41]. Examples for enzymes used as food processing aids are represented in Table 2.

5.3. Enzymes Used as Food Additives

Enzymes used in food industry as food additives should provide: 1) safety when used and their use must not mislead the consumer; 2) technological advantages for their use; 3) benefits to the consumer; 4) a better use of raw materials and thereby less wastage of food in the production process; and 5) no technological effect on the final product [42]. The food enzyme regulation only covers enzymes that are added during food processing to perform a technological function in the manufacture, processing, preparation, packaging, transport, and storage of foods [42]. Some permitted food enzymes used in industrial application are listed in Table 3.

Table 2. Industrial application of enzymes used as food processing aids [22]

Enzymes	Application
Proteases	Used in biscuit manufacturers to lower the protein level of flour.
Trypsin	Used to predigest baby foods.
Cellulases, pectinases	Used to clarify fruit juices.
Papain	Used to tenderize meat for cooking.
Rennin (from stomachs of young ruminant)	Used in dairy industry to manufacture of cheese to hydrolyze proteins.
Lipases	Implemented during production of Roquefort cheese to enhance ripening of the blue-mold cheese.
Lactases	Break down lactose to glucose and galactose.
Glucose isomerase	Converts glucose into fructose in production of high-fructose syrups from starchy materials.
Amylases	Used in production of sugars from starch such as in making high-fructose corn syrup.

Table 3. Some permitted enzymes used as food additives in the food industry [43, 44]

Industrial field	Enzyme	Role
Fruit juice	Pectinases Cellulases & hemicellulases Naringinases and limoninases	Degrade pectin and facilitate juice filtration. Result in clear and high-quality juice. Acting on soluble pectin hydrolysis and lowering viscosity. Acting on compounds that cause bitterness in citrus juices.
Starch	Amylases, amyloglucosidases and glucoamylases, Glucose isomerase	Convert starch to glucose and other sugar syrups. Converts glucose syrups into fructose syrups.
Dairy Industry	Rennin from stomachs of young ruminant Lipases ß-glactosidases and Lactases	Manufacture of cheese. Enhance ripening of blue-mold cheeses. Break down lactose to glucose and galactose to avoid lactose intolerance.
Baking Industry	α-amylase, ß-xylanase Lipases Proteinases Asparaginases	Catalyzes the breakdown of starch in flours. Used in the manufacture of bread to improve shelf life of bread and cakes. Improves the characteristics and rising of bread. Modify the natural lipids in flour to strengthen the dough. Reduce the protein in flour. Used in biscuit manufacture. Reduce the amount of acrylamide formed during baking.
Brewing Industry	Amylases, glucanases, proteinases Amyloglucosidase, ß-glucanases	Split polysaccharides and proteins in the malt. Low Calorie beer production. Improves filtration characteristics.

5.4. Enzymes Used in Dairy Technology

Application of enzymes in the processing of milk has a long tradition. In ancient times, calf rennet (contained the enzyme chymosin) was used for coagulation during cheese production. Nowadays there are many industrially produced chymosin products or similar proteases available as substitutes. Lipases are used mainly in cheese ripening. Proteases are used to accelerate cheese ripening, modify the functional properties of cheese, and modify milk proteins to reduce allergenic properties of some dairy products [45]. Therefore, attention should be paid to the food technological processes that may eliminate or reduce allergenicity of cow's milk proteins. Composition and hygienic quality of milk determine its nutritional value and processing suitability. However, biological quality of milk depends on the content of bioactive components which due to its health promoting properties have a positive effect on human health [46]. Biologically active substances present in cow's milk include proteins, peptides, amino acids, sugars, vitamins, enzymes, sterols, phospholipids and fatty acids. Among these components, several proteins (lactoferrin, β-lactoglobulin, bovine serum albumin and casein) that inhibit cancer cell growth, deserve special attention. Cow's milk contains approximately 30 potentially allergenic proteins. The most common bovine milk allergens, whey proteins and casein fraction, naturally not present in human breast milk [46]. Strategies for destroying or modifying these allergens to eliminate milk allergy either for adults or infants are reviewed and reported by scientists all over the world [47, 48].

5.5. Enzymes Used in Baking Technology

Enzymes are usually added in bread manufacture to modify dough rheology, gas retention, and crumb softness as well as reducing acrylamide formation in bakery products. It has been reported that the amount of the potentially carcinogenic substance acrylamide is relatively high in a number of cereal- and potato-based products like biscuits, crackers, crisp bread, French-fries, and potato crisps. Acrylamide is a substance that is formed at high temperatures when the amino acid asparagine reacts with a reducing sugar like glucose, through the Maillard reaction [49]. This reaction, known as non-enzymatic browning, is typically occur at temperatures above 100°C. Maillard reactions are responsible for important color and flavor development in fried and baked starchy products. Thus, the same reactions that make fried and baked foods tasty give rise to a potential carcinogen [39]. To address this issue, the enzyme asparaginase has been developed to reduce formation of acrylamide. By adding asparaginase before baking or frying the food, asparagine is converted to aspartic acid and ammonium. With the use of an asparaginase, formation of acrylamide can be reduced by up to 90% in a range of food products [49].

6. Strategies for Improving Properties of Microbial Enzymes

The continuously expanding application of enzymes is creating a growing demand for biocatalysts that exhibit improved or new properties. Although enzymes have favorable turnover numbers, they do not necessary fulfill all process requirements and need further fine tuning to achieve industrial scale production. Genetic modification is very important and the recombinant DNA techniques have increased production by 100-fold [6]. Development of new/improved biocatalysts is a challenging and a complex task. To improve enzyme performance, different approaches can be considered such as: *a*) screening the biota for suitable enzymes; *b*) production of enzyme variants through protein engineering; and *c*) enzyme immobilization, typically for improved stability [50]. Production of improved enzymes variants is possible either through a rational approach involving site-directed mutation, or through a random mutagenesis directed evolution approach [50].

7. Nanotechnology in Food Industry

The food processing techniques utilize enzymes to modify food components and improve their flavor and nutritional quality. Nanoparticles are used as a source to immobilize these enzymes and enhance their activity. Application of nanotechnology in food has gained great importance in recent years [51, 52, 53]. Food nanotechnology opens up new possibilities for developing innovative food products. Application of nanotechnology in food includes improvement of packaging, processing, nano-additives, cleaning and sensors for detection of contaminants. The main driving principle behind these developments seems to be aimed at enhancing uptake of nano-sized nutrients, improving taste, consistency, stability, and texture of food products [54]. Nanotechnology can be used in all phases of food production chain, from the agricultural production to the industrial processing and finally the packaging of foods material. Nanoparticles can also be used in food to enhance the nutritional aspects by means of nano-supplements and nano-delivery systems for bioactive compounds [54].

Nanotechnology has the potential to improve the nutritional value of foods, generate new novel food products, and new safe food packaging to extend storage. However, this new technology also raises concern about risk assessment. Nanotechnology risks include environmental risks from release of nanoparticles into the environment, safety risks from nanoparticles for workers and consumers [55, 56]. To control such risks, it will be necessary to systematically identify those critical issues which should be looked at in more detail. Risks of nanoparticles are related to the free nanoparticles which in most applications will be embedded in the final product [55]. Release of immobilized nanoparticles during use of products will be of a great impact on the environment and human health where diffusion, transport, and transformation in air, soil, and water may be occurred.

8. Safety Regulation and Future Needs

The majority of industrial enzymes available at present is used in food industry. However, apart from the problems with enzymes during 35 years ago is that, enzymes have neither been a focal issue for regulators nor for consumer groups or general public. In the beginning of the 1990s this started to change when consumer and environmental groups were alarmed by the increasing use of genetically modified microorganisms for enzyme production [2]. This eventually led to review of safety concerns and regulatory aspects, especially in Germany,

Switzerland and Austria. Safety regulations of food enzymes differ among countries [57]. Safety regulations of foods associated with enzymes focus on allergenic properties, as it is well established that enzymes are potent inhalation sensitizers and can cause allergic reactions including asthma [42]. Safety concerns for consumers of using genetically modified microorganisms for enzyme production is largely focus on bacterial toxins and mycotoxins that might be present in the enzyme preparation. In addition, toxic by-products might be transferred to foodstuff and, thus, constitute a safety risk to consumers. Like all proteins, enzymes can cause allergic reactions when people have been sensitized through exposure to large quantities [42].

Attitudes toward safety, efficacy, and values about what is important in food and life will be important in determining future needs. Demonstrating efficacy requires clinical studies with well-defined products and rigorous experimental designs, and the studies must be replicable [58]. Basic and more applied challenges will continue well into the future. Much remains to be learned about the effects of bioactive constituents such as flavonoids in foods and dietary supplements on health outcomes. More and better biomarkers need to be developed and their associations with health outcomes should be clarified [58]. Assessments of the safety, quality, and efficacy of nutrients and bioactive compounds are needed to provide the scientific information that regulators need to improve the public health.

Conclusion

Industrial enzymes are playing a significant role in today's commercial status of biotechnology and will assist more and novel applications. During the last four decades, enzyme applications is continuously increasing and covering various scientific and technological issues. Enzymes contributed to more environmentally adapted clean and green technology due to their biodegradable nature and are currently used in several industrial processes. Common to most applications, introduction of enzymes as effective biocatalysts working under mild conditions results in significant savings in resources such as energy and water for benefits of both the industry and environment. Microorganisms provide an impressive amount of catalysts with a wide range of applications across several industries. Recent advances in genomics, metagenomics, proteomics, and emerging recombinant DNA techniques have facilitated the discovery of new microbial enzymes from nature.

References

[1] Avendano, K., et al. (2016). Microbial enzymes: Applications in food processing. *Agro FOOD Ind Hi Tech.*, *27*: 63-67.

[2] Buchholz, K., Kasche, V. & Bornscheuer, U. (2012). Biocatalysts and enzyme technology. In: *Introduction to Enzyme Technology*. Published by WILEY-VCH Verlag GmbH & Co. KGaA, Weinheim.

[3] Rigoldi, F., et al. (2018). Review: Engineering of thermostable enzymes for industrial applications. *APL Bioeng.*, *2*: 011501-011517.

[4] *Global Market Insights.* (2018).

[5] Kelly, D. (2018). Biotechnological and industrial applications of enzymes produced by extremophilic bacteria. *A Mini Review*. Doi:10.20944/preprints201801.0198.v1.

[6] Jose, L. and Arnold, L. (2014). Microbial enzymes: Tools for biotechnological processes. *Biomolecules*, *4*: 117-139.

[7] Martinez, M., Rafael, B. and Manuel, F. (2017). Metagenomics and the search for industrial enzymes. In: *Biotechnology of Microbial Enzymes*, pp. 167-184. DOI: 10.1016/B978-0-12-803725-6.00007-8.

[8] Thapa, S., et al. (2017). Metagenomics prospective in bio-mining the microbial enzymes. *J Gen Proteins*, *1*: 1-5.

[9] Tyson, J. (2012). *Industrial microbiology*. (www.jtyson.com).

[10] Pasteur, L. (1858). On Alcoholic Fermentation, *Weekly Reports of the Sessions of the Academy of Science*. *25*, janvier 1858, XLVI, pp. 179-180.

[11] Doran, P. (1995). *Bioprocess engineering principles*. Academic Press, Elsevier, New York.

[12] Parales, R., Bruce, N., Schmid, A. and Wackett, L. (2002). Biodegradation, biotransformation, and biocatalysis. *Appl Env Microbiol.*, *68*: 4699-4709.

[13] Scragg, A. (1991). *Bioreactors in biotechnology: A practical approach*. Ellis Horwood Series in Biochemistry and Biotechnology, New York.

[14] Thackray, A. (1998). *Biotechnology and the rise of the molecular sciences.* Philadelphia: University of Pennsylvania Press, ISBN 0-8122-3428-6.

[15] Ran, N., et al. (2008). Recent applications of biocatalysis in developing green chemistry for chemical synthesis at the industrial scale. *Green Chem.*, *10*: 361-372.

[16] Kaushik, N., Biswas, S. and Singh, J. (2014). Biocatalysis & Biotransformation Processes – An Insight. *The Scitech J.*, *1*: 15-22.

[17] Wohlgemuth, R. (2010). Biocatalysis—Key to sustainable industrial chemistry. *Curr Opin Biotechnol.*, *21*: 713-724.

[18] Truppo, M. (2017). Biocatalysis in the pharmaceutical industry: The Need for Speed. *ACS Med Chem Lett.*, *8*: 476–480.

[19] Greg, H. & Lewis, J. (2018). Introduction: Biocatalysis in industry. *Chem Rev, 118*: 1-3.

[20] Carla, C. (2016). Whole cell biocatalysts: essential workers from nature to the industry. In: *Microbial Biotechnology*. Published by John Wiley & Sons Ltd and Society for Applied Microbiology.

[21] Faber, K. (1997). *Biotransformations, in organic chemistry: A textbook, 3rd Ed.*; Springer-Verlag: Berlin, Germany.

[22] Duza, M. and Mastan, S. (2013). Micobial enzymes and their applications. *Indo Amr J Pharm Res.*, *3*: 6208-6219.

[23] https://www.boundless.com. Enzymes used in industry. *Boundless Microbiology. Boundless*, 27 Jun, 2014.

[24] Tim, J. (2006). *Enzymes: Lock and Key Model*. Washington University.

[25] Leschine, S. (1995). Cellulose degradation in anaerobic environments. *Ann Rev Microbiol.*, *49*: 399-426.

[26] Roberts, M. et al. (1995). *Introduction to biocatalysis using enzymes and microorganisms*, 195 pp. Cambridge, UK: Cambridge University Press.

[27] Thomas, L. and Chamberlin, G. (1980). *Colorimetric chemical analytical methods.* Tintometer Ltd, Salisbury, United Kingdom.
[28] Bairoch, A. (2000). The enzyme database 2000. *Nucleic Acids Res.*, *28*: 304-305.
[29] Jaeger, K. and Eggert, T. (2004). Enantioselective biocatalysis optimized by directed evolution. *Curr Opin Biotechnol.*, *15*: 305-313.
[30] Fersht, A. (1985). *Enzyme structure and mechanism.* San Francisco: Publisher W.H. Freeman & Co Ltd. 475 pages. ISBN 0-7167-1615-1.
[31] Price, N. (1979). What is meant by competitive inhibition? *Tren Biochem Sci.*, *4*: N272-N273.
[32] Yoshikawa, S. and Caughey, W. (1990). Infrared evidence of cyanide binding to iron and copper sites in bovine heart cytochrome c oxidase. Implications regarding oxygen reduction. *J Biol Chem.*, *265*: 7945-7958.
[33] Dunaway-Mariano, D. (2008). Enzyme function discovery. *Structure*, *16*: 1599-1600.
[34] Brian, E., et al. (2006). Ever-fluctuating single enzyme molecules: Michaelis-Menten equation revisited. *Nature Chem Biol.*, *2*: 87-94.
[35] Radzicka, A. and Wolfenden, R. (1995). A proficient enzyme. *Sci.*, *267*: 90-93.
[36] Author=fullofstars. (2007). *Michaelis-Menten saturation curve of an enzyme reaction.* http://en.wikipedia.org/wiki/Image:MM_curve_v3.png (PD).
[37] Ghaffari-Moghaddam, M., et al. (2014). Industrial applications of enzymes. *Rev. J. Chem.*, *4*: 341-361.
[38] Health Canada. (2014). *Policy for Differentiating Food Additives and Processing Aids.* www.hc-sc.gc.ca.
[39] Novozymes, A. S. (2013). *Enzymes at work.* www.novozymes.com, Denmark.
[40] Jha, A., Ajeet, S. and Hemant, K. (2014). Enhancement of functional properties of cholesterol oxidase and glucose oxidase enzymes through co-immobilization on organic support. *Int. J. Curr. Microbiol. App. Sci.*, *3*: 249-257.
[41] Silva, A., Tomotani, E. and Michele, V. (2011). Invertase, glucose oxidase and catalase for converting sucrose to fructose and gluconic acid through batch and membrane-continuous reactors. *Braz J Pharm Sci.*, *47*: 399-407.
[42] Armin, S. (2006). Safety regulations of food enzymes. *Food Technol Biotechnol.*, *44*: 197-209.
[43] David, J., Moreland, D. and Peter, L. (2011). *Biocatalysts: Industrial enzymes and the exploitation of microorganisms.* University of Auckland, Private Bag 90219, New Zealand.
[44] Li, S., et al. (2012). Technology prospecting on enzymes: Application, markting and engineering. *Computat. Struct. Biotechnol. J.*, *2*, doi.org/10.5936/ csbj. 201209017.
[45] Bindslev-Jensen, C., et al. (2006). Investigation on possible allergenicity of 19 different commercial enzymes used in the food industry. *Food Chem Toxicol.*, *44*:1909-1915.
[46] Jan, M., et al. (2013). Characteristics of cow's milk proteins including allergenic properties and methods for its reduction. *Polish Ann Med.*, *20*: 69-76.
[47] Bu, G., Luo, Y., Chen, K. and Zhu, T. (2013). Milk processing as a tool to reduce cow's milk allergenicity: a mini-review. *Dairy Sci Technol.*, *93*: 211-223.

[48] Safri, M., et al. (2015). Cow's milk protein sensitization on the lymphocyte Th-1 and Th-2 activity in relation to wheezing in the first year of life. *Intr J Bioflux Soc.*, *7*: 31-37.

[49] Whitehurst, R. and Barry, L. (eds.). (2002). *Enzymes in Food Technology*, Sheffield Academic Press, CRC Press.

[50] Pedro, F. (2014). Improving Enzyme Performance in Food Applications. *Max Rubner Conference 2014*. Max Rubner – Institute.

[51] Gokila, T. and Thiruvengadam, S. (2014). Nanotechnology in food industry. A review. *Intr J Chem Tech Res.*, *6*: 4096-4101.

[52] Singh, T., et al. (2017). Application of nanotechnology in food science: Perception and overview. *Front. Microbiol.*, *8*: 1501-1507.

[53] Husain, Q. (2018). Nanocarriers immobilized proteases and their industrial applications: An overview. *J Nanosci Nanotechnol.*, *18*: 486-499.

[54] Chaudhry, Q., Castle, L. and Watkins, R. (2010). *Nanotechnologies in Food.* Royal Society of Chemistry Publishers, Cambridge, UK.

[55] www.nano.gov. (2015). *Opportunities and risks of nanotechnologies.*

[56] Eric, K. (1986). *Drexler Engines of Creation: The Coming Era of Nanotechnology*, Anchor Books 1986 ISBN 0-385-19973-2.

[57] Food.gov.uk. (2014). *Food Additives Legislation Guidance to Compliance.* Food Standards Agency.

[58] Dwyer, J., Paul, M. and Smith, M. (2018). Dietary supplements: Regulatory challenges and research resources. *Nutrients*, *10*: 41-64, doi:10.3390/nu10010041.

In: Microbial Catalysts. Volume 1
Editors: S. M. Abdel-Aziz, N. Garg, A. Aeron et al.
ISBN: 978-1-53614-679-0

Chapter 10

POTENTIAL APPLICATION OF ENZYMES OF LACTIC ACID BACTERIA IN THE FOOD INDUSTRY

Akash Mathew[1] and Smriti Gaur[2,*]
[1]NanoBiotechnology Centre, TERI, Darbari Seth Block,
IHC Complex, New Delhi, India
[2]Department of Biotechnology,
Jaypee Institute of Information Technology Noida,
Uttar Pradesh, India

ABSTRACT

Lactic acid bacteria have been employed in food processing since historical times. Indeed, many of these practices have carried on to modern times. However, the underlying mechanisms, and systems that make lactic acid bacteria suitable for this application have only recently come under scientific scrutiny. This review discusses some of the recent findings regarding lactic acid bacteriaand the various underlying metabolic pathways and enzymes that function to provide LAB with many of its beneficial qualities. Specifically, this review is centered around three principle pathways, namely,lactic acid metabolism, microbial inhibition via bacterocin synthesis and the molecular basis of the pro-biotic activity attributed to LAB. Furthermore, this review goes on to discuss possible future directions that LAB research can take, and the benefits of elucidating the underlying mechanisms of these microbial catalysts.

Keywords: lactic acid bacteria, probiotic, bacteriocin

1. INTRODUCTION

Lactic acid bacteriaor (LAB) for short, are a diverse set of organisms that are unique in their ability to synthesize lactic acid carbon and energy [1]. However, the enzymatic pathways

* Author's Email: smriti.gaur@jiit.ac.in.

for the production of lactic acid have been found to vary within this group and under different environmental conditions [1]. LAB are particularly relevant to the food-technology sector, due to their ability to prevent spoilage of food via elimination or limitation of growth of pathogenic organisms [2], and to alter the texture and taste of certain foods to make them more palatable and nutritious [3]. Further study of the enzymatic mechanisms that provide LAB with these beneficial characteristics will help in the development of modern foods and preservatives.

Lactic acid Bacteria comprise a number of diverse organisms. The first detailed study and classification of these organisms was carried out by Orla-Jensen in the early nineteen hundreds [1]. The basis of classification, still in use today, stems from the morphology rods or cocci [4] and the substrate specificity and reaction end products, heterofermentative or homofermentative [5] which as the names imply, indicate the ability of a given LAB to utilize one or more than one substrate for lactic acid productionor results in a single or multiple end products.

LAB have been the subject of much scrutiny and study not only due to their immense potential for application in the food preservation and processing industry, but also due to their perceived benefits to human health. Thus, studying the enzymes that make these organisms so beneficialwill serve to refine existing technologies applying LAB and better adapt them for industrial applications. Furthermore, the current practice of employing LAB is riddled with challenges [6] chief among them is to do with the sensitivity of LAB to environmental conditions [7]. Changes in nutrient composition, pH or temperature, has been known to alter the expression and activation of the LAB enzymatic pathways [6]. There have been several studies aimed at overcoming this obstacle by employing metabolic engineering to produce a more hardy strain of LAB [8]. Studying these enzymes may lead to the development of new enzymatic tools that do not suffer from these constraints.

Bacteriocin production, lactic acid metabolism, and the probiotic activity of LAB all have enormous potential for commercial application and would benefit greatly from a detailed study. In order to understand the significance of the pathways and enzymes discussed further in this chapter, it is important to first study some of the well known benefits of lactic acid bacteria to human health.

2. Lactic Acid Bacteria in Human Health

In human health, lactic acid bacteria has been shown to effective in treating a number of conditions ranging from Lactose intolerance to cancer and inflammatory bowel disease. However, the most promising role of LAB with regard to human well being stems from its application as a microbial catalyst in the food processing industry. The application of LAB to enhance food quality and shelf life is a historical application and has been reported to be in practice since as early as 2000 B.C.

Another important function of lactic acid bacteria is their probiotic action. There have been several studies that show the efficacy of these organisms in maintaining a healthy gut environment [9-11]. The ability of these organisms to interact with mucosal lining of the intestines has generated considerable interest as a potential drug delivery system. Indeed, *L. lactis* recently became the first recombinant transgenic organism to be used to successfully treat Chron's disease in humans [12]. As a probiotic, lactic acid bacteria are already commercialized and are generally regarded safe for consumption. Experiments on mice have shown the benefits of commercially available probiotics containing LAB in alleviating lower gut issues such as colitis [13].

Consumption of LAB has been found to enhance the body's immune response against infection [14]. *L. casei* and *B. Longum* have been reported to strengthen the natural immune response against *E. coli* in gastrointestinal tract [15]. Studies have shown that the mechanism of their action mainly comprises stimulation of the phagocytic activity of macrophages. Apart from their role in inducing macrophage activity, some species of LAB have also been reported to exert anti-tumor effects [16]. Dietary supply of LAB has also been shown to promote the IgA response and assist in the transportation antigens through Peyer's patches for modulation of immune response [17].

3. Existing Applications of Lactic Acid Bacteria in the Food Industry

Though the application of LAB in the food industry has been in practice for such a significant period, many of the benefits of this microbial catalyst has only recently come to light. In addition to the ability of LAB to alter and preserve foodstuff, recent research has led to the discovery of bacterocin-like antimicrobial peptides that are produced by certain species of LAB [2]. Studying the enzymatic pathways that produce these anti-microbials may lead to the development of alternative strategies for combating the rising threat of antibiotic resistant infections.

With regard to food processing, LAB has become a mainstay in adding to the nutritive value, palatability as well as preservation of many major food products. Most of these applications employ LAB directly to catalyze the conversion of complex sugars and other components present in food stuff, to metabolic by-products that in addition to enhancing the quality of the food, act as a bio-preservatives due to inherent enzymatic products produced by LAB during the course of its metabolism.

There are a large number of species that have been classified as lactic acid bacteria, however, only a few of these species have been employed in the food-technology industry [3]. Among which *Streptococcus thermophylus* and *Lactobacillus bulgaricus* [18] used in the manufacture of yogurtand *Lactococcus lactis* traditionally used in the manufacture of cheese and buttermilk have been well studied [19].

4. The Major Enzymatic Pathways and Products of LAB

4.1. Lactic Acid Metabolism

As previously mentioned, LAB comprises of a diverse set of organisms that can be classified on the basis of their ability to convert a given substrate to lactic acid [4]. The enzymatic pathways that comprise this process can be classified along the same lines into hexose, pentose and disaccharide metabolism pathways. Lactic acid synthesis is relevant to the food industry due to its ability to limit the growth of pathogenic bacteria while increasing the palatability of the food product itself [3].

4.1.1. Hexose Metabolism

Hexose metabolism can be further classified into homolactic and heterolactic fermentation. The homolactic pathway employs the glycolysis pathway found in all species of LAB. The major steps involved in this direction involve activation of the sugar via phosphorylation followed by the synthesis of dihydroxyacetone phosphate (DHAP) and glyceraldehydes-3-phosphate (GAP) which in turn are converted to pyurvate. Pyurvate is then further reduced to lactic acid via the action of the enzyme Lactate dehydrogenase. One of the steps that have shown a level of variation within the LAB group is the initial activation of the sugar via phosphorylation. This is dependent on the breaking of a high energy phosphate bond but can be carried out by two separate enzymes namely, the ATP-dependent glucokinase and the sugar phosphotransferase system (PTS). Though the method of phosphorylation can occur via either pathway the basis of sugar activation is conserved. This pathway results in the release of a single molecule of lactic acid and two molecules of ATP for every molecule of glucose consumed. An important caveat to this pathway [4, 20] is that lactic acid production in these species is triggered only under high concentrations of sugar or oxygen starvation.

In contrast heterolactic fermentation results in the formation of 1 molecule of acetic acid, 1 molecule of lactic acid and 1 molecule of CO_2 in addition to 1 ATP [20]. This type of fermentation relies on the Pentose-phosphoketolase pathway which is characterized by an initial decarboxylation step following activation of the hexose moiety. The heterolactic fermentation pathway has become more relevant today due to its ethanol production capability which is particularly relevant for application in the field of bio-fuel production [21].

4.1.2. Disaccharide Metabolism

Disaccharide metabolism specifically lactose fermentation is a particularly well studied pathway in species of LAB. This type of metabolism has two modes of action: lactose can either be enzymatically cleaved to its component monosaccharides glucose and galactose that are then subject to the pathways discussed in the previous section or can be first phosphorylated via PTS and then cleaved by the enzyme phospho-β-D galactosidase into one phosphorylated galactose and non-phosphorylated glucose molecule [4].

In case of Lactate metabolism, the cleavage step results in the synthesis of one glucose-6-phosphate molecule which enters glycolysis and a molecule of galactose-6-phosphate which enters the tagtose-6-phosphate pathway [4]. In *L. lactis*, this pathway has generated considerable interest due to the propensity for this species of LAB to dephosphorylate and secrete the dephosphorylated galactose. Galactose has been associated with several traits in fermented products including the effect of 'browning' which is observed in certain fermented dairy products [22, 23]. *L. lactis* in this context is important as it is one of the most widely used organisms for the fermentation of dairy products [3]. There have been attempts to metabolically engineer strains of *L. lactis* that are more efficient in lactose metabolism and less likely to secrete galactose [24]. Elucidating the functional and biochemical aspects of the enzymes of this pathway may open up the possibility of engineering new enzymes that can be applied in the dairy industry.

4.1.3. Pentose Metabolism

Apart from a few exceptions all LAB can readily metabolize pentoses via the pentose phosphoketolase pathway. However, the requirement of a decarboxylation step is rendered redundant thus ablating the production of CO_2. Thus, this pathway produces equimolar quantities of lactic and acetic acid at the expense of a pentose molecule [4].

4.2. Bacteriocin Production in LAB

Bacteriocins have gained a singular significance in the modern world, where there is a rapid rise in strains of pathogens that have developed resistance to conventional antibiotics [25]. Based on the range of activity exhibited by these molecules, bacteriocins can be either be narrow or broad range. As the terms imply narrow range bacteriocins act against a smaller range of microbes often closely related to the producerwhereas broad rage bacteriocins serve to inhibit a broader range of microbes [26].

Bacteriocins have been conventionally classified into two main groups: Class I bacteriocins that comprise of lantibiotic bacteriocins, and are generally small peptides, and Class II bacteriocins that contain non-lantibiotic bacteriocins, and are generally larger thermostable proteins. Class II bacteriocins are further classified into five sub-groups based on particular structural and functional features [27].

One of the earliest and most comprehensive studies carried out on the LAB bacteriocins was disseminated by Todd R. Klaenhammer in 1987. In his paper, Klaenhammer went on to discuss the properties of bacteriocins from a number of LABs such as *L. Helviticus, L. acidophilus, L. plantarum* and several species of streptococci [28]. This paper serves to elucidate, and showcase the wide variety of bacteriocins that are produced by LAB. Indeed, there have been many recent discoveries that have further extended the list of bacteriocins produced in LAB to include anti-gram negative bacteriocins as well [27].

An important broad spectrum bacterocin found in LABand widely used in the food industry as a preservative is Nisin [28]. *S. lactis* has been known to natively produce large amounts of nisinvia a plasmid system and is known to contribute to the highly adaptive nature of this species in competitive environments. Nisin has been extensively characterized and comprises of a 3500 kDa protein that exists as a dimer or a tetramer and contains several unusual amino acids [28].

These bacteriocins have been shown to be synthesized as precursors from ribosomes that undergo extensive post-translational modifications before maturation [29]. Studying the process of synthesis of and characterizing the bacteriocins themselves has enormous potential for designing effective food preservatives for industrial and commercial applications.

4.3. Probiotic Action of LAB

A 'Probiotic' by definition is something that assists in maintaining a healthy microenvironment and encourages the growth of beneficial microbiota within the human gut. Lactic acid bacteria have been well established in this role and have been extensively studied in this context.

The probiotic activity of LAB is multi-faceted. Secretion of anti-microbial bacteriocins and modulation of the gut pH prevent the growth of pathogenic bacteria [30] while direct immune-stimulatory effects [31-33] and anti-tumor effects of LAB [34, 35, 16] on the host serve to maintain intrinsic vitality. Furthermore, the ability of these bacteria to break down complex sugars serve to assist in the digestive process and have been shown to alleviate symptoms of lactose intolerance in humans [36].

Further effects of LAB on pathological conditions such as diabetes and obesity have recently come to light, and are currently under study [37]. Studying these effects and elucidating the specific pathways that may be responsible for modulating these activities in LAB will provide a basis for developing targeted treatments for the aforementioned conditions. Deodan is one such molecule, a product of the *L. bulgaricus,* has been shown to enhance levels of IL-12 when orally administered in mice [38]. Thus, studying the enzymes and compounds inherent to LABhas the potential to lead to the discovery of new methodologies and novel applications of these aspects in the food technology industry.

FUTURE PERSPECTIVES

As discussed in the aforementioned sections, LAB have been shown to have several beneficial effects for human health. In order to refine the usage of these organisms and possibly increase the scope of application of these benefits, it is imperative to gain a deeper understanding of the enzymes native to this group of microorganisms. The potential of LAB in the food industry is already apparent [39, 3] indeed the use of LAB in processing food to increase its palatability, nutritive value and shelf life has already been explored and applied commercially. However, many of these functions are still not clearly understood. Furthermore, using live organisms in food technology is subject to its own set of limitations and challenges dictated by the requirements of the organism itself [6].

There have been several studies that have employed metabolic engineering and adaptive laboratory evolution to engineer new more applicable strains of LAB [24, 8]. Elucidation of the enzymatic pathways and characterization of the enzymes of LAB is a task that would simplify this problem and pave the way for developing more specific LAB tailored to the needs of the industry and the application in question. Large scale proteomic and genomic analysis combined with organisms level studies will prove to be useful in developing these technologies.

REFERENCES

[1] Benninga, H. (1990). *A history of lactic acid making: A chapter in the history of biotechnology*. Netherlands: Kluver Academic Publishers.

[2] Vandenbergh, P. A. (1993). "Lactic acid bacteria, their metabolic products and interference with microbial growth". *FEMS Microbiol Rev., 12*: 221 - 237.

[3] Das, D. and Goyal, A. (2012). Lactic acid bacteria in food industry. In: *Microorganisms in Sustainable Agriculture and Biotechnology,* pp. 757-772. Springer Science & Business Media.

[4] Axelsson, L. (2004). Lactic acid bacteria: Classification and physiology. In: *Lactic acid bacteria: Microbiological and functional aspects.* Taylor & Francis eBooks *3*: 1-66.

[5] Okano, K., Zhang, Q., Shinkawa, S., et al. (2009). Efficient production of optically pure d-lactic acid from raw corn starch by using a genetically modified l-lactate dehydrogenase gene-deficient and Î±-amylase-secreting Lactobacillus plantarum strain. *Appl Env Microbiol., 75*: 462-467.

[6] Mattila-Sandholm, T., Myll ñrinen, P., Crittenden, R., et al. (2002). Technological challenges for future probiotic foods. *Intr Dairy J. NIZO Dairy Conference on Food Microbes, 12*: 173-182.

[7] Conway, P. L., Gorbach, S. L. and Goldin, B. R. (1987). Survival of lactic acid bacteria in the human stomach and adhesion to intestinal cells. *J Dairy Sci., 70*: 1-12.

[8] Kleerebezem, M. and Hugenholtz, J. (2003). Metabolic pathway engineering in lactic acid bacteria". *Cur Opin Biotechnol., 14*: 232-237.

[9] García-Ruiz, A., González de Llano, D., Esteban-Fernández, A., et al. (2014). Assessment of probiotic properties in lactic acid bacteria isolated from wine. *Food Microbiol., 44*: 220-225.

[10] Lamari, F., Sadok, K., Bakhrouf, A., et al. (2014). Selection of lactic acid bacteria as candidate probiotics and in vivo test on Artemia nauplii. *Aquac Intr., 22*: 699-709.

[11] Rubio, R., Jofré, A., Martín, B., et al. (2014). Characterization of lactic acid bacteria isolated from infant faeces as potential probiotic starter cultures for fermented sausages. *Food Microbiol., 38*: 303-311.

[12] Bermudez-Humaran, L. G. (2009). Lactococcus lactis as a live vector for mucosal delivery of therapeutic proteins. *Vaccines, 5*: 264-267.

[13] Yoda, K., Miyazawa, K., Hosoda, et al. (2014). SLactobacillus GG-fermented milk prevents DSS-induced colitis and regulates intestinal epithelial homeostasis through activation of epidermal growth factor receptor. *Eur J Nutr., 53*: 105-115.

[14] Sato, K., Saito, H., Tomioka, H., et al. (1988). Enhancement of host resistance against listeria infection by Lactobacillus casei: Efficacy of cell wall preparation of Lactobacillus casei. *Microbiol Immunol., 32*: 1189-1200.

[15] Romond, M. B., Haddou, Z., Mielcareck, C., et al. (1997). Bifidobacteria and human health: Regulatory effect of indigenous bifidobacteria on Escherichia coli Intestinal colonization. *Anaerobe, 3*: 131-136.

[16] Kato, I. (2000). *Antitumour activity of lactic acid bacteria.* Fuller, R. and Perdigon, G. (Eds) 115-138.

[17] Kagnoff, M. F. and Eckmann, L. (1997). Epithelial cells as sensors for microbial infection. *The Journal of Clinical Investigation, 100*: 6-10.

[18] Sozzi, T. and Smiley, M. B. (1980). Antibiotic resistances of yogurt starter cultures Streptococcus thermophilus and Lactobacillus bulgaricus. *Appl Envi Microbiol., 40*: 862-865.

[19] Terzic-Vidojevic, A., Mihajlovic, S., Uzelac, G., et al. (2014). Characterization of lactic acid bacteria isolated from artisanal travnik young cheeses, sweet creams and sweet kajmaks over four seasons. *Food Microbiol., 39:* 27-38.

[20] Kandler, O. (1983). Carbohydrate metabolism in lactic acid bacteria. *Antonie Van Leeuwenhoek, 49*: 209-224.

[21] Berezina, O., Zakharova, N., Brandt, A., et al. (2010). Reconstructing the clostridial n-butanol metabolic pathway in lactobacillus brevis". *Appl Microbiol Biotechnol.*, *87*: 635-646.

[22] De Vin, F., Radstrom, P., Herman, L., et al. (2005). Molecular and biochemical analysis of the galactose phenotype of dairy Streptococcus thermophilus strains reveals four different fermentation profiles. *Appl Env Microbiol., 71*: 3659-3667.

[23] Mukherjee, K. K. and Hutkins, R. W. (1994). Isolation of galactose-fermenting thermophilic cultures and their use in the manufacture of low browning mozzarella Cheese. *J Dairy Sci.*, *77*: 2839-2849.

[24] De Vos, W. (1996). Metabolic engineering of sugar catabolism in lactic acid bacteria. *Antonie Van Leeuwenhoek*, *70*: 223-242.

[25] Hassan, M., Kjos, M., Nes, I. F., et al. (2012). Natural antimicrobial peptides from bacteria: Characteristics and potential applications to fight against antibiotic resistance. *J Appl Microbiol.*, *113*: 723-736.

[26] Riley, M. A. and Wertz, J. E. (2002). Bacteriocins: Evolution, ecology, and application. *Ann Rev Microbiol.*, *56*: 117-137.

[27] De Vuyst, L. and Leroy, F. (2007). Bacteriocins from lactic acid bacteria: Production, purification, and food applications. *J Mol Microbiol Biotechnol.*, *13*: 194-199.

[28] Klaenhammer, T. R. (1988). Bacteriocins of lactic acid bacteria. *Biochimie, 70*: 337-349.

[29] Nes, I., Diep, D., HÃ¥varstein, L., et al. (1996). Biosynthesis of bacteriocins in lactic acid bacteria. *Antonie Van Leeuwenhoek, 70*: 113-128.

[30] Heyman, M. (2000). Effect of lactic acid bacteria on diarrheal diseases. *J Amr College Nutr., 19*: 137S-146S.

[31] Chen, C. Y., Tsen, H. Y., Lin, C. L., et al. (2013). Enhancement of the immune response against salmonella infection of mice by heat-killed multispecies combinations of lactic acid bacteria. *J Med Microbiol.*, *62*: 1657-1664.

[32] Maeda, M., Shibata, A., Biswas, G., et al. (2014). Isolation of lactic acid bacteria from kuruma shrimp (Marsupenaeus japonicus) intestine and assessment of immunomodulatory role of a selected strain as probiotic. *Marine Biotechnol.*, (New York, NY), *16*: 181-192.

[33] Yoshida, T., Fujiwara, W., Enomoto, M., et al. (2013). An increased number of CD4+CD25+ cells induced by an oral administration of lactobacillus plantarum NRIC0380 are involved in antiallergic activity. *Intr Arch Allergy Immunol.*, *162*: 283-289.

[34] De Moreno, De LeBlanc, A., Matar, C., et al. (2005). Effects of milk fermented by Lactobacillus helveticus R389 on a murine breast cancer model. *Breast Cancer Research: BCR, 7*: 477-486.

[35] Kim, Y., Oh, S., Yun, H. S., et al. (2010). Cell-bound exopolysaccharide from probiotic bacteria induces autophagic cell death of tumour cells. *Lett App Microbiol.*, *51*: 123-130.

[36] Martini, M. C., Lerebours, E. C., Lin, W. J., et al. (1991). Strains and species of lactic acid bacteria in fermented milks (yogurts): Effect on in vivo lactose digestion. *The Amr J Clin Nutr., 54*: 1041-1046.

[37] Remely, M., Aumueller, E., Jahn, D., et al. (2014). Microbiota and epigenetic regulation of inflammatory mediators in type 2 diabetes and obesity. *Beneficial Microbes, 5*: 33-43.

- The industrial production of enzymes for use in food processing dates back to 1874 when Danish scientist Christian Hansen extracted rennin (chymosin and pepsin) from calves' stomachs for use in cheese manufacturing [6]. Chymosin is now produced from microorganisms that contain the bovine prochymosin gene introduced through recombinant deoxyribonucleic acid (rDNA) techniques. Bovine chymosin expressed in *Escherichia coli* K-12 became the first recombinant microbial catalyst approved for use in food by the US Food and Drug Administration (FDA) [7].
- In 1833, French chemist Anselme Payen discovered the first enzyme, diastase. In 1835, the hydrolysis of starch by diastase was acknowledged as a catalytic reaction by another Swedish scientist Jons Jacob Berzelius [8].
- In 1846 the activity of invertase was demonstrated by Dubonfout [3].
- In 1871 when Louis Pasteur discovered that heating juices to a certain temperature would kill off bad bacteria, affecting wine and fermentation. This process was applied to milk processing, heating milk to a certain temperature to improve food hygiene [1].
- In 1894, Jokichi Takamine, a Japanese scientist is often quoted as the father of industrial enzymology since he pioneered the use of fungal amylases for treatment of pancreatic disorders in amylase deficient people. Indeed, *Aspergillus* amylase is sometimes known as Taka-amylase [3].
- Pectinases were used for juice clarification in the 1930s, and for a short period during World War II, invertase was also used for the production of invert sugar syrup in a process that pioneered the use of immobilized enzymes in the sugar industry.
- In 1946, the first patent was accepted for enzymatic hydrolysis of starch. Still, the large-scale application of enzymes only became really established in the 1960s, when the traditional acid hydrolysis of starch was replaced by an approach based on the use of amylases and amyloglucosidases (glucoamylases), a cocktail, that some years later would include glucose (xylose) isomerase.
- In the1960s, more new enzymes were discovered, in particular the alkaline protease of *Bacillus licheniformis* which had almost perfect properties for maintaining activity when included in washing powders. In the late 1970s *B. licheniformis* again came to prominence because its amylase was temperature stable and had more sophisticated genetics. The enzyme producers were able to increase yield to make this enzyme competitive in price terms with the less temperature stable enzyme from *B. amyloliquefaciens* which naturally secreted large amounts of extracellular enzymes, particularly α- amylase [9].
- In 1980s, the technology to convert starch into fructose syrups (high fructose corn syrups or HFCS) for use as a sweetener emerged, but this proved too competitive for the sugar beet farmers of the EU and the process was effectively banned from the EU by the imposition of high tariffs [10].
- Developments in the latter part of the 20th century were centered on the contribution of modern molecular biology to provide exceptionally high yields of enzymes and modify enzyme molecules to tailor enzymes for the particular market requirement. Perhaps the impact of modern biotechnology has nowhere been more widely felt than in the application of industrial microbial catalysts produced by large scale fermentation of microorganisms. Over the past 20 years, microbial catalysts have been developed to play an ever expanding role in food processing industries.

1.2. World Market

Microbial catalysts, as well as other types of enzymes, are very well established products in biotechnology based markets. The business worth of US $1.3 billion in 2002 skyrocketed to US $7 billion in 2013 with an increasing global demand of around 6% yearly [2, 11]. In this growing market, a large number of applications [12] like baking, brewing, dairy, dietary supplements as well as fats and oils production etc. are related to food and feed industries. A recent survey on world sales of enzymes ascribes 31% to food enzymes, 6% to feed enzymes and the remaining to technical enzymes [13].

A good number of companies from Europe such as Novozymes (Denmark), Danisco (Denmark), DSM (The Netherlands) and BASF (Germany) are dominating the world market of microbial enzymes[13, 14]. However, it is widely predicted that India and China would join in this emerging markets in near future [15, 16].

After World War II, the use of microbial catalysts brought a radical change in many sectors including food industries. Thus, the production and distribution of microbial catalysts became an established global industry, though food grade enzymes represented only about one third of this market [5].

2. Microbial Catalyst and Food Biotechnology

2.1. Definition

Enzymes are biological macromolecules originating in living organisms and catalyzing important chemical interconversions that are required to sustain life [7, 17]. "Microbial catalyst" is a type of enzymes which have heterologous genes and are collected from microorganisms (i.e., fungal or bacterial hosts). Recently, the use of recombinant DNA technology has made it possible to manufacture microbial catalysts that are tailored to function in specific food processing conditions. It has also been termed as "biocatalysts" [2] or "microbial enzymes" [3].

Biotechnology is defined in accordance with the Convention on Biological Diversity (CBD), i.e., "any technological application that uses biological systems, living organisms, or derivatives thereof, to make or modify products or processes for specific use." [18]. According to the Organization of Economic Co-operation and Development (OECD), biotechnology is the application of scientific and engineering principles for the processing of materials by biological agents to provide goods and services' [19].

Food biotechnology comprises the use of microorganisms for a number of applications such as preservation of food, microbial cultures, production of microbial catalysts, enrichment of different dietary supplements and food ingredients, and enhancement of flavor compounds. Therefore, the integrative culture of safe microorganisms with optimum yield is the prime objective in modern food biotechnology.

The improvement of microbial cultures to be applied in food processing sectors so far used two methodologies [20]. The first is direct evolution of microbial catalyst, through traditional biotechnology such as random mutagenesis and recombination, where the environmental adaptation is reproduced in-vitro in a much accelerated timescale. In order to control the pathway of the process, either a screening test for the assessed feature is performed after each round of modification or selective pressure is applied [21, 22]. This methodology, which allows for a high throughput, has been extensively applied, aimed for more efficient catalysts [23]. Examples of this traditional biotechnological advancement in the food and feed processing include the enhancement of the activity of hyperthermostable glucose (xylose) isomerase from *Thermotoga neapolitana* at relatively low temperature and pH without hampering thermostability [2] and the improvement of the thermostability of maltogenic amylase, amylosucrase, glucoamylase, phytase and xylanase from *Thermus* sp., *Neisseria polysaccharea*, *Aspergillus niger*, *Escherichia coli*, and *Bacillus subtilis* respectively [24-28]. Hybridization is another example of classical biotechnology applied to the improvement of yeasts such as *Saccharomyces cerevisiae* strains. This field has been scoured for improved fermentation, processing and biopreservation abilities and for capacities to increase the wholesomeness and sensory quality of wine [29]. Bovine chymosin expressed in *Escherichia coli* K-12 became the first recombinant microbial catalysts approved for use in food by the US Food and Drug Administration (FDA) [30]. Since then the recombinant gene technology is widely employed in R&D for strain improvement. For example, alpha-amylases with improved heat stability were developed through DNA sequence modifications of the alpha-amylase genes and phospholipase A1 gene from *Fusarium venenatum* were expressed in GM *Aspergillus oryzae* to produce the phospholipase A1 enzyme which was used for cheese manufacturing to improve process efficiencies and cheese yields [7]. The key component in evaluating safety of microbial catalysts derived from native and recombinant microorganisms is the safety assessment of the production strain, in particular, its pathogenic and toxigenic potential [31]. Although the safety evaluation of food processing microbial catalysts from recombinant microorganisms has been extensively discussed in many literature, in the United States, FDA regularly reviews the Generally Recognized As Safe (GRAS) affirmation petitions for the microbial catalysts preparations and gives the final decision regarding safety issue of the catalysts mentioned in the petitions [31-33].

The second methodology states that rational pinpoint modification can be carried out in the structure of one or more amino acids indispensable to a targeted enzyme function. Bioinformatics provides data on amino acid propensities and protein sequences. Adequate computational processing of the data through molecular simulations or developing a model based on known enzyme structures with homologous sequences enables the output of generalized rules predicting the effect of mutations on enzyme properties and the potential molecular functions [20, 34]. This methodology is actually a computer aided modeling technique to produce a site-directed mutant. This strategy mostly aims to improve thermal stability, catalytic efficiency, the effective range of pH/temperature of the catalysts to function.

Enzymes used in food processing industry are sold as enzyme preparations which typically contain the target enzyme and several other substances suitable for use in food such as diluents, preservatives, and stabilizers. The preparations may also contain other enzymes and metabolites from the organisms and the residues. However, all these materials are expected to be of appropriate purity consistent with current good manufacturing practice (cGMP) [7].

2.2. Relevant Microbial Catalysts

According to the enzyme commission (provide EC number) the enzymes are divided into the 6 parts as given below [3, 35].

(I) Oxidoreductase (EC 1)
(II) Transferase (EC 2)
(III) Hydrolase (EC 3)
(IV) Lyase (EC 4)
(V) Isomerase (EC 5)
(VI) Ligase (EC 6)

2.2.1. Oxidoreductase

This type of enzymes causes oxidation in biological systems either by the transfer of electrons from one molecule, the reductant (electron donor), to another, the oxidant (electron acceptor) or by the removal of hydrogen from the substrate molecule and this is why they are also called dehydrogenases. As for example, alcohol dehydrogenase catalyzes alcohols to aldehydes as shown in the following reactions.

$$R\text{–}CH_2OH + A \rightarrow R\text{–}CHO + H_2A$$

Where, A is an acceptor molecule. If A is oxygen, the relevant enzymes are called oxidases or laccases; if A is hydrogen peroxide, the relevant enzymes are called peroxidases. Oxidoreductases are further classified into 22 subclasses.

2.2.2. Transferase

These enzymes catalyze the transfer of specific functional groups (e.g., a methyl or glycosyl group) from one molecule (called the donor) to another (called the acceptor). Based on the type of biochemical group transferred, transferases can be divided into ten categories that comprise over 450 different unique enzymes [36, 37].

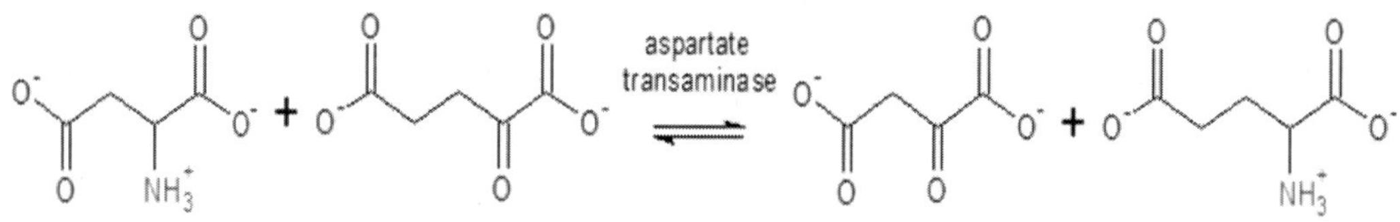

Figure 1. Aspartate transaminase reversibly transfers an amino group from one molecule to the other.

2.2.3. Hydrolase

Hydrolases catalyze hydrolysis, the cleavage of substrates by water. These reactions include the cleavage of peptide bonds in proteins, glycosidic bonds in carbohydrates, and ester bonds in lipids. In general, larger molecules are broken down to smaller fragments by hydrolases. Based upon the bonds they act upon, hydrolases are further classified into 13 subclasses.

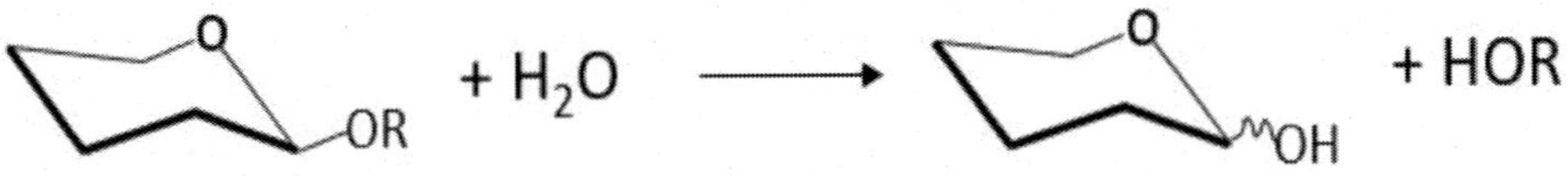

Figure 2. Glycosidases catalyze the synthesis and breakage of glycosidic bonds.

2.2.4. Lyase

Lyases catalyze the addition of groups to double bonds or the formation of double bonds through the removal of groups. Thus bonds are cleaved using a principle different from hydrolysis. Lyases can be further classified into seven subclasses.

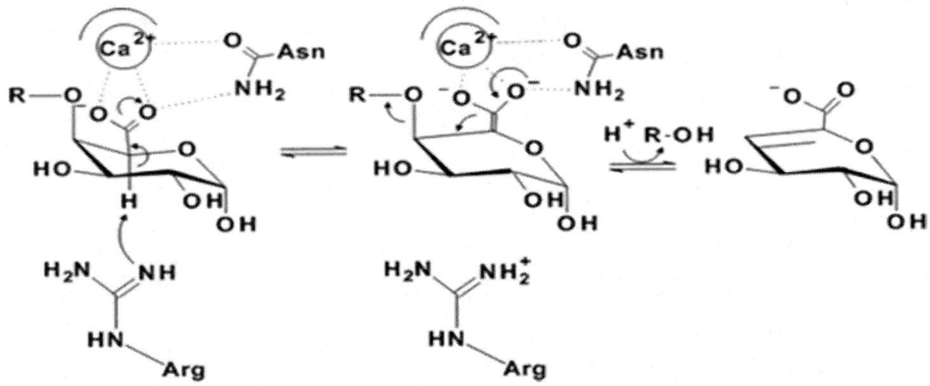

Figure 3. General β-elimination reaction mechanism for catalysis by pectate lyases.

2.2.5. Isomerase

Isomerases catalyze the transfer of groups from one position to another in the same molecule. In other words, these enzymes change the structure of a substrate by rearranging its atoms. Isomerases are further classified into six subclasses.

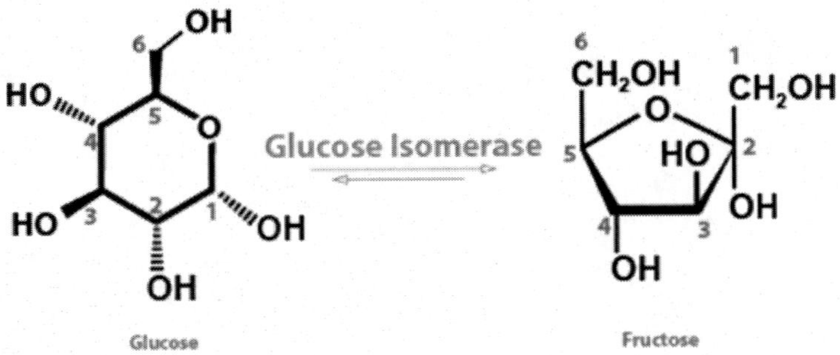

Figure 4. Isomerization of glucose and fructose.

2.2.6. Ligase

Ligases join molecules together with covalent bonds. These enzymes participate in biosynthetic reactions where new groups of bonds are formed. Such reactions require the input of energy in the form of cofactors such as ATP. Ligases are further classified into six subclasses.

3. Food Microbial Catalysts

The commonly used various food and feed microbial catalysts are tabulated below [3, 38].

Table 1. An overview of enzymes used in food and feed processing [2]

Class	Enzyme	Role
Oxido-	Glucose oxidase	Dough strengthening
reductases	Laccases	Clarification of juices, flavor enhancer (beer)
	Lipoxygenase	Dough strengthening, bread whitening
	Cyclodextrin	Cyclodextrin production
	Glycosyltransferase	
Transferases	Fructosyltransferase	Synthesis of fructose oligomers
	Transglutaminase	Modification of viscoelastic properties, dough processing, meat processing
		Starch liquefaction and sachcarification
		Increasing shelf life and improving quality by retaining moist,
	Amylases	elastic and soft nature
		Bread softness and volume, flour adjustment, ensuring uniform
		yeast fermentation
		Juice treatment, low calorie beer
	Galactosidase	Viscosity reduction in lupins and grain legumes used in animal feed, enhanced digestibility
	Glucanase	Viscosity reduction in barley and oats used in animal feed, enhanced digestibility
	Glucoamylase	Saccharification
	Invertase	Sucrose hydrolysis, production of invert sugar syrup
	Lactase	Lactose hydrolysis, whey hydrolysis
Hydrolases		Cheese flavor, in-situ emulsification for dough conditioning,
	Lipase	support for lipid digestion in young animals, synthesis of aromatic
		molecules
		Protein hydrolysis, milk clotting, low-allergenic infant food
	Proteases (chymosin, papain)	formulation, enhanced digestibility and utilization, flavor
		improvement in milk and cheese, meat tenderizer, prevention of
	Pectinase	chill haze formation in brewing
	Peptidase	Mash treatment, juice clarification
		Hydrolysis of proteins (namely, soy, gluten) for savoury flavors,
	Phospholipase	cheese ripening
	Phytases	In-situ emulsification for dough conditioning
		Release of phosphate from phytate, enhanced digestibility
	Pullulanase	Saccharification
	Xylanases	Viscosity reduction, enhanced digestibility, dough conditioning
Lyases	Acetolactate decarboxylase	Beer maturation
Isomerases	Xylose (Glucose) isomerase	Glucose isomerization to fructose

3.1. Carbohydrases- Microbial Catalysts for Carbohydrate Hydrolysis

3.1.1. Amylases

Amylases are one of the main microbial catalysts used in food industries. Amylases catalyze the chemical conversion of starch into simple sugars. Amylase is abundantly present in human saliva. It is also present in seeds containing starch as a food reserve and is secreted by many fungi. Starch containing foods like rice and potato taste little sweet in the mouth because amylase present in saliva turns some of their starch into sugar. Amylase present in the pancreas hydrolyzes dietary starch into disaccharides and trisaccharides which are further converted by other enzymes to glucose to supply the body with energy.

The history of amylases began in 1811 when the first starch degrading enzyme was discovered by Kirchhoff [39]. Amylase was the first catalyst isolated by Anselme Payen in 1833 [3]. In1930, Ohlsson first suggested the classification of starch digestive enzymes in the malt as α and β-amylases according to the anomeric type of sugars produced by the enzyme reaction [39]. All amylases are glycoside hydrolases and act on glycosidic bonds. It is widely used in industries and has nearly 25% of the enzyme market [40, 41]. Today amylase has almost replaced chemical hydrolysis of starch in starch processing industry. The amylases obtained from microorganisms have a broad spectrum of industrial uses as they are more stable than plant and animal amylases. The two major reasons for using microorganisms for the production of amylases are possibility to produce economically in bulk and capacity to tailor microbes using genetic engineering to derive enzymes of desired nature [42, 43].

3.1.1.1. Types of Amylases

Amylase based on the bonding type, are divided into three sub classes which are as α, β and γ amylase.

3.1.1.2. α-Amylase (1, 4-α-D-Glucan Glucanohydrolases)

The correct name of α-amylase family is glycoside hydrolase family 13. α-Amylase (E.C.3.2.1.1) is an endo-hydrolase enzyme that catalyzes the hydrolysis of internal α-1, 4-glycosidic bonds in polysaccharides with the retention of α-anomeric configuration in low molecular weight products, such as glucose, maltose and maltotriose units [44]. As an endo-hydrolases it acts on the interior of the starch substrate molecule, and neither cleave the terminal non reducing ends of glucose residues nor α-1,6-linkages [45]. α-amylases are known to be metalloenzymes [39]. Enhancement of amylase activity is based on the ability of Ca^{2+} ions to interact with negatively charged amino acid residues such as aspartic and glutamic acids, which resulted in stabilization as well as maintenance of enzyme conformation [43, 46]. In addition, calcium also has a substrate binding capacity [47]. Other divalent cations such as Ba^{2+}, Mg^{2+} and Sr^{2+} can also act as activators for the α-amylases, however, with less activity. All the metal chelators are strong inhibitors of amylases as they are metalloenzymes and can be removed by treating with metal chelators such as EDTA, EGTA, etc. [48]. The inhibition usually increases at high temperatures and acidic pH. Heavy metal ions such as Hg^{2+}, $Pb^{2+}$and Al^{3+} are also susceptible inhibitors for amylases.

The amylase has a three-dimensional structure composed of 512 amino acids in a single oligosaccharide chain with a molecular weight of 57.6 kDa [46]. The protein contains 3 domains: A, B, and C. A domain, the largest one, is a typical barrel shaped 8 (α/β) or TIM barrel super structure. B domain is in between the A and C domains and attached to A domain by disulphide bond. C domain has a β sheet structure linked to A domain by a simple polypeptide chain. Calcium (Ca^{2+}) is situated in a long cleft of about 3 nm, between the carboxyl end of the A and B domains [44].

α-Amylase has become an enzyme of crucial importance due to its starch hydrolysis activity. The optimum pH for activity is found to be 7.0. One such activity is the production of glucose and fructose syrup from starch. α-Amylase catalyzes the first step in this process. Previously, starch was hydrolyzed into glucose by acid hydrolysis. But this method has drawbacks like the operating conditions are of highly acidic nature and required high temperatures. These limitations are overcome by enzyme hydrolysis of starch to yield high fructose syrup. Hydrolysis patterns of starch are controlled by the structure of starch granules, maturity of granules and how the amylases adsorb to a granule. Minor constituents such as surface proteins and lipids also influence enzyme adsorption and thus the hydrolysis of starch. The starch granules are degraded either by exocorrosion or by endocorrosion. SEM studies on wheat and barley starch granules revealed that the initial amylase attack was on the starch granule forming discrete tunnels into the granule and finally degrade the central portion [50]. α-Amylase is also used in ethanol production to break starches in grains into fermentable sugars. The first step in the production of high-fructose corn syrup is the treatment of cornstarch with α-amylase, producing shorter chains of sugars called oligosaccharides.

3.1.1.3. β-Amylase

β-amylase (EC 3.2.1.2) (alternative names: 1,4-β-D-glucan maltohydrolase; glycogenase; saccharogen amylase) occurs in plants (mostly in the seeds of higher plants and sweet potatoes) and some bacteria but it is not yet produced commercially from microorganisms. It is classified in glycoside hydrolase family 14. It is an exo-acting enzyme that hydrolyzes amylose chains from the non-reducing ends releasing two glucose units in the β-configuration (maltose) at a time. Since it is unable to cleave branched linkages in branched polysaccharides such as glycogen or amylopectin, the hydrolysis remains incomplete [43, 44].

β-amylase helps to add sweetness to the ripe fruit by breaking down its starch into maltose. β-amylase is present in an inactive form prior to germination in seeds. Animal tissues do not contain β-amylase, although it may be present in microorganisms contained within the digestive tract. The optimum pH for β-amylase is 4-5. It can be used for structural studies of starch and glycogen molecules produced by various methods. In the industry, it is used for fermentation in the brewing of malted barley and distilling industry. Also, it is used to produce high maltose syrups [51].

3.1.1.4. γ-Amylase

γ-Amylase (EC3.2.1.3) (alternative names: Glucan 1,4-α-glucosidase; amyloglucosidase; Exo-1,4-α-glucosidase; glucoamylase; lysosomal α-glucosidase; 1,4-α-Dglucan glucohydrolase) will cleave α(1-6) glycosidic linkages, as well as the last α (1-4) glycosidic linkages at the nonreducing end of amylose and amylopectin to finally produce glucose [3]. This enzyme is a member of glycoside hydrolase family 15. This enzyme is almost invariably derived from fungi such as *Aspergillus niger and Aspergillus oryza* and has typical low pH (3.5 to 5.0) and temperature optima (few are active above 60°C) of fungal enzymes [43, 51].

3.1.2. Pullulanase (α-Dextrin 6-Glucanohydrolase)

Pullulanase is an extracellular amylolytic exoenzyme. Pullulanase, an important debranching microbial catalyst, has been widely utilized to hydrolyze the α-1,6 glucosidic linkages in starch, amylopectin, pullulan, and related oligosaccharides, which enables a complete and efficient conversion of the branched polysaccharides into small fermentable sugars during saccharification process. Pullulanase is used to increase the final glucose concentration with reduced amount of glucoamylase in saccharification process. Thus, the reversion reaction that involves resynthesis of saccharides from glucose molecules is prevented. To date, five groups of pullulanase enzymes have been reported, that is, (i) pullulanase type I,

(ii) amylopullulanase, (iii) neopullulanase, (iv) isopullulanase, and (v) pullulan hydrolase type III. [53]. The enzyme has been studied most extensively in the gram-negative bacterium *Klebsiella pneumoniae*, however, they are usually collected from *Bacillus acidopullulyticus* or from the mold *Aureobasidium pullulans* [54, 55].

3.1.3. Glucose Isomerase

Glucose isomerase (GI) (D-xylose ketol-isomerase; EC. 5.3.1.5) is a cytoplasmic enzyme that catalyzes the reversible isomerization of D-glucose and D-xylose to D-fructose and D-xylulose respectively [56]. The isomerization of glucose is a reversible reaction that reaches equilibrium at about 50% fructose. This intracellular catalyst is commercially produced from GM *Streptomyces.* As it is an expensive catalyst, they are normally immobilized by being entrapped within the producer cell cross-linked with glutaraldehyde. Alternatively, the enzyme may be attached to a solid support of cellulose fibers or similar materials [57]. The catalyst is used industrially to convert glucose to fructose in the manufacture of high-fructose corn syrup. Since fructose tastes almost twice as sweet as sucrose, fructose syrups produced from this catalyst are used to sweeten foods and lemonade and cola beverages [58].

3.1.4. Cellulases

Cellulase is the common name of any of several enzymes that catalyze the decomposition of cellulose and some related polysaccharides; like the hydrolysis of the 1,4-β-D-glycosidic linkages in cellulose, hemicellulose, lichenin, and cereal beta-D-glucans. Cellulases break down the cellulose molecule into simple sugars such as β-glucose, or shorter polysaccharides and oligosaccharides [59]. Cellulose hydrolysis is mediated by a combination of three main types of cellulases: endoglucanases, exoglucanases and β-glucosidase. Endoglucanase randomly cleaves β-1,4-bonds of cellulose chains to generate glucose, soluble cellodextrins or insoluble cellulose fragment. However, some endoglucanases can hydrolyze crystalline cellulose and generate the major products as cellobiose or longer cellodextrins. Exoglucanases act on the reducing or nonreducing ends of cellulose polysaccharide chains, liberating either cellobiose or glucose as major products The cellobiose released is then hydrolyzed to glucose by the β-glucosidases and slowly the substrate is converted to glucose [60, 61].

Cellulase is produced mainly by symbiotic bacteria in the ruminating chambers of herbivores that allow them to digest the cellulose from their vegetable diet. Cellulases are also produced by a few other types of organisms such as some termites. However, commercial cellulose is produced mostly from *Trichoderma reesei* fungi, *cellumonas* and *cytophaga* bacteria and from some actinomycetes [62]. Cellulase is commercially used for processing coffee beans during producing coffee from them [63].

3.1.5. β-Glucanase

Glucanases are enzymes that break down a glucan, a polysaccharide made of several glucose sub-units. 1,3-1,4-β-Glucans are polysaccharides particularly abundant in the endosperm cell walls of commercially valuable cereals such as barley, rye, sorghum, oats and wheat [64]. These polysaccharides are made up of up to 1,200 β-D-glucosyl residues linked through β-1,3 and β-1,4 glycosyl bonds. The proportions of β-1,3 and β-1,4-linkages vary around 30 and 70% respectively in the length of the mixed-linked segments [65]. They can act both as endo and exo hydrolases. They are further classified into β- 1,3, β-1,6 and β- 1,4-glucanases on the basis of the type of glycosidic bonds cleaved by them [66].

They are mostly produced from *Trichoderma* sp., however, several other microorganisms including *Bacillus* sp., *Trichoderma* sp., *Talaromyces emersonii*, and *Rhizobium* sp. are also commercially used to produce β-glucanases [67]. The β-glucans are not crystalline rather they dissolve in water to form viscous solutions. During the preparation of beer, a fungal β-glucanase can reduce the viscosity of the beer and improve the efficiency of filtration. In other circumstances β-glucan is important, for example, oat β-glucan is associated with lowering cholesterol and consequently, the food producer would want to preserve the β-glucan content of oatmeal [67].

3.1.6. Pectinase

Pectinase catalyze the breakdown process of pectin (a polysaccharide) found in plant cell walls of several fruits which include pectolyase esterase, pectozyme and polygalacturonase, a pectin depolymerizing enzymes [68]. Pectin comprises chains of α-1,4-linked galacturonic acid units. In high methoxy pectins (polymethylgalacturonic acid), many of the residues are esterified at the carboxyl groups with methanol. When few or no carboxyl groups are esterified, the result is pectic or polygalaturonic acid. There are basically two types of pectic enzymes, pectin and pectin. Pectinses are secreted by many fungi such as *Aspergillus niger* and is used in several food applications, in particular, fruit juice and vegetable processing.

3.2. Protein Hydrolysis Enzymes

3.2.1. Proteases

Proteases are enzymes of metabolic as well as commercial importance, that catalyze the hydrolysis of the peptide and isopeptide bonds that join amino acids within proteins, on top of that, the breakdown process of protein is known as proteolysis. Based on their catalytic mechanism, they are divided into two types: exopeptidases and endopeptidases. Exopeptidases remove the terminal amino acids from the protein chain such as aminopeptidases, carboxypeptidase, etc. and endopeptidases attack an internal peptide bond, as in trypsin, chymotrypsin, pepsin, elastase and papain, etc. [69, 70]. Among the proteases, trypsins are highly specific in nature and will hydrolyze only those peptide bonds preceded by a lysine or arginine residue. Other proteases are far less specific and will hydrolyze a broad range of peptide linkages.

Proteases produced from microbial sources are currently the most widely used commercial protease because their desired characteristics are more suitable in biotechnological applications than those from plant and animal sources [71]. Most commercial proteases are mainly produced from the bacterial genus of *Bacillus.* A wider variety of fungi than that of bacteria produce proteases. However, the common strain is *Aspergillus oryzae.* Moreover, some viral proteases are now also produced to be used in some biomedical applications [72]. Based on the functional group present at the active site, proteases are further classified into four prominent groups, i.e., serine proteases, aspartic proteases, cysteine proteases, and metalloproteases [73, 74]. There are a few miscellaneous proteases which do not precisely fit into the standard classification, e.g., ATP-dependent proteases which require ATP for activity [41]. Proteases are also classified into different families based on their amino acid sequences [75].

3.2.2. Serine Proteases

Serine proteases are characterized by the presence of a serine residue at their active site. They are very common and found in all organisms ranging from viruses to humans, indeed the archetypal serine proteases are the animal enzymes trypsin and chymotrypsin. Based on their structural similarities, the serine proteases tag have been assigned to more than 20 families emphasizing the enormous diversity and separate evolutionary roots of these enzymes. They may be exopeptidases, oligopeptidase, and omega peptidase, while, the common catalysts are endopeptidases [76]. They are optimally active in the alkaline pH range from neutrality to pH 10.5 or higher and are often referred to as the alkaline proteases. The principal industrial microbial serine catalysts are derived from *Bacillus licheniformis* and alklaliphilic *Bacillus* species such as *B. clausii*. They have high pH optima, temperature optima around 60°C, and broad specificity. Whereas subtilisin enzymes are representatives of this group, similar enzymes are also produced by other bacteria such as *Thermus caldophilus*, *Desulfurococcus mucosus* and *Streptomyces* and by the genera *Aeromonas* and *Escherichia coli* [77].

The serine protease (trypsin) enzyme preparation is intended for use as a food additive, in the manufacture of partially or extensively hydrolyzed proteins for applications in food and beverages, protein-fortification, and for emulsification or flavor enhancement. However, the activity of the enzyme is terminated by high temperatures. Therefore no reaction products are formed during the production or storage of the enzyme treated foods [78].

3.2.3. Metalloproteases

The metalloproteases are another enormous collection of proteolytic enzymes with the unique feature of requiring a divalent metal ion for activity; usually zinc. They have a pH range of around neutrality, hence their alternative name neutral proteases arises from [41]. Microbial Thermolysin is one of the major commercial metalloproteases used to produce low calorie sweetener; aspartame. Thermolysin is also used with other proteases in producing protein hydrolysate, enhancing flavor peptides, accelerating the ripening of dry sausages, brewing low calorie beer and improving the filtration of beer [79]. Recently it has been published that they can be used to hydrolyze whey proteins in making cheese [80]. It is commercially produced from the thermophilic bacterium *Bacillus stearothermophilus*. Other well-known metalloproteases are collagenase and some snake venoms.

3.2.4. Cysteine or Thiol Proteases

Thiol proteases use a sulphydryl group at their active site which is provided by the amino acid cysteine. The pH range for these enzymes is rather broad generally with greater activity below neutrality [41]. Papain, derived from the papaya plant is a thiol protease that is commonly used in meat tenderizing and also in other food preparations. The microbial cystain proteases are produced from the species *Streptococcus*, thermophilic *Bacillus* and *Pyococcus sp.* The major cysteine protease, L-cysteine is obtained industrially by hydrolysis of poultry feathers or human hair. However, the synthetic route involves fermentation using a mutant of *E. coli* [81]. Cysteine is one of the 599 additives used to manufacture cigarettes. In food technology, they are used to improve clarity in beer and Marie–type biscuits [82].

3.2.5. Aspartic Proteases

Aspartic endopeptidases depend on the aspartic acid residues for their catalytic activity [41]. The acid or aspartic proteases are related to the human stomach enzyme pepsin and chymosin from the digestive tract of the young cow. Acid proteases have an optimum pH of around 2.5 to 3.5. An acid protease rennin is traditionally used to coagulate casein of milk in the preparation of cheese. However, the limited supply of natural calf rennin has inspired researchers to search for alternative sources of rennin such as microbial origin (known as rennet) and the production of recombinant bovine chymosin. Filamentous fungi are the major producers of microbial aspartic proteases, although a few more examples of bacterial source have been published [83]. *Rhizomucor miehei* and *Rhizomucor pusillus* are the most common industrial source of rennet [84, 85].

3.3. Lipid Hydrolysis Enzymes; Lipases

Simple fats or oils (lipids) are composed of fatty acid chains esterified to a glycerol backbone in the form of a triglyceride. Lipases (EC 3.1.1.3) catalyze both the hydrolysis and the synthesis of esters formed from glycerol and long-chain fatty acids. These reactions usually proceed with high regio- and enantioselectivity. The ability of lipase to hydrolyze ester linkages in lipids to obtain fatty acids and glycerol is called lipolysis. Generally, they release two of the fatty acids producing a monoglyceride and two free fatty acids. These free fatty acids produce more flavor than the triglyceride however they are reactive to oxygen leading to rancidity.

The two main categories in which lipase-catalyzed reactions may be classified are as follows [86]:

(i) Hydrolysis:
$RCOOR' + H_2O \Leftrightarrow RCOOH + R'OH$

(ii) Synthesis:
Reactions under this category can be further separated as:
(a) Esterification
$RCOOH + R'OH \Leftrightarrow RCOOR' + H_2O$

(b) Interesterification
$RCOOR' + R''COOR^* \Leftrightarrow RCOOR^* + R''COOR'$
(c) Alcoholysis
$RCOOR' + R''OH \Leftrightarrow RCOOR'' + R'OH$
(d) Acidolysis
$RCOOR' + R''COOH \Rightarrow R''COOR' + RCOOH$

The last three reactions are often grouped together into a single term, viz., transesterification. Several other types of lipase activities exist in nature, such as phospholipases, lipoxygenase and sphingomyelinases; however, these are usually treated separately from "conventional" lipases. Lipoxygenase is a metalloenzyme that catalyzes the reaction of oxygen with certain unsaturated fatty acids, common substrates being linoleic and linolenic acids. These enzymes catalyze the addition of oxygen to the double bonds in these molecules giving rise to hydroperoxides which is responsible for the bleaching effect of some flours.

Lipases played important roles in human practices as ancient as yogurt and cheese fermentation. In the chocolate industry, the free fatty acid contributed to the flavor of milk chocolate, caramels, toffees, and buttercreams. Biolipolysis is already being used for the production of fat-free meats and fish. Lipase helps to retard bread staling. Lipases are also used in soybean milk preparation, smoked crab processing and vegetable fermentation [87-89]. A number of lipases have been produced commercially, with the majority of them originating from microbial flora comprising bacteria, fungi and yeast [90].

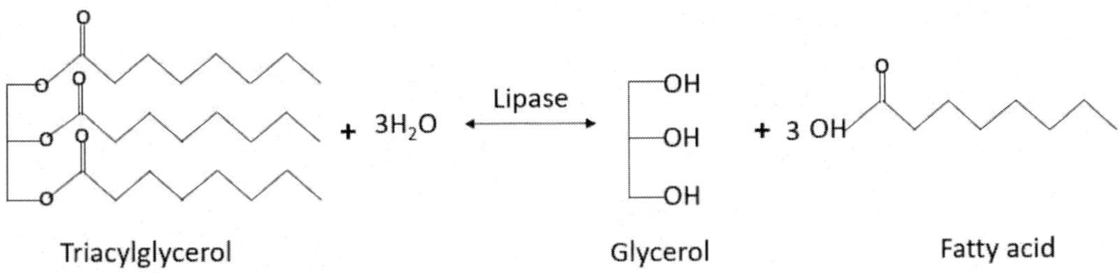

Figure 5. The catalytic action of lipases. A triglyceride can be hydrolyzed to form glycerol and fatty acids, or the reverse (synthesis) reaction can combine glycerol and fatty acids to form the triglyceride.

3.4. Tannase

Tannase or tannin acyl hydrolase catalyzes the breakdown processes of hydrolyzable tannins and gallic acid esters. It converts tannic acids into glucose gallic acid and some galloyl esters [92]. Gallic acid is used for the synthesis of food preservatives as antioxidants such as gallates and pyrogallol, propyl gallate [93]. Although tannase is present in plants, animals and microorganisms, it is mainly produced by fungi and few bacterial species such as *Bacillus sp.*, *Corinebacterium sp.*, *Lactobacillus sp.* and *Serratia sp.* [92]. The most prominent use of tannase in the food industry is as an antioxidant. It is commonly used in fruit juices to retard the oxidation of vitamin C [93]. After that, the major food applications of tannase is in the manufacturing of instantaneous tea [94, 95]. Tannase's applications in food and beverage industrial products are to remove the undesirable effects of tannins as well. Also, it can be used in the preparation of animal feed [96].

3.5. Transglutaminase

A transglutaminase (TGases) catalyzes an acyl transfer reaction between the γ-carboxyamide group of a peptide-bound glutaminyl residue (acyl donors) and a variety of primary amines (acyl acceptors), including the amino group of lysine. Bonds formed by transglutaminase exhibit high resistance to proteolytic degradation (proteolysis) [97]. The microbial transglutaminase (MTGase) are produced from the genetic manipulation using host microorganisms such as *Escherichia coli*, *Bacillus*, yeast, or *Aspergillus*, or *Corynebacterium glutamicum*. MTGase are used in various food processing industries. MTGase can produce restructured meat by binding together small pieces of meat [98, 99]. Both endogenous fish TGase and exogenous MTGase could improve the efficacy of fish raw materials by increasing crosslinking [100]. Milk casein, which does not gel, even when heated, is a very good substrate for TGases.

4. Sources of Food Microbial Catalysts

Sources of enzymes used in food technology are diverse. They can be isolated from a number of plants, different parts of various animal species and a vast range of microorganisms. However, there are two major reasons for the increase in demand for catalysts from microorganism sources. The first one is the scope of bulk and rapid growth of microorganisms leading to lower cost, lower space requirement during production and higher productivity of microbial catalysts compared to that of animals and plants. Secondly, modern genetic engineering can tailor the host strain for the optimum quality and yield [43, 101]. The microbial catalysts commercially used in food processing are mostly derived from two major microbial sources, bacteria of the genus *Bacillus* and various molds, in particular, *aspergilla* [44].

4.1. Bacterial Sources

Bacillus is one of the oldest and largest rod-shaped aerobic endospore-forming bacteria whose more than 12 genera including *Bacillus*, *Brevibacillus*, *Paenibacillus* and *Thermobacillus* has been recognized with the advent of molecular taxonomic methods [102, 103]. These bacteria are commonly terrestrial microbial origin; found in the soil and in decomposing plant materials, however, the marine environment has also been identified as a source for microbial organisms [104, 105]. Among the genus, *Bacillus anthracis* causes anthrax in animals and man while *Bacillus cereus* is responsible for mild food poisoning. Other than these two species of the genus, the majority of species are harmless to human [106, 107].

Bacillus bacteria have several benefits which can be exploited in the food biotechnology industry. Most of them are safe, non-pathogenic, robust and inexpensive. Amylases and proteases were among the first enzymes to be used extensively in the food and related industries to hydrolyze macromolecules to the smaller products such as sugars from starch or peptides from proteins. There is a variety of species like thermophilic (maximum growth temperature around 80°C), psychrophiles (minimum growth temperature around -5°C), alkaliphilic species able to grow at pH 10 or higher, acidiphiles etc. to fit for the extreme conditions. For example, thermostable amylolytic catalysts such as *Bacillus subtilis*, *Bacillus stearothermophilus*, *Bacillus licheniformis*, and *Bacillus amyloliquefaciens* are used for the production of crystalline dextrose, glucose, maltose, dextrose syrup, and maltodextrins in starch processing industries [108, 109]. Most commercial proteases, mainly neutral and alkaline, are produced by organisms belonging to the genus *Bacillus* [41]. Some of the lipase-producing bacterial genera include *Bacillus*, *Pseudomonas*, and *Burkholderia* [3].

Table 2. List of some common enzymes found from different species [3]

Source	Enzyme	Microorganism
Fungal	Amylase	*Aspergillus oryzae*
	Glucosidases	*A. flavus*
	Proteases	*A. niger*
	Pectinases	*A. niger*
	Glucose oxidase	*Penicillium notatum*
	Catalase	*A. niger*
Bacterial	Amylase Protease Penicillinase	*Bacillus subtilis*
Yeast	Invertase	*Saccharomyces cerevisiae*
	Lactase	*S. fragilis*

4.2. Fungal Sources

Fungi have a great importance in several food industries. Approximately 200 species of *Aspergillus* like *Aspergillus flavus, A. niger, A. oryzae* are applied in food biotechnology. However, *Aspergillus niger* is of great commercial interest due to its applications in several and diverse reactions [110]. Its cell mass is also used as a component in animal feed [77]. They used to secrete extracellular hydrolases such as amylases, glucanases, cellulases and pectinases.

The common black mold on bread is actually asexual *Aspergillus. Aspergillus oryzae*, however, has been used by the Japanese for many centuries in the production of various fermented materials. It is known as the *koji mold* which is used to convert the starch in rice into sugars suitable for subsequent fermentation. Amylase is secreted by the mold which converts the starch to sugar. Indeed, *A. oryzae* is used in the production of many different oriental foods such as soy sauce and miso, a form of fermented soybean used in soups. *Rhizomucor* is a cosmopolitan filamentous fungus found in soil and decaying fruit and vegetables. *Rhizomucor* sp. is often isolated from fermenting and composting organic matter and is very rarely pathogenic.

Glucoamylase is produced industrially by *Aspergillus, Rhizopus,* and *Endomyces* species. Amylolytic yeasts are potentially valuable in the use of starch substrates for direct conversion into single-cell protein. These are known to produce at least two types of amylases, α-amylase and glucoamylase. The nature and amount of extracellular amylases produced by the yeast species cause considerable cost savings. These include *Candida tsukubaensis, Filobasidium* capsuligenum, *Lipomyces kononenkoae, Saccharomycopsis bispora, Trichosporon pulMans* and *Lipomyces spe*are nonfermentative yeasts. *Saccharomycopsis fibuligera* has been used to ferment premalted 20% wheat starch. *Schwanniomyces affuvius* is an alcohol-producing yeast, 'O' and it may be used for single- cell protein and industrial alcohol production. The amylases secreted by *Schwunniomyces castellii* have been used in the production of low-carbohydrate beer.

Fungi produce a wider variety of proteases than do bacteria. For example, *Aspergillus oryzae* produces acid, neutral, and alkaline proteases [41]. Other major lipase producing fungi are *Mucor, Candida, Penicillium, Rhizopus, Geotrichum, Rhizomucor, Aspergillus, Humicola,* and *Rhizopus* [111].

5. Potential Applications in Food Biotechnology

In the twentieth century, microbial catalysts have become the most important commercial enzymes used in the food industry. Modern biotechnology is consistently developing the methods for optimizing microorganisms for the production of microbial catalysts. The major concern about these methods was whether they were safe for food use or not. In this respect, it is noticeable that the genetic modification is applied directly to the production of microorganism which is the source of the food catalysts; however, no modification is practiced onto the catalysts produced from them [7]. The catalysts produced through biotechnology are identical to those found in nature.

The microbial catalysts used in the food and related industries are observed to be employed mostly for two purposes. The most commercial use of them is in the starch processing industries and the other important use is to improve or modify the flavor or texture of foods.

5.1. Starch Processing Industries

Starch is a polymer of glucose linked to one another through the oxygen, known as the glycosidic bond. This glycosidic bond is stable at high pH and hydrolyzes at low pH. At the end of the polymeric chain, an aldehyde group is present, which is called the reducing end. Two types of glucose polymers are present in starch: (i) amylose and (ii) amylopectin. Amylose is a linear polymer consisting of up to 6000 glucose units with α-1-4 glycosidic bonds. The number of glucose residues in starch varies with the origin. The average amylose content in starches typically varies between 20–25%. Amylopectin consists of short α-1-4 linked linear chains of 10–60 glucose units and α-1-6 linked side chains with 15–45 glucose units. The average number of branching points in amylopectin is 5% which varies with the botanical origin. The complete amylopectin molecule contains on average about 2000000 glucose units and it is the largest molecules in nature [10]. Amylopectin is soluble in water, however, amylase and the starch granule are insoluble in cold water [112]. Of the common starch sources, corn (maize) starch is the most important. At first, the raw material for the preparation of the starch was wheat. Currently main starch sources are, maize (in America) – 70%, potatoes (in Europe) – 12%, wheat - 8%, tapioca - 9%, rice, sorghum and other - 1% [113]. Around 5% of the annual world corn production is used for the manufacture of starch.

The starch processing industry traditionally used hydrochloric acid to hydrolyze starch into dextrins and simple sugars [10]. However, the use of amylases in starch hydrolysis dates back to the 9th century AD when malt was used to convert arrowroot starch to sweetener and in 1833, precipitated malt extract was first observed to produce a sugar [114]. The conversion of starch to sugars has undergone a very impressive growth from the very beginning. The first major technological breakthrough occurred in 1946 when Dale and Langois patented the use of commercially available enzymes and hydrolyzed starch in the manufacture of commercial corn syrups of unique properties [115]. Thus, enzymes established a more controlled industrial hydrolysis process of the starch. There are basically four groups of starch-converting enzymes: (i) endoamylases; (ii) exoamylases; (iii) debranching enzymes; and (iv) transferases, however, among them the amylases are considered as the most important [112]. The enzymatic conversion of all starch includes gelatinization, which involves the dissolution of starch granules, thereby forming a viscous suspension; liquefaction, which involves partial hydrolysis and loss in viscosity; and saccharification, involving the production of glucose and maltose via further hydrolysis [109].

5.1.1. Starch Liquefaction and Production of Maltodextrins

The starch containing grains are milled to release the starch from the grain. The milled grains are then mixed with water and alkali to raise the pH to 6-6.5 and calcium ions (about 50 ppm). Starch slurry about 30% by weight is prepared and temperature stable α-amylase of *B. licheniformis* added prior to gelatinization obtained by heat treatment. The amylase is added prior to the heat treatment to initiate hydrolysis before degradation of the enzyme at the high temperature. The typical gelatinization temperature range is 62 to 70ºC. However, continuous cooking may employ relatively low temperatures (105ºC) for 5-7 minute, medium heat (118-121ºC for 5-7 min) or ultrahigh temperature (138-149ºC) for seconds. The higher temperatures have the advantages of improved starch gelatinization and sterility is almost assured. Pressure batch cooking involves the addition of enzyme, and heating by direct steam injection to about 150ºC for a brief period followed by flash cooling to 85-95ºC, followed by a second addition of enzyme to complete the starch dextrinization [4].

Using traditional α-amylases, the pH has to be adjusted to an undesirable high level and calcium must be added to stabilize the enzyme. Recently developed α-amylases have offered optimized properties, such as enhanced thermal stability, acid tolerance and ability to function without the addition of calcium [116-118]. There has been interest in developing amylases that liquefy raw starch granules thus avoiding the energy-intensive gelatinization process. However, effective granule-degrading enzymes have yet to be identified. If the dextrin obtained is allowed to continue further hydrolysis, then maltodextrins (DE 5-15) are obtained. These powders have high viscosity, low hygroscopicity, bland flavor and low sweetness. They are commonly included as an inexpensive additive to thicken food products and in ice cream to retard crystallization. More recently, maltodextrin of low DE (largely maltopentaose and maltohexaose) are included as extenders in various foods and as fat replacements because of their creamy texture [119].

5.1.2. Maltose and Related Syrups Prepared by Saccharification

Further hydrolysis of the maltodextrins using various microbial catalysts produce various maltose/glucose syrups of different sweetness.

Table 3. Sweetness of various sugar syrups at 15% (weight/volume) and 15ºC

Sugar	Relative sweetness
Sucrose	100
Glucose	75
HFCS (42%)	100
Maltose	30-50
Lactose	20

5.1.3. Glucose Syrups

Glucose syrups are prepared using glucoamylase that hydrolyzes consecutive 1,4-a bonds and the 1,6-o-bonds in dextrins to yield P, D-glucose. Sometimes, a specific debranching enzyme is used to speed up the reaction and improve the yield of glucose [10]. This fungal enzyme has lower pH and temperature optima than the bacterial liquefying enzymes. The inefficient hydrolysis of α-1,6-bonds by the glucoamlyse is aided by the addition of pullulanase. At the high sugar levels used, transglycosylation reactions lead to the formation of undesirable maltose and isomaltose. However, under optimal conditions, the maltodextrins can be converted to 97 to 99% glucose. If the syrup is destined for isomerization, glucose isomerase is used [112]. Engineering efforts have also been undertaken to develop improved versions of the enzymes used later in the process (i.e., glucoamylase and glucose isomerase) [120, 121].

5.1.4. Maltose Syrups

Maltose syrups are produced using a combination of enzymes to obtain desired ratios of glucose, maltose and higher oligosaccharides in the syrups. If the fungal enzymes are used then a reduction in pH is required [114]. Maltose syrups are of three principal types such as high maltose syrups, very high maltose syrups and high conversion syrups. High maltose syrups are produced by hydrolyzing the liquefied starch using a combination of fungal α-amylase and maltogenic α-amylase. In the absence of a debranching enzyme, about 40% maltose is reached with about 1% glucose and the remainders are maltotriose and higher oligosaccharides. Very high maltose syrups (VHMS) are obtained by hydrolyzing the liquefied starch with maltogenic α-amylase and glucoamylase. In this way, 90% maltose syrups can be achieved. High conversion syrups are obtained by a careful balance of the ratio of glucoamylase to fungal α-amylase. High conversion syrups have much higher glucose concentrations, typically 35 to 40% and are obtained by limited hydrolysis of a lightly liquefied starch [10].

Sugar syrups are used in baking, soft drink and confectionery industries as well as in various dried food preparations such as sauces. They have low viscosity and mild sweetness so are useful as extenders and contributors to taste. The VHMS have a mild sweetness that is often more desirable in cakes and candies than sucrose. The brewing industry uses various syrups as adjuncts to the fermentable sugars (largely maltose) derived from malted barley. High maltose syrups are desirable because they mirror more closely the sugars derived from malted barley during mashing. Hydrogenation of maltose to form maltitol results in a low calorie sweetener that is readily metabolized [122].

5.1.5. High Fructose Corn Syrups (HFCS)

In the late 1960s, the commercialization of fructose syrups derived from corn starch hydrolysis started in Japan [123, 124]. Fructose is a very sweet tasting sugar that occurs in several natural fruits and in honey. The sweetness value is equal to or greater than that of sucrose depending on concentration, a 42% solution is of about the same sweetness as a sucrose solution, 55% sucrose is about 5% sweeter and 90% fructose is sweeter by some 20-30%. The demand for fructose based sweetener by the beverage and food industry has been the driving force behind the dramatic increase in global production capacity of HFCS [39, 125].

The isomerization of fructose from glucose required the enzyme glucose isomerase. They are resistant to thermal denaturation and will act at very high substrate concentrations, which have the additional benefit of substantially stabilizing the enzymes at higher operational temperatures. Glucose isomerases, being a cytoplasmic enzyme, are used in an immobilized form to provide rapid, efficient and economical conversion. The corn syrup is then converted to fructose in a batch process to make 42% fructose syrup. For many purposes, a 42% fructose syrup is perfectly satisfactory for use but it does not match the exacting criteria of the quality which soft drink manufacturers demand as a replacement for sucrose in acidic soft drinks. For use in the better colas, 55% fructose is required. This is produced by using vast chromatographic columns of zeolites or the calcium salts of cation exchange resins to adsorb and separate the fructose from the other components. High fructose corn syrups are classified according to the fructose content (i.e., 42%, 55%, 90%) [125, 126].

With the advancement of biotechnology, the enhancement of the thermostability of maltogenic amylase from *Thermus* sp. IM6501, of amylosucrase from *Neisseria polysaccharea* and of glucoamylase from *Aspergillus niger* have also been achieved [2]. Furthermore, the development of hyperthermostable glucose (xylose) isomerase from *Thermotoga neapolitana* shows activity at a relatively low temperature and pH [127]. The enzyme from the parent strain is highly active at 97◦C, but it retains only 10% of its activity at 60◦C. Therefore, there was an interest in selecting a catalyst able to operate efficiently at temperatures close to those currently used but at a lower pH. The mutant glucose isomerase 1F1 obtained by Sriprapundh et al. (2003) displayed a roughly 5-fold higher activity at 60◦C and pH 5.5 when compared with the parent *T. neapolitana* isomeraseor or even with the wild type isomerase [128].

5.2. Brewery

Beer brewing and winemaking are old technologies and archaeological evidence suggests that this technique was used in most emerging civilizations including ancient Egypt and Mesopotamia. In simple terms, beer brewing involves malting the barley in a malt house followed by the preparation and fermentation of the wort in the brewery, while wine making requires the extraction of juice from grapes and subsequent fermentation of the juice by yeast. Microbial catalyst plays a central role in both of these processes.

5.2.1. Beer Brewing

Traditionally beer is made from a hot water infusion of milled malted barley together with some other form of starch (so-called adjuncts) such as maize (corn), rice or barley. After mashing which is the process by which the malt enzymes convert the starch in the cereals to maltose and other fermentable sugars, the sweet wort is recovered from the mash through a mash filter or lauter tun. The wort is then boiled in the presence of hops, cooled and fermented. The green beer fresh from the fermentation must be aged for flavor maturation and yeast removal before being packaged for trade.

During the mashing process, the enzymes of the malted barley (α- and β-amylases) convert the starch into fermentable sugar (glucose and predominantly maltose) for consumption by the yeast. The uses of high levels of adjunct and un-malted or low quality barley which have low amylolytic activity will result in low extract and an unacceptably high level of non-fermentable dextrin will remain in the beer. A second problem associated with high levels of adjunct is insufficient proteolytic hydrolysis during mashing giving rise to nitrogen starvation of the yeast in the fermentation. Yeast is weakly proteolytic and unable to hydrolyze barley proteins to provide its amino acid requirements [129]. The polysaccharides, β-glucan (70-75%) and arabinoxylan, (20-23%) originate from endosperm cell wall of un-malted or poor hydrolysis barley, which are responsible of the wort viscosity [130]. These form gels during the brewing process and leads to poor filtration of the wort, slow run-off times, low extract and yields and development of haze in the final product. To overcome these problems, microbial α-amylase, neutral protease and β-glucanase/pentosanase are either added at a malt equivalent blend at the mashing in stage or separately as required. Bacterial α-amylases added to the mash permits simpler starch liquefaction and shorter process times with an overall increase in productivity. Addition of a neutral protease to the mash will alleviate the protein hydrolyse problem. Addition of microbial β-glucanase/pentosanase preparations will enhance hydrolysis of these cell wall polysaccharides and greatly assist wort separation and beer filtration by reducing viscosity [131-133]. Traditional maturation of beer at low temperature requires long periods (several months) but this has now been reduced to weeks or even days by industrial microbial catalyst. One of the key maturation reactions is the removal of a yeast metabolite; α-acetolactate. α-acetolactate is excreted by yeast during growth but being unstable chemically undergoes decarboxylate reaction to form diacetyl in the fermenting wort. The diacetyl is rapidly reduced by the yeast to acetoin and finally is converted to butanediol. Acetoin and butanediol are essentially tasteless compared to diacetyl and has buttery flavor and aroma. By adding the enzyme acetolactate decarboxylase at the beginning of the primary fermentation, α-acetolactate is converted to acetoin and its accumulation is prevented. This greatly reduces maturation times and the brewery enjoys greater fermentation and maturation capacity.

There are no sources in the current document.

5.2.2. Light and Low Carbohydrate Beers

Low carbohydrate beers, commonly referred to as “low carb” beers, can be defined as having 30% fewer calories than normal beers. Most light beers are made from highly fermentable wort that leave very little carbohydrate in the finished beer. Along mash, the addition of amylases, particularly, glucoamylase with its debranching activity and high adjunct ratios are some of the common methods used to produce light beer. The low carb beers are produced from light beer by adjusting the amounts of grains and sources of carbohydrate, the mashing temperature and conversion. The rest is to ensure that the complex carbohydrates are broken down into sugars that the yeast will metabolize to alcohol, and this is supplemented with glucoamylase. However, to get to the 30% lower calorie count is generally necessary to dilute the beer with water but by starting with very high alcohol content, the final alcohol concentration is maintained.

5.2.3. Wine Production

The first step in winemaking involves crushing grapes and subsequent extraction of the juice for fermentation. This is a biotechnological process in which both yeast cells and enzymes play a key role. Three main exogenous enzymes used in wine production are pectinases, b-glucanases and hemicellulases [132, 133]. During mashing and extraction of the juice the grape’s own enzymes assist the degrading the protopectin and pectin. Commercial pectinase from Aspergillus, which contained varying amounts of pectin esterase, polygalacturonase, pectin lyase and small amounts of hemicellulase are widely used. Addition of pectinase degrades the pectin complex, promotes the release of juice and various flavoring compounds, improves juice extraction, reduces the clarification time and increases the terpene content of wine [132, 133]. But the pectin active enzymes have no effect on the complex carbohydrates of the grape cell wall, particularly those found in red grapes. Significant and reproducible improvements in grape pressability, settling rate and total juice yield were achieved using a combination of macerating enzymes compared to using pectinase alone [134]. Therefore a correct balance of exogenous pectinolytic, cellulolytic and hemicellulolytic enzymes have been developed to give efficient maceration of black grape skins and help to optimize the extraction of valuable tannins, anthocyanins and aroma compounds. The wines produced from red and white grapes with such enzyme mixture also show better performance. In fact, the enzyme technology offers enormous benefits to the wine industry, but new enzymes with better properties are expected to emerge and provide further benefits to both wine producers and consumers.

Table 4. Microbial catalyst in brewery and wine biotechnology (Collected from [131])

Enzyme/micro-organism	Function	Application
β-Glucanase/glucanolytic yeast	Hydrolysis of β-1,3, and β -1,4 glucan; reducing the viscosity and releasing reducing sugars during primary fermentation	Improvement in primary fermentation, filtration and quality of beer
Pectin esterase	De-esterification and gelling of pectins	Improvement in the clarification of cider
Macerating enzymes (cellulases, hemicellulases and pectinases)	Hydrolysis of plant cell wall polysaccharides	Improvement in skin maceration and colour extraction of grapes; quality, stability, filtration and clarification of wines
β-Glucosidase	Modification of aromatic residues	Improvement in the aroma of wines

5.3. Extraction and Clarification of Fruit and Vegetable Juices

Extraction, clarification and stabilization are the three main unit process involve in the production of fruit and vegetable juices. Microbial catalyst has been used in the fruit juice manufacturing business for more than 65 years. At the beginning of juices production, juice was produced without enzyme which faced many difficulties such as low yield, high viscosity hampering the filtering etc. [129]. Pectin is an important gelling molecule in jams and other semisolid fruit based foods but causes difficulties with haze formation and viscosity in fruit juices. The development of industrially suitable pectinases with other enzymes such as cellulases and hemicellulases from food-grade micro-organisms (*Aspergillus niger* and *Trichoderma sp.*) helped to overcome these difficulties [135]. Currently, a combination of pectinases (pectin lyase, pectin methylesterase, endo and exo-polygalacturonases, pectin acetylesterase, rhamnogalacturonase, endo- and exo-arabinases), cellulases (endoglucanases, exoglucanases and cellobiases) and hemicellulases (endo- and exo-xylanases, galactanases, xyloglucanases and mannanases) collectively called macerating enzymes are used in the extraction and clarification of fruit and vegetable juices [131]. In addition, a-amylase and amyloglucosidase, active at acidic pH, were used to process starch containing fruits, especially apples harvested during the early stages in order to prevent haze formation [131]. Sparkling clear juices (apple pear and grape) and cloudy juices (apple, citrus and tomato) are the two basic types of fruit juices.

5.3.1. Sparkling Juices

During the extraction of juice from fruits such as apples and pears, the whole fruit is washed, sorted and crushed to pulp mash. The crushed fruits are generally stored in a holding tank for a short period of time (15-20 minutes) so that pectinases enzyme inhibitors (polyphenols) are oxidized by endogenous polyphenol oxidases. Macerating enzymes are then added to aid the extraction of the juice from the fruits pulp about 15 min to 2 hours depending on the process and the fruits variety [132]. The macerating enzymes enhances yield percentage and process performance without additional capital investment. It is very crucial to quantity enzymes and treatment time for good quality juices, if the pulp is exposed to excess macerating enzymes it becomes soggy and difficult to press. The enzymes are generally used in two steps: (1) after crushing, to macerate the fruit pulp either to partial or complete liquefaction, which not only increases the juice yield and reduces the processing time, but also improves the extraction of valuable fruit components, and (2) after the juice extraction, whereby pectinases are used for its clarification, thereby lowering the viscosity of fruit juice prior to concentration and increasing the filtration rate and stability of the final product [131]. If a cloudy juice is required the juice is pasteurized immediately after pressing and packaged. The clear juice requires a clarification stage in which the juice is centrifuged to remove any large particulate. Other sparkling juices such as cranberry, raspberry and blackberry are produced in very similar ways with minor modifications to the processing according to the structure of the fruit.

5.3.2. Cloudy Juices

The cloudy juices of citrus fruits such as grapefruit and orange have high nutritional value and are produced as cloudy rather than clear. The pectin of oranges is mainly esterase rather than methylated ones in apples that lead to high levels of polygalaturonic acid (demethylated pectin). This charged molecule forms insoluble calcium pectate in the presence of calcium ions and produced undesirable precipitation of the juice due to the formation of coagulated complexes. Heat treatment of juice can prevent the precipitation but it impairs the flavor. On the other hand, microbial catalyst polygalacturonase can depolymerize polygalacturonate before significant complexing with calcium can take place. Pectinases enzymes are also used in the extraction phase or pulp wash process of cloudy juice production to reduce viscosity and increase yields in much the same way as the enzymes are added in the initial phase of apple juice production [132].

The formation of calcium pectate complexes is important to improve the quality of some cooked or blanched vegetables, especially tinned vegetable. The texture and firmness of tinned tomatoes are greatly enhanced by adding calcium chloride. The gelation properties of the calcium pectate lead to firmer and plumper tomatoes in the tin. Owing to the gelation properties of calcium pectate, pure pectin methylesterase is incorporated into foods such as yogurts, ice cream and pastries. Pectin demethylation of these small fruit pieces and treatment with calcium firms up the calcium pectate and improves the texture and appearance of the fruit pieces after processing.

5.3.3. Production of Fruit Nectars and Purees

The production and preservation of fruit nectars and purees are of tremendous commercial importance to attract a wide range of consumers. Many tropical fruits are not easily pressable. They are either too acidic or too strongly flavored to be used to produce pleasant beverages without dilution, blending or both. Nevertheless, the juice from tropical fruits is delicious after dilution or blending with other fruit juices. Some good examples of these are apricot, peach, pear, plum, mango, guava, papaya and banana. Use of macerating enzymes not only improves the cloud stability, texture, and facilitates easy concentration of nectars and purees, but also decreases their viscosity rapidly [135]. Hence, a suitable combination of macerating enzymes is expected to be ideal for the production of fruit nectars and purees.

5.3.4. Infusion of Microbial Catalyst to Alter the Sensory Properties of Fruits and Vegetables

Enzyme infusion has the potential to alter the texture, flavor and other sensory properties of foods. Vacuum infusion of pectinases to ease the peeling of citrus fruits has been commercialized [131, 136]. The potential applications of pectinase infusion include, (1) production of freshly peeled citrus fruits and salads; (2) reduction of excessive bitterness in citrus peels; (3) restoration of the flavor lost during drying; and (3) improvement of the firmness of peaches and processed pickles [136-139]. The infusion of pectinases and b-glucosidases enhances the aroma and volatile characteristics of specific fruits and vegetables [140-143]. Thus, enzyme infusion to alter the sensory attributes of fruits, vegetables and other foods has immense potential in food biotechnology.

5.4. Baking Industry

Flour contains protein in the form of gluten which forms a large network and this gives the baking dough the elasticity and the crumb structure in bread after baking. Normal hard wheat flour is composed of about 82% starch, 13% protein, 3.5% fiber and 1.5% lipid with trace amounts of minerals whereas, soft wheat flours have less protein (about 8-10%) and correspondingly more starch [144]. Flour also contains about 4% non-starch polysaccharides in the form of pentosans. These macromolecules are responsible for absorbing about 20 to 25% of the water in the dough. The dough used in baking industry for different products consists of hard flour, water, yeast, salt, possibly sugar and fat in various proportions. As soon as the yeast is added it ferments low molecular sugars to produce the carbon dioxide that raises the dough.

Flour naturally contains variable amounts of β-amylase but virtually no α-amylase. β-amylase releases a small amount of fermentable sugar, which is often insufficient to promote the efficient raising of the dough. Traditionally, malted barley was used to supplement the flour; however, α-amylase supplementation in flour is the present trend in baking. α-amylase increases the rate of fermentation, reduces the viscosity and generates additional sugar in the dough which ultimately improves the volume, texture, taste, crust color and toasting qualities of the product [145, 146]. Recently, intermediate temperature stable (ITS) α-amylases have been used which become inactive much before the completion of the baking process and remove excessive dryness of bread [39]. Interesting improvement happens in the use of lipolytic enzymes such as phospholipases to substitute or supplement traditional emulsifiers which degrade polar wheat lipids to emulsifying lipids *in situ*.

All undesirable changes together are called staling for bread, and this causes a considerable economic loss in the baking industry as it reduces the shelf life of baked products. Several additives so far have been tried to reduce staling effect in bread, however, use of α-amylases and xylanases show that water-binding capacity and retention in the starch and hemicellulose fractions of the bread, being the substrates of α-amylases and xylanases, respectively, to be critical for maintaining softness and elasticity [147]. The recently determined three-dimensional structure of the widely applied amylase for antistaling (Novamyl™) is probably capable of degrading amylopectin to a degree that prevents re-crystallization after gelatinization, without completely degrading the amylopectin network which provides the bread with elasticity. This provided further insight into the mechanism of enzyme action [5,

148]. Gluten development in the dough can be modified by the judicial use of neutral protease. Supplementation of flour with protease will reduce the gluten content giving rise to softer doughs with reduced mixing time and improved flow characteristics. These qualities are particularly valuable for biscuits and cookies where crumb structure is not a desired feature. Doughs from particularly hard wheat may also benefit from protease supplementation since they would otherwise result in bread with course crumb structure and texture [41]. Oxidases like glucose oxidase have been recommended as replacements for the chemical oxidants such as bromates, azodicaronamide and ascorbic acid used before to strengthen the gluten networks in the dough.

Table 5. Enzymes used in baking industry (adapted from [149])

Enzyme	Effects
Fungal α-amylase	Maximizes the yeast fermentation giving even crumb structure and high loaf volume
Maltogenic α-amylase	Improves shelf life
Glucose oxidase	Oxidases free sulfhydryl groups in gluten to make the dough stronger and more elastic
Lipase	Improves dough by producing more uniform, smaller crumb structure and a smoother texture
Lipoxygenase	Bleaches and strengthens dough
Protease	Weakens gluten, provides plastic properties to dough for cookies and biscuits
Xylanase	Improves dough leading to easier handling and improved crumb structure

5.5. Enzymes in Dairy Industry

With growing urbanization, demand for processed dairy foods has increased considerably, in particular demand for different cheese varieties and low-lactose milk due to increasing intolerance of human beings to lactose in milk and other milk products. In order to improve the quality of milk and milk products, a number of different enzymes from microbial as well as from nonmicrobial sources have been used in dairy processing. The application of exogenous enzymes in the processing of milk has a long tradition going back some 8000 years and is based on animal proteases, in particular, calf rennin. The rennet enzyme has the largest single application of enzymes in dairy food processing. Recently, proteinases have widely used in the acceleration of cheese ripening and preparation of the dairy products. Lipases and β-galactosidases have important and growing applications in dairy food industry. On the other hand, glucose oxidase, superoxide dismutase, sulphydryl oxidase, etc. have limited applications [150]. Microbial rennin substitutes referred to as rennets were developed in 1960 owing to increase in demand of cheese and declining availability of calf rennin. At present, microbial rennet is used for one third of the entire cheese produced worldwide.

5.5.1. Cheese Making

At first, the fat and protein contents of the milk are made uniform by blending the milk. Often, the milk is centrifuged to modify fat content. Membrane filtration is carried out to alter protein content. The milk is then pasteurized at 72ºC for about 15 seconds and cooled to 30ºC before adding a starter culture of lactic acid bacteria for fermentation and rennet to promote milk protein clotting. Rennin is a mixture of chymosin and pepsin which has ideal properties for milk coagulation in the production of cheese with good flavor and texture. This is extracted from the gastric mucosa of the abomasums of young mammals (calves and lambs). Rennin acts on the milk protein in two stages, by enzymatic and by nonenzymatic action, resulting in coagulation of milk. In the enzymatic phase, the resultant milk becomes a gel due to the influence of calcium ions and the temperature used in the process [151]. Bovine milk proteins comprise caseins, a group of phosphoproteins assigned to four major families; αs1-, αs2-, β- and κ-casein. The milk proteins form casein micelles which are stabilized at their surface by κ-casein molecules. The limited hydrolysis of the κ-casein molecules by calf chymosin destroys the stability of the micelles by removing the exposed regions of the molecules and keeps the micelles dispersed. Once the change is removed, the micelles aggregate and coagulation of the milk ensues. Chymosin is particularly effective since it acts only on κ-casein and lactoglobulin and not on the other casein molecules to any extent. In detail, it hydrolyzes the phenylalanine (105)-methionine (106) bond to produce para κ-casein and macropeptides. This is referred to as the primary phase of milk clotting and it is enhanced by high temperature [152].

Many microorganisms are known to produce rennet like proteinases, which can substitute the calf rennet. Microorganisms like *Rhizomucor pusillus*, *R. miehei*, *Endothia parasitica*, *Aspergillus oryzae*, and *Irpex lactis* are used extensively for rennet production in cheese manufacture [152, 153]. Different strains of species of *Mucor* are often used for the production of microbial rennets. Whereas the best yields of the milk-clotting protease from *Rhizomucor pusillus* are obtained from semisolid cultures containing 50% wheat bran. During growth, lipase is secreted together with the protease. Therefore, the lipase activity has to be destroyed by reducing the pH, before the preparation can be used as cheese rennet. About 30 minutes after rennet addition to the milk, it is cut or stirred to promote syneresis. Heating also promotes the association of micelles and coagulation. When coagulation is complete, the curd is pressed and the whey is removed. Although the proteases produced from these fungi are different in their molecular properties from calf chymosin, the rennets from microbial sources are more proteolytic in nature in comparison to rennet from animal sources. One major drawback is the formation of bitter flavors in the non-ripened as well as the ripened cheese due to the high proteolytic activity and the formation of bitter peptides [154]. However, judicious application of the enzymes overcomes this problem.

Hence, attempts have been made to clone the gene for calf chymosin, and to express it in selected bacteria, yeasts, and molds. Because of calf chymosin is an ideal enzyme for manufacturing cheese from bovine milk, gene from calf chymosin was one of the first genes from mammalian enzymes that was cloned and expressed in microorganisms [155]. The first cloned chymosin gene expressed in yeast became available in the early 1990s and gained acceptance later that decade. It is now registered for food use in the USA, Europe and several other countries. Around 90% of the hard style cheeses produced in the UK is made with recombinant rennets [41]. No major differences have been detected between cheeses made with recombinant chymosins or natural enzymes, regarding cheese yield, texture, smell, taste and ripening. The recombinant chymosins are identical to calf rennet according to the report on biochemical and genetic evidence [156]. Such cheeses have been accepted for consumption even by vegetarians.

5.5.2. Flavor Development

Cheese ripening is a complex process mediated by biochemical and biophysical changes during which a bland curd is developed into a mature cheese with characteristic flavor, texture and aroma. Natural cheese ripening requires large storage space and controlled temperatures and is consequently expensive. Accelerating the ripening process can save large sums of money and vast research has been carried out to produce the amino acids more quickly so that the microorganisms can convert them to various flavor compounds. Cheese ripening is essentially an enzymatic process which can be accelerated by augmenting the activity of the key enzymes. This has the advantage of initiating more specific action for flavor development compared to the use of elevated temperatures that can result in accelerating undesirable nonspecific reactions, and consequently off-flavor development. Fresh curd is composed of casein, fats, some carbohydrate and minerals and has a very mild taste. The desirable cheeses are produced by the partial and gradual breakdown of carbohydrates, lipids and protein during ripening mediated by several agents, viz. (i) residual coagulants, (ii) starter bacteria and their enzymes, (iii) nonstarter bacteria and their enzymes, (iv) indigenous milk enzymes, especially proteinases, and (v) secondary inoculate with their enzymes. Enzymes from the milk and the starter culture develop flavor during the ripening stage. Lipases from the milk convert the milk fat into fatty acids with a strong flavor. Fermentation of the lactose by lactic acid bacteria enhances the flavor largely through limited proteolysis of the casein. Variations in the microflora contribute to the huge variety of cheeses available. The pathways leading to the formation of flavor compounds are largely unknown, and therefore the use of exogenous enzymes to accelerate ripening is mostly an empirical process [157]. So, although there is huge potential for the addition of industrial enzymes at this stage, controlling their activity is difficult and consequently, quality control is problematic and they have not been widely adopted. Nevertheless, enzyme modified cheeses have been developed as highly flavored food ingredients in which cheese flavor is required without the texture and composition of a true cheese.

5.6. Processing of Fats and Oils

Since 1980s immobilized lipases have been used in the fat and oil industries for the interesterification of triglycerides, for example, in the production of margarine. The high cost of immobilization of the catalyst is the main obstacle in this regard. A new process for immobilizing lipases based on the granulation of silica has dramatically lowered process costs to produce fats and oils with no content of *trans*-fatty acids [158]. Use of a highly selective microbial phospholipase in the removal of phospholipids in vegetable oils, which is called 'de-gumming', has enabled an energy and water saving benefit in the industries [159].

During storage, one of the most important changes that occur in the lipid fraction is the hydrolysis of triacylglycerols, catalyzed by lipase retaining non-esterified fatty acids, which are very important for the characteristic flavor of these products [160]. To prevent hydrolysis, specific immobilized or free lipases have recently been applied successfully. Buisman et al. used immobilized lipases from *C. antarctica* (CAL-B), *C. cylindracea AY*30, *Humicola lanuginosa*, *Pseudomonas* sp. and *Geotrichum candidum* for the esterification of functionalized phenols for the synthesis of lipophilic antioxidants to be used in sunflower oil [90, 161].

High quality olive oil is usually obtained from freshly picked and slightly immature fruit under cold pressing conditions. Though processing of fully ripened fruit at higher than ambient temperatures gives high yields of oil, it produces oil with high acidity, rancidity and poor aroma [132]. Hence, to develop an improved method for the extraction of high quality olive oil, a commercial enzyme Olivex, which is a pectinase from *Aspergillus aculeatus,* was first used

[162]. Soon after the first enzymatic extraction process, it is noticed that no single enzyme was adequate for the efficient maceration and extraction of oil from olives. A combination of enzymes, consisting of pectinases (from *Aspergillus*), cellulases and hemicellulases (from *Trichoderma*) was then applied and they performed better [163]. The use of a commercial enzyme Cytolase O later increased the oil yield on an average by 1–2 kg per ton of olives and the use of macerating enzymes increased the anti-oxidants in extra-virgin olive oil reducing the chance of rancidity [132].

5.7. Flavors and Aromas

Flavor and aroma of food is usually due to the presence of many volatile and nonvolatile components in it. The nonvolatile compounds contribute mainly to the taste and the volatile ones influence both taste and aroma [164]. A number of compounds are actually responsible for the aroma of the food products such as alcohols, aldehydes, esters, dicarbonyls, short to medium- chain free fatty acids, methyl ketones, lactones, phenolic compounds and sulphur compounds [165, 166]. For thousands of years, microorganisms have been used for the purpose of adding flavor to food and beverages. The use of microorganisms in the production of wine and beer in Egypt and Assyria dates back to 3500 B.C. [167]. More than 150 years ago, benzaldehyde was the first flavor compound identified [168] and after that vanillin was isolated, identified and synthesized [169]. From the very beginning, flavor compounds have been extracted from plant sources. After the elucidation of their structure, synthetic flavors were produced by chemical synthesis [164]. Another route for flavor synthesis is based on microbial biosynthesis or bioconversion and this method is equally popular like the plant cell culture method [169-171]. Though the flavoring of food has started from an ancient age, however, the first publication regarding microbial flavor came out in 1923 [172]. Microorganisms can synthesize flavors as secondary metabolites during fermentation on nutrients such as sugars and amino acids. This capability may be used in two different ways, one is *in situ* flavor generation, as an integral part of food or beverage production processes (i.e., cheese, yogurt, beer, wine) and another one is microbial cultures specifically designed to obtain aroma compounds to be isolated later on and used as additives in food manufacture [164].

With the development of beer, wine, cheese, soy sauce, and related fermented products, microbial processes have traditionally played an integral role in adding aromas in the food. Flavors produced using microorganisms currently compete with those from traditional plant sources. Nowadays, more than 100 commercial aroma chemicals are derived using biotechnology either through the screening for overproducers, the elucidation of metabolic pathways and precursors or through the application of conventional bioengineering [173]. Recombinant DNA technologies have also enhanced efficiency in the production of non-nutritive sweeteners such as aspartame and thaumatin [18]. Moreover, the market development has been particularly dynamic for flavor and aroma compounds [174]. Thousands of tons per year of non-volatile flavors, such as sweeteners, acidulants, and savory compounds, are produced by means of biotechnology, while bioprocesses for volatile flavors have emerged only recently [175]. A number of microbial catalysts (i.e., lipases, proteases, glucosidases) produce aroma-related compounds from precursor molecules with the advantages of providing higher stereoselectivity than chemical routes. Moreover, the products obtain the legal status of natural substances which has high demand in the market [164, 176].

The microbial lipases from *Mucor miehei* and *Rhizopus arrhizus* were more selective for long-chain acids and acetates, which is a valuable method for the production of flavor inducing esters [177]. Immobilized lipases from *M. miehei* and *C. antarctica* are used for synthesis of short-chain flavor inducing thioesters in solvent-free medium [178]. The addition of bacterial lipase to fermented sausages increases the quantity of saturated nonesterified fatty acids, *Pseudomonas* lipase in cod liver oil converts triacylglycerols into diacylglycerols and monoacylglycerols which further turn into non-esterified fatty acids and fatty acid propan-2-yl esters. *C. rugosa* lipase in cattle, sheep and goat milk fats increases the amounts of volatile branched-chain fatty acids to help fragrance development in dairy products [90, 179, 180].

Short oligopeptides play an important role in the sensory appreciation of food and different sequences can be used to recreate the four basic taste sensations (sweet, bitter, sour and salty) [181, 182]. The fungal protease catalyzed hydrolysis of whey protein does not taste bitter and is applied in milk, however, hydrolysis of proteins often results in the formation of hydrophobic bitter peptides, which limits the use of protein hydrolysates in food products. This bitterness can further be reduced by hydrolysis with exopeptidases [183, 184]. Protease treated crayfish was observed to obtain an increase in the concentration of flavoring agents such as benzaldehyde and pyrazines [185]. The use of enzyme preparations including proteases, as well as peptidases, lipases and esterases, to produce enzyme-modified flavor-enhanced cheese has been found to increase the flavor up to 30 fold compared to the flavor of natural cheese [186]. Glutamic acid in the form of monosodium glutamate (MSG) is an important "meat" flavor and the most widely used flavor enhancer derived from protein. Flavor enhancer glutamate is produced by fermentation of sugar sources such as molasses, sucrose or glucose using high-performance strains of *Corynebacterium glutamicum* and *Escherichia coli* [187].

Glucosidases can be used to free the glycosidic bond present in natural terpenols in grapes and thus transform glycosidically bound odorless terpenols to flavor-full volatile terpenes to increase the taste of wine. Commercially, b-glucosidases isolated from numerous sources including *Vitis, Saccharomyces, Oenococcus, Aspergillus* or *Candida* is now being used by wine companies [188]. A hydroxylase of monooxygenase type were observed to produce nootkatone, a grapefruit flavor compound [189]. The very popular flavoring agent vanillin has also been synthesized from vanillylamine using amine oxidase from *Aspergillus niger* and a continuous commercial production process using immobilized enzyme has also been reported [190].

Artificial sweetener aspartame consists of L-aspartic acid and the methyl ester of L-phenylalanine is approved by the FDA, USA. The L-configurations of the two amino acids are responsible for the sweet taste of it. The enzymatic synthesis of it using proteases may be applied, however, immobilized thermolysin from *Bacillus thermoprotyolyticus* are industrially used for the production of aspartame [191].

5.8. Food Preservation

Food is traditionally preserved using heat and sunlight to destroy the microbes present in them which may cause putrefaction, adulteration or simply wastage of food. The common preservative agents include chemicals like weak organic acid (acetic, lactic, benzoic and sorbic acid), hydrogen peroxide, chelators, small organic biomolecule and spices. However, some cell wall perturbing antimicrobials from microbial sources have recently been studied and many antifungal enzymes of microbial origin including those originating from the biocontrol fungus, *Trichoderma harzianum* [192, 193] have been used extensively in recent times. Individual enzyme activities have also been characterized and in some cases these enzymes have been purified and used successfully [194].

5.9. Food Colorants/Carotenoid Extraction

Carotenoids are the main group of food colorants available in nature being responsible for many plant colors from red to yellow [195]. Though many carotenoids are being commercially produced by chemical synthesis, however, several microorganisms, including fungi, bacteria, and algae, are able to produce carotenoids naturally [196]. Biotechnological synthesis of certain carotenoids, by either homologous or heterologous production, may become more and more attractive [197]. As for example, red yeast *Xanthophyllomyces dendrorhous*, which was formerly known as *Phaffia rhodozyma*, produces the carotenoid astaxanthin [198]. *X. dendrorhous* mutants with higher carotenoid production capacities have been obtained by chemical mutagenesis or by recombinant DNA technology [199, 200]. The use of pectinase and cellulase enzymes disrupts the cell wall of orange peel, sweet potato and carrot, and releases carotenoids in the chloroplasts and in cell fluids. These pigments remain in their natural state as being bonded with proteins which prevents pigment oxidation and also affects color stability [201, 202].

5.10. Few Other Common Fermented Food Products

5.10.1. Soy Sauce

Soy sauce production involves a two-step fermentation process that makes use of koji inoculants from *Aspergillus oryzae* in the initial phase followed by moromi inoculants in the second phase. The initial phase of the fermentation involves hydrolyzing soy proteins. The second phase of the fermentation process, referred to as moromi fermentation, involves the addition of brine solution to the koji. *Saccharomyces rouxii,* a salt-tolerant yeast, is the predominant micro-organism in this phase of the fermentation [203].

5.10.2. Fermented Pork Sausage (nham)

Nham is an indigenous Thai fermented pork sausage produced in Southeast Asia. It is prepared from ground pork, pork rinds, garlic, cooked rice, salt, chili, sugar, pepper and sodium nitrite. Fermentation of the product occurs during transportation from the manufacturer to the point of retail. Traditionally produced nham is considered having high risk. Nham is fermented either spontaneously or by using starter cultures. *Lactobacillus plantarum* BCC 9546 is one of the commercial Nham starter cultures available in the Thai market [204].

5.10.3. Fermented Fish Paste (Som Fug)

Som Fug is a traditional fermented minced fish cake in Southeast Asia, which consists of 84 percent (by weight) minced freshwater fish (mud carp, *Cirrhina microlepis*), 8 percent garlic, 4 percent water, 2 percent salt, 1 percent boiled rice, 0.1 percent sucrose and black pepper. It is considered a healthy and highly nutritious product and is an excellent source of protein. It is produced using a spontaneous microbial fermentation process for about 2–4 days at ambient temperature. The garlic-fermenting lactic acid bacteria associated with Som Fug fermentations belonged to *Lactobacillus pentosus* and *Lact. Plantarum* [205].

5.10.4. Alkaline-Fermented Beans

Dawadawa is an important condiment in West/Central African Savannah region, produced by alkaline fermentation of African fermented locust bean (Steinkraus, 1995). Similar fermented food products can be found throughout Africa with regional differences in the raw materials used as processing inputs or in post-processing operations. Foods produced by alkaline fermentation in other parts of the world include natto in Japan, thua noa in Thailand and kinema in India. They are mainly used to enhance or intensify meatiness in soups, sauces and other prepared dishes. The production of dawadawa involves spontaneous fermentation of the softened beans is subsequently allowed to take place over 2–4 days. Micro-organisms associated with the fermentation include *Bacillus subtilis*, *B. pumilus*, *B. licheniformis* and *Staphylococcus saprophyticus* [18].

5.10.5. Amasi

Amasi is the common word for fermented milk in South Africa. It tastes like cottage cheese or plain yogurt. Amasi is traditionally prepared by fermenting unpasteurized cow's milk commercially using *Lactococcus lactis* subspecies *lactis* and *L. lactis* subspecies *cremoris* [206].

5.10.6. Miso

Miso is a traditional Japanese seasoning produced by fermenting soybeans with salt and the fungus *Aspergillus oryzae*, known in Japanese as *kōjikin*, and sometimes rice, barley, or other ingredients. The result is a thick paste used for sauces and spreads, pickling vegetables or meats and soup as a Japanese culinary staple [207].

5.10.7. Tofu

Tofu, a typical soybean curd product, is prepared by the coagulation of soybean proteins with the addition of Ca^{2+} and Mg^{2+} and glucono-δ-lactone. The addition of MTGase enables the smooth texture of retorted tofu to be maintained for a long time [97].

5.10.8. Coffee and Tea Fermentation

Pectinase treatment accelerates tea fermentation and also destroys the foam forming property of instant tea powders by destroying pectins. They are also used in coffee fermentation to remove mucilaginous coat from coffee beans [208].

6. Recent Advancement in Food Biotechnology

The isolation of commercial and stable microbial catalyst suitable for extremely harsh industrial condition such as high temperature and pH is in demand. In this regard, one appropriate method for such operation is to look for novel microbial strains from extreme environments such as hydrothermal vents, salt and soda lakes, and brine pools [209, 210] and the approach was successfully used by a number of research groups to unearth a thermostable catalysts pullulanase from *Fervidobacterium penna vorans* [211]. Another invention involves finding out potential enzymes to use the nucleotide or amino acid sequence of the conserved domains in designing degenerated PCR primers to screen microbial genomes for the presence of genes putatively encoding the enzyme of interest. This approach found a novel thermostable isoamylase enzyme from two *Sulfolobus* species and *Rhodothermus marinus* [212]. The third method of commercial engineering of microbial catalyst is to find out what specific regions are of importance for a given property. Hybrids of two homologous enzymes can be generated such as a hybrid of *B. licheniformis* and *B. amyloliquefaciens* α-amylase. [213]. The introduction of disulfide bonds in the enzyme can also lead to improved stability. By altering amino acids prone to oxidation, such as methionine, tryptophane, cysteine, histidine, of tyrosine, by an amino acid less oxidative, one can increase the stability in the presence of bleach, peracids, or chloramines [214]. A more recent approach for the development of microbial catalysts is random protein mutagenesis coupled with high-throughput screening [215]. Because of the randomness of this method, a large collection of mutants needs to be screened to find those that are of interest and the beneficial mutations found by random mutagenesis were combined with the already known ones.

Genome sequence information along with advanced molecular techniques have established mechanisms of various host-defensive pathogen's counter-defensive strategies to design healthy and safe food by optimizing the effect of probiotic bacteria, the design of starter culture bacteria and functional properties. For example, nucleotide sequences of the genomes of many important food microbes are available now. *Saccharomyces cerevisiae* was the first food microbe for which a complete genome sequence was characterized [216].

Food biotechnology helps to produce fresher, better-tasting and healthier foods for human. The initial developments in food biotechnology were mostly on the genetic modification (GM) of microorganisms to enhance production of enzymes to be used in food processing and still this application is the most extensively used in food production. Biotechnologically improved foods may help to combat chronic diseases by providing more healthful compounds, including increased levels of antioxidants, vitamins and decreased amounts of unhealthy fats. These foods are being improved by modern biotechnology to combat certain food allergies and pathogens so that every people may be able to consume previously allergenic and non-safe foods safely. The food biotechnologists aim to produce and use microbial catalysts more effectively and in diversified ways while keeping in mind efforts to minimize the risk of microbial contamination. The immobilization of enzymes has also been a key supporting tool for rendering these proteins easily fit for industrial application. All these strategies either individually or in an integrated way have been used to improve existing processes by the food scientists.

CONCLUSION

The developments in molecular biology brought many changes in the production of microbial catalysts in recent years. Innovative steps such as cloning of genes in host microorganisms in large-scale industrial fermentation, tailoring properties such as food-processing temperature or pH of catalyst by modifying their amino acid sequences and developing well-characterized nonpathogenic and nontoxigenic microbial strains have been adopted by modern food biotechnology. Moreover, advances in food biotechnology revealed new methods like classical mutagenesis and recombinant DNA techniques. These methods helped to substantially reduce the toxigenic potential of few microorganisms such as *A. oryzae* or *A. niger* fungal species used in food industries. Genetically modified microbial host strains improved catalysts production, for example, it is now possible to produce specific proteases deficient bacterial and fungal host strains to obtain better target catalysts. The present trend of building knowledge of green technology for the food industries is undoubtedly expanding more than ever before. It can be concluded that the progress in genetic techniques have started a new era of using microbial catalysts in food biotechnology.

REFERENCES

[1] Campbell-Platt, G. *Food science and technology,* John Wiley & Sons, 2011.

[2] Fernandes, P. (2010). Enzymes in Food Processing: A Condensed Overview on Strategies for Better Biocatalysts. *Enz Res., 2010*: 1-19.

[3] Gurung, N., Ray, S., Bose, S., Rai, V. A. (2013). Broader View: Microbial Enzymes and Their Relevance in Industries, Medicine, and Beyond. *BioMed Res Intr., 2013*: 1-18.

[4] Cherry, J. R. and Fidantsef, A. L. (2003). Directed evolution of industrial enzymes: an update. *Cur Opin Biotechnol., 14*: 438-443.

[5] Kirk, O., Borchert, T. V. and Fuglsang, C. C. (2002). Industrial enzyme applications. *Cur Opin Biotechnol., 13*: 345-351.

[6] Nielsen, A. H., Hagemann, A., Svenstrup, B., et al. (1994). Angiotensin II Receptor Density in Bovine Ovarian Follicles Relates to Tissue Renin and Follicular Size. *Clin Exp Pharmacol Physiol., 21*: 463-469.

[7] Olempska-Beer, Z. S., Merker, R. I., Ditto, M. D., et al. (2006). Food-processing enzymes from recombinant microorganisms—a review. *Regulatory Toxicol Pharmacol.,* 45: 144-158.

[8] Payen, A. and Persoz, J. F. (2000). Memoir on diastase, the principal products of its reactions, and their applications to the industrial arts. In: *Annales de Chimie et de Physique, 1833*: 73-92.

[9] Underkofler, L., Barton, R. and Rennert, S. (1958). Production of microbial enzymes and their applications. *Appl microbiol., 6*: 212-221.

[10] Nigam, P. and Singh, D. (1995). Enzyme and microbial systems involved in starch processing. *Enz Microbial Technol., 17*: 770-778.

[11] Godfrey, T. and West, S. (1996). *Industrial enzymology,* Grove's Dictionaries Inc, Avon.

[12] Binod, P., Singhania, R., Soccol, C., et al. (2008). Industrial enzymes. In: *Advances in Fermentation Technology*. pp. 291-320.

[13] Berka, R. M. and Cherry, J. (2006). Enzyme biotechnology. In: *Basic Biotechnology*, (C. Ratledge and B. Kristiansen, Eds.), pp. 477-498.

[14] Ogawa, J. and Shimizu, S. (2002). Industrial microbial enzymes: their discovery by screening and use in large-scale production of useful chemicals in Japan. *Cur Opin Biotechnol., 13*: 367-375.

[15] Chandel, A. K., Rudravaram, R., Rao, L. V., et al. (2007). Industrial enzymes in bioindustrial sector development: an Indian perspective. *J Com Biotechnol., 13*: 283-291.
[16] Carrez, D. and Soetaert, W., (2005). Looking ahead in Europe: white biotech by 2025. *Ind Biotechnol., 1*: 95-101.
[17] Smith, A., Datta, S. P., Smith, G. H., Campbell, P. N., et al. (2000). *Oxford dictionary of biochemistry and molecular biology,* Oxford University Press (OUP).
[18] B.f.A. (2011). Development, Current Status and Options for Biotechnologies in Food Processing and in Food Safety in Developing Countries. In: *Food and Agriculture Organization of the United Nations*. Rome, Italy, pp. 244-246.
[19] Bud, R. (1994). *The uses of life: a history of biotechnology.* Cambridge University Press.
[20] Dalby, P. A. (2007). Engineering enzymes for biocatalysis. *Recent patents on biotechnology, 1*: 1-9.
[21] Tracewell, C. A. and Arnold, F. H. (2009). Directed enzyme evolution: climbing fitness peaks one amino acid at a time. *Cur Opin Chem Biol., 13*: 3-9.
[22] Sen, S., Venkata Dasu, V., et al. (2007). Developments in directed evolution for improving enzyme functions. *Appl Biochem Biotechnol., 143*: 212-223.
[23] Adamczak, M. and Krishna, S. H. (2004). Strategies for improving enzymes for efficient biocatalysis. *Food Technol Biotechnol., 42*: 251-264.
[24] Kim, Y. W., Choi, J. H., Kim, J. W., et al. (2003). Directed evolution of thermus maltogenic amylase toward enhanced thermal resistance. *Appl env microbiol., 69*: 4866-4874.
[25] Emond, S., André, I., Jaziri, K., et al. (2008). Combinatorial engineering to enhance thermostability of amylosucrase. *Protein Sci., 17*: 967-976.
[26] Wang, Y., Fuchs, E., da Silva, R., et al. (2006). Improvement of *Aspergillus niger* Glucoamylase Thermostability by Directed Evolution. *Starch - Stärke, 58*: 501-508.
[27] Kim, M. and Lei, X. (2008). Enhancing thermostability of Escherichia coli phytase AppA2 by error-prone PCR. *Appl Microbiol Biotechnol., 79*: 69-75.
[28] Miyazaki, K., Takenouchi, M., Kondo, H., et al. (2006). Thermal stabilization of *Bacillus subtilis* family-11 xylanase by directed evolution. *J Biol Chemis., 281*: 10236-10242.
[29] Pretorius, I. and Bauer, F. (2002). Meeting the consumer challenge through genetically customized wine-yeast strains. *Trends Biotechnol., 20*: 426-432.
[30] Flamm, E. L. (1991). How FDA approved chymosin: a case history. *Nature Biotechnol., 9*: 349-351.
[31] Pariza, M. and Johnson, E. (2001). Evaluating the Safety of Microbial Enzyme Preparations Used in Food Processing: Update for a New Century. *Regulatory Toxicol Pharmacol., 33*: 173-186.
[32] Jonas, D., Antignac, E., Antoine, J., et al. (1996). The safety assessment of novel foods: guidelines prepared by ILSI Europe Novel Food Task Force, Food and Chemical. *Toxicol., 34*: 931-940.
[33] I.F.B.C. (IFBC), (1990). Biotechnologies and Food: Assuring the safety of foods produced by genetic modification, regulatory toxicology and pharmacology. *12*: S1-S196.
[34] Vieceli, J., Müllegger, J. and Tehrani, A. (2006). Computer-assisted design of industrial enzymes: The resurgence of rational design and in silico mutagenesis. *Ind Biotechnol., 2*: 303-308.

[35] Webb, E. C. (1992). *Enzyme nomenclature 1992: recommendations of the Nomenclature Committee of the International Union of Biochemistry and Molecular Biology on the nomenclature and classification of enzymes*. Published for the International Union of Biochemistry and Molecular Biology by Academic Press,.

[36] Transferase, Encyclopædia Britannica, in, *Encyclopædia Britannica,* Inc., 2014.

[37] *EC2 Transferase Namenclature,* in, School of Biological & Chemical Sciences at Queen Mary, University of London. Nomenclature Committee of the International Union of Biochemistry and Molecular Biology (NC-IUBMB). 2014.

[38] Sarrouh, B., Santos, T., Miyoshi, A., et al. (2012). Up-to-date insight on industrial enzymes applications and global market. *J Bioprocess Biotech.*, S4:002. Doi:10.4172/2155-9821.S4-002.

[39] Gupta, R., Gigras, P., Mohapatra, H., et al. (2003). Microbial α-amylases: a biotechnological perspective. *Process Biochem.*, *38*: 1599-1616.

[40] Sidhu, G., Sharma, P., Chakrabarti, T., et al. (1997). Strain improvement for the production of a thermostable α-amylase. *Enz Microb Technol.*, *21*: 525-530.

[41] Rao, M., Tanksale, A., Ghatge, M., et al. (1998). Molecular and biotechnological aspects of microbial proteases. *Microbiol Mol Biol Rev.*, *62*: 597-635.

[42] Lonsane, B. and Ramesh, M. (1990). Production of bacterial thermostable a-Amylase by solid-state fermentation: A potential tool for achieving economy in enzyme production and starch hydrolysis. *Adv Appl Microbiol.*, 35: 1-56.

[43] Sundarram, A. and Murthy, T. (2014). α-Amylase Production and Applications: A Review. *J Appl Env Microbiol.*, *2*: 166-175.

[44] Mobini-Dehkordi, M. and Javan, F. (2012). Application of alpha-amylase in biotechnology. *J Biol.*, *1*: 39-50.

[45] Singh, S., Sharma, V., Soni, M., et al. (2011). Biotechnological applications of industrially important amylase enzyme. *Intr J Pharma Bio Sci.*, *2*: 486-496.

[46] Monteiro, P. and Magalhaes, P. (2010). Application of microbial α amylases in industry–A review. *Braz J Microbiol.*, *41*: 850-861.

[47] Belitz, H., Grosch, W. and Schieberle, P. (2009). *Food chemistry,* 4th revised and extended edn, in, Springer, Heidelberg,.

[48] Bush, D., Sticher, L., Van Huystee, R., et al. (1989). The calcium requirement for stability and enzymatic activity of two isoforms of barley aleurone alpha-amylase. *J Biol Chem.*, 264: 19392-19398.

[49] Payan, F. (2004). Structural basis for the inhibition of mammalian and insect α-amylases by plant protein inhibitors. *Biochim Biophys Acta.*, *1696*: 171-180.

[50] Dronzek, B. L., Hwang, P. and Bushuk, W. (1972). Scanning Electron Microscopy of Starch from Sprouted Wheat. *Cereal Chem.*, *49*: 232-239.

[51] Sivaramakrishnan, S., Gangadharan, D., Nampoothiri, K., et al. (2006). α-Amylases from microbial sources–an overview on recent developments. *Food Technol Biotechnol.*, 44: 173-184.

[52] Rejzek, M., Stevenson, C., Southard, A., et al. (2011). Chemical genetics and cereal starch metabolism: structural basis of the non-covalent and covalent inhibition of barley β-amylase. *Mol BioSystems*, 7: 718-730.

[53] Hii, S., Tan, J., Ling, T., et al. (2012). Pullulanase: role in starch hydrolysis and potential industrial applications. *Enz Res.*, *2012*: 1-14.

[54] Bender, H. and Wallenfels, K. (1966). Pullulanase (an amylopectin and glycogen debranching enzyme) from *Aerobacter aerogenes. Methods Enzymol.*, 8: 555-559.
[55] Douglas C. and Mitchinson, C. (1997). Enzymes involved in the processing of starch to sugars. *Trends Biotechnol., 15*: 349-352.
[56] Bhosale, S., Rao, M. and Deshpande, V. (1996). Molecular and industrial aspects of glucose isomerase, *Microbiol Rev., 60*: 280-300.
[57] Glucose isomerase, in, *GMO Compass,* 2014.
[58] Katz, A., Li, X., Carrell, H., et al. (2006). Locating active-site hydrogen atoms in D-xylose isomerase: Time-of-flight neutron diffraction. *Proceedings of the National Academy of Sciences, 103*: 8342-8347.
[59] Bayer, E., Chanzy, H., Lamed, R., et al. (1998). Shoham, Y. Cellulose, cellulases and cellulosomes, *Curr opin struct biol, 8*: 548-557.
[60] Zhang, X. and Zhang, Y. (2013). Cellulases: Characteristics, sources, production, and applications. *Bioprocessing Technologies in Biorefinery for Sustainable Production of Fuels, Chemicals, and Polymers,* 131-146.
[61] Yang, S., El-Ensashy, H., Thongchul, N. (eds.), (2013). *Bioprocessing technologies in biorefinery for sustainable production of fuels, chemicals, and polymers*, John Wiley & Sons.
[62] Ali, S. (2003). Biotechnological application of alkaline cellulase In food technology, *Ass. Univ Bull Env Res, 6*: 117-124.
[63] Kuhad, R., Gupta, R. and Singh, A. (2011). Microbial cellulases and their industrial applications. *Enz Res., 2011*:1-10.
[64] Stone, B. and Clarke, A. (1992). *Chemistry and Biology of (1, 3)-β-Glucans*. Victoria: La Trobe University Press.
[65] Planas, A. (2000). Bacterial 1,3-1,4-β-glucanases: structure, function and protein engineering. *Biochim Biophys Acta., 1543*: 361-382.
[66] Watanabe, H. and Tokuda, G. (2001). Animal cellulases. *Cel Mol Life Sci CMLS., 58*: 1167-1178.
[67] Celestino, K., Cunha, R. and Felix, C. (2006). Characterization of a β-glucanase produced by Rhizopus microsporus var. microsporus, and its potential for application in the brewing industry. *BMC Biochem., 7*: 23-31.
[68] Pectinase, in, *Wikipedia*, the free encyclopedia, 2014.
[69] Fox, J., Shannon, J. and Bjarnason, J. (1991). Proteinases and their inhibitors in biotechnology. Enzymes in biomass conversion. In: *ACS Symp. Ser,* pp. 62-79.
[70] Kalisz, H. M. (1988). Microbial proteinases. In: *Enzyme Studies,* Springer, pp. 1-65.
[71] Fullbrook, P., Godfrey, T. and Reichelt, J. (1996). *Industrial enzymology*. In: MacMillan Press, London.
[72] Rawlings, N. D. and Barrett, A. J. (1993). Evolutionary families of peptidases, *Biochem J., 290*: 205-218.
[73] García-Carreño, F. (1991). Proteases in food technology, *Biotechnol Education, 2*: 150-153.
[74] Hartley, B. (1960). Proteolytic enzymes. *Ann Rev Biochem., 29*: 45-72.
[75] Argos, P. A. (1987). Sensitive procedure to compare amino acid sequences. *J Mol Biol., 193*: 385-396.
[76] Barrett, A. J. (1994). *Proteolytic enzymes: serine and cysteine peptidases*. Academic Press.

[77] Soares, I., Távora, Z., Barcelos, R., Baroni, S. (2012). Microorganism-Produced Enzymes in the Food Industry. In: *Scientific, health, and social aspects of the food industry*. Valdez, B. (ed.). InTech.
[78] Srinivasan, J. R. (2011). Serine Protease (Trypsin) from Fusarium Oxysporum Expressed *in Fusarium Venenatum*, pp. 1-7.
[79] Boye, J. and Arcand, Y. (2012). *Green technologies in food production and processing*. Springer, 2012.
[80] Sorensen, T., Madkor, S. and Mims, S. (2003). *Hydrolysing whey proteins by a metalloprotease; food product especially cheese,* in, Google Patents.
[81] Cysteine, in, *Wikipedia*, the free encyclopedia, 2014.
[82] Whitehurst, R. and Van Oort, M. (2009). *Enzymes in food technology,* Wiley Blackwell.
[83] Yegin, S. and Fernandez-Lahore, M. (2013). Production of extracellular aspartic protease in submerged fermentation with Mucor mucedo DSM 809. *Afr J Biotechnol.*, 9: 6380-6386.
[84] Ismail, A., El-Aassar, S. and Abdel-Fattah, A. (1984). Production of milk-clotting and proteolytic enzymes by fungi. *Agr wastes,* 10: 95-102.
[85] Preetha, S. and Boopathy, R. (1994). Influence of culture conditions on the production of milk-clotting enzyme from Rhizomucor. *World J Microbiol Biotechnol.,* 10: 527-530.
[86] Gandhi, N. N. (1997). Applications of lipase. *J Amr Oil Chemists' Society,* 74: 621-634.
[87] Nielsen, T. (1985). Industrial application possibilities for lipase. *Fette, Seifen, Anstrichmittel*, 87: 15-19.
[88] Seitz, E. W. (1974). Industrial application of microbial lipases: A review. *J Amr Oil Chemists' Society,* 51: 12-16.
[89] Wiseman, A. (1975). Industrial practice with enzymes: Applications and sources of industrial enzymes. *Handbook Enz Biotechnol.*, pp. 252-259.
[90] Pandey, A., Benjamin, S., Soccol, C., (1999). The realm of microbial lipases in biotechnology. *Biotechnol Appl Biochem.*, 29: 119-131.
[91] Jaeger, K. and Reetz, M. (1998). Microbial lipases form versatile tools for biotechnology. *Trends Biotechnol.*, 16: 396-403.
[92] Aguilar-Zárate, P., Cruz-Hernández, M., Montañez, J., (2014). Bacterial tannases: Production, properties and applications. *Revista Mexicana de Ingeniería Química*, 13: 63-74.
[93] Srivastava, A. and Kar, R. (2009). Characterization and application of tannase produced by *Aspergillus niger* ITCC 6514.07 on pomegranate rind. *Braz J Microbiol.,* 40: 782-789.
[94] Belmares, R., Contreras-Esquivel, J., Rodriguez-Herrera, R., et al. (2004). Microbial production of tannase: an enzyme with potential use in food industry. *LWT-Food Sci Technol.*, 37: 857-864.
[95] Pometto, A., Shetty, K., Paliyath, G., et al. (2005). *Food Biotechnology,* CRC Press.
[96] Mueller-Harvey, I. (2001). Analysis of hydrolysable tannins. *Animal Feed Sci Technol.*, 91: 3-20.
[97] Yokoyama, K., Nio, N. and Kikuchi, Y. (2004). Properties and applications of microbial transglutaminase. *Appl Microbiol Biotechnol.,* 64: 447-454.
[98] Takeoka, G., Teranishi, R., Williams, P., et al. (1996). *Biotechnology for improved foods and flavors*. American Chemical Society.

[99] Kuraishi, C., Sakamoto, J. and Soeda, T. (1996). The usefulness of transglutaminase for food processing. In: *Biotechnology for improved foods and flavors*. Amr Chem Soc. pp. 29-38.

[100] Kumazawa, Y., Sakamoto, H., Kawajiri, H., et al. (1996). Determination of ε-(γ-glutamyl) lysine in several fish eggs and muscle proteins. *Fisheries sci.*, 62: 331-332.

[101] Hasan, F., Shah, A. and Hameed, A. (2006). Industrial applications of microbial lipases, *Enz Microbial Technol., 39*: 235-251.

[102] Konsoula, Z. and Liakopoulou-Kyriakides, M., (2007). Co-production of α-amylase and β-galactosidase by *Bacillus subtilis* in complex organic substrates. *Biores Technol., 98*: 150-157.

[103] Pandey, A., Nigam, P., Soccol, C., et al. (2000). Advances in microbial amylases. *Biotechnol Appl Biochem., 31*: 135-152.

[104] Shahidi, F. and Kamil, Y. (2001). Enzymes from fish and aquatic invertebrates and their application in the food industry. *Trends Food Sci Technol., 12*: 435-464.

[105] Rasmussen, R. and Morrissey, M. (2007). Marine biotechnology for production of food ingredients. *Adv food Nutr Res., 52*: 237-292.

[106] *Bacterial Food Poisoning,* in: Call Us, California Poison Control System, 2010.

[107] Anthrax, in, *Wikipedia*, the free encyclopedia, 2014.

[108] Asgher, M., Asad, M., Rahman, S., et al. (2007). A thermostable α-amylase from a moderately thermophilic *Bacillus subtilis* strain for starch processing. *J Food Eng., 79*: 950-955.

[109] Gomes, I., Gomes, J. and Steiner, W. (2003). Highly thermostable amylase and pullulanase of the extreme thermophilic eubacterium *Rhodothermus marinus*: Production and partial characterization. *Biores Technol., 90*: 207-214.

[110] Andersen, M., Nielsen, M. and Nielsen, J. (2008). Metabolic model integration of the bibliome, genome, metabolome and reactome of *Aspergillus niger*. *Mol Systems Biol., 4*: 178-190.

[111] Sharma, R., Chisti, Y. and Banerjee, U. (2001). Production, purification, characterization, and applications of lipases. *Biotechnol adv., 19*: 627-662.

[112] Van Der Maarel, M., Van Der Veen, B., Uitdehaag, J., et al. (2002). Properties and applications of starch-converting enzymes of the α-amylase family. *J Biotechnol., 94*: 137-155.

[113] Starch production, in, *Wikipedia*, the free encyclopedia, 2014.

[114] Muralikrishna, G. and Nirmala, M., (2005). Cereal α-amylases—An overview. *Carbohyd polymers, 60*: 163-173.

[115] *Enzyme,* in: U.S.P.a.T. Office (Ed.), 1940.

[116] Bisgaard-Frantzen, H., Svendsen, A., Norman, B., et al. (1999). Development of Industrially Important. ALPHA.-Amylases. *J Appl Glycosci., 46*: 199-206.

[117] Shaw, A., Bott, R. and Day, A. (1999). Protein engineering of α-amylase for low pH performance. *Curr Opin Biotechnol., 10*: 349-352.

[118] Declerck, N., Machius, M., Wiegand, G., (2000). Probing structural determinants specifying high thermostability in *Bacillus licheniformis* α-amylase. *J mol biol., 301*: 1041-1057.

[119] What Is Maltodextrin?, in, *WiseGeek*, 2014.

[120] Sauer, J., Sigurskjold, B., Christensen, U., (2000). Glucoamylase: structure/function relationships, and protein engineering. *Biochimica et Biophysica Acta (BBA)-Protein Struct Mol Enz., 1543*: 275-293.

[121] Hartley, B., Hanlon, N., Jackson, R., (2000). Rangarajan, M. Glucose isomerase: insights into protein engineering for increased thermostability. *Biochimica et Biophysica Acta-Protein Struct Mol Enzymol., 1543*: 294-335.

[122] *Enzyme Technology,* in, 2014.

[123] Yamanaka, K. (1966). [104] d-Xylose isomerase. *Meth enzymol.,* 9: 588-593.

[124] Chen, W. P. (1980). Glucose Isomerase (A Review). *Process Biochem., 15*: 30-35.

[125] *High fructose corn syrup,* in, Wikipedia, the free encyclopedia, 2014.

[126] Parker, K., Salas, M. and Nwosu, V. (2010). High fructose corn syrup: production, uses and public health concerns. *Biotechnol Mol Biol Rev., 5*: 71-78.

[127] Sriprapundh, D., Vieille, C., Zeikus, J., et al. (2003). Directed evolution of thermotoga neapolitana xylose isomerase: High activity on glucose at low temperature and low pH, *Protein Eng., 16*: 683-690.

[128] Rubin-Pitel, S., and Zhao, H. (2006). Recent advances in biocatalysis by directed enzyme evolution. *Combinatorial chemistry high throughput screening, 9*: 247-257.

[129] Linsmaier-Bednar, E. M., Uhlig, H. *Industrial enzymes and their applications,* John Wiley & Sons, 1998.

[130] Canales, A., Garza, R., Sierra, J., et al. (1988). The application of a beta-glucanase with additional side activities in brewing. *Technical quarterly-Master Brewers Association of the Americas* (USA).

[131] Bhat, M. (2000). Cellulases and related enzymes in biotechnology. *Biotechnol Adv., 18*: 355-383.

[132] Galante, Y., De Conti, A. and Monteverdi, R. (1998). Application of Trichoderma enzymes in the food and feed industries. *Trichoderma and Gliocladium, 2*: 327-342.

[133] Harman, G. E., Kubicek, C. P. (2002). *Trichoderma and Gliocladium: Enzymes, biological control and commercial applications,* CRC Press.

[134] Harbord, R., Simpson, C. and Wegstein, J. (1990). Winery scale evaluation of macerating enzymes in grape processing. *Wine industry J., 5*: 134-137.

[135] Grassin, C. and Fauquembergue, P. (1996). Fruit juices. *Ind enzymol.,* 2: 225-264.

[136] Baker, R. and Wicker, L. (1996). Current and potential applications of enzyme infusion in the food industry. *Trends Food Sci Technol., 7*: 279-284.

[137] Roe, B. and Bruemmer, J. (1977). Treatment requirements for debittering and fortifying grapefruit and stable storage of the product. In: *Proc Fla State Hortic Soc.,* pp. 180-182.

[138] Crocco, S. (1976). New way to modify flavor. *Food Eng., 48*: 6-8.

[139] Javeri, H., Toledo, R. and Wicker, L. (1991). Vacuum Infusion of Citrus Pectinmethylesterase and Calcium Effects on Firmness of Peaches. *J Food Sci., 56*: 739-742.

[140] Humpf, H. and Schreier, P. (1991). Bound aroma compounds from the fruit and the leaves of blackberry (Rubus laciniata L.). *J Agr Food Chem., 39*: 1830-1832.

[141] Krammer, G., Winterhalter, P., Schwab, M., et al. (1991). Glycosidically bound aroma compounds in the fruits of Prunus species: apricot (*P. armeniaca*, L.), peach (*P. persica*, L.), yellow plum (*P. domestica*, L. ssp. Syriaca). *J Agr Food Chem., 39*: 778-781.

[142] Marlatt, C., Ho, C. and Chien, M. (1992). Studies of aroma constituents bound as glycosides in tomato. *J Agr Food Chem., 40*: 249-252.

[143] Pabst, A., Barron, D., Etievant, P., et al. (1991). Studies on the enzymic hydrolysis of bound aroma constituents from raspberry fruit pulp. *J Agr Food Chem., 39*: 173-175.
[144] Chu, M. Wheat Flour. In, *Cooking for Engineers,* 2014.
[145] El-Fallal, A., Dobara, M., El-Sayed, A., Omar, N. (2012). *Starch and Microbial α-Amylases: From Concepts to Biotechnological Applications*.InTech Publication.
[146] Chang, C. F. (2012). *Carbohydrates-Comprehensive Studies on Glycobiology and Glycotechnology,* InTech Publication.
[147] Andreu, P., Collar, C. and Martínez-Anaya, M. (1999). Thermal properties of doughs formulated with enzymes and starters. *Eur Food Res Technol., 209*: 286-293.
[148] Dauter, Z., Dauter, M., Brzozowski, A., (1999). X-ray Structure of Novamyl, the Five-Domain “Maltogenic” α-Amylase from *Bacillus stearothermophilus*: Maltose and Acarbose Complexes at 1.7 Å Resolution. *Biochem., 38*: 8385-8392.
[149] *Enzymes at work.* In: Nozomyes, Denmark, 2013.
[150] Neelakantan, S., Mohanty, A. and Kaushik, J. (1999). Production and use of microbial enzymes for dairy processing. *Curr Sci.,* 77: 143-148.
[151] Boopathy, R. (1994). Enzyme technology in food and health industries. *Indian Food Ind., 13*: 22-22.
[152] Fox, P. (1993). Exogenous Enzymes in Dairy Technology—A Review. *J Food biochem., 17*: 173-199.
[153] Green, M. L. (1977). Milk coagulants. *J Dairy Res.,* 44: 159-188.
[154] Davies, F. L. and Law, B. A. (1984). *Advances in the microbiology and biochemistry of cheese and fermented milk,* Elsevier Applied Science Publishers Ltd.
[155] Nishimori, K., Shimizu, N., Kawaguchi, Y., et al. (1984). Expression of cloned calf prochymosin cDNA under control of the tryptophan promoter. *Gene,* 29: 41-49.
[156] Teuber, M. (1990). Production of chymosin (EC 3.4.23.4) by microorganisms and its use for cheesemaking, *Bull Intr Dairy Federation, 251*: 3-15.
[157] Fox, P, McSweeney, P., Cogan, T., et al. (2004). *Cheese: Chemistry, Physics and Microbiology: General Aspects,* Academic Press.
[158] Christensen, M., Andersen, L., Kirk, O., et al. (2001). Enzymatic interesterification of commodity oils and fats: approaching the tonnes scale. *Lipid Technol News, 7*: 33-37.
[159] Clausen, K. (2001). Enzymatic oil-degumming by a novel microbial phospholipase. *Eur J Lipid Sci Technol., 103*: 333-340.
[160] Heidt, M., Bornscheuer, U. and Schmid, R. (1996). Studies on the enantioselectivity in the lipase-catalyzed synthesis of monoacylglycerols from isopropylidene glycerol. *Biotechnol Techniques, 10*: 25-30.
[161] Buisman, G., Van Helteren, C., Kramer, G., et al. (1998). Enzymatic esterifications of functionalized phenols for the synthesis of lipophilic antioxidants. *Biotechnol Lett., 20*: 131-136.
[162] Fantozzi, P., Montedoro, G. and Petruccioli, G. (1977). Enzymatic preparation treatment of olive pastes submitted to single-pressing extraction. Effects exerted by cultivars, harvesting time and preservation. *Rivista Italiana delle Sostanze Grasse* (Italy).
[163] Galante, Y., Monteverdi, R., Inama, S., et al. (1993). New applications of enzymes in wine making and olive oil production. *Italian Biochem Soc Trans, 4*: 34-40.
[164] Longo, M. and Sanromán, M. (2006). Production of food aroma compounds: microbial and enzymatic methodologies. *Food Technol Biotechnol., 44*: 335-353.

[165] Urbach, G. (1997). The flavour of milk and dairy products: II. Cheese: contribution of volatile compounds. *Intr J Dairy Technol., 50*: 79-89.

[166] Gatfield, I. L. (1988). Production of flavor and aroma compounds by biotechnology. *Food Technol., 42*: 110-169.

[167] Hausler, A. and Munch, T. (1997). Microbial production of natural flavors, *ASM News-American Society for Microbiolog, 63*: 551-559.

[168] Wöhler, F. and Liebig, J. (1837). Notiz über die Bildung des Bittermandelöls. (Note about the formation of bitter almond oil). *Annalen der Pharmacie, 21*: 96-97.

[169] Krings, U. and Berger, R. (1998). Biotechnological production of flavours and fragrances, *Appl Microbiol Biotechnol., 49*: 1-8.

[170] Vandamme, E. and Soetaert, W. (2002). Bioflavours and fragrances via fermentation and biocatalysis. *J Chem Technol Biotechnol.,* 77: 1323-1332.

[171] Aguedo, M., Ly, M., Belo, I., et al. (2004). The use of enzymes and microorganisms for the production of aroma compounds from lipids. *Food Technol Biotechnol., 42*: 327-336.

[172] Omelianski, V. (1923). Aroma-producing microörganisms. *J Bacteriol., 8*: 393-419.

[173] Berger, R. G. (2009). Biotechnology of flavours—the next generation. *Biotechnol Lett., 31*: 1651-1659.

[174] Leuchtenberger, W., Huthmacher, K. and Drauz, K. (2005). Biotechnological production of amino acids and derivatives: current status and prospects. *Appl Microbiol Biotechnol., 69*: 1-8.

[175] Hagedorn, S. and Kaphammer, B. (1994). Microbial biocatalysis in the generation of flavor and fragrance chemicals. *Ann Rev Microbiol., 48*: 773-800.

[176] Kashi, F., Fooladi, J. and Bayat, M. (2008). The Use of Biotechnology for the Production of Flavor and Fragrance. *Biotechnol., 7*: 194-199.

[177] Langrand, G., Rondot, N., Triantaphylides, C., et al. (1990). Short chain flavour esters synthesis by microbial lipases. *Biotechnol Lett., 12*: 581-586.

[178] Cavaille-Lefebvre, D. and Combes, D. (1997). Lipase synthesis of short-chain flavour thioesters in solvent-free medium. *Biocatal Biotransformation, 15*: 265-279.

[179] Zuyi, L. and Ward, O. (1993). Stability of microbial lipase in alcoholysis of fish oil during repeated enzyme use. *Biotechnol Lett.,* 15: 393-398.

[180] Zalacain, I., Zapelena, M., Astiasaran, I., et al. (1995). Dry fermented sausages elaborated with lipase from *Candida cylindracea*: Comparison with traditional formulations. *Meat sci., 40*: 55-61.

[181] Nishimura, T. and Kato, H. (1988). Taste of free amino acids and peptides. *Food Rev Intr., 4*: 175-194.

[182] Maehashi, K., Matsuzaki, M., Yamamoto, Y., et al. (1999). Isolation of peptides from an enzymatic hydrolysate of food proteins and characterization of their taste properties. *Biosci Biotechnol Biochem., 63*: 555-559.

[183] Stevenson, D., Ofman, D. and Fenton, G. (1998). Protease-catalysed condensation–oligomerisation of hydrophobic peptides as a means of flavour modification. *J Mol Catalysis B: Enzymati, 5*: 39-44.

[184] Sawhill, J. W. *Peptide product,* in, Google Patents, 2004.

[185] Baek, H. and Cadwallader, K. (1996). Volatile compounds in flavor concentrates produced from crayfish-processing byproducts with and without protease treatment. *J Agr Food Chem., 44*: 3262-3267.

[186] Kilcawley, K., Wilkinson, M. and Fox, P. (1998). Enzyme-modified cheese. *Intr Dairy J., 8*: 1-10.

[187] Nandakumar, R., Yoshimune, K., Wakayama, M., et al. (2003). Microbial glutaminase: biochemistry, molecular approaches and applications in the food industry. *J Mol Catalysis B: Enzymatic, 23*: 87-100.

[188] Aryan, A., Wilson, B., Strauss, C., et al. (1987). The properties of glycosidases of Vitis vinifera and a comparison of their β-glucosidase activity with that of exogenous enzymes. An assessment of possible applications in enology. *Amr J Enol Viticult., 38*: 182-188.

[189] de Kraker, J., Schurink, M., Franssen, M., et al. (2003). Hydroxylation of sesquiterpenes by enzymes from chicory (*Cichorium intybus* L.) roots. *Tetrahedron, 59*: 409-418.

[190] Yoshida, A., Takenaka, Y., Tamaki, H., et al. (1997). Vanillin formation by microbial amine oxidases from vanillylamine. *J Fermen Bioeng., 84*: 603-605.

[191] Martin, L., Ebert, C., Gardossi, L., et al. (2001). High isolated yields in thermolysin-catalysed synthesis of Z-l-aspartyl-l-phenylalanine methyl ester in toluene at controlled water activity. *Tetrahedron Lett., 42*: 3395-3397.

[192] Fuglsang, C., Johansen, C., Christgau, S., et al. (1995). Antimicrobial enzymes: applications and future potential in the food industry, *Trends Food Sci Technol., 6*: 390-396.

[193] Bom, I., Dielbandhoesing, S., Harvey, K., et al. (1998). A new tool for studying the molecular architecture of the fungal cell wall: one-step purification of recombinant Trichoderma β-(1-6)-glucanase expressed in Pichia pastoris. *Biochim Biophys Acta - General Subjects, 1425*: 419-424.

[194] Brul, S. and Coote, P. (1999). Preservative agents in foods: mode of action and microbial resistance mechanisms. *Intr J Food Microbiol., 50*: 1-17.

[195] Çinar, I. (2005). Effects of cellulase and pectinase concentrations on the colour yield of enzyme extracted plant carotenoids. *Process Biochem., 40*: 945-949.

[196] Johnson, E. A. and Schroeder, W. A. (1996). Microbial carotenoids. In: *Downstream processing biosurfactants carotenoids,* Springer, pp. 119-178.

[197] Mijts, B. N. and Schmidt-Dannert, C. (2003). Engineering of secondary metabolite pathways, *Curr Opin Biotechnol., 14*: 597-602.

[198] Golubev, W. I. (1995). Perfect state of Rhodomyces dendrorhous (Phaffia rhodozyma). *Yeast, 11*: 101-110.

[199] An, G., Schuman, D. and Johnson, E. (1989). Isolation of Phaffia rhodozyma mutants with increased astaxanthin content. *Appl Env Microbiol., 55*: 116-124.

[200] Verdoes, J., Sandmann, G., Visser, H., et al. (2003). Metabolic engineering of the carotenoid biosynthetic pathway in the yeast Xanthophyllomyces dendrorhous (Phaffia rhodozyma). *Appl Env Microbiol., 69*: 3728-3738.

[201] Bassi, R., Pineau, B., Dainese, P., et al. (1993). Carotenoid-binding proteins of photosystem II. *Eur J Biochem., 212*: 297-303.

[202] Fennema, O. (1985). *Food Chemistry,* In: Marcel Dekker, Inc.

[203] Valyasevi, R. and Rolle, R. (2002). An overview of small-scale food fermentation technologies in developing countries with special reference to Thailand: scope for their improvement. *Intr J Food Microbiol., 75*: 231-239.

[204] Jaichumjai, P., Valyasevi, R., Assavanig, A., et al. (2010). Isolation and characterization of acid-sensitive *Lactobacillus plantarum* with application as starter culture for Nham production. *Food Microbiol., 27*: 741-748.

[205] Paludan-Müller, C., Valyasevi, R., Huss, H., et al. (2002). Genotypic and phenotypic characterization of garlic-fermenting lactic acid bacteria isolated from som-fak, a Thai low-salt fermented fish product. *J Appl Microbiol., 92*: 307-314.

[206] *Amasi,* in, Wikipedia, the free encyclopedia, 2014.

[207] *Miso,* in, Wikipedia, the free encyclopedia, 2014.

[208] *Pectinase database*, in, 2014.

[209] Sunna, A., Moracci, M., Rossi, M., et a. (1997). Glycosyl hydrolases from hyperthermophiles. *Extremophiles, 1*: 2-13.

[210] Vieille, C. and Zeikus, G. (2001). Hyperthermophilic enzymes: sources, uses, and molecular mechanisms for thermostability. *Microbiol Mol Biol Rev., 65*: 1-43.

[211] Bertoldo, C., Duffner, F., Jorgensen, P., et al. (1999). Pullulanase type I from Fervidobacterium pennavorans Ven5: cloning, sequencing, and expression of the gene and biochemical characterization of the recombinant enzyme. *Appl Env Microbiol., 65*: 2084-2091.

[212] Tsutsumi, Y., Matsubara, H., Masaki, H., et al. (1999). Angiotensin II type 2 receptor overexpression activates the vascular kinin system and causes vasodilation. *J Clin Investig., 104*: 925-935.

[213] Suzuki, Y., Ito, N., Yuuki, T., et al. (1989). Amino acid residues stabilizing a Bacillus alpha-amylase against irreversible thermoinactivation. *J Biolog Chem., 264*: 18933-18938.

[214] Barnett, C. C., Mitchinson, C., Power, S. D., et al. (1998). Requadt, C. A., Solheim, L. P. *Oxidatively stable alpha-amylase,* in, Google Patents.

[215] Chen, L., Tatum, V., Glauert, H., et al. (2001). Chow, C. K. Peroxisome proliferator perfluorodecanoic acid alters glutathione and related enzymes. *J Biochem Mol Toxicol., 15*: 107-113.

[216] Goffeau, A., Barrell, B., Bussey, H., et al. (1996). Life with 6000 genes. *Science,* 274: 546-567.

SECTION E. STRUCTURAL MICROBIOLOGY

In: Microbial Catalysts. Volume 1
Editors: S. M. Abdel-Aziz, N. Garg, A. Aeron et al.
ISBN: 978-1-53614-679-0

Chapter 12

FUNCTIONAL, STRUCTURAL, AND THERMODYNAMIC CHARACTERISTICS OF ENZYMES FROM DEEP-SEA MICROORGANISMS

Eiji Ohmae[1,*], Yurina Miyashita[1,†] and Chiaki Kato[2,‡]
[1]Department of Mathematical and Life Sciences, Graduate School of Science, Hiroshima University, Higashi-Hiroshima, Japan
[2]Department of Marine Biodiversity Research, Japan Agency for Marine-Earth Science and Technology (JAMSTEC), Yokosuka, Japan

ABSTRACT

Hydrostatic pressure analysis is an ideal approach for studying protein dynamics and hydration because hydrostatic pressure affects the size and shape of cavities and the hydration state of proteins in solution. Furthermore, modern developments in submersibles capable of exploring the full depth of the ocean, and in molecular biology techniques, now allow us to investigate enzymes from deep-sea microorganisms at the molecular level. The deep-sea environment, including the lack of light, a very low temperature across almost of it, other than extremely high temperatures within specific confined locations, and high hydrostatic pressure across the entire environment, have caused deep-sea enzymes to evolve special hyperthermophilic, psychrophilic, and piezophilic characteristics that can serve as useful tools in biotechnological and industrial applications. Although research in deep-sea enzymes began only 40 years ago, work in the field has increased exponentially since, with the whole genome sequence of many deep-sea microorganisms as well as the crystal structure of several deep-sea proteins being determined one after another in recent years. However, the molecular adaptation mechanisms that deep-sea enzymes have evolved to exist in these environments are still unclear, although the primary and tertiary

* Author's Email: ohmae@hiroshima-u.ac.jp.
† Author's Email: ymiya@rikkyo.ac.jp.
‡ Author's Email: kato_chi@jamstec.or.jp.

structures of these enzymes are often quite similar to, and sometimes essentially the same as, the counterpart homologues in organisms living in moderate temperature and atmospheric pressure conditions. This review provides an overview of several deep-sea enzymes, and suggests corresponding environmental adaptation mechanisms, particularly from the perspective of hydrostatic pressure.

Keywords: deep-sea microorganisms, hydrostatic pressure, hyperthermophilic enzymes, psychrophilic enzymes, piezophilic enzymes

1. INTRODUCTION

The deep sea is an extreme environment containing no light, usually low (1–4°C) but in some locations extremely high (up to 400°C) temperatures, and high hydrostatic pressure. Despite such an unfavorable environment to life, many microorganisms, especially bacteria and archaea, are adapted to live and thrive in the deep sea. The enzymes produced by these microorganisms, so-called "deep-sea enzymes," must function at the same temperature and hydrostatic pressure conditions as the environment in which they exist, and adaptation mechanisms to such conditions need to be a part of the enzymes themselves, because temperatures and hydrostatic pressures are the same on the inside and outside of these microorganisms' cells. Thus, the extreme environments of the deep sea enable deep-sea enzymes to be used as special and useful tools for biotechnological and industrial applications. Furthermore, modern developments in submersibles capable of exploring the full depth of the ocean, and in molecular biology techniques, now allow us to investigate enzymes from deep-sea microorganisms at the molecular level.

ZoBell and Kim [1] began deep-sea enzyme research only 40 years ago; work in the field has increased exponentially ever since. Deep-sea enzymes can generally be classified into three categories based on the living environment of the source organism: 1) hyperthermophilic or thermophilic enzymes, 2) psychrophilic or psychrotolerant enzymes, and 3) piezophilic or piezotolerant enzymes, although several enzymes in the third category can also be included in either the first or second category. Most research interest has centered on hyperthermophilic, psychrophilic, and piezophilic enzymes, because of their potential applications.

Hyperthermophilic enzymes are predominantly obtained from archaea (and rarely bacteria) living in deep-sea hydrothermal vents and adjacent fields that optimally grow at temperatures greater than 80°C [2]. The most widely recognized application of such a hyperthermophilic enzyme is the polymerase chain reaction's (PCR) use of a DNA polymerase from the hyperthermophilic archaea *Thermococcus litoralis* [3, 4] and *Pyrococcus* sp. strain GB-D [5], which are commercially sold by New England Biolabs Inc. (Ipswich, MA, US) as "Vent" and "Deep Vent" DNA polymerases, respectively. Several hydrolytic enzymes, including amylases, proteases, and lipases, from deep-sea hyperthermophilic microorganisms have also been investigated for potential industrial use, because they can speed up reaction processes by using high-temperature conditions.

Furthermore, psychrophilic enzymes have recently captured the interest of many researchers, because this type of enzyme can save on the energy requirements, and consequently the cost of reaction processes, by using low-temperature conditions. Psychrophilic enzymes are predominantly obtained from bacteria, optimally growing at temperatures less than 15°C and unable to grow above 20°C, living in deep-sea water or sediments, or in the tissues of multicellular organisms, such as deep-sea fish and shrimp [6].

Although hyperthermophilic and psychrophilic enzymes can be obtained from organisms living in environments at atmospheric pressure, such as in hot springs and polar ices, respectively, those from the deep sea may possess unique characteristics that guard against the excessive hydrostatic pressure of such an environment. Piezophilic microorganisms grow faster under pressurized conditions than at atmospheric pressure [7], but piezophilic enzymes can be obtained from both piezophilic and piezotorelant microorganisms. However, studies on piezophilic enzymes are still quite limited, because of the technical need for the additional pressurization system on measuring instruments.

The whole genome sequence of many deep-sea microorganisms and the crystal structure of several deep-sea proteins have recently been determined, one after another. However, the primary and tertiary structures of deep-sea enzymes are very similar to their counterpart homologues obtained from organisms living in moderate-temperature and atmospheric pressure conditions. Furthermore, the specific molecular adaptation mechanisms of deep-sea enzymes to the environment in which they exist are still unclear. This review provides an overview of deep-sea enzymes and suggests several environmental adaptation mechanisms, especially from the perspective of hydrostatic pressure.

2. Functional Characteristics of Enzymes from Deep-Sea Microorganisms

2.1. Hyperthermophilic and Thermophilic Enzymes

The reaction rate catalyzed by enzymes is generally increased by an increase in temperature; however, enzymes are unfolded at too high of a temperature and consequently lose catalytic ability. Thus, specific optimal temperature will exist and can be evaluated in the temperature dependence of the enzyme activity. In this context, enzymes with high thermal stability show thermophilic characteristics and a high optimal temperature, whereas those that have low thermal stability show psychrophilic characteristics and a low optimal temperature. Therefore, the thermal stability of enzymes should be discussed within the context of thermodynamic measurement results, but unfolded enzymes often aggregate at high temperature in the working solution, and cannot be described as the equilibrium unfolding process. Furthermore, the modification of enzymes by components in the solution, such as the oxidation or reduction of sulfhydryl groups, and the self-digestion of proteases should not be ignored in many cases. Therefore, the time needed to achieve a 50% loss of enzymatic activity at a fixed reaction condition, the so-called "half-life," is used as the usual guide for the thermal stability of enzymes.

The optimal temperatures and half-lives of several enzymes from deep-sea hyperthermophilic microorganisms, along with their collection location and depth, are listed in Figure 1. In the case of archaeal enzymes, Michels and Clark [8] reported a hyperthermophilic

Source organism	Collection location (depth)	Enzyme	Optimum temperature (at pH)	Half-life (at temperature)	Reference
Archaeal Enzymes					
Methanocaldococcus jannaschii	East Pacific Rise (2,600 m)	Protease	116°C (6.5)	45 min (116°C) 7 min (125°C)	[8]
		NADH oxidase	>95°C (7.0)	ND [a]	[9]
		Adenylosuccinate synthetase	>85°C (6.5)	150 min (70°C) 60 min (80°C)	[49]
		20S proteasome	55–75°C (7.5)	ND [a]	[61]
Pyrococcus abyssi GE5	North Fiji Basin (2,000 m)	Protease	95°C (9.0)	165 min (100°C) 7.3 min (110°C)	[10]
		Nitrilase	80°C (7.2)	9h (80°C) 6h (90°C)	[47]
		Alkaline phosphatase	70°C (11.0)	18h (100°C) 5h (105°C)	[62]
P. holikoshii OT3	Okinawa Trough (1,395 m)	Protease	80°C (8.5)	ND [a]	[63]
		Endoglucanase	>97°C (5.6)	>3h (97°C)	[64]
		Inorganic pyrophosphatase	89°C (8.0)	5h (95°C) 4h (100°C)	[65]
		UDP-glucose 4-epimerase	65°C (6.5)	>1h (90°C)	[66]
		Glutamate decarboxylase	>97°C (8.0)	3h (95°C)	[67]
		Aminoacylase	90°C (7.5)	217 min (75°C) 119 min (85°C)	[44]
		Exo-β-D-glucosamidase	90°C (6.0)	9h (90°C)	[68]
P. yayanosii GH1	Mid-Atlantic Ridge (4,100 m)	Amidase	85°C (7.6)	4h (70°C) 110 min (80°C)	[46]
		Amylopullulanase	95°C (5.0 or 6.0) [b]	2.5h (100°C) 1h (110°C)	[12]
Themococcus fumicolans	North Fiji Basin (2,000 m)	DNA polymerase I	72°C (9.0)	7h (92°C) 2h (100°C)	[69]
T. siculi HJ21	Mid-Okinawa Trough (1,394 m)	α-Amylase	95°C (5.0)	5h (90°C)	[11]
Bacterial Enzymes					
Aerophyrum camini SY1	Izu-Bonin Arc (1,385 m)	Hydrogenase	85°C (8.5)	48h (90°C)	[13]
Bacillus sp. SCSIO 15121	Indian Ocean (3,412 m)	α-Amylase	75°C (7.5)	1h (85°C)	[14]
Geobachillus sp. EPT9	East Pacific (3,191 m)	Lipase	55°C (8.5)	1h (70°C)	[15]
G. sp. MT-1	East Pacific (3,191 m)	Xylanase	70°C (7.0)	50 min (65°C)	[70]

[a] Not determined. [b] The optimal pH for amylase activity is 5.0, while that for pullulanase activity is 6.0.

Figure 1. Enzyme examples from deep-sea hyperthermophilic microorganisms.

protease from *Methanocaldococcus jannaschii* whose optimal temperature is a surprising 116°C, and its half-life at the same temperature is 45 min. Case et al. [9] also reported a hyperthermophilic NADH oxidase from the same archaeon whose optimal temperature is more than 95°C. Dib et al. [10] reported a hyperthermophilic protease from *Pyrococcus abyssi* whose optimal temperature is 95°C and its half-life at 100°C is 165 min. Wang et al. [11] reported a hyperthermophilic α-amylase from a *Thermococcus* sp. strain HJ21 whose optimal temperature is 95°C, with a half-life at 90°C of 5h. Jiao et al. [12] reported a hyperthermophilic amylopullulanase from the same archaeon whose optimal temperature is also 95°C, with a half-life at 100°C of 2.5 h.

Bacterial deep-sea enzymes exhibit a trend in that their optimal temperatures are lower than those of the hyperthermophilic archaeal enzymes, although only limited examples have been reported thus far. Nishimura and Sako [13] reported a hyperthermophilic hydrogenase whose optimal temperature is 85°C, with a half-life at 90°C of 48 h. This enzyme was obtained from *Aerophyrum camini* strain SY1 isolated from a deep-sea hydrothermal vent chimney sample collected at the Suiyo Seamount in the Izu-Bonin Arc at a depth of 1,385 m. Li et al. [14] reported a thermophilic α-amylase from *Bacillus* sp. SCSIO 15121 isolated from Indian Ocean sediment at a depth of 3,412m, whose optimal temperature is 75°C with a half-life at 85°C of 1h. Zhu et al. [15] reported a mesophilic lipase whose optimal temperature is 55°C from a *Geobacillus* sp. strain EPT9 isolated from a sediment sample collected at a deep-sea hydrothermal field in the east Pacific at a depth of 3,191m. Other hyperthermophilic enzymes, including those that did not originate from deep-sea microorganisms, have been summarized in a previous review [16].

2.2. Psychrophilic and Psychrotolerant Enzymes

Most microorganisms living in the deep sea are psychrophilic, because the temperature of the deep sea is usually quite low, 1–4°C. The enzymes obtained from such psychrophilic microorganisms are frequently psychrophilic, with optimal temperatures less than 50°C, and with catalytic activity being retained even at 0°C in many cases. This characteristic is very useful for applications that need to avoid high temperatures in the food, laundry, pharmaceutical, and medical industries. The optimal temperatures and half-lives of several enzymes from deep-sea psychrophilic microorganisms, along with their collection location and depth are listed in Figure 2. Zhang et al. reported two psychrophilic lipases from the psychrotrophic bacteria *Psychrobacter* sp. strain 7195 and *Pseudomonas* sp. strain 7323, both collected from a sediment sample in the Antarctic Ocean at a depth of 900m; the optimal temperature of both enzymes is 30°C [17, 18]. Zhang and Zeng [19] also reported a psychrophilic α-amylase from *Nocardiopsis* sp. strain 7326 collected at the same location, whose optimal temperature is 35°C that maintained 38% of its highest activity at 0°C. Zhang and coworkers actively searched for psychrophilic proteases. They reported three cold-adapted proteases, MCP-01, MCP-02, and MCP-03, from the deep-sea psychrotolerant bacterium *Pseudoalteromonas* sp. strain SM9913 [20–22]. In particular, MCP-01, named "deseasin," was fully characterized [20, 23–27]. The optimal temperature and half-life of this protease are dependent on buffer systems, because the decrease in its activity at high temperatures is mainly caused by its self-digestion [20]. They also reported two collagenolytic cold-adapted proteases, named "myroilysin" and "myroicolsin," from the deep-sea bacterium *Myroides profundi* strain D25 isolated from a sediment sample taken from the southern Okinawa Trough, at a depth of 1,245m, whose optimal temperatures are 60 and 50°C, respectively [28, 29].

Source organism	Collection location (depth)	Enzyme	Optimum temperature (at pH)	Half-life (at temperature)	Reference
Bacterial Enzymes					
Alcanivirax sp. L27	Eastern Pacific	5-Enopyruvylshikimate-3-phosphate synthase	50°C (7.0)	ND [a]	[71]
Myroides profundi D25	Okinawa Trough (1,245 m)	Protease (Myroicolsin)	60°C (8.5)	10 min (60°C)	[29]
		Protease (Myroilysin)	50°C (9.0)	28 min (40°C) 5 min (45°C)	[28]
Moritella sp. 5710	Suruga Bay (2,220 m)	Malate dehydrogenase	35°C (7.0)	10 min (42°C)	[72]
Moritella sp. 2D2	Northwest Pacific (6,000 m)	Malate dehydrogenase	40°C (7.0)	10 min (45°C)	[73]
M. profunda	Sierra Leone Rise (2,815 m)	Aspartate carbamoyltransferase	30–35°C (7.5)	>60 min (55°C) 10 min (60°C)	[74]
		Dihydrofolate reductase	34°C (7.0)	25 min (45°C)	[75]
M. abyssi	Sierra Leone Rise (2,815 m)	Aspartate carbamoyltransferase	30–35°C (9.0)	8 min (50°C) 4 min (60°C)	[76]
Nocardiopsis sp. 7326	Prydz Bay (900 m)	α-Amylase	35°C (8.0)	40 min (35°C)	[19]
Pseudoalteromonas sp. DY3	East Pacific (5,000 m)	Cellulase	40°C (6–7)	120 min (35°C) 60 min (40°C)	[77]
P. issachenkonii	Aleutian deep-sea (3,310 m)	Protease	35°C (7.0)	ND[a]	[78]
Pseudomonas sp. SM9913	Okinawa Trough (1,855 m)	Protease (Deseasin, MCP-01)	30–35°C (9.0) [b] 40–45°C (8.5) [c] 45–50°C (6.5) [d] 50–55°C (7.5) [e]	15 min (40°C) [b] 48 min (40°C) [c] 58 min (40°C) [d] 75 min (40°C) [e]	[20]
		Protease (MCP-02)	57°C (8.0)	90 min (55°C) 23 min (60°C)	[21]
		Protease (MCP-03)	45°C (8.0)	30 min (50°C)	[22]
Pseudomonas sp. 7323	Prydz Bay (900 m)	Lipase	30°C (9.0)	4.5h (30°C)	[18]
Pseudomonas sp. MS300	Mariana Trench (10,897 m)	α-Maltotetraohydrolase	50°C (7.0)	30 min (50°C)	[79]
Pseudomonas sp. DY-A	East Pacific (5,225 m)	Protease	40°C (10.0)	>24h (30°C)	[80]
Pseudomonas sp. MM15	Southern Okinawa Trough	Endoglucanase	30°C (4.5)	1h (50°C)	[52]
Psychrobacter sp. 7195	Prydz Bay (900 m)	Lipase	30°C (9.0)	5h (30°C) 15 min (60°C)	[17]
Metagenome	Edison Seamount (1,440 m)	Lipase	25°C (8.0)	>30 min (35°C)	[30]
	Central Pacific (5,886 m)	Esterase	20°C (7.5)	30 min (30°C)	[31]

[a] Not determined. [b-d] The buffers used were carbonate (b), Tris (c), phosphate (d), and borate (e) buffers.

Figure 2. Enzyme Examples from Deep-Sea Psychrophilic Microorganisms.

It is noteworthy that psychrophilic enzymes can now be obtained not only from isolated microorganisms, but also from metagenomic samples. Jeon et al. [30] reported a novel cold-active lipase having an optimal temperature of 25°C cloned from a deep-sea sediment metagenomic sample at a depth of 1,440m. Jiang et al. [31] reported nine new cold-active esterases cloned from deep-sea sediment metagenomic samples in the Pacific Ocean at a depth of 5,886m, one of which showed an optimum temperature of only 20°C and a half-life at 30°C of only 30 min.

2.3. Piezophilic and Piezotolerant Enzymes

The rate of chemical reactions can be accelerated or decelerated by an increase in hydrostatic pressure depending on the activation volume of the reaction. This is because both negative and positive values are acceptable for the activation volume, whereas only positive values are acceptable for the activation energy. Enzyme reactions consist of multiple steps—the binding of substrates, the actual chemical reaction, and the release of products—and the apparent reaction rate depends on the rate-determining step within these multiple steps. If the activation volume of the rate-determining step under conditions of atmospheric pressure is positive, then the enzyme's activity is decelerated by the pressurization, and the optimal pressure of such enzymes is 0.1 MPa, which is the atmospheric pressure. However, if the activation volume of the rate-determining step under conditions of atmospheric pressure is negative, then the reaction is accelerated by the pressurization. Furthermore, another reaction step, which is not a rate-determining step under conditions of atmospheric pressure, could be decelerated by pressurization because of its positive activation volume, and therefore, become the rate-determining step at the optimal pressure of the enzyme reaction. In this case, additional pressurization decelerates the activity of the enzyme. In other cases, an enzyme can be unfolded by pressurization, similar to heating. Enzymes show piezophilic activities in these cases.

Source organism	**Collection location (depth)**	**Enzyme**	**Optimum pressure, MPa (at pH and temperature)**	**Reference**
Archaeal Enzyme				
Methanocaldococcus jannaschii	East Pacific Rise (2,600 m)	Protease	>50 (6.5, 115°C)	[8]
Bacterial Enzymes				
Moritella profunda	Sierra Leone Rise (2,815 m)	Dihydrofolate reductase	50 (7.0, 25°C)	[34]
M . yayanosii DB21MT-5	Mariana Trench (10,897 m)	Dihydrofolate reductase	0.1–75 (7.0, 25°C) [a]	[35]
Photobacterium profundum SS9	Sulu Sea (2,551 m)	Tryptophan indole-lyase	60 (8.0, 25°C)	[36]
Shewanella violacea DSS12	Ryukyu Trench (5,110 m)	Dihydrofolate reductase	100 (7.0, 25°C)	[32]
*S.*benthica DB21MT-2	Mariana Trench (10,897 m)	Dihydrofolate reductase	50 (7.0, 25°C)	[33]

[a] Piezotolerant enzyme.

Figure 3. Piezophilic and Piezotolerant Enzyme Examples from Deep-Sea Microorganisms.

The optimal pressure of a few piezophilic and piezotolerant enzymes from deep-sea microorganisms along with their collected location and depth are listed in Figure 3. There are only few reports on piezophilic or piezotolerant enzymes from deep-sea microorganisms, because of the technical difficulties involved in measuring enzymatic activity under pressurized conditions. We found three piezophilic dihydrofolate reductase (DHFR) from deep-sea bacteria, whose optimal pressures are 50–100 MPa [32–34], and one piezotolerant DHFR from an obligatory piezophilic bacterium *Moritella yayanosii* strain DB21MT-5 [35], as shown in Figure 3. Two other piezophilic enzymes, tryptophan indole-lyase from *Photobacterium profindum* strain SS9 [36] and a protease from *Methanocaldococcus jannaschii* [8], whose optimal pressures are 60 and more than 50 MPa, respectively, were also reported. As discussed in section 4.3, hydrostatic pressure affects the cavity and hydration states of proteins in solution, thus piezophilic enzymes are a very good model for investigating relationships between structural dynamics and the function of enzymes. Future research of these important enzymes is anxiously awaited.

3. Structural Characteristics of Proteins from Deep-Sea Microorganisms

3.1. Primary Structure

Many deep-sea microorganisms have been isolated since ZoBell [37] initially reported the direct observation of bacteria in a deep-sea sample using a microscope. Although some deep-sea microorganisms belong to new genera, most of them belong to known genera that include species living in atmospheric pressure conditions, suggesting that these organisms adapted to

Species	Collection location (depth)	Genome size, bp	Reference
Archaea			
Archaeoglobus profundus Av18	Guaymas Basin (2,000 m)	1,563,423	[81]
Methanocaldococcus jannaschii DSM2661	East Pacific Rise (2,600 m)	1,739,927	[39]
Methanocaldococcus sp. FS406-22	Juan de Fuca Ridge (1,525 m)	1,773,136	CP001901 and CP001902 in GenBank
M. okinawaensis IH1	Iheya Ridge (972 m)	1,677,455	CP002792 and CP002793 in GenBank
Pyrococcus abyssi GE5	North Fiji Basin (2,000 m)	1,765,118	[82]
P. holikoshii OT3	Okinawa Trough (1,395 m)	1,738,505	[83]
P. yayanosii CH1	Mid Atlantic Ridge (4,100 m)	1,716,818	[84]
Pyrococcus sp. NA2	Western Pacific (1,650 m)	1,861,320	[85]
Pyrococcus sp. ST04	Juan de Fuca Ridge (2,290 m)	1,736,885	[86]
Thermococcus balophilus MP	Mid Atlantic Ridge (3,550 m)	2,064,237	[87]
T. eurythermalis A501	Guaymas Basin (2,007 m)	2,122,535	[88]
T. gammatolerans EJ3	Guaymas Basin (2,616 m)	2,045,438	[89]
T. nautili 30-1	East Pacific Rise (2,633 m)	1,976,356	[90]
T. onnurineus NA1	East Manus Basin (1,650 m)	1,847,607	[91]
Thermococcus sp. 4557	Guaymas Basin (2,000 m)	2,011,320	[92]
Thermococcus sp. AM4	East Pacific Rise (2,600 m)	2,086,428	[93]
Thermococcus sp. CL1	Juan de Fuca Ridge	1,950,313	[94]
Thermococcus sp. ES1	Juan de Fuca Ridge	1,957,742	[95]
Bacteria			
Deferribacter desulfuricans SSM1	Izu-Bonin Arc (1,385 m)	2,542,933	[96]
Desulfurobacterium thermolithotrophum BSA	Mid Atlantic Ridge (3,500 m)	1,541,968	[97]
Glaciecola agarilytica 4H-3-7+YE-5	Suruga Bay (755 m)	5,393,591	[98]
Idiomarina loihiensis L2TR	Lō'ihi Seamount (1,296 m)	2,839,318	[40]
Marinithermus hydrothermalis T1	Izu-Bonin Arc (1,385 m)	2,269,167	[99]
Marinitoga piezophila KA3	East Pacific Rise (2,630 m)	2,244,793	[100]
Thiomicrospira crunogena XCL-2	Galapagos Rift	2,427,734	[101]
Oceanobacillus iheyensis HTE831	Iheya Ridge (1,050 m)	3,630,528	[102]
Persephonella marina EX-H1	East Pacific Rise (2,507 m)	1,983,966	[103]
Photobacterium profundum SS9	Sulu Sea (2,551 m)	6,403,280	[104]
P. phosphoreum ANT-2200	Mediterranean Sea (2,200 m)	5,107,216	[105]
Psychromonas sp. CNPT3	North Pacific Ocean (5,700 m)	3,052,410	[106]
Shewanella violacea DSS12	Ryukyu Trench (5,110 m)	4,962,103	[41]
S. loihica PV-4	Lō'ihi Seamount (1,325 m)	4,602,594	CP000606 in GenBank
Thermodesulfatator indicus CIR29812	Central Indian Ridge (2,420 m)	2,322,224	[107]
Thermosediminibacter oceani JW/IW-1228P	Peru Margin (252 m)	2,280,035	[108]
Thermovibrio ammonificans HB-1	East Pacific Rise (2,500 m)	1,759,526	[109]
Zunongwangia profunda SM-A87	Southern Okinawa Trough (1,245 m)	5,128,187	[110]

Figure 4. List of Deep-Sea Microorganisms Whose Genome Sequence has been Completely Determined.

the deep-sea environment after differentiation from other genera under atmospheric pressure conditions. Therefore, the primary structure of deep-sea enzymes is more similar to that of counterpart homologues from congeneric species, versus homologous deep-sea enzymes from other genera [38]. Furthermore, the complete genome sequence of more than 30 deep-sea microorganisms has been determined, including the hyperthermophilic *Methanocaldococcus jannaschii* strain DSM2661 [39], the psychrophilic *Idiomarina loihiensis* strain L2TR [40], and the piezophilic *Shewanella violacea* strain DSS12 [41]. A list of deep-sea microorganisms whose genome sequences have been completely determined as of this writing is shown in Figure 4. The sequence similarity between deep-sea enzymes from these microorganisms and their non-deep-sea homologues is usually 30−60%, which is sufficient to estimate their function and expected tertiary structure based on homology.

3.2. Tertiary Structure

The three-dimensional structure of many recombinant proteins originating from deep-sea microorganisms has been determined and deposited in the Protein Data Bank (PDB) database [42], especially those from hyperthermophilic archaea, including *Pyrococcus horikoshii* (323 structures), *Methanocaldococcus jannaschii* (95 structures), and *Pyrococcus abyssi*

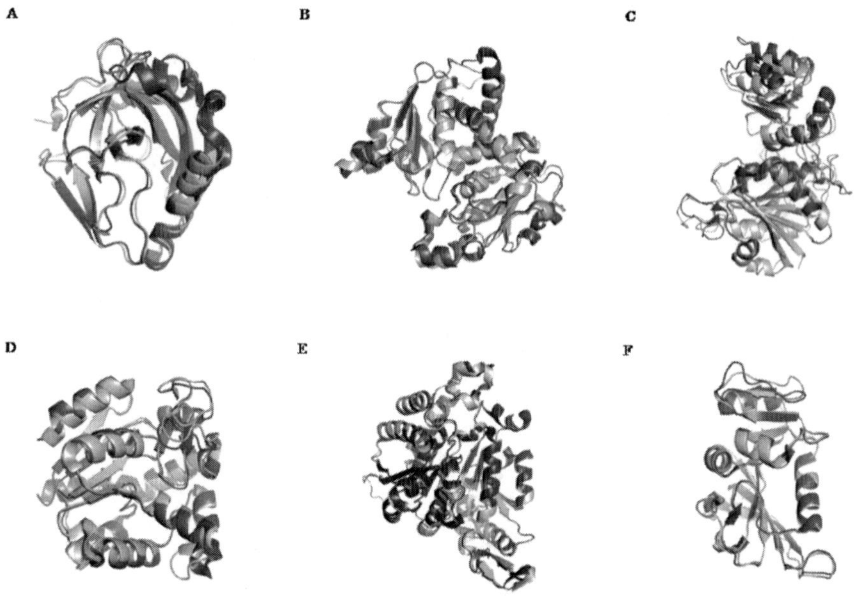

Figure 5. Backbone structure superpositioning of deep-sea enzymes (black) and their non-deep-sea homologues (gray). The single subunits drawn in each panel were produced with PyMol [http://www.pymol.org/]. (A): Inorganic pyrophosphatase from *Pyrococcus horikoshii* (PDB code: 1ude [111]) and *Escherichia coli* (PDB code: 1jfd [112]). (B): Aspartate transcarbamylase from *Pyrococcus abyssi* (PDB code: 1ml4 [113]) and *Escherichia coli* (PDB code: 1ekx [114]). (C): L-asparaginase from *Pyrococcus horikoshii* (PDB code: 1wls [115]) and *Escherichia coli* (PDB code: 3eca [116]). (D): Adenylate kinase from *Photobacterium profundum* (PDB code: 4k46, no article) and *Escherichia coli* (PDB code: 1ake [117]). (E): 3-Isopropylmalate dehydrogenase from *Shewanella benthica* (PDB code: 3vmk [118]) and *Shewanella oneidensis* (PDB code: 3vmj [118]). (F): Dihydrofolate reductase from *Moritella profunda* (PDB code: 2zza [34]) and *Escherichia coli* (PDB code: 1rx2 [119]).

(28 structures), although the functions of some of these proteins remain unknown. Furthermore, the number of protein structures deposited in the PDB database originating from deep-sea psychrophilic microorganisms has also recently increased: *Shewanella loihica* (19 structures), *Idiomarina loihiensis* (6 structures), and *Photobacterium profundum* (3 structures). However, only a very limited number of protein structures from deep-sea piezophilic microorganisms have been deposited in the database: four structures of two enzymes, DHFR and aspartate carbamoyltransferase, from *Moritella profunda*, and three structures of two enzymes, 3-isopropylmalate dehydrogenase and dihydropicolinate synthase, from *Shewanella benthica*. The backbone structure of these deep-sea enzymes and their non-deep-sea homologues is usually very similar, or essentially the same in some cases, although the number of subunits in a biological assembly is sometimes different. Typical examples are shown in Figure 5. Other deep-sea proteins that have not had their tertiary structures determined experimentally have been predicted to have very similar structures as their non-deep-sea homologues using homology-modeling approaches [15, 21, 43–46]. Therefore, the stability and functional characteristics of deep-sea enzymes probably need to be attributed to other factors, because the tertiary structure of deep-sea enzymes and their counterpart non-deep-sea homologues are so very similar, at least in all cases observed so far.

4. Thermodynamic Characteristics of Deep-Sea Enzymes

4.1. Hyperthermophilic Enzymes

Proteins from hyperthermophilic microorganisms are generally more thermostable than proteins originating from mesophilic microorganisms. This is reflected in the fact that recombinant proteins cloned from these organisms and then overexpressed in *Escherichia coli* do not aggregate and are easily separated from host proteins upon heating. However, as mentioned in section 2.1, the thermal stability of hyperthermophilic enzymes is usually measured by a half-life, which is based on the remaining activity of the protein measured at a standard reaction condition after incubation at a fixed temperature higher than the standard condition. Moreover, because enzymes can be refolded to an active structure and reactivated by reducing the temperature, a more realistic measure of thermal stability would be accomplished using thermodynamic methods, such as differential scanning calorimetry, or structure-based optical methods, such as circular dichroism (CD) or fluorescence spectroscopies. Structure-based methods are also applicable for measuring stability under conditions of varying hydrostatic pressure, and in the presence of chemical denaturants, such as guanidine hydrochloride or urea.

Although structure-based measurements are still not extensively used with hyperthermophilic enzymes, Mueller et al. [47] reported that the midpoint temperature (T_m) of a nitrilase from *Pyrococcus abyssi* measured using differential scanning calorimetry was 112.7°C, which is one of the highest protein melting temperatures known. Kleeb et al. [48] reported that the T_m and midpoint guanidine hydrochloride concentration (C_m) of a prephenate dehydratase from *Methanocaldococcus jannaschii* measured by CD were 94°C and 4.54 M, respectively. Mehrotta and Balaram [49] reported that the T_m of an adenylosuccinate synthetase from the same archaeon, also measured by CD, was 85°C. This enzyme showed a three-state unfolding process in the presence of guanidine hydrochloride, including a defined intermediate. The C_m values measured by CD and fluorescence were 2.1 and 4.6 M for native to intermediate and intermediate to unfolded states, respectively. It is noteworthy that the secondary and tertiary structures of this enzyme do not collapse even in the presence of an 8 M urea solution. Miyazono et al. [50] also reported that the C_m value of an acylphosphatase from *Pyrococcus horikoshii* measured by CD was 5.7 M. Overall, hyperthermophilic enzymes seem more stable against heat and chemical denaturants than their mesophilic homologues, although further comparative studies are needed.

4.2. Psychrophilic Enzymes

Enzymes from psychrophilic deep-sea bacteria generally show lower thermal stabilities than those of their counterpart mesophilic and hyperthermophilic homologues. He et al. [25] reported that the T_m value of protease MCP-01 (deseasin) from *Pseudoalteromonas* sp. SM9913 measured by CD was 49.5°C, which is much lower than its mesophilic homologue, subtilisin Carlsberg (62.5°C). This value increases by 10.5°C, becoming comparable to its mesophilic homologue, by the addition of 1 M trimethylamine N-oxide (TMAO), which is a natural osmolyte in many marine animals. Xie et al. [21] reported that the T_m values of protease MCP-02 from the same bacteria measured by fluorescence and CD were 64 and 65°C, which is 11 and 7°C, respectively, lower than those of its mesophilic homologue, pseudolysin. However, it is noteworthy that the optimal temperatures for mesophilic and thermophilic enzymes are consistent with their unfolding temperatures, whereas those for psychrophilic enzymes are sometimes less than their unfolding temperatures [51]. Yang and Dang [52] reported that the T_m value of an endo-β-1,4-glucanase from *Pseudomonas* sp. MM15 measured by CD was approximately 46°C, which is 16°C higher than its enzymatic optimal temperature, 30°C. Although the thermal unfolding of this enzyme measured by CD follows a two-state unfolding process, the active-site structure is disrupted and enzymatic activity is lost before the overall structure is unfolded. We previously reported that the secondary structure of a psychrophilic DHFR from *Moritella profunda* measured by CD did not unfold at temperatures up to 80°C, although mesophilic DHFR from *Escherichia coli* unfolds at ~50°C [34]. These results clearly indicate that the structural stability of deep-sea psychrophilic enzymes is not necessarily less than their mesophilic and hyperthermophilic homologues; however, the active sites of these enzymes are quite flexible and seem to be less stable than other parts of the enzymes.

4.3. Piezophilic Enzymes

According to Le Chatelier's principle, pressurization produces a shift in the equilibrium of a system in such a direction that the total volume of the system is decreased. However, any conformational change of proteins in solution cannot affect the total volume of the system according to Archimedes' principle. As a result, pressurization produces a conformational change in proteins that decreases cavities within them, and enhances the hydration state of the solution; this produces an unfolded conformation. Thus, hydrostatic pressure directly affects the hydration state of proteins, and offers an ideal approach for investigating the effect of hydration on the stability and function of enzymes [53].

The pressure dependence of the standard Gibbs free energy change between native and unfolded states of proteins in solution (ΔG^o, the so-called "protein stability") decreases in a nearly linear fashion with pressurization; therefore, it is unlikely that high-pressure stabilizes the native state of proteins. Hence, there are no reports of deep-sea proteins that have piezophilic stability. However, pressure-induced refolding of proteins can be observed at a temperature range slightly above the T_m, because pressure-temperature phase diagrams of protein stability often show elliptical shapes [54–56]. Furthermore, a co-solute can contribute to protein stability through preferential interactions between the protein and the co-solute, because the ΔG^o value is attributed to the free energy change of not only the protein but also the whole solution. In this context, chemical denaturants are such co-solutes that will reduce the ΔG^o value of thermal stability in response to the experimental temperature. As mentioned in section 4.2, TMAO enhances the thermal stability of the deep-sea cold-adapted protease, deseasin [25]. Yancey and coworkers also reported that TMAO stabilizes the lactate dehydrogenases of deep-sea fish from pressure-induced inactivation [57, 58]. Furthermore, the T_m value of double-stranded DNA is usually increased by pressurization, although pressurization itself induces the melting of the same double-stranded DNA [59, 60]. These results clearly indicate that the structural stability of macromolecules in solution depends not only on the macromolecules themselves, but also on the hydration state of the macromolecules and on preferential interactions between macromolecules and other solutes. Enzymes have adapted to deep-sea environments by substituting amino acid side chains without significantly altering their backbone structure from that of their non-deep-sea homologues, because protein hydration states and preferential interactions are dependent on amino acid side chains.

CONCLUSION

This review discusses many of the enzymes that have been obtained and investigated from hyperthermophilic, psychrophilic, and piezophilic deep-sea microorganisms. Research indicates that these enzymes have adapted to the extreme environment of the deep sea, although primary and tertiary structures are very similar to their counterpart homologues obtained from the organisms living in moderate temperatures and atmospheric pressure conditions. An elucidation of the molecular adaptation mechanisms of these deep-sea enzymes can provide novel information regarding principles of structural stability and function in enzymes. Because hydration is a key component in the stability and function of enzymes, these deep-sea enzymes offer an ideal model for the study of this phenomenon, and hydrostatic pressure is the ideal approach to clarify the problem. Although research on these deep-sea enzymes began only 40 years ago, work in the field has since increased exponentially, and will become even more important in future.

REFERENCES

[1] ZoBell, C. E. and Kim, J. (1972). Effects of deep-sea pressures on microbial enzyme systems. *Symp Soc Exp Biol., 26*: 125–146.

[2] Stetter, K. O. (1996). Hyperthermophilic procaryotes. *FEMS Microbiol Rev., 18*: 149–158.

[3] Cariello, N. F., Swenberg, J. A., and Skopek, T. R. (1991). Fidelity of *Thermococcus litoralis* DNA polymerase (Vent) in PCR determined by denaturing gradient gel electrophoresis. *Nucleic Acids Res., 6*: 4193–4198.

[4] Mattila, P., Korpela, J., Tenkanen, T., et al. (1991). Fidelity of DNA synthesis by the *Thermococcus litoralis* DNA polymerase--an extremely heat stable enzyme with proofreading activity. *Nucleic Acids Res., 6*: 4967–4973.

[5] Perler, F., Kumar, S., and Kong, H. (1996). Thermostable DNA polymerases. In *Enzymes and proteins from hyperthermophilic microorganisms* vol. 48, edited by Adams, M. W. W., 377–435. San Diego:Academic Press.

[6] Cavicchioli, R., Siddiqui, K. S., Andrews, D., et al. (2002). Low-temperature extremophiles and their applications. *Curr Opin Biotechnol., 13*: 253–261.

[7] Kato, C. and Nogi, Y. (2001). Correlation between phylogenetic structure and function: examples from deep-sea *Shewanella*. *FEMS Microbiol Ecol., 35*: 223–230.

[8] Michels, P. and Clark, D. S. (1997). Pressure-enhanced activity and stability of a hyperthermophilic protease from a deep-sea methanogen. *Appl Env Microbiol., 63*: 3985–3991.

[9] Case, C. L., Rodriguez, J. R., and Mukhopadhyay, B. (2009). Characterization of an NADH oxidase of the flavin-dependent disulfide reductase family from *Methanocaldococcus jannaschii*. *Microbiology, 155*: 69–79.

[10] Dib, R., Chobert, J., Dalgalarrondo, M., et al. (1998). Purification, molecular properties and specificity of a thermoactive and thermostable proteinase from *Pyrococcus abyssi*, strain st 549, hyperthermophilic archaea from deep-sea hydrothermal ecosystem. *FEBS Lett., 431*: 279–284.

[11] Wang, S., Lu, Z., Lu, M., et al. (2008). Identification of archaeon-producing hyperthermophilic α-amylase and characterization of the α-amylase. *Appl Microbiol Biotechnol., 80*: 605–614.

[12] Jiao, Y. L., Wang, S. J., Lv, M. S., et al. (2011). A GH57 family amylopullulanase from deep-sea *Thermococcus siculi*: expression of the gene and characterization of the recombinant enzyme. *Curr Microbiol., 62*: 222–228.

[13] Nishimura, H. and Sako, Y. (2009). Purification and characterization of the oxygen-thermostable hydrogenase from the aerobic hyperthermophilic archaeon *Aeropyrum camini*. *J Biosci Bioeng., 108*: 299–303.

[14] Li, L., Yang, J., Li, J., et al. (2014). Role of two amino acid residues' insertion on thermal stability of thermophilic α-amylase AMY121 from a deep sea bacterium *Bacillus* sp. SCSIO 15121. *Bioprocess Biosyst Eng.,* doi: 10.1007/s00449-014-1330-2.

[15] Zhu, Y., Li, H., Ni, H., et al. (2015). Molecular cloning and characterization of a thermostable lipase from deep-sea thermophile *Geobacillus* sp. EPT9. *World J Microbiol Biotechnol., 31*: 295–306.

[16] Bouzas, T. M., Barros-Velázquez, J., and Villa, T. G. (2006). Industrial applications of hyperthermophilic enzymes: A review. *Protein and peptide Lett., 13*: 645–651.
[17] Zhang, J. W., Lin, S., and Zeng, R. Y. (2007). Cloning, expression, and characterization of a cold-adapted lipase gene from an Antarctic deep-sea psychrotrophic bacterium, *Psychrobacter* sp. 7195. *J Microbiol Biotechnol., 17*: 604–610.
[18] Zhang, J. W. and Zeng, R. Y. (2008). Molecular cloning and expression of a cold-adapted lipase gene from an Antarctic deep sea psychrotrophic bacterium *Pseudomonas* sp. 7323. *Marine Biotechnol., 10*: 612–621.
[19] Zhang, J. W. and Zeng, R. Y. (2008). Purification and characterization of a cold-adapted α-amylase produced by *Nocardiopsis* sp. 7326 isolated from Prydz Bay, Antarctic. *Marine Biotechnol., 10*: 75–82.
[20] Chen, X. L., Sun, C. Y., Zhang, Y. Z., et al. (2002). Effects of different buffers on the thermostability and autolysis of a cold-adapted protease MCP-01. *J Protein Chem., 21*: 523–527.
[21] Xie, B. B., Bian, X. L., Chen, X. L., et al. (2009). Cold adaptation of zinc metalloproteases in the thermolysin family from deep sea ice bacteria revealed by catalytic and structural properties and molecular dynamics: New insights into relationship between conformational flexibility and hydrogen bonding. *J Biol Chem., 284*: 9257–9269.
[22] Yan, B. Q., Chen, X. L., Hou, X. Y., et al. (2009). Molecular analysis of the gene encoding a cold-adapted halophilic subtilase from deep-sea psychrotorelant bacterium *Pseudoalteromonas* sp. SM9913: cloning, expression, characterization and function analysis of the C-terminal PPC domains. *Extremophiles, 13*: 725–733.
[23] Chen, X. L., Xie, B. B., Lu, J. T., et al. (2007). A novel type of subtilase from the psychrotolerant bacterium *Pseudoalteromonas* sp. SM9913: catalytic and structural properties of deseasin MCP-01. *Microbiology, 153*: 2116–2125.
[24] Zhao, G. Y., Chen, X. L., Zhao, H. L., et al. (2008). Hydrolysis of insoluble collagen by deseasin MCP-01 from deep-sea *Pseudoaltelomonas* sp. SM9913. *J Biol Chem., 283*: 36100–36107.
[25] He, H. L., Chen, X. L., Zhang, X. Y., et al. (2009). Novel use for the osmolyte trimethylamine N-oxide: Retaining the psychrophilic characters of cold-adapted protease deseasin MCP-01 and simultaneously improving its thermostability. *Marine Biotechnol., 11*: 710–716.
[26] Wang, Y. K., Zhao, G. Y., Li, Y., et al. (2010). Mechanistic insight into function of the C-terminal PKD domain of the collagenolytic serine protease deseasin MCP-01 from deep sea *Pseudoaltelomonas* sp. SM9913: Binding of the PKD domain to collagen results in collagen swelling but does not unwind the collagen triple helix. *J Biol Chem., 285*: 14285–14291.
[27] Ran, L. Y., Su, H. N., Zhao, G. Y., et al. (2013). Structural and mechanistic insights into collagen degradation by a bacterial collagenolytic serine protease in the subtilisin family. *Mol Microbiol., 90*: 997–1010.
[28] Chen, X. L., Xie, B. B., Zhao, G. Y., et al. (2009). Ecological function of myroilysin, a novel bacterial M12 metalloprotease with elastinolytic activity and a synergistic role in collagen hydrolysis, in biodegradation of deep-sea high-molecular-weight organic nitrogen. *Appl Environ Microbiol., 75*: 1838–1844.

[29] Ran, L. Y., Su, H. N., Zhou, M. Y., et al. (2014). Characterization of a novel subtilisin-like protease myroicolisin from deep sea bacterium *Myroides profundi* D25 and molecular insight into its collagenolytic mechanism. *J Biol Chem., 289*: 6041−6053.

[30] Jeon, J. H., Kim, J. T., Kim, Y. J., et al. (2009). Cloning and characterization of a new cold-active lipase from a deep-sea sediment metagenome. *Appl Microbiol Biotechnol., 81*: 865−874.

[31] Jiang, X., Xu, X., Huo, Y., et al. (2012). Identification and characterization of novel esterases from a deep-sea sediment metagenome. *Arch Microbiol., 194*: 207−214.

[32] Ohmae, E., Kubota, K., Nakasone, K., et al. (2004). Pressure-dependent activity of dihydrofolate reductase from a deep-sea bacterium *Shewanella violacea* strain DSS12. *Chem Lett., 33*: 798−799.

[33] Murakami, C., Ohmae, E., Tate, S., et al. (2011). Comparative study on dihydrofolate reductase from *Shewanella* species living in deep-sea and atmospheric-pressure environments. *Extremophiles, 15*: 165−175.

[34] Ohmae, E., Murakami, C., Tate, S., et al. (2012). Pressure dependence of activity and stability of dihydrofolate reductase of the deep-sea bacterium *Moritella profunda* and *Escherichia coli. Biochim Biophys Acta, 1824*: 511−519.

[35] Murakami, C., Ohmae, E., Tate, S., et al. (2010). Cloning and characterization of dihydrofolate reductases from deep-sea bacteria. *J Biochem., 147*: 591−599.

[36] Phillips, R. S., Ghaffari, R., Dinh, P., et al. (2011). Properties of tryptophan indole-lyase from a piezophilic bacterium, *Photobacterium profundum* SS9. *Arch. Biochem Biotechnol., 506*: 35−41.

[37] ZoBell, C. E. (1952). Bacterial life at the bottom of the Philippine Trench. *Science, 115*: 507−508.

[38] Ohmae, E., Gekko, K., and Kato, C. (2015). Environmental adaptation of dihydrofolate reductase from deep-sea bacteria. In: *High pressure bioscience – Basic concepts, applications and frontiers*, edited by Akasaka, K. and Matsuki, H. 423–442. New York:Springer.

[39] Bult, C. J., White, O., Olsen, G. J., et al. (1996). Complete genome sequence of the methanogenic archaeon, *Methanococcus jannaschii. Science, 273*: 1058−1073.

[40] Hou, S., Saw, J. H., Lee, K. S., et al. (2004). Genome sequence of the deep-sea gamma-proteobacterium *Idiomarina loihiensis* reveals amino acid fermentation as a source of carbon and energy. *Proc Natl Acad Sci USA., 101*: 18036−18041.

[41] Aono, E., Baba, T., Ara, T., et al. (2010). Complete genome sequence and comparative analysis of *Shewanella violacea*, a psychrophilic and piezophilic bacterium from deep sea floor sediments. *Mol Biosyst., 6*: 1216−1226.

[42] Berman, H., Henrick, K., and Nakamura, H. (2003). Announcing the worldwide Protein Data Bank. *Nat Struct Biol., 10*: 980.

[43] Saito, R., Kato, C., and Nakayama, A. (2006). Amino acid substitutions in malate dehydrogenases of piezophilic bacteria isolated from intestinal contents of deep-sea fishes retrieved from the abyssal zone. *J Gen Appl Microbiol., 52*: 9–19.

[44] Tanimoto, K., Higashi, N., Nishioka, M., et al. (2008). Characterization of thermostable aminoacylase from hyperthermophilic archaeon *Pyrococcus horikoshii. FEBS J., 275*: 1140−1149.

[45] Sayed, A., Ghazy, M. A., Ferreira, A. J., et al. (2014). A novel mercuric reductase from the unique deep brine environment of Atlantis II in the Red Sea. *J Biol Chem., 289*: 1675–1687.

[46] Fu, L., Li, X., Xiao, X., et al. (2014). Purification and characterization of a thermostable aliphatic amidase from the hyperthermophilic archaeon *Pyococcus yayanosii* CH1. *Extremophiles, 18*: 429–440.

[47] Mueller, P., Egorova, K., Vorgias, C. E., et al. (2006). Cloning, overexpression, and characterization of a thermoactive nitrilase from the hyperthermophilic archaeon *Pyrococcus abyssi*. *Protein Expr Purif., 47*: 672–681.

[48] Kleeb, A. C., Kast, P., and Hilvert, D. (2006). A monofunctional and thermostable prephenate dehydratase from the archaeon *Methanocaldococcus jannaschii*. *Biochemistry, 45*: 14101–14110.

[49] Mehrotra, S. and Balaram, H. (2008). *Methanococcus jannaschii* adenylosuccinate synthetase: Studies on temperature dependence of catalytic activity and structural stability. *Biochim Biophys Acta, 1784*: 2019–2028.

[50] Miyazono, K., Sawano, Y., and Tanokura, M. (2005). Crystal structure and structural stability of acylphosphatase from hyperthermophilic archaeon *Pyrococcus horikoshii* OT3. *Proteins,* 61: 196–205.

[51] Feller, G. (2013). Psychrophilic enzymes: from folding to function and biotechnology. *Scientifica,* doi: 10.1155/2013/512840.

[52] Yang, J. and Dang, H. (2011). Cloning and characterization of a novel cold-active endoglucanase establishing a new subfamily of glycosyl hydrolase family 5 from psychrophilic deep-sea bacterium. *FEMS Microbiol Lett., 325*: 71–76.

[53] Ohmae, E., Miyashita, Y., and Kato, C. (2013). Thermodynamic and functional characteristics of deep-sea enzymes revealed by pressure effects. *Extremophiles, 17*: 701–709.

[54] Ravindra, R. and Winter, R. (2003). On the temperature--pressure free-energy landscape of proteins. *Chemphyschem, 4*: 359–365.

[55] Scharnagl, C., Reif, M., and Friedrich, J. (2005). Stability of proteins: temperature, pressure and the role of the solvent. *Biochim Biophys Acta, 1749*: 187–213.

[56] Meersman, F., Smeller, L., and Heremans, K. (2006). Protein stability and dynamics in the pressure-temperature plane. *Biochim Biophys Acta, 1764*: 346–354.

[57] Yancey, P. H. and Siebenaller, J. F. (1999). Trimethylamine oxide stabilizes teleost and mammalian lactate dehydrogenases against inactivation by hydrostatic pressure and trypsinolysis. *J Exp Biol., 202*: 3597–3603.

[58] Yancey, P. H., Rhea, M. D., Kemp, K. M., et al. (2004). Trimethylamine oxide, betaine and other osmolytes in deep-sea animals: depth trends and effects on enzymes under hydrostatic pressure. *Cell Mol Biol.,* (Noisy-le-grand) *50*: 371–376.

[59] Macgregor, R. B. Jr. (1998). Effect of hydrostatic pressure on nucleic acids. *Biopolymers, 48*: 253–263.

[60] Takahashi, S. and Sugimoto, N. (2013). Effect of pressure on thermal stability of g-quadruplex DNA and double-stranded DNA structures. *Molecules, 18*: 13297–13319.

[61] Frankenberg, R. J., Anderson, M., and Clark, D. S. (2003). Effect of temperature and pressure on the proteolytic specificity of the recombinant 20S proteasome from *Methanococcus jannaschii*. *Extremophiles, 7*: 353–360.

[62] Zappa, S., Rolland, J. L., Flament, D., et al. (2001). Characterization of a highly thermostable alkaline phosphatase from the Euryarchaeon *Pyrococcus abyssi*. *Appl Environ Microbiol., 67*: 4504–4511.

[63] Zhan, D., Bai, A., Yu, L., et al. (2014). Characterization of the PH1704 protease from *Pyrococcus horikosii* OT3 and the critical functions of Tyr120. *PLOS one, 9*: e103902. doi: 10.1371/journal.pone.0103902.

[64] Ando, S., Ishida, H., Kosugi, Y., et al. (2002). Hyperthermostable endoglucanase from *Pyrococcus horikoshii*. *Appl Env Microbiol., 68*: 430–433.

[65] Lu, D., Xie, G., and Gao, R. (2014). Cloning, purification, and characterization of inorganic pyrophosphatase from the hyperthermophilic archaea *Pyrococcus horikoshii*. *Protein Expr Purif., 99*: 94–98.

[66] Chung, S. K., Ryu, S. I., and Lee, S. B. (2012). Characterization of UDP-glucose 4-epimerase from *Pyrococcus horikoshii*: Regeneration of UDP to produce UDP-galactose using two-enzyme system with trehalose. *Bioresour Technol., 110*: 423–429.

[67] Kim, H. W., Kashima, Y., Ishikawa, K., et al. (2009). Purification and characterization of the first archaeal glutamate decarboxylase from *Pyrococcus horikoshii*. *Biosci Biotechnol Biochem., 73*: 224–227.

[68] Liu, B., Li, Z., Hong, Y., Ni, J., et al. (2006). Cloning, expression and characterization of a thermostable exo-β-D-glucosaminidase from the hyperthermophilic archaeon *Pyrococcus horikoshii*. *Biotechnol Lett., 28*: 1655–1660.

[69] Cambon-Bonavita, M. A., Schmitt, P., Zieger, M., et al. (2000). Cloning, expression, and characterization of DNA polymerase I from the hyperthermophilic archaea *Thermococcus fumicolans*. *Extremophiles, 4*: 215–225.

[70] Wu, S., Liu, B., and Zhang, X. (2006). Characterization of a recombinant thermostable xylanase from deep-sea thermophilic *Geobacillus* sp. MT-1 in East Pacific. *Appl Microbiol Biotechnol., 72*: 1210–1216.

[71] Zhang, Y., Yi, L., Lin, Y., et al. (2014). Characterization and site-directed mutagenesis of a novel class II 5-enopyruvylshikimate-3-phosphate (EPSP) synthase from the deep-sea bacterium *Alcanivorax* sp. L27. *Enz Microbiol Technol., 63*: 64–70.

[72] Ohkuma, M., Ohtoko, K., Takada, N., et al. (1996). Characterization of malate dehydrogenase from deep-sea psychrophilic *Vibrio* sp. strain 5710 and cloning of its gene. *FEMS Microbiol Lett., 137*: 247–252.

[73] Saito, R. and Nakayama, A. (2004). Differences in malate dehydrogenases from the obligately piezophilic deep-sea bacterium *Moritella* sp. strain 2D2 and the psychrophilic bacterium *Moritella* sp. strain 5710. *FEMS Microbiol Lett., 233*: 165–172.

[74] De Vos, D., Xu, Y., Hulpiau, P., et al. (2007). Structural investigation of cold activity and regulation of aspartate carboxytransferase from the extreme psychrophilic bacterium *Moritella profunda*. *J Mol Biol., 365*: 379–395.

[75] Xu, Y., Feller, G., Gerday, C., et al. (2003). *Moritella* cold-active dihydrofolate reductase: Are there natural limits to optimization of catalytic efficiency at low temperature? *J Bacteriol., 185*: 5519–5526.

[76] Xu, Y., Zhang, Y., Liang, Z., et al. (1998). Aspartate carbamoyltransferase from a psychrophilic deep-sea bacterium, *Vibrio* strain 2693: properties of the enzyme, genetic organization and synthesis in *Escherichia coli*. *Microbiology, 144*: 1435–1441.

[77] Zeng, R., Xiong, P., and Wen, J. (2006). Characterization and gene cloning of a cold-active cellulose from a deep-sea psychrophilic bacterium *Pseudoalteromonas* sp. DY3. *Extremophiles, 10*: 79–82.
[78] Xiong, H., Song, L., Xu, Y., et al. (2007). Characterization of proteolytic bacteria from the Aleutian deep-sea and their proteases. *J Ind Microbiol Biotechnol., 34*: 63–71.
[79] Kobayashi, H., Takaki, Y., Kobata, K., et al. (1998). Characterization of α-maltotetraohydrolase produced by *Pseudomonas* sp. MS300 isolated from the deepest site of the Mariana Trench. *Extremophiles, 2*: 401–407.
[80] Zeng, R., Zhang, R., Zhao, J., et al. (2003). Cold-active serine alkaline protease from the psychrophilic bacterium *Pseudomonas* strain DY-A: enzyme purification and characterization. *Extremophiles, 7*: 335–337.
[81] von Jan, M., Lapidus, A., Del Rio, T. G., et al. (2010). Complete genome sequence of *Archaeoglobus profundus* type strain (AV18). *Stand Genomic Sci., 2*: 327–346.
[82] Cohen, G. N., Barbe, V., Flament, D., et al. (2003). An integrated analysis of the genome of the hyperthermophilic archaeon *Pyrococcus abyssi. Mol Microbiol., 47*: 1495–1512.
[83] Kawarabayasi, Y., Sawada, M., Horikawa, H., et al. (1998). Complete sequence and gene organization of the genome of a hyper-thermophilic archaebacterium, *Pyrococcus horikoshii* OT3. *DNA Res., 30*: 55–76.
[84] Jun, X., Lupeng, L., Minjuan, X., et al. (2011). Complete genome sequence of the obligate piezophilic hyperthermophilic archaeon *Pyrococcus yayanosii* CH1. *J Bacteriol., 193*: 4297–4298.
[85] Lee, H. S., Bae, S. S., Kim, M. S., et al. (2011). Complete genome sequence of hyperthermophilic *Pyrococcus* sp. strain NA2, isolated from a deep-sea hydrothermal vent area. *J Bacteriol., 193*: 3666–3667.
[86] Jung, J. H., Lee, J. H., Holden, J. F., et al. (2012). Complete genome sequence of the hyperthermophilic archaeon *Pyrococcus* sp. strain ST04, isolated from a deep-sea hydrothermal sulfide chimney on the Juan de Fuca Ridge. *J Bacteriol., 194*: 4434–4435.
[87] Vannier, P., Marteinsson, V. T., Fridjonsson, O. H., et al. (2011). Complete genome sequence of the hyperthermophilic, piezophilic, heterotrophic, and carboxydotrophic archaeon *Thermococcus barophilus* MP. *J Bacteriol., 193*: 1481 –1482.
[88] Zhao, W. and Xiao, X. (2015). Complete genome sequence of *Thermococcus eurythermalis* A501, a conditional piezophilic hyperthermophilic archaeon with a wide temperature range, isolated from an oil-immersed deep-sea hydrothermal chimney on Guaymas Basin. *J Biotechnol., 193*: 14–15.
[89] Zivanovic, Y., Armengaud, J., Lagorce, A., et al. (2009). Genome analysis and genome-wide proteomics of *Thermococcus gammatolerans*, the most radioresistant organism known amongst the Archaea. *Genome Biology, 10*: R70. doi: 10.1186/gb-2009-10-6-r70.
[90] Oberto, J., Gaudin, M., Cossu, M., et al. (2014). Genome sequence of a hyperthermophilic archaeon, *Thermococcus nautili* 30-1, that produces viral vesicles. *Genome Announc., 2*: e00243-14. doi: 10.1128/ genomeA.00243-14.
[91] Lee, H. S., Kang, S. G., Bae, S. S., et al. (2008). The complete genome sequence of *Thermococcus onnurineus* NA1 reveals a mixed heterotrophic and carboxydotrophic metabolism. *J Bacteriol., 190*: 7491–7499.
[92] Wang, X., Gao, Z., Xu, X., et al. (2011). Complete genome sequence of *Thermococcus* sp. strain 4557, a hyperthermophilic archaeon isolated from a deep-sea hydrothermal vent area. *J Bacteriol., 193*: 5544–5545.

[93] Oger, P., Sokolova, T. G., Kozhevnikova, D. A., et al.. (2011). Complete genome sequence of the hyperthermophilic archaeon *Thermococcus* sp. strain AM4, capable of organotrophic growth and growth at the expense of hydrogenogenic or sulfidogenic oxidation of carbon monoxide. *J Bacteriol., 193*: 7019–7020.

[94] Jung, J. H., Holden, J. F., Seo, D. H., et al. (2012). Complete genome sequence of the hyperthermophilic archaeon *Thermococcus* sp. strain CL1, isolated from a *Paralvinella* sp. polychaete worm collected from a hydrothermal vent. *J Bacteriol., 194*: 4769–4770.

[95] Jung, J. H., Kim, Y. T., Jeon, E. J., et al. (2014). Complete genome sequence of hyperthermophilic archaeon *Thermococcus* sp. ES1. *J Biotechnol., 174*: 14–15.

[96] Takaki, Y., Shimamura, S., Nakagawa, S., et al. (2010). Bacterial lifestyle in a deep-sea hydrothermal vent chimney revealed by the genome sequence of the thermophilic bacterium *Deferribacter desulfuricans* SSM1. *DNA Res., 17*: 123–137.

[97] Göker, M., Daligault, H., Mwirichia, R., et al. (2011). Complete genome sequence of the thermophilic sulfur-reducer *Desulfurobacterium thermolithotrophum* type strain (BSA^T) from a deep-sea hydrothermal vent. *Stand Genomic Sci., 5*: 407–415.

[98] Klippel, B., Lochner, A., Bruce, D. C., et al. (2011). Complete genome sequence of the marine cellulose- and xylan-degrading bacterium *Glaciecola* sp. strain 4H-3-7+YE-5. *J Bacteriol., 193*: 4547–4548.

[99] Copeland, A., Gu, W., Yasawong, M., et al. (2012). Complete genome sequence of the aerobic, heterotroph *Marinithermus hydrothermalis* type strain ($T1^T$) from a deep-sea hydrothermal vent chimney. *Stand Genomic Sci., 6*: 21–30.

[100] Lucas, S., Han, J., Lapidus, A., et al. (2012) Complete genome sequence of the thermophilic, piezophilic, heterotrophic bacterium *Marinitoga piezophila* KA3. *J Bacteriol., 194*: 5974-5975.

[101] Scott, K. M., Sievert, S. M., Abril, F. N., et al. (2006). The genome of deep-sea vent chemolithoautotroph *Thiomicrospira crunogena* XCL-2. *PLoS Biol., 4*: e383. doi: 10.1371/journal.pbio.0040383.

[102] Takami, H., Takaki, Y., and Uchiyama, I. (2002). Genome sequence of *Oceanobacillus iheyensis* isolated from the Iheya Ridge and its unexpected adaptive capabilities to extreme environments. *Nucleic Acids Res., 30*: 3927–3935.

[103] Reysenbach, A. L., Hamamura, N., Podar, M., et al. (2009). Complete and draft genome sequences of six members of the *Aquificales*. *J Bacteriol., 191*: 1992–1993.

[104] Vezzi, A., Campanaro, S., D'Angelo, M., et al. (2005). Life at depth: *Photobacterium profundum* genome sequence and expression analysis. *Science, 307*: 1459–1461.

[105] Zhang, S. D., Barbe, V., Garel, M., et al. (2014). Genome sequence of luminous piezophile *Photobacterium phosphoreum* ANT-2200. *Genome Announc., 2*: e00096-14. doi: 10.1128/genomeA.00096-14.

[106] Lauro, F. M., Stratton, T. K., Chastain, R. A., et al. (2013) Complete genome sequence of the deep-sea bacterium *Psychromonas* strain CNPT3. *Genome Announc., 1*: e00304-13. doi: 10.1128/ genomeA.00304-13.

[107] Anderson, I., Saunders, E., Lapidus, A., et al. (2012). Complete genome sequence of the thermophilic sulfate-reducing ocean bacterium *Thermodesulfatator indicus* type strain ($CIR29812^T$). *Stand Genomic Sci., 6*: 155–164.

[108] Pitluck, S., Yasawong, M., Munk, C., et al. (2010). Complete genome sequence of *Thermosediminibacter oceani* type strain (JW/IW-1228P). *Stand Genomic Sci., 3*: 108–116.

[109] Giovannelli, D., Ricci, J., Pérez-Rodríguez, I., et al. (2012). Complete genome sequence of *Thermovibrio ammonificans* HB-1^T, a thermophilic, chemolithoautotrophic bacterium isolated from a deep-sea hydrothermal vent. *Stand Genomic Sci., 7*: 82−90.
[110] Qin, Q. L., Zhang, X. Y., Wang, X. M., et al. (2010). The complete genome of *Zunongwangia profunda* SM-A87 reveals its adaptation to the deep-sea environment and ecological role in sedimentary organic nitrogen degradation. *BMC Genomics, 11*: 247. doi: 10.1186/1471-2164-11-247.
[111] Liu, B., Bartlam, M., Gao, R., et al. (2004). Crystal structure of the hyperthermophilic inorganic pyrophosphatase from the archaeon *Pyrococcus horikoshii. Biophys J., 86*: 420−427.
[112] Avaeva, S., Kurilova, S., Nazarova, T., et al. (1997). Crystal structure of *Escherichia coli* inorganic pyrophosphatase complexed with SO_4^{2-}. Ligand-induced molecular asymmetry. *FEBS Lett., 410*: 502−508.
[113] Van Boxstael, S., Cunin, R., Khan, S., et al. (2003). Aspartate transcarbamylase from the hyperthermophilic archaeon *Pyrococcus abyssi*: thermostability and 1.8Å resolution crystal structure of the catalytic subunit complexed with the bisubstrate analogue N-phosphonacetyl-L-aspartate. *J Mol Biol., 326*: 203−216.
[114] Endrizzi, J. A., Beernink, P. T., Alber, T., et al. (2000). Binding of bisubstrate analog promotes large structural changes in the unregulated catalytic trimer of aspartate transcarbamoylase: implications for allosteric regulation. *Proc Natl Acad Sci USA., 97*: 5077−5082.
[115] Yao, M., Yasutake, Y., Morita, H., et al. (2005). Structure of the type I L-asparaginase from the hyperthermophilic archaeon *Pyrococcus horikoshii* at 2.16 angstroms resolution. *Acta Crystallogr., D 61*:2 94−301.
[116] Swain, A. L., Jaskólski, M., Housset, D., et al. (1993). Crystal structure of *Escherichia coli* L-asparaginase, an enzyme used in cancer therapy. *Proc Natl Acad Sci USA., 90*: 1474−1478.
[117] Müller, C. W. and Schulz, G. E. (1992). Structure of the complex between adenylate kinase from *Escherichia coli* and the inhibitor Ap5A refined at 1.9Å resolution. A model for a catalytic transition state. *J Mol Biol., 224*: 159−177.
[118] Nagae, T., Kato, C., and Watanabe, N. (2012). Structural analysis of 3-isopropylmalate dehydrogenase from the obligate piezophile *Shewanella benthica* DB21MT-2 and the nonpiezophile *Shewanella oneidensis* MR-1. *Acta Crystallogr., F 68*: 265–268.
[119] Sawaya, M. R. and Kraut, J. (1997). Loop and subdomain movements in the mechanism of *Escherichia coli* dihydrofolate reductase: crystallographic evidence. *Biochemistry, 36*: 586–603.

In: Microbial Catalysts. Volume 1
Editors: S. M. Abdel-Aziz, N. Garg, A. Aeron et al.
ISBN: 978-1-53614-679-0

Chapter 13

SOLVING THE STRUCTURE OF AMIDASE-03

Ravi Datta Sharma* and Garima Bhardwaj
Centre of Microbial and Plant Genetics, Leuven, Belgium
Amity Institute of Integrative Science and Health, Amity University Haryana, Panchgaon (Manesar), Haryana, India

ABSTRACT

Autolysins are bacteriolytic enzymes that are capable of hydrolyzing the cell wall peptidoglycan. These potentially lethal autolysins are important in several processes including cell growth, cell-wall turnover, peptidoglycan maturation, cell division, separation, protein secretion, differentiation and pathogenicity. In this chapter, firstly we focused on peptidoglycan amidases, their classification and secondly, we studied the structural stability of amidase-03 from *Bacillus antharacis*. Unlike other amidases, amidase-03 in *B. anthracis* plays an important role in innate immunity by decreasing the inflammatory response in host by hydrolyzing the bond between glycan and peptide. So the structure stability of amidase-03 is important to assess its function and role during infection. The chapter includes the prediction of amidase-03 structure using bioinformatics tools and structural analysis using molecular dynamics simulation.

Keywords: autolysins, amidase, bacillus antharacis

INTRODUCTION

The hydrolysis of C-N bond containing compounds such as organic cyanides or nitriles (R-C≡N), inorganic cyanides (H-C≡N) and amides (for example; acid amides R-C(=O)-NH2], secondary amides [R-C(=O)NH-R'] and N-carbamyl amides [R-NH-C(=O)-NH2]) are found in both prokaryotes and eukaryotes. The hydrolysis of these compounds is done by nitrilase family enzymes [1], signature amidases [2] [3], N-terminal nucleophile hydrolases, amidotransferases [4] and hydrazidase [5]. Using a conserved cysteine residue, nitrilase family

* Corresponding Author's Email: ravidattasharma@gmail.com.

enzymes catalyze either the cyano carbon of a linear nitrilase substrate or the planar carbon of an amide substrate. Signature amidases possess highly conserved Ser-Ser-Lys catalytic triad that is used for amide hydrolysis and a conserved stretch of 130 amino acids known as the AS sequence. However the catalytic mechanism among the family member is highly substrate specific. Recently found hydrazidase, hydrolyze the C-N bond of acylated nitrogen. The hydrazidase contains the catalytic Ser-Ser-Lys motif like Signature amidases but they lack the catalytic Ser residue that is conserved among members of the amidase signature family. N-terminal nucleophile hydrolases and amidotransferases [4] act on C-N bond-containing substrates but differ in mode of action. Amidotransferase catalyses the formation of Gln-tRNA(Gln) where tRNA(Gln) is misacylated with Glu first and then, this Glu-tRNA(Gln) is converted into Gln-tRNA(Gln) by the process of amidolysis.

In general, these enzyme act on linear amides where the mode of action include two substrate (monocarboxylic acid amide and H_2O) and production of monocarboxylate and NH_3 (Figure 1).

In both eukaryotic and prokaryotic organisms, amidases are important in order to produce substances such as auxin and biotin in plants, in nutrient metabolism, degradation of toxic cyanogenic compounds [6] and in the post-translational modification of amino acids, proteins and other compounds [7]. In metabolism pathway, they found to be taking part in urea cycle and metabolism of amino groups, phenylalanine metabolism, tryptophan metabolism, cyanoamino acid metabolism, benzoate degradation via coa ligation, and styrene degradation.

The biological function of amidases varies from peptidoglycan biosynthesis to industrial catalyst, treatment of waste-water and in inflammatory response to host. The main function of amidases in prokaryotes varies from; fixation of the carbon or nitrogen through hydrolysis of amides, the generation of properly charged tRNAGln by transamidation of misacylated Glu (in eubacteria) or by amidolysis of Glutamine (in bacteria) [8], the formation of indole-3-acetic acid (IAA) in Agrobacterium. Amidases are widely used in production of pharmaceutical chemicals such as carboxylic acids, hydroxamic acids [9], [10], hydrazide, acrylic acid, *p*-aminobenzoic acid, pyrazinoic acid, nicotinic acid, etc. [11] and treatment of sewage water [12], [13]. Some studies have suggested their role in survival of bacteria by decreasing the inflammatory response in host [14].

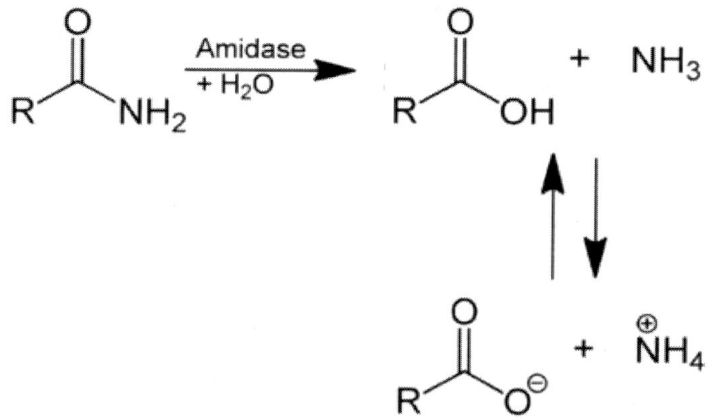

Figure 1. The formation of monocarboxylate and NH_3 by amidase in presence of a water molecule and monocarboxylic acid amide.

Most microorganisms have a major cell wall component called peptidoglycan [15]. Generally, peptidoglycan is consist of a carbohydrate backbone of alternating units of N-acetylglucosamine (GlcNAc) and N-acetylmuramic acid (MurNAc) that are connected by b-(1,4)-glycosidic bonds. The peptidoglycan layer constantly undergoes modification by hydrolysis through the transfer of new molecules and by the release of old molecules. The hydrolysis activity is essential for the correct cell wall assembly and other metabolic activities. The enzymes performing peptidoglycan cell wall hydrolysis are known as amidase (E.C. 3.5.1.4) or peptidoglycan amidohydrolase or systematic name *acylamide amidohydrolase.* These enzymes belong to the family of hydrolases and catalyze the hydrolysis of amide bonds (CO-NH2) in peptidoglycan. Structural information combined with biochemical studies suggest that peptidoglycan amidases are almost specific for every bond found in peptidoglycan even though they might share the similar fold but cleave different bonds and vice-versa, cleave the same bond despite the fact of entirely different folds [16].

In eukaryotes, peptidoglycan amidases play an important role in innate immunity, which is first line of defense against microorganisms [17][18]. Eukaryotic-peptidoglycan amidases are generally known as peptidoglycan recognition proteins (PGRPs) that share homology to T7 lysozyme. Thomas Werner et al., [19] showed PGRP-SC1B in *Drosophila* to be an N-acetylmuramoyl-l-alanine amidase (NAMLAA), an enzyme that cleaves the lactylamide bond between muramic acid and the peptide chain in peptidoglycan. Later, authors in [20] found M.mPGRP-L in mammalian sera that had N-acetylmuramoyl-l-alanine amidase activity and were able to degrade peptidoglycan from both *Escherichia coli* and *Staphylococcus aureus*. Also, the catabolism of neuromodulatory fatty-acid-amides catalyzed by amidases is very well known in mammals [21], [3].

CLASSIFICATION OF AMIDASES

The peptidoglycan layer is made up of glycan strands that consists of alternate **N**-acetylglucosamine (GlcNAc) and **N**-acetylmuramyl (MurNAc) residues [22]. The peptidoglycan layer continues to undergo hydrolysis and then, forms new layer by the process of polymerization of glycan strands using glycosyltransferases and cross-linking of stem peptides using DD-transpeptidases [22]. Since the activity of the peptidoglycan hydrolases include the cell lysis in response to various stresses or antibiotics, these enzymes are also known as autolysins. A number of enzymes with peptidoglycan hydrolytic activities have been discovered so far. The function of the these enzyme vary a lot since they are found in great extent in any bacteria and may play different roles, for example, *E. coli* has five N-acetylmuramoyl-L-alanine amidases, three peptidoglycan endopeptidase and six membrane bound lytic transglycosylases. All of which have role in cleavage of the septum during cell division, release of turnover product during cell growth and autolysin activity [23]. These enzymes are generally classified as *N-acetylmuramoyl-l-alanine amidases, N–acetyl-beta-D-muramidase, N-acetylglucosaminidases, endopeptidases* based on their substrate specificity [24], [23]. Other names in common use include acetylmuramyl-L-alanine amidase, N-acetylmuramyl-L-alanine amidase, N-acylmuramyl-L-alanine amidase, acetylmuramoyl-alanine amidase, N-acetylmuramic acid L-alanine amidase, acetylmuramyl-alanine amidase, N-acetylmuramylalanine amidase, N-acetylmuramoyl-L-alanine amidase type I, and N-acetylmuramoyl-L-alanine amidase type II. Autolysins and some phage lysins are examples of N-acetylmuramoyl-L-alanine amidases.

N-acetylmuramoyl-l-alanine amidases is the most studied enzyme in family amidases including pathogenic bacteria such as *Bacillus anthracis* [25]. This enzyme activity is also reported in human [26] where it play a vital role in human innate immunity by causing bacterial cell lysis. These enzymes are commonly known as peptidoglycan N-acetylmuramoylhydrolase or autolytic N-acetylmuramoylhydrolase (EC 3.2.1.17). The enzymes perform the hydrolysis of amide bond between MurNAc and the N-terminal L-alanine residue in peptidoglycan. Unlike other bacteriophage peptidoglycan amidases that are endolysins, these enzymes are highly substrate specific for example, in *Streptococcus faecium* the enzyme contains an autolytic peptidoglycan hydrolase activity of only a single specificity towards P-1,4-N-acetyl-muramoyl-hydrolase. This enzyme activity has specific affinity for binding to the cell wall of this species. But in disrupted cells, these enzymes are found to be associated with the insoluble cell wall fraction [27]. *Several MurNAc-LAAs can be found in bacterial proteome for example, in E. coli, there are AmiA, AmiB, AmiC, AmiD and AmiE* [28], [29], [23].

N–Acetyl-Beta-D- Muramidase (Lysozymes and Lytic Transglycosylases)

N–acetyl-beta-D- muramidases are either Lysozymes or lytic transglycosylases depending on the mode how they hydrolyze the glycosidic bond, however, both cleave the beta 1-4 glycosidic bond between MurNAc and GlcNAc. Lysozymes are the most studied class of enzymes that found in phages, bacteria, fungi, vertebrates and invertebrates. Out of nine classes, only four are proven to have hydrolytic activity against peptidoglycan but all of them cleave beta 1-4 glycosidic bonds [23]. Some examples include; two lysoszymes (SF muramidase and pesticin) that are found in *Enterococcus faecium* ATCC9790 [27], [30], cellosyl produced by *Streptomyces coelicolor* [31], the autolysin LytC from *S. pneumonia* [32]. Lytic transglycosylases also cleave the β1,4-glycosidic bond between MurNAc and GlcNAc but the process include a ring formation (1,6-anhydro) at the MurNAc residue [33]. Enzyme Slt70, MltA, MltB, MltC, MltD, YfhD and EmtA from *E. coli* are such example of Lytic transglycosylases [23], [34], [35], [36].

N-acetylglucosaminidases are known as N-acetylmuramoyl-glycoprotein N-acetyl-D-glucosamine hydrolase (EC 3.2.1.B37) or N-acetylmuramoyl-glycoprotein endo-beta-N-acetylglucosaminidase. N-acetylglucosaminidases possess hydrolase acitivty as well as lytic activity on the peptidoglycan in many Gram-positive bacteria including *Clostridium difficile* [37]. This enzyme hydrolyses peptidoglycan bonds between N-acetylglucosamine and N-acetylmuramic acid. In *E. Faecalis*, N-acetylglucosaminidase (AtlA) has been reported to require a LysM peptidoglycan-binding domain for full activity [38]. In *Lactococcus lactis*, there are three known N-acetylglucosaminidase (AcmA, AcmB and AcmC) and one hypothetical (AcmD) [39], [40], [41], [23].

Endopeptidases and Carboxypeptidases

Endopeptidases and carboxypeptidases cleave amide bonds between amino acids in peptidoglycan or its soluble fragments [42], [43], [23]. Depending on their specificity, peptidoglycan peptidases are divided into two categories, first carboxypeptidases that remove a C-terminal amino acid and second, endopeptidase that cleave within the peptide. In both the case, DD-Peptidases cleave between two D-amino acids while LD- or DL-peptidases cleave between an L- and a D-amino acid [23]. PBP4 in *E. coli* and PBP4a in *B. Subtilis* are example of DD-endopeptidase [44], [45]. However, MepA from *E. coli* can hydrolyze both DD and LD-peptide cross-links [46], [47], [48]. Other examples are CwlK and LytH in *B. Subtilis*. LdcA is an example of LD-carboxypeptidases and PBP5 of DD-carboxypeptidases from *E. coli* (reviewed in [23]).

Another class of enzyme, *hexosaminides* (EC 3.2.1.52) is found to be involved in hydrolysis of terminal non-reducing N-acetyl-D-hexosamine residues in N-acetyl-beta-D-hexosaminides. Human β-N-acetyl-D-hexosaminidase, which is able to hydrolyze the peptidoglycan of bacterial cell wall, showed mycobactericidal effect that play an important role in host defense [49]. *N-acetylmuramidases* is systematically known as peptidoglycan beta-N-acetylmuramidase (EC 3.2.1.92) that hydrolyze the terminal and non-reducing N-acetylmuramic residues. N-acetylmuramidases has synonymous exo-beta-N-acetylmuramidases and exo-beta- acetylmuramidase. The activity has been well described in *Bacillus subtilis* [50], *Streptococcus faecali and in Streptococcus faecium* [51].

Functions of Bacterial Peptidoglycan Amidases

Although, there have been number of studies reporting amidases, but the area seems to be unexplored that needs to be investigated in more detail. Bacterial peptidoglycan hydrolases not only play important role in bacterial cell physiology but also work as determinant in bacterial population. Moreover, they release the fragments of peptidoglycan that, in turn, are recognized by the host immune system, i.e., host-pathogen interaction. Peptidoglycan hydrolases properties of amidase are used in genetic transformation, lysis of cell during development, biofilm formation and also, for the digestion of peptidoglycan of prey cells. Bacterial peptidoglycan hydrolases can affect the cell physiology by regulating the cell wall growth, by enlarging the size of sacculus in peptidoglycan and by participating in cell separation during or after the cell division. Interestingly, the turnover product during hydrolysis of peptidoglycan can be reutilized in synthesis of peptidoglycan, itself.

Solving Structure of the Amidases-03

The etiologic agent of anthrax, *Bacillus anthracis* contains two virulent plasmids pXO1 and pXO2 causing cutaneous, gastrointestinal and inhalation anthrax. *B. anthracis* produces protective antigen (83 kDa), edema factor (84) kDa) and lethal-factor (83kDa) causing inflammatory response with the release of toxic compounds and mediators [53]. The genome sequencing and annotation shows the presence of several amidases like proteins in the

B. anthracis genome. The most studied amidase gene from *B. anthracis* is amiA belonging to class amidase-03 of autolysins [54]. Many autolysins usually increase the inflammatory response. However, amidase-03 decreases the inflammatory response in host by hydrolyzing the bond between glycan and peptide thereby reducing the amplification to response [54]. So the structure stability of amidase-03 is important to assess its function and role during infection. Therefore, it is important to construct its protein structure for amidase-03 to glean functional information using modeling methods described elsewhere [54], [55], [56]. The advancement in bioinformatics approaches allows the prediction of a protein from its protein sequence using homology modeling.

The homology modeling approach include the search for suitable template, building of model, validation of model and finally energy minimization of the structure. Template selection was done using BLASTp for the query sequence against PDB (Protein Data Bank) available at NCBI [57]. BLOSUM62 was applied as matrix. The target subsequently selected was the x-ray crystal structure with PDB code 1XOV with resolution 1.8 Å and R-value 0.122. Sequence alignment for template 1XOV and query (Figure 2) was done by using python script align2d.py in Modeller [58]. Sequence alignment was used to generate several 3D models of amidase-03 using auto-model class. Five models were built by Modeller default routine in which initial coordinates are randomized and evaluated by pseudo-energy parameter. If several models are calculated for the same target, the best model can be selected in several ways. We evaluated the model with the lowest value of the Modeller objective function and PROCHECK statistics, which is also recognized by ADIT tool at PDB Server. The stereo-chemical properties were assessed using PROCHECK [59]. Energy minimization of the selected model was done by using steepest descent method for removal of the bad vander-waal contacts from the model. The calculations were done by using GROMACS 3.2 [60] software with G43a1 force field with flexible SPC water model in cubic box of size 1 Angstrom instead of rigid SPC model. Structure analysis and visualization were carried out with Visual Molecular Dynamics (VMD) [61] software. Structural alignment and RMSD calculation with template and several models were also done with VMD.

```
                  10        20        30        40        50        60        70        80
          ....*....|....*....|....*....|....*....|....*....|....*....|....*....|....*....|
28843       6 LHAGHNSIVQGANYGNRKEHIMDRQVKDAVVAKLRALGHTVyDDTDEVGTTQaqNLNNIVSKTNSHDVDLVVSFHLNS-Y 84
pfam01520   3 IDPGHGGKDPGAVGNGLQEKDVTLDIALRLRKLLRAQGANV-ILTRTSDTFV--DLYERVAKANAANADLFVSIHANAaP 79
                  90       100       110       120       130       140       150       160
          ....*....|....*....|....*....|....*....|....*....|....*....|....*....|....*....|
28843      85 DTRANGVEVLYYDQQA---------LSAKIAAQLSKDIGWSNRGAKErKDLYVLSNTKAPAILIELGFIDNEADMAKWN- 154
pfam01520  80 DGSARGTEVYYYSNKGaasarakalLADELVDSLQKVTILDSRGVKE-ANLYVLRNTTMPSVLVELGFISNPEDARLLKs 158
                 170
          ....*....|....*..
28843     155 ---PDKIANSIVYALTG 168
pfam01520 159 akyRQKLAQAIAKGILR 175
```

Figure 2. Sequence alignment of amidase-03 with pfam database (amidase family) as a BLASTp output.

BLAST results had shown higher similarity with the crystal structure of 1XOV and 1JWQ, while the template was selected on the basis of greater identity. 1XOV template showed 35% sequence identity, 47 conserved residues and 55% of sequence similarity with amidase-03, while the 1JWQ showed 26% sequence identity and 55% sequence similarity. A putative domain was also found against pfam database entry pfam01520, a family of hydrolase enzyme consisting of catalytic domain, which cleaves peptidoglycan of the bacterial cell wall by MurNAc-L-alanine linkage and has 51 conserved regions in BLASTp results (Figure 2). So, the 245 amino acids protein from *B. anthracis* genome encoded by GenBank accession NP_844822 consists of amidase-03 domain known to have catalytic functions to hydrolyze the cell wall. This protein in this study is referred as amidase-03 hereon.

The template structure 1XOV consists of two domains, a C- terminus and N-terminus while we used only N- terminus of the protein to build the model. Figure 1 shows the alignment of 171 residues of amidase-03 with 316 residues of 1XOV. After 175 initial residues, Modeller put a long gap with C-terminal of the template as we used only N-terminus of the protein to build models of amidase-03. Due to the high target and template similarity, there are only two gaps inserted at 96th and 97th position in the N- terminus of the sequence alignment. Using crystal structure for Homology modeling provides high quality structure alignment for structure prediction [62]. Phi and Phi torsion angles of predicted models were checked by PROCHECK. In Ramachandran plot analysis, one residue of each model was found in disallowed region. For a conformational asset, a PROCHECK analysis was done with template 1XOV, which also confirmed one residue in disallowed region of template. So, the probable error can be negotiated here.

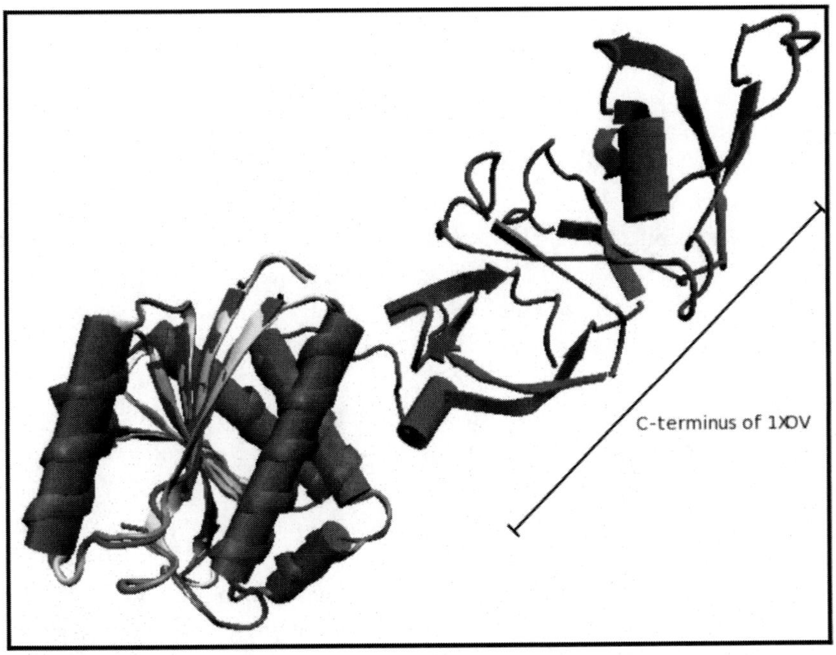

Figure 3. Superimposition of structures of amidase-03 (Red-new cartoon selection) on N-terminal of 1XOV (Ribbon selection).

The total number of residues in selected model was found as: 94.3% residues in most favored region, 3.8% in additional allowed region and 1.3% in generously allowed region of the Ramachandran plot. Structural alignment was done using VMD to calculate the structure deviation between template (1XOV) and selected model of amidase- 03 of *B. anthracis* (Figure 3). The RMSD between template and selected model showed a value of 0.0976 nm, which is an indicative of a good quality predicted model. However, all predicted models had less than 0.15 nm RMSD values. The predicted model of amidase-03 consists of six α-helices and six β-sheets. All β-sheets are surrounded with helices forming inner core of the protein and providing the amidase moiety. The overall conformation of the model was found similar to N-terminal of 1XOV (Figure 3), which is an enzymatically active autolysin as reported in *Listeria monocytogenes* bacteriophage PSA endolysin PlyPSA [63]. These autolysins have unusual substrate specificity and high activity towards degradation of the murein layers and lysis of susceptible cells [63]. Based on their antimicrobial properties and substrate specificity, these enzymes show great potential to be effective therapeutic agents [64]. The increasing resistance to the known antibiotics has driven attention towards finding newer ways to combat pathogenic organisms and thus the investigation of the structures for the development of new drugs/drug targets from the molecules like amidases have become reasonably essential especially for pathogenic organisms like *B. anthracis.*

ESTABLISHING STABILITY OF AMIDASES-03

Amidases-03 (503 residues) is a well-studied enzyme protein encoded by the gene *amiA* present in the plasmid pXO2. AmiA consists of two functional domains, of which the N-terminal one consists of three SLH (S-layer homology domain) repeats (50 residues). The remaining part of the protein is composed of two sub-domains (one domain of 100 residues with unknown structure and the other N-acetylmuramoyl-L-alanine amidase with similarity to pfam01520 family) [65]. The amidase-03 domain shows enzymatic activity of hydrolyzing the peptidoglycan at MurNAc-L-alanine linkage [65]. It decreases the inflammatory response and hence favoring the multiplication of *B. anthracis* in the host. However, some peptidoglycan hydrolases increase the inflammatory response by releasing the muramyl peptide [53], [66]. Hence this amidase is of important value. So, the stability of amidase-03 domain (GenBank accession NP_844822) was done using data obtained from MD simulations and homology modeling.

Protein folding and unfolding dynamics are still under scrutiny by the scientific community using *in vivo*, *in vitro* and *in silico* approaches [67], [68], [69], [70]. Protein unfolding is one of the most fundamental biochemical processes and it is very difficult to characterize using *in vivo* methods. However, several theoretical methods with the use of computing power have been used in recent years to address this issue. We use molecular dynamics (MD) simulations to characterize the unfolding process of an N acetylmuramoyl- L-alanine amidase (PDB ID-2IR0), a cell wall hydrolase enzyme from *Bacillus anthracis* at high temperature. We used high temperature for this study because it is known to accelerate the unfolding path during simulation [71].

Molecular dynamics simulation uses classical Newton mechanics to generate trajectories to study molecular motion using force fields. Molecular trajectories are coordinates formed by recording structural changes over time showing the positions of each atom along a series of small time step (usually 2 fs). The G43a1 force field with explicit SPC water and cubic boundary conditions are used to parameterize system interactions. The system consists of 8614 water molecules surrounding the protein, in a box of edge length 65.69 Å. This was subjected to an isobaric- isothermal MD simulation at 1 atom through a Berendsen thermostat-barostat [60] with coupling constants of 0.1 ps and 0.5 ps, respectively. The real space coulomic and short range contributions were truncated at 9 Å and 14 Å, respectively. The net negative charge on the protein was offset by replacing two water molecules with two sodium ions. This resulted in 8612 water molecules and 2 Na+ surrounding the protein. The resulting box had a density of 1.02 g/cc. All simulations were performed using a 2 fs time step.

The amidase-3 model (PDB ID: 2IR0) was used as a starting structure. In an initial equilibration, the protein amidase-03 was kept fixed, whereas the solvent and Na+ were subjected to an energy minimization using steepest descent avoiding any steric energy conflicts. The minimum energy was achieved in 475 steps. Electrostatics is calculated using the Particle Mesh Ewald (PME) method [72]. All rigid bonds were monitored using the Lincs method [60]. The software package used for simulation is GROMACS v 3.3.2 [73]. All simulations were performed in a desktop AMD64 4300+ processor, with 2GB RAM, using Fedora Core 8 as operating system. The protein was simulated at two temperatures - 300 K and 498 K. Simulation data at high and low temperatures provide measurements for studying protein conformation stability [67]. Proteins are usually stable at 300K and often undergo unfolding at 498 K [74]. We used the root-mean-square-displacement (RMSD) parameter to analyze the unfolding of the protein during simulation. The RMSD of the backbone and different secondary structures of the protein were computed with respect to the initial structure of the respective trajectories after a least square fit of the non-hydrogen atoms. A cut- off value of 5 ns was used for equilibration simulation. The precompiled Dictionary of Protein secondary structure (DSSP) [75] was subsequently used to analyze the secondary structure unfolding, while Visual Molecular Dynamics (VMD) software [61] was used for structure visualization.

Table 1. Structures Designation

S. No.	Residues		Structure	Name
	Start	End		
1	2	12	β-sheet	B1
2	24	42	α-helix	H1
3	45	51	β-sheet	B2
4	57	71	α-helix	H2
5	75	82	β-sheet	B3
6	89	93	β-sheet	B4
7	96	113	α-helix	H3
8	117	122	β-sheet	B5
9	127	131	α-helix	H4
10	136	143	β-sheet	B6
11	147	154	α-helix	H5
12	158	170	α-helix	H6

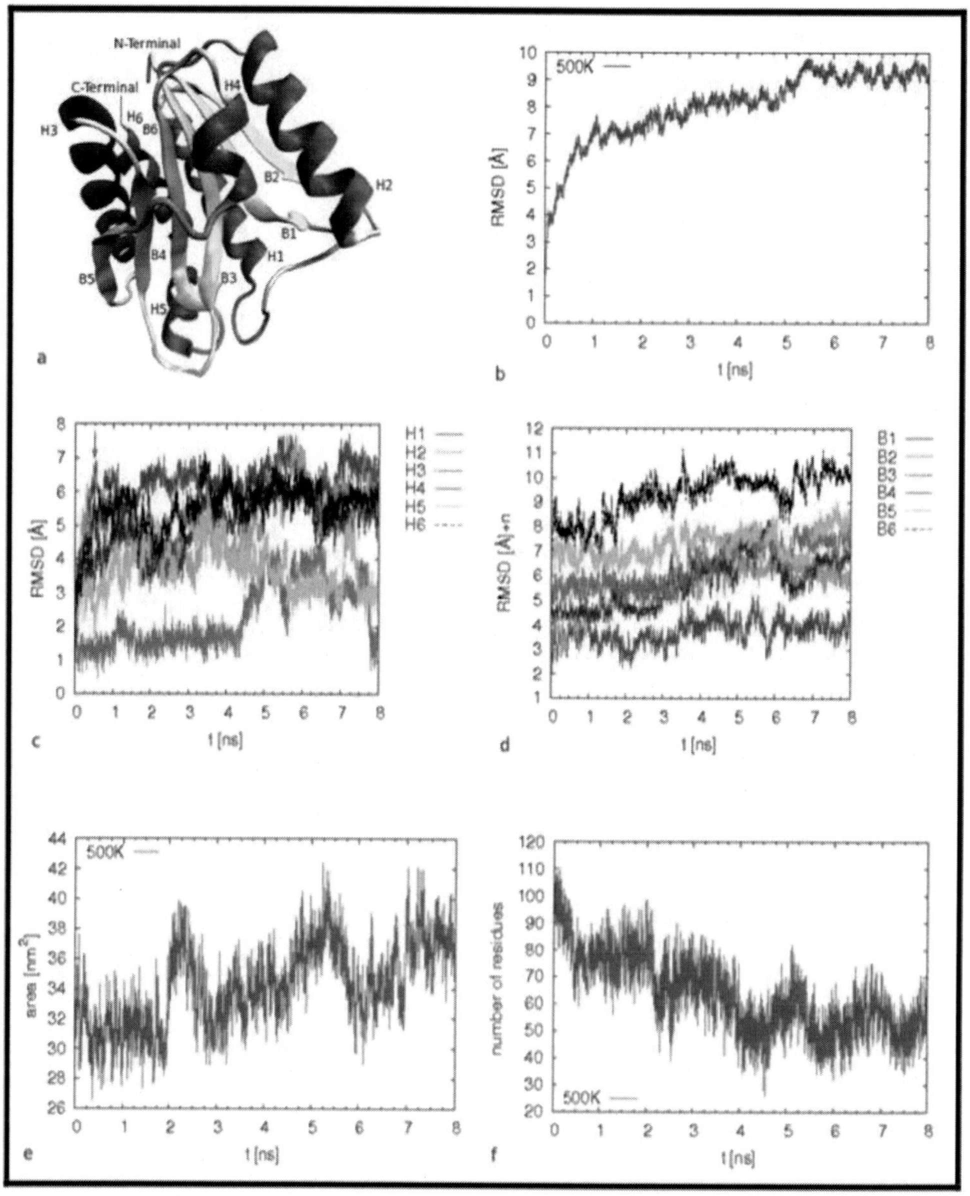

Figure 4. (a) Structure snapshot of 2ir0. The N-terminus and C-terminus, with β -sheets named as B1, B2, B3, B4, B5, B6 (forming the core inner part of the protein) surrounded by α-helices named likewise as H1, H2, H3, H4, H5, H6; (b) Evolution of Root-mean-square-deviation (Å) of the 2ir0 protein backbone with respect to its initial structure at 498 K. (c) Structural changes observed in terms of RMSD in α-helices from its native structure with time at 498 K. (d). Structural changes observed in terms of RMSD in β-sheets from its native structure with time at 498 K; for clarity all RMSD values are incremented by n, where n is the no. of the β-sheet. (e). Variation of solvent accessible surface area correlates with time at 498 K. (f). Decrement in total secondary structure including α-Helix, β -Sheet, B-bridge, turn at 498 K confirms unfolding.

The amidase-03 enzyme from *B. anthracis* is composed of a central core made of six β-sheets and an outer sheath of six α-helices (Table 1). The RMSD of the protein backbone was found to be stable in the range of 2-3 Å at 300 K, but increased to 6 Å after 1 ns at 498 K (Figure 4b). The RMSD of three α-helices (H1, H2 and H6) was similarly increased to 6 Å in 0.5 ns time step at 498 K. Most of the helices except H4 and H6 uncoil in 1 ns. The short helix H4 (residue 127 to 131) remains stable up to 4.3 ns and H6 near the inner core is stable throughout the simulation. The unfolding of the helices leads to the opening of the core to the solvent. A transition state is observed after equilibration until unfolding of α-helices at high temperature and is described as the Transition State Ensemble (TSE) [76]. The RMSD for β-sheets increases after 2 ns until end of simulation (Figure 4d) and hence β- sheets are recorded to have greater stability than α-helices. The collapse in the core consisting of β-sheets is accompanied by increment in backbone RMSD and is indicative of protein denaturation. This is similar to the two state reversible denaturation process described elsewhere [77]. The native structure showed RMSD ≤ 2 Å, the intermediate (partial unfolded) state showed ≥ 2 Å RMSD ≤ 8 Å (Figure 4b) during the transition state. The high RMSD value (≥ 8 Å) is associated with unfolded state.

REFERENCES

[1] Pace, H. C. and Brenner, C. (2001). The nitrilase superfamily: classification, structure and function. *Genome Biol.*, *2*: reviews0001.1–reviews0001.9.

[2] Chebrou, H., Bigey, F., Arnaud, A., et al. (1996). Study of the amidase signature group. *Biochim Biophys Acta*, *1298*: 285–293.

[3] Patricelli, M. P. and Cravatt, B. F. (2000). Clarifying the catalytic roles of conserved residues in the amidase signature family. *J. Biol. Chem.*, *275*: 19177–19184.

[4] Zalkin, H. and Smith, J. (1998). Enzymes utilizing glutamine as an amide donor. *Adv Enzymol Relat Areas Mol Biol.*, *72*: 87–144.

[5] Oinuma, KI., Takuwa, A., Taniyama, K., et al. (2015). Hydrazidase, a novel amidase signature enzyme that hydrolyzes acylhydrazides. *J Bacteriol.*, *197*: 1115–1124.

[6] Piotrowski, M., Schönfelder, S. and Weiler, E. (2001). The Arabidopsis thaliana isogene NIT4 and its orthologs in tobacco encode beta-cyano-L-alanine hydratase/nitrilase. *J. Biol. Chem.*, *276*: 2616–2621.

[7] Sharma, M., Sharma, N. N. and Bhalla, T. (2009). Amidases: versatile enzymes in nature. *Rev Env Sci Bio/Technol.*, *8*: 343–366.

[8] Curnow, A. W., Hong, K., Yuan, R., et al. (1997). Glu-tRNAGln amidotransferase: a novel heterotrimeric enzyme required for correct decoding of glutamine codons during translation. *Proc Natl Acad Sci.*, U. S. A. *94*: 11819–11826.

[9] Fournand, D., Bigey, F. and Arnaud, A. (1998). Acyl transfer activity of an amidase from *Rhodococcus sp.* strain R312: formation of a wide range of hydroxamic acids. *Appl Env Microbiol.*, *64*: 2844–52.

[10] Holmes, L. B. (1996). Hydroxamic acid: a potential human teratogen that could be recommended to treat ureaplasma. *Teratology*, *53*: 227–229.

[11] Banerjee, A., Sharma, R. and Banerjee, U. C. (2002). The nitrile-degrading enzymes: current status and future prospects. *Appl Microbiol Biotechnol.*, *60*: 33–44.

[12] Nawaz, M. S., Khan, A. A. Bhattacharayya, D., et al. (1996). Physical, biochemical, and immunological characterization of a thermostable amidase from *Klebsiella pneumoniae* NCTR 1. *J Bacteriol.*, *178*: 2397–23401.

[13] Madhavan N., Roopesh, K., Chacko, S., et al. (2005). Comparative study of amidase production by free and immobilized *Escherichia coli* cells. *Appl Biochem Biotechnol.*, *120*: 97–108.

[14] Sharma, R.D., Goswami, N., Lynn, R., et al. (2009). A modeled structure for amidase-03 from *Bacillus anthracis*. *Bioinformation*, *4*: 242–244.

[15] Fukushima, T., Kitajima, T., Yamaguchi, H., et al. (2008). Identification and characterization of novel cell wall hydrolase CwlT: a two-domain autolysin exhibiting n-acetylmuramidase and DL-endopeptidase activities. *J Biol Chem.*, *283*: 11117–11125.

[16] Gavin, AC., Aloy, P., Grandi, P., et al. (2006). Proteome survey reveals modularity of the yeast cell machinery. *Nature*, *440*: 631–636.

[17] Girardin, S. E., Philpott, D. J. and Lemaitre, B. (2003). Sensing microbes by diverse hosts. Workshop on pattern recognition proteins and receptors. *EMBO Rep.*, *4*: 932–936.

[18] Guan, R. and Mariuzza, R. A. (2007). Peptidoglycan recognition proteins of the innate immune system. *Trends Microbiol.*, *15*: 127–134.

[19] Werner, T., Liu, G., Kang, D., et al. (2000). A family of peptidoglycan recognition proteins in the fruit fly Drosophila melanogaster. *Proc Natl Acad Sci.,* U. S. A. *97*: 13772–13777.

[20] Gelius, E., Persson, C., Karlsson, J., et al. (2003). A mammalian peptidoglycan recognition protein with N-acetylmuramoyl-L-alanine amidase activity. *Biochem Biophys Res Commun.*, *306*: 988–994.

[21] Cravatt, B. F., Giang, D. K., Mayfield, S. P., et al. (1996). Molecular characterization of an enzyme that degrades neuromodulatory fatty-acid amides. *Nature*, *384*: 83–87.

[22] Mesnage, S., Chau, F., Dubost, L., et al. (2008). Role of N-acetylglucosaminidase and N-acetylmuramidase activities in *Enterococcus faecalis* peptidoglycan metabolism. *J Biol Chem.*, *283*: 19845–19853.

[23] Vollmer, W., Joris, B., Charlier, P., et al. (2008). Bacterial peptidoglycan (murein) hydrolases. *FEMS Microbiol. Rev.*, *32*: 259–286.

[24] Fujita, M., Tsuchida, A., Hirata, A., et al. (2011). Glycoside hydrolase family 89 alpha-N-acetylglucosaminidase from *Clostridium perfringens* specifically acts on GlcNAc alpha1,4Gal beta1R at the non-reducing terminus of O-glycans in gastric mucin. *J Biol Chem.*, *286*: 6479–6489.

[25] Martinez-Fleites, C., Korczynska, J. E., Davies, G. J., et al. (2009). The crystal structure of a family GH25 lysozyme from *Bacillus anthracis* implies a neighboring-group catalytic mechanism with retention of anomeric configuration. *Carbohydr Res.*, *344*: 1753–1757.

[26] Huang, P., Li, W., Xie, J., et al. (2011). Characterization and expression of HLysG2, a basic goose-type lysozyme from the human eye and testis. *Mol Immunol.*, *48*: 524–531.

[27] Kawamura, T. and Shockman, G. D. (1983). Purification and some properties of the endogenous, autolytic N-acetylmuramoylhydrolase of *Streptococcus faecium*, a bacterial glycoenzyme. *J Biol Chem.*, *258*: 9514–9521.

[28] Heidrich, C., Templin, M., Ursinus, A., et al. (2001). Involvement of N-acetylmuramyl-l-alanine amidases in cell separation and antibiotic-induced autolysis of *Escherichia coli*. *Mol Microbiol.*, *41*: 167–178.

[29] Uehara, T. and Park, J. T. (2007). An anhydro-N-acetylmuramyl-L-alanine amidase with broad specificity tethered to the outer membrane of *Escherichia coli*. *J Bacteriol.*, *189*: 5634–5641.

[30] Barrett, J. F., Dolinger, D. L., Schramm, V. L. et al. (1984). The mechanism of soluble peptidoglycan hydrolysis by an autolytic muramidase. A processive exodisaccharidase. *J Biol Chem.*, *259*: 11818–11827.

[31] Vollmer, W. (2008). Structural variation in the glycan strands of bacterial peptidoglycan. *FEMS Microbiol Rev.*, *32*: 287–306.

[32] Garcia, P., Gonzalez, M., Garcia, E., et al. (1999). The molecular characterization of the first autolytic lysozyme of Streptococcus pneumoniae reveals evolutionary mobile domains. *Mol Microbiol.*, *33*: 128–138.

[33] Holtje, J. V., Mirelman, D., Sharon, N., et al. (1975). Novel type of murein transglycosylase in *Escherichia coli*. *J Bacteriol.*, *124*: 1067–1076.

[34] Thunnissen, A. M., Isaacs, N. W. and Dijkstra, B. W. (1995). The catalytic domain of a bacterial lytic transglycosylase defines a novel class of lysozymes. *Proteins*, *22*: 245–58.

[35] Romeis, T., Vollmer, W. and Holtje, J. V. (1993). Characterization of three different lytic transglycosylases in *Escherichia coli*. *FEMS Microbiol Lett.*, *111*: 141–146.

[36] Ursinus, A. and Holtje, J. V. (1994). Purification and properties of a membrane-bound lytic transglycosylase from *Escherichia coli*. *J Bacteriol.*, *176*: 338–343.

[37] Dhalluin, A., Bourgeois, I., Pestel-Caron, M., et al. (2005). Acd, a peptidoglycan hydrolase of Clostridium difficile with N-acetylglucosaminidase activity. *Microbiology*, *151*: 2343–2351.

[38] Eckert, C., Lecerf, M., Dubost, L., et al. (2006). Functional analysis of AtlA, the major N-acetylglucosaminidase of *Enterococcus faecalis*. *J Bacteriol.*, *188*: 8513–8519.

[39] Huard, C. (2003). Characterization of AcmB, an N-acetylglucosaminidase autolysin from *Lactococcus lactis*. *Microbiology*, *149*: 695–705.

[40] Huard, C., Miranda, G., Redko, Y., et al. (2006). Analysis of the peptidoglycan hydrolase complement of *Lactococcus lactis*: identification of a third N-acetylglucosaminidase, AcmC. *Appl Environ Microbiol.*, *70*: 3493–3499.

[41] Steen, A., Buist, G., Horsburgh, G., et al. (2005). AcmA of Lactococcus lactis is an N-acetylglucosaminidase with an optimal number of LysM domains for proper functioning. *FEBS J.*, *272*: 2854–68.

[42] Huysen, H., J-MHakenbeck, R. and Shockman, G. (1994). Microbial peptidoglycan (murein) hydrolases. *Bacterial Cell Wall*, Elsevier, Amsterdam, pp. 131–166,.

[43] Höltje, J.V. (1995). From growth to autolysis: the murein hydrolases in *Escherichia coli. Arch. Microbiol.*, 164: 243–54.

[44] Sauvage, E., Herman, R., Petrella, S., et al. (2005). Crystal structure of the Actinomadura R39 DD-peptidase reveals new domains in penicillin-binding proteins. *J. Biol. Chem.*, *280*: 31249–56.

[45] Kishida, H., Unzai, S., Roper, D., et al. (2006). Crystal structure of penicillin binding protein 4 (dacB) from *Escherichia coli*, both in the native form and covalently linked to various antibiotics. *Biochem.*, *45*: 783–92.

[46] Keck, W. and Schwarz, U. (1979). *Escherichia coli* murein-DD-endopeptidase insensitive to beta-lactam antibiotics. *J Bacteriol.*, *139*: 770–774.

[47] Keck, W., van Leeuwen, A., Huber, M., et al. (1990). Cloning and characterization of mepA, the structural gene of the penicillin-insensitive murein endopeptidase from *Escherichia coli. Mol Microbiol.*, *4*: 209–19.

[48] Engel, H., van Leeuwen, A., Dijkstra, A., et al. (1992). Enzymatic preparation of 1,6-anhydro-muropeptides by immobilized murein hydrolases from *Escherichia coli* fused to staphylococcal protein A. *Appl Microbiol Biotechnol.*, *37*: 772–83.

[49] Koo, IC., Ohol, YM., Wu, P., et al. (2008). Role for lysosomal enzyme beta-hexosaminidase in the control of mycobacteria infection. *Proc Natl Acad Sci.*, U. S. A., *105*: 710–715.

[50] Rio, L. A. and Berkeley, R. C. (1976). Exo-beta-N-acetylmuramidase - A Novel Hexosaminidase Production by Bacillus subtilis B, Purification and Characterization. *Eur J Biochem.*, *65*: 3–12.

[51] Shockman, G. and Cheney, M. (1969). Autolytic enzyme system of *Streptococcus faecalis* V. Nature of the autolysin-cell wall complex and its relationship to properties of the autolytic enzyme of *Streptococcus faecalis*. *J Bacteriol.*, *98*: 1199–1207.

[52] Smith, T., Blackman, S. and Foster, S. (2000). Autolysins of *Bacillus subtilis*: multiple enzymes with multiple functions. *Microbiology*, *146*: 249–262.

[53] Gupta, D., Jin, Y. and Dziarski, R. (1995). Peptidoglycan induces transcription and secretion of TNF-alpha and activation of lyn, extracellular signal-regulated kinase, and rsk signal transduction proteins in mouse macrophages. *J Immunol.*, *155*: 2620–2630.

[54] Michielin, O., Luescher, I. and Karplus, M. (2000). Modeling of the TCR-MHC-peptide complex. *J Mol Biol.*, *300*: 1205–1235.

[55] Paine, G. and Scheraga, H. (1987). Prediction of the native conformation of a polypeptide by a statistical-mechanical procedure. III. Probable and average conformations of enkephalin. *Biopolymers*, *26*: 1125–1162.

[56] Abagyan, R., Batalov, S., Cardozo, T., et al. (1997). Homology modeling with internal coordinate mechanics: deformation zone mapping and improvements of models via conformational search. *Proteins*, *29* : 29–37.

[57] Altschul, S., Gish, W., Miller, W., et al. (1990). Basic local alignment search tool. *J Mol Biol.*, *215*: 403–410.

[58] Martí-Renom, M., Stuart, A., Fiser, A., et al. (2000). Comparative protein structure modeling of genes and genomes. *Ann Rev Biophys Biomol Struct.*, *29*: 291–325.

[59] Ramachandran, G., Ramakrishnan, C. and Sasisekharan, V. (1963). Stereochemistry of polypeptide chain configurations. *J Mol Biol.*, *7*: 95–99.

[60] Hess, B., Bekker, H., Berendsen, H., et al. (1997). LINCS: A linear constraint solver for molecular simulations. *J Comput Chem.*, *18*: 1463–1472.

[61] Humphrey, W., Dalke, A. and Schulten, K. (1996). VMD: visual molecular dynamics. *J Mol Graph.*,*14*: 33–38.

[62] Chumchua, V., Pornputtapong, N., Thammarongtham, C., et al. (2008). Homology modeling of Mycoplasma pneumoniae enolase and its molecular interaction with human plasminogen. *Bioinformation*, *3*: 18–23.

[63] Korndörfer, I., Danzer, J., Schmelcher, M., et al. (2006). The crystal structure of the bacteriophage PSA endolysin reveals a unique fold responsible for specific recognition of Listeria cell walls. *J Mol Biol.*, *364*: 678–689.

[64] Fischetti, V. A. (2005). Bacteriophage lytic enzymes: novel anti-infectives. *Trends Microbiol.*, *13*: 491–496.

[65] Mesnage, S. and Fouet, A. (2002). Plasmid-encoded autolysin in Bacillus anthracis: modular structure and catalytic properties. *J Bacteriol.*, *184*: 331–334.

[66] Milohanic, E., Jonquières, R., Cossart, P., et al. (2001). The autolysin Ami contributes to the adhesion of *Listeria monocytogenes* to eukaryotic cells via its cell wall anchor. *Mol Microbiol.*, *39*: 1212–1224.

[67] Alonso, DO. and Daggett, V. (1995). Molecular dynamics simulations of protein unfolding and limited refolding: characterization of partially unfolded states of ubiquitin in 60% methanol and in water. *J Mol Biol.*, *247*: 501–520.

[68] Vijayakumar, S., Vishveshwara, S., Ravishanker, G., et al. (1993). Differential stability of beta-sheets and alpha-helices in beta-lactamase: a high temperature molecular dynamics study of unfolding intermediates. *Biophys J.*, *65*: 2304–2312.

[69] Bolhuis, P. G. (2005). Kinetic pathways of beta-hairpin (un)folding in explicit solvent. *Biophys J.*, *88*: 50–61.

[70] Fersht, A. R. and Daggett, V. (2002). Protein folding and unfolding at atomic resolution. *Cell*, *108*: 573–82.

[71] Day, R., Bennion, B., Ham, S., et al. (2002). Increasing temperature accelerates protein unfolding without changing the pathway of unfolding. *J Mol Biol.*, *322*: 189–203.

[72] Essmann, U., Perera, L., Berkowitz, M., et al. (1995). A smooth particle mesh Ewald method. *J Chem Phys.*, *103*: 8577. Doi.org/10.1063/1.470117.

[73] Berendsen, H., van der Spoel, D. and van Drunen, R. (1995). GROMACS: A message-passing parallel molecular dynamics implementation. *Comput Phys Commun.*, *91*: 43–56.

[74] Settanni, G. and Fersht, A. (2008). High temperature unfolding simulations of the TRPZ1 peptide. *Biophys J.*, *94*: 4444–4453.

[75] Kabsch, W. and Sander, C. (1983). Dictionary of protein secondary structure: pattern recognition of hydrogen-bonded and geometrical features. *Biopolymers*, *22*: 2577–2637.

[76] De Jong, D., Riley, R., Alonso, D., et al. (2002). Probing the energy landscape of protein folding/unfolding transition states. *J Mol Biol.*, *319*: 229–242.

[77] Itzhaki, L., Otzen, D. and Fersht, A. (1995). The Structure of the transition state for folding of chymotrypsin inhibitor 2 analysed by protein engineering methods: Evidence for a nucleation-condensation mechanism for protein folding. *J Mol Biol.*, *254*: 260–288.

[78] Tsai, J., Levitt, M. and Baker, D. (1999). Hierarchy of structure loss in MD simulations of src SH3 domain unfolding. *J Mol Biol.*, *291*: 215–225.

In: Microbial Catalysts. Volume 1
Editors: S. M. Abdel-Aziz, N. Garg, A. Aeron et al.
ISBN: 978-1-53614-679-0

Chapter 14

AN OVERVIEW ABOUT OUR UNDERSTANDING OF PROTEIN ACTIVITY, SPECIFICITY, STRUCTURE AND FUNCTION

Amro Abd Al Fattah Amara*

Protein Research Department,
Genetic Engineering and Biotechnology Research Institute,
City for Scientific Research and Technological Applications,
Universities and Research Center District, New Borg El-Arab,
Alexandria, Egypt

ABSTRACT

Enzymes are proteins responsible for conducting different cell activities. Some enzymes work together in a pathway to produce one or more products. These different proteins are built of an amino acid chain. Their difference in structure and function is based on their difference in amino acid sequence, thus in their 3D structure. An effective change in a protein structure and function is well represented by changes in its active site, which in turn changes its activity and/or its substrate specificity. Accumulated changes could lead to change in the entire protein function. Similar enzymes have conserved amino acid residues. Conserved amino acids are an interesting subject of research. One amino acid in protein active site could dramatically affect the activity and the specificity of an enzyme (protein). Sickle cell anaemia represents a famous case in which, the change in a single amino acid is the cause of such degenerative disease. The image is not that simple. Changes in amino acids other than the active residues are also affecting the protein activity, specificity, polarity, and so on. That is because such changes affect the 3D structure of the protein, which interferes with the catalytic residue configurations and their fitting to their specific substrate. This is a clear prove that any change in an amino acid will lead to a change in protein's activity. Changes in DNA are permanent. Changes caused by environmental factors such as pH, temperature, ions, and so on are temporary (in case of correct denature). The environment where a protein exists could be optimized to match the better structure and function. This chapter aims to highlight the different criteria involved

* Prof. Head of Protein Research Department; Email: amroamara@web.de.

in protein structure and function. It covers some important historic stations in the progress of protein science.

Keywords: protein activity, structure and function, Sickle cell anaemia, insulin

INTRODUCTION

The biological system is rich with uncountable number of molecules and macromolecules. Each structure has its own criteria based on the criteria of each unit plus the effect of the external surrounding factors [1-4]. Even most proteins are different from one another, but they can be classified into groups. Each group has some sort of similarity [5-9]. What are the criteria, which govern such similarities and differences? Simply, it is protein chemical structure and any other factor that could affect its structure, function and specificity. The structure is responsible for the function. Our understanding of protein structure will give us a better understanding of its function. Any of the biological system molecules and macromolecules is built from certain elements. Such elements exist in the native nature form. But, oxygen, hydrogen, carbon and nitrogen are more representative. The name of a molecule is given for two or more atoms chemically bonded. The new molecules react as one unit, such as O_2. Their atoms bond with one another by chemical bonds through the interactions of the electrons in their outer shells. Compounds are bigger than molecules. Macromolecules are further bigger; usually they exist in polymeric form, such as protein, DNA, RNA, Lipid, poyhydroxyalkanoate and so on. Bio-macromolecules are polymers composed of repeated homogenous or heterogeneous units, which are named homo or hetero polymer respectively. Some forms of chemical bonds are more common, such as covalent, ionic and hydrogen bonds [10-12]. Other natural forces exist such as van der Waals forces, which occur between two, or more atoms existed closely (3-4 Å) (such as substrates and enzymes binding) [10-12]. Protein is a polymer that has long backbone, so, the interactions between their amino acids represent different internal structures such as helix, coil, turn and so on. Another force is the hydrophobic interaction generated on the molecule or the macromolecules in the water. Water repels molecules that tend to cluster in an aqueous environment leading to associating of the nonpolar parts. Such a phenomenon can be found in the phospholipids' membrane, protein folding, and substrate/enzyme binding [13-18]. The non-bonds forces contribute in the biological system while they show the importance of the water and its contents of ions, charges and molecules in protein 3D structure [19]. One protein exists in uncountable number of 3D structures particularly as a response to its existence in the different environments. The monomeric backbone of the biopolymers and the macromolecules is important. One should put into his account that the monomeric structure finally build the 3D structure is responsible for the function. Moreover, such biopolymer and macromolecules on degradation will be probably divided to their monomeric constituents. Monomeric structures in their single form (for example amino acid) are important. Food, which contains cysteine, will be better than that which contains cystine. The cell membrane transports cysteine ten times faster than cystine. Our understanding of such properties will enable us to select the type of our foods [20]. None of the components of our biological system is functionless; but we still have limited knowledge. Even cancer could be controlled by controlling their cells supplement of amino acids [21]. An old approach, was recalled and re-evaluated [21]. This chapter concerns with one type of polymer, the protein.

PROTEINS ARE CODED IN DNA

After solving the relationship between both of DNA and protein codes, DNA and protein, sequences enrich our knowledge about the behaviour of any protein under investigation. A protein constituent of amino acids enables the understanding of its function. Protein function is the functions of its amino acids. Even, some amino acids are functioning-less, but they are affecting protein 3D structure and its fitness to the substrate. Besides the knowledge about each amino acid in protein backbone, many tactics for DNA mutagenesis enable changes. By relating the change in a protein function to the change in its amino acids, the role of each amino acid becomes clearer. Establishing protein, DNA and RNA sequences, PCR, primer synthesis, and so on, steps taking us to the random and site directed mutagenesis protocols. Both are protocols for mutant(s) induction and controlled by another selection and screening protocols. To characterize the changes, happen in a gene, one should first do sequencing to specify and prove that new change is able to change protein properties. If the mutant is translated to a new amino acid, this is a new protein different from the wild type protein. By measuring the change in protein function, folding, 3D structure, activity, and so on, the role of the new amino acid(s) as well the original amino acids become clearer. This is not the complete story. Scientists are interested with each single idea and opportunity that can deep their understanding to protein such as, structure, function, specificity, interaction with the substrate, folding properties, 3D structure, homology with other proteins, classification, substrate docking, specifying the active sites and the amino acids involved in catalysis, and so on. The progress in molecular biology, bioinformatics, chemistry, physics, computer science, mathematics, mechanical engineering, and perhaps some applied and even theoretical sciences enriched protein science. Such different branches of science were playing significant roles in improving, and supporting our understanding to the different proteins by enabling better tools for their studies [22-26].

ENZYMES

Enzymes occur in each cell, hence in any microbe. As biological catalyst, enzymes show ability to work faster than any other known chemical catalyst whenever there is a substrate and suitable conditions. Enzymes have two extraordinary properties can be distinguished from any other protein and chemical catalyst by: 1) they are highly efficient catalysts able to accelerate reaction by as much as 10^{17} fold 2) and they are exquisitely selective, being able to discriminate among closely related substrates [27-31]. Although enzymes are protein in their nature and are usually non-toxic, they could harm if they are used incorrectly. Being safe and eco-friendly there is a rising interest in their use over the chemical catalyst. They can maintain their activities for long time if used correctly. Immobilized enzymes prove that. Few amino acids are responsible for catalysis. Those amino acids if changed, the enzyme function could be modified or impaired. Sometimes, the change will affect its activity and in other cases, it will affect its specificity. Other changes in one enzyme backbone (amino acids other than the catalytic ones) could cause change in its activity because of the change in its 3D configuration. These changes might affect the enzyme activity but seldom, could affect its specificity. As a dynamic chemical structure, enzymes' surrounding environments are important. Enzymes present in folding 3D forms and can be changed to uncountable structures (configurations). Enzymes could react singly or in combination with other biological forms including other types of enzymes to produce single function or single product. Enzymes could react inside the cell (*in vivo*) or outside the cell (*in vitro*) whenever the conditions are suitable. They were classified based on their functions to six main groups 1) Oxidoreductases 2) Transferases 3) Hydrolases 4) Lyases 5) Isomerases and 6) Ligases.

STRUCTURES REACT AUTO-MECHANICALLY

Analogs are structures similar to their related compounds with minor differences. Some analogs are more stable or even indegradable. However, analogs are recognized as correct substrates and can activate the biological system, they are not usually degradable. IPTG can be given as a good example. Analogs can show us that even though our biological system is perfect but finally the chemical and the physical laws are the governor. So, similar compounds to a certain substrate could react as analogs and interfere with their related enzymes and/or wrongly activate/inactivate certain biopathway(s). Even such a phenomenon could help in the research field but might be harmful in real life. Where, no one can guarantee the type and the total physicochemical effect of such analogs could enter our bodies. Understanding the effect of the structure on the function, will let us avoiding improbable compounds could wrongly act as analogs and activate mistakenly some bio pathways. Or, we can control their use to activate or block certain points in a certain pathway to overcome certain deficiencies or illness. However, I am personally, recommended analogs derived from natural products particularly those who are derived from the edible plants. The structure of the analogs could be built and designed for the better result. Some medications are based on blocking proteins active sites or changing protein configuration to be functionless. For that, there are two possible ways for discovering new drugs could inhibit particular protein either by conducting *in silico* drug design, or by *in vivo* and *in vitro* bioassay(s) to screen for such inhibitors. The bioassay in each case is critical. However, the type of the screened materials in their better case is from natural resources. Natural structures are perfect and specific. While synthetic structures are made through chemical reactions, which during the industrial process may incorporate incorrect structures or isomers. Biological system guarantee 100% *livo* rotator structure. Additionally, incomplete synthesized forms are existed. What could such forms do for us? Particularly, those come with structures not belong to our biological system.

In our bodies, the degraded products of incorrect structures depend on the activated enzyme types and the existed chemical compounds subjected to the degradation process, also depend on which compounds activate the cytochrome P450 (CYTP450) and type as well as mode of action of activated enzymes. Different CYP450 enzymes have different degradation products for one compound, which upon degradation could be converted to toxins and carcinogens.

Even though, some of the natural products are toxic, but we still have the opportunity to reduce their toxicity for appropriate use.

One should differentiate between low molecular weight bioactive compounds and the high molecular weight or polymeric structures. High molecular weight molecules even those with safe physiological activity in their mother source might become antigenic and induce the immune system if they enter our bodies without being degraded to their fine monomeric forms. Any foreign protein could be an example. There is a major shift towards the use of natural active compounds particularly those derived from edible plants. Compounds derived from edible plants proved historically to be safe for our health.

Edible plants are the most products subjected to continuous open clinical trials and prove to be safe. The human memory and experiences should not be neglected [32].

From the above facts, some important criteria about protein function, structure, length, bioavailability, antigenicity, route of administration, its degraded products, toxicity, physiological effect, substrate analogs, inhibitors and others points were described. Those criteria are important for any protein, but finally any protein is a chemical structure, so it will do its work auto-mechanically, whenever, there is a suitable condition or based on the conditions where it exists.

PROTEIN DEFINITION BASED ON ITS STRUCTURE AND FUNCTION

Protein name is derived from the Greek word prǒtos, meaning "first" or "foremost". Proteins are macromolecules (Macromolecules are very large complex molecules; found only in plants and animals) that consist of distributed twenty amino acids (amino acids are organic compounds containing an amino group and a carboxylic acid group). Protein is simply described in dictionaries as "a large group of nitrogenous organic compounds that are essential constituents of living cells; consist of polymers of amino acids; essential in the diet of animals for growth and for repair of tissues; can be obtained from meat, eggs, milk, and legumes." However, protein definition could be larger and more specific. Protein is a biopolymer (mostly hetero-biopolymer) that is essential constituent of living cells and among their macromolecules, protein is responsible for most of the cells functions. Its code is saved in another macromolecule, DNA, which contains different genes; each gene is transcribed to RNA, which will be translated to one protein. Few numbers of amino acids are responsible for catalysis but the overall number, types and distribution of the amino acids in protein backbone are responsible for the 3D structure, hence the function. Protein is a dynamic molecule folded to (fold: An angular or rounded shape made by folding) uncountable number of configurations based on the surrounding factors affecting the process. However, collectively it could be folded either to a native form or a denatured form. Being denatured reversibly, protein can return to its correct native form under correct refolding conditions. Protein constituent of amino acids has the major impact on its function. However, many other factors such as pH, buffers (if they exist), temperature, salt concentration, other existed macro and micro molecules, the different charges in protein environment and many others factors affect protein's structure and function. However, the most effective criterion that can govern our judgment about certain protein is its ability to do certain function(s) correctively (its activity). There is no sense to produce a perfectly pure protein, either as crystals or as a folded protein in solution but without any activity. Structure study should be related to protein activity. That is the correct way to understand a protein structure and function. However, the available process to get a certain protein structure is through crystallography. From protein sequence or gene sequence or gene/protein sequence to a recombinant protein and crystals, there are many considerations to get (as a result) a correct crystal, which can be turned to correct active and folded protein representing correct case study! This is an industrial and medicinal demand. Change in protein structure could lead to a new function. For example, about protein specificity, the differences between DNA and RNA polymerase can be dictated, where DNA polymerase reaction is catalyzed by two-metal ion mechanism in the 3'-exonuclease reaction. In contrast, RNA polymerase does function that is more complicated while, it works in one ribonucleic acid strand only. They can start RNA synthesis without requiring a primer oligonucleotide; exhibiting an abortive initiation phase and they are targeting host regulatory proteins (activators, inhibitors, terminators and anti-terminators) that modulate gene expression.

HISTORIC STATIONS

The Swedish chemist Jöns Jakob Berzelius was the first to describe protein as a term in 1838. James B. Sumner (1926) showed that urease is a protein. Protein and the recombinant protein story is excited. It is full of knowledge, events, discoveries, prices and challenges.

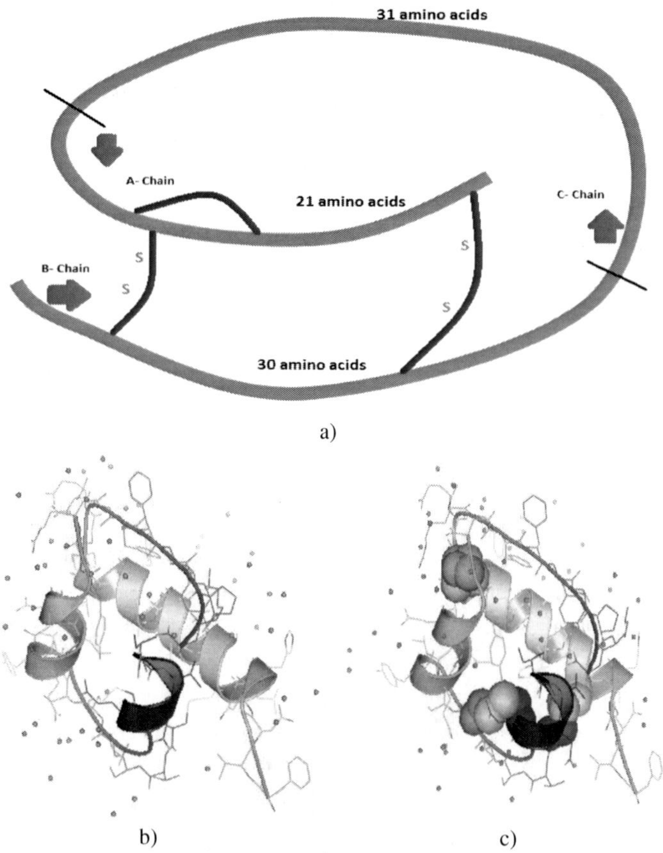

Figure 1. a) Pro-insulin with its three fragments (chains) a, b and c; and its disulfide bonds, b) 3D structure of Insulin fragment A and B, c) same like (a) but with the disulfide bonds.

The smallest known protein can fold into many structures. Each structure could be different in many properties. By understanding the steps from the gene, which lead to a protein, many recombinant proteins were produced. However, the history of protein teaches us that a safe way to clone a correct protein is by purifying it and then decoding its N and C terminal amino acids, and then designing primers to amplify its responsible gene. Alternately, by doing gene library. That because protein could be modified through its gene, under very specific condition, which might not have existed in the expression system or by using recombinant tools. Insulin, which is existed in native form as Pro-insulin contains three fragments A, B and C. Fragment C is existed between fragment A and B (Figure 1). To be able to turn to a correct insulin, fragment C must be removed. However, it must be removed only after the Pro-insulin is correctly folded. Such remove keeps both of fragment A and B in correct folded form attached to one another only by two disulfide bonds existed between four cystine, two in each fragment. Singer has solved the insulin sequence using protein amino acids sequencing rather than using its genes. Insulin was the first protein to be sequenced, by Frederick Sanger, who won the Nobel Prize for this achievement in 1958. It was a complicated process especially in the presence of disulfide bonds between A and B fragments (Figure 1). However, Sanger [33, 34] has identified the importance of the insulin 3D structure and its function and specificity. It takes a life of many researchers. What an effort! This two short peptide is essential for the human life! That is only one example out of thousands.

The X-Ray Analysis for the First Protein [Sperm Whale's Meat Myoglobin]

The German physicist Max von Laue devised the technique of the x-ray crystallography. At Cambridge, William Bragg and his son Lawrence in 1910s, worked out the atomic structures of many inorganic crystals. In 1930s J. Desmond Bernal and Dorothy Hodgkin in the Cavendish Lab., showed that protein crystals could yield sharp diffraction pattern, raising the prospect that their structures could be decoded. Perutz started to work as Bernal's student in 1937, and a year later Lawrence Bragg becomes a director of the Cavendish. Bragg was fascinated by the idea that the powers of x-ray analysis might be extended to the giant molecules, which form the crystals of living cells. Preutz suggested that Kendrew start to work and compare the adult and fetal sheep hemoglobin a work published after 22 years of effort in 1960 [35-38]. However, the first protein structure was something else and was published at 1958. Kendrew decided to work on a smaller protein, the myoglobin, which is only a quarter, the size of hemoglobin. This represents one major point in protein structure. In addition, Protein has irregular 3D structure, so how it could be arranged correctly to a crystal can be solved to its structure! Myoglobin is a protein that store oxygen temporarily in cells before transferring it to hemoglobin, the oxygen-ferrying molecule in the blood stream. Kendrew could not get crystals large enough from horse myoglobin. As a scientist, he realized that diving mammal such as wheals have much more myoglobin in their muscles, because they need a big oxygen store. His efforts on the wheals' myoglobin and he succeeds to crystallize it, adds an essential step. He adds very heavy metal atoms, the uranium, which scatters x-ray strongly and thus provide a set reference point in the diffraction pattern. If the pattern from heavy-atom crystals is compared to that from native crystals, the phase problem can be solved [37]. The 3D structures of both proteins were determined by x-ray diffraction analysis. Based on their pioneering work, they were awarded Nobel Prize in Chemistry in 1962 [35-39].

PROTEIN FOLDING

The primary structure of a protein is its linear sequence of amino acids and usually appreciated as 1D. The secondary structure of a protein is the local spatial arrangement of a polypeptide's backbone atoms without regard to the conformations of its side chains and usually appreciated as 2D. In 1957 by his unique experiments, Christian Anfinsen proves clearly that protein primary structure dictates its three dimensional (3D) structure. Anfinsen was working on ribonuclease A (RNase A) which contains eight Cys. By denaturing RNase A, using 8 M urea followed by removing the urea; the RNase A did not lose its activity, while it does if oxidised at denatured condition with activity equal to 1%. This 1% active protein is called "scrambled protein." If it is exposed to trace of 2-mercaptoethanol, which breaks the improper disulfide bonds and protein allowed to refold again, protein will be again fully active. Anfinsen explained that, the 1% activity comes from that the eight Cys have made the disulfide bond incorrectly where the probability could be calculated as 1/7*1/5*1/3*1/1=1/105 or 1:105 which equal to 1%. This is clearly proved that protein primary structure has enough information for its 3D and could rearrange itself in the native correct folding form. While many proteins composed of two or more polypeptide chains, loosely called subunits. Protein's quaternary structure refers to the spatial arrangement of its subunits. Proteins work together to achieve a particular function, and they often associate to form stable complexes [4, 17, 19, 40-46].

Another hypothesis about protein 3D structure named "Thermodynamic hypothesis." This hypothesis states that the 3D structure of a native protein in its normal physiological milieu (solvent, pH, ionic strength, presence of other components such a metal ions or prosthetic groups, temperature, and so on) is the one in which the Gibbs free energy of the whole system is lowest. That is, because native conformation is determined by the totality of interatomic interactions and hence by the amino acid sequence, in a given environment [47-49].

ALLELES ENABLE VARIATIONS BUT DISABLE SPECIES ALTERATION!

Recently, Amara (2013), highlighted that degenerative diseases which cause severe illness because of mutants cause protein alteration, hence cause illness could be avoided [50]. Costly research and medication spent every year to find a solution for such illness but with nearly no success. That because such change happened in the patient genomes (somatic cells); so it is hard to be repaired [50]. At least that is not in the coming near years. But the solution is so simple, it can start before the marriage happens [50]. Simply by avoiding or controlling the marriage from the relatives. The biological system is sensitive for the chemical structures. An enzyme could be so specific. Other protein forms could be also sensitive. Red Blood Cells (RBCs) could differentiate between O_2 and CO_2. The conditions and the structure draw the function of the RBCs; where they gain O_2 and where they lose it; where they gain CO_2 and where they lose it. Such specificity is not only in the heam active amino acids but also in the atoms of O_2 and CO_2. Minor changes in the RBCs are responsible for serious diseases. Sickle cell anaemia is an example.

When oxygen is removed from Sickle hemoglobin, those molecules change their shapes and combine with one another. The red blood cell structure changes from ring to sickle shape in absence of oxygen. This causes blood to clot and deprives vital organs from their supply of blood, resulting in pain, intermittent illness, and often a shortened life span. The only difference between normal and Sickle cell hemoglobin is that in each β- chain, one glutamic acid is replaced by one valine. Valine, unlike glutamic acid, contains a nonpolar group. The result is a hydrophobic "sticky" region that can interact with the hydrophobic region on neighbouring molecules, producing the observed clumping (Figure 2). A slight change in the β- the neighbouring subunits of the hemoglobin. The HBB gene provides instructions for making β-globin. Various versions of β-globin result from different mutations in the HBB gene. One particular HBB gene mutation produces an abnormal version of β-globin known as hemoglobin S (HbS). Other mutations globin 3D structure will induce a change in its configuration when it interacts within the HBB gene lead to additional abnormal versions of β-globin such as hemoglobin C (HbC) and hemoglobin E (HbE). HBB gene mutations can also result in an unusually low level of β-globin; this abnormality is called β-thalassemia.

Out of nearly three billion base pair representing the human genome, only single base change could cause fatal diseases. The β-globin molecule contains 146 amino acids. Humans carry two copies of each gene. One allele is inherited from each parent (Figure 4). The existence of one Sickle cell allele in a person will not cause an apparent serious problem, however if two of the Sickle cell alleles exist in one person they will develop, the symptoms of Sickle cell anemia (Figures 3 and 4). Where the normal β-globin contains in its fragment sequence is [Valine – Histidine – Leucine – Threonine – Proline - Glutamic acid - Glutamic acid - Lysine]

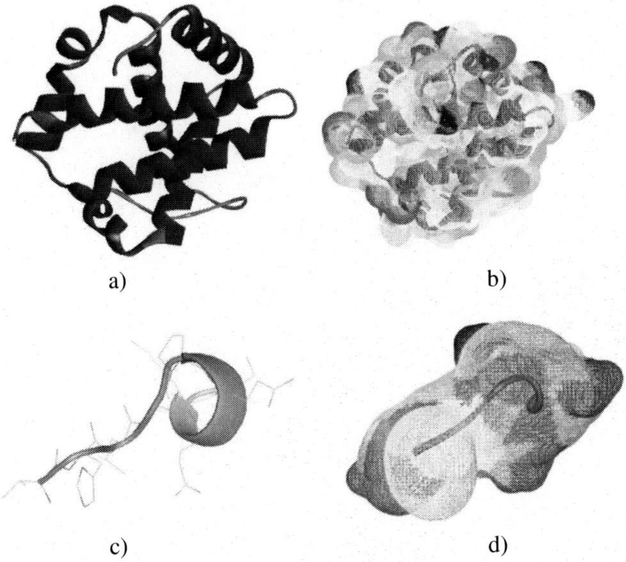

Figure 2. a) The 3D structure of correct β-globin; b) amino acids energy of the correct β-globin; c) the correct fragment (eight amino acids: of the correct β-globin [Valine-Histidine-Leucine-Threonine-Proline-Glutamic acid-Glutamic acid-Lysine]; d) the energy around the amino acids in the correct β-globin fragment.

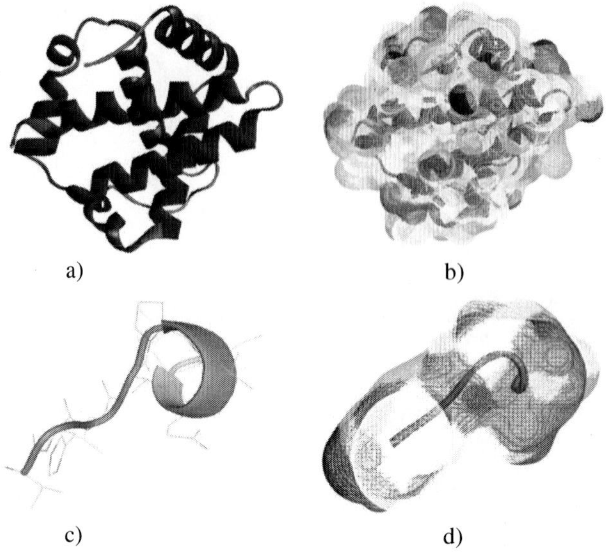

Figure 3. a) The 3D structure of sickle β-globin; b) amino acids energy of the sickle β-globin; c) the correct fragment (eight amino acids: of the sickle β-[Valine-Histidine-Leucine-Threonine-Proline-Valine-Glutamic acid-Lysine]; d) the energy around the amino acids in the sickle β-globin fragment.

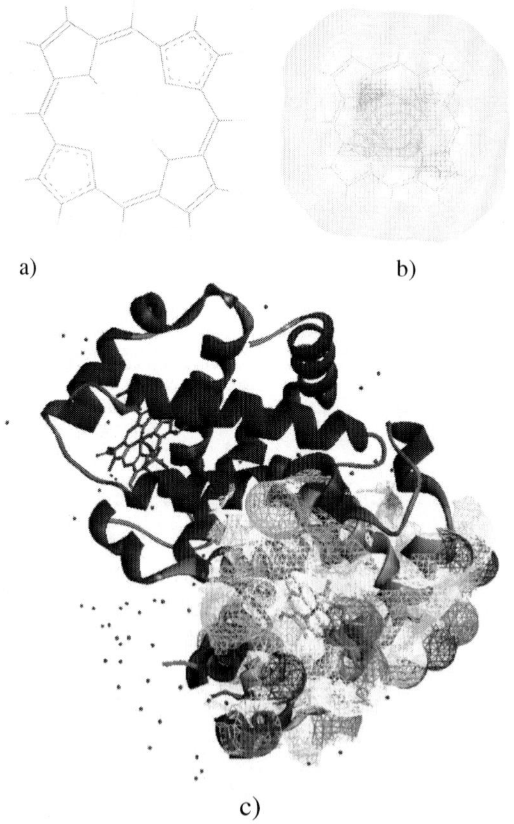

Figure 4. a) porphyrin ring; b) energy around the porphyrin ring, c) 3D structure for two β-globin units with energy around the amino acids surrounded to a porphyrin ring in one unit.

and the Sickle β-globin fragment sequence is [Valine – Histidine – Leucine – Threonine – Proline – Valine - Glutamic acid - Lysine]. Existence of two alleles could dilute many genetic disorders if one is correct. For that the marriage outside the family, save a lot and protect from the degenerative diseases and many other genetic related illness [50]. In contrast, marriage from the same genetic pool increases the genetic disorder probability. Verses to Darwin theory, alleles differences enable variations (colour, length. and so on) but disable species alteration. It becomes clear that Darwin has clearly succeeded to explain the role of the genetic material in the variation happened in the offspring phenotypes. However, he has failed in relating that to the species alteration. That simply because, alleles, if they are subjected to genes change(s) it will cause variation but if they are subjected to mutational change they will cause degenerative illness including cancer, and other degenerative diseases. The animals have such selectivity. The female for breading will accept the strongest, healthiest and the best male. The issue is not only, science, but also the behaviour, and many other factors. Our attempt to change the genetic elements should be under high restriction and precautions. While most of the genes are existed in pair each in one chromosome, it enables variation, also enable saving the host if one gene is defected. Even so but if the two essential genes are defected, that cause fatal disease such as the degenerative disease and put a big question mark about our understanding to the evolution. Genes constituents have narrow range of change if compared by their mobility in the same species, and particularly within the sexually reproduced (have x and y chromosome) organisms including humans. Narrow range of change enables variation but mutational change for essential amino acid(s) is restricted, that safe the whole species feature. Direct mating between an apparently related species induces a forced change. The product is mostly sterile or degenerative!

PROTEIN ENGINEERING

The understanding of protein active sites (catalytic residues) was achieved by elevating the site directed mutagenesis protocol. As in the above-mentioned part, changing protein composition caused by change(s) happened in its DNA, which is a natural phenomenon [51-53]. Post-translation modification happen either before protein is used in the cell, or as a part of a control mechanism are other examples. Logical approach or Rational designs enable manipulating protein through its gene [54-60]. Rational design was improved by the improvement in the site directed mutagenesis methods, protein 2D and 3D configurations, structure modelling, protein-protein interaction and their related software [18, 61-63]. The concept of protein 3D modelling is based on relating a structure to the possible expected function(s). Conserved amino acids within a group of enzymes do the same job. So, those amino acids could have an essential function in the catalysis. This is an easy method to specify the nucleophilic residue. The most important ones will be those, which have physicochemical properties. A change in an amino acid of a protein could happen naturally. Protein engineering started in 1978 when a site directed mutagenesis was introduced in the laboratory of Michael Smith. Michael Smith and his co-workers were the first to introduce synthetic oligonucleotides mediate mutations. The first modified enzyme; a tyrosyl-transfer RNA synthetase was obtained in 1982. Through site directed mutagenesis, a cysteine was replaced by a serine altering protein's substrate binding characteristics. His revolutionary work in this field earned Michael

Smith the Nobel Prize in 1993. Many alterations in a protein can be generated by making amino acids replacements at a specific site in the polypeptide backbone. There are many tools that specify the active amino acid(s) residue(s) such as: 1) through understanding protein structure and function 2) aligning similar proteins with same or nearly same function to specify the conserved regions 3) crystallography and x-ray analysis 4) NMR analysis 5) study the shift of amino acids or their configuration after adding the substrate. This could be done by crystallizing both of the substrate and the amino acid(s) or by using the NMR with or without the substrate 6) In silico or even using normal computer to do protein substrate docking 7) by targeting directly the amino acids with physicochemical properties .

The probabilities in need to change a protein using site directed mutagenesis is high. Proteins, which are composed of 20 different amino acids, need many sites directed mutagenesis steps to alter their structure and function. The random mutagenesis becomes a better choice when there is a shortage of the existed information about protein structure and function and the role of active residues. The random mutagenesis is applied to a gene coded for a protein. The process passes through a selection protocol that enables selecting newly mutated genes by detecting a difference in the expressed protein. A gene could be a subject of more than one cycle of mutagenesis. The random mutagenesis usually seems to be time-consuming process, but alternative methods will be impossible if there is not enough knowledge about protein structure and function. Random mutagenesis mimics what could happen naturally. That will satisfy all the criteria which lead to mutate a certain gene; even randomly but with more force and directed tactics. Random mutagenesis can be used *in vivo* using mutator strains or *in vitro* using different kinds of PCR.

Facts versus Big Words: Concerning Protein Structure and Function

We still need to know more about protein structure; and its effect on protein function. We have nowadays a lot of important information; however, such information is still unable to build the correct model for a single protein. This is mainly because protein structures are sensitive to any change in its surrounding environment. Such changes could be physical, chemical, or even biological ones. Those three main categories, which affect protein structure, include many factors but some are more famous than others; such as pH, temperature, ions, protein content and arrangement of the amino acids in its backbone. Any change in the surrounding conditions for a protein will affect its structure hence its function. Protein is not that simple macromolecule. For me it is a chance to highlight some weakness in the research concerning protein. Protein appears as one clear band after purification and running a PAGE stained with comassiee blue, might be one of many if stained with silver staining. Even appearing as one band does not mean that it is a single protein. Any contamination will affect protein structure and it will give a wrong crystal. Contaminating protein crystal will give wrong x-ray analysis hence will give wrong structure. The expert knows that to be sure from a protein purity, it should give one band on PAGE, and in the 2D PAGE, as well as gives a correct C and N terminal amino acids. Recently, we get in our protein lab, a new 2D protein analysis instrument with sophisticated software linked with the Internet. The students, who found the classical and basic protein purification methods timely, some of them, are happy with such an instrument.

However, they were depressed when they were told that they need (after all of that) to be sure from the C and N terminal of any purified protein particularly new ones. Yes, there is the database for the 2D solved protein, but no one could guarantee that the condition is similar in two 2D gels made by 2D machines (in different places).

SOME REMARKS AND REMINDERS ABOUT PROTEIN AND PROTEIN RESEARCH

Our current understanding to protein structure and function comes through many different routes of information, knowledge, experiments, results and theories collectively build our understanding to structure and function of protein. Proteins have unique properties and features. Here I try to highlight some common misunderstood points about protein mainly because of a technical incorrect start practice.

- A protein must be a mirror for a correct DNA gene that has a start codon and a single stop codon in its end. One could find in protein database, a protein sequence which if reverse translated it will show a stop codon in its backbone. That did not mean wrong cloning but wrong sequencing. Scientists and Researchers should give high concern about DNA sequence; most of the high sophisticated machines for DNA sequencing give wrong sequences. At molecular level, wrong sequence, means even one wrong nucleotide. This will not allow a correct judgment for the sequence, which should be 100% accurate. I believe that only naked eye judgment for the four bases, A, T, C and G could do that. Should we go back to the traditional method for sequencing! There are still some machines in the market allow eye judgment and visualize each base. It might be a chance to highlight that, during my PhD, I made some work on site directed mutagenesis and during sequencing I found that the sequence contains both the wild and the replaced base. That because, the colony I used contains both wild and mutagenized gene (in two coexisted plasmids).
- There are other common mistakes used to be done by many researchers, they are designing a primer from DNA database instead of reverse translating the first amino acids in the N and C terminal of the purified protein.
- Many researchers nowadays are interested in bypassing the sophisticated step for standardizing a method for purifying a protein. They jump to the affinity purification methods by modifying the original gene, such as adding six Histidine either in C or in N terminal, which helps a lot in protein purification. However, such change will affect protein structure hence its function.
- Using strong promoter could lead to high expression. However, one should be sure from RNA polymers in the used cells because strong polymerization with RNA polymers with low fidelity could lead to a mistake in RNA hence in protein.
- The above evidence is impotent for anyone goes to do structure research. If a real purified protein is obtained and correctly crystallized, it is the time of the machine to do the further job. Machine is not perfect as well. One should know that most of the data taken from the machine is subjected to calculation and most of them are based on

a modelling research for facts and knowledge in our hand, which need more adjustment and optimization.

- A protein is a sequence of 20 amino acids coming through translating RNA of its related gene. The properties of each single amino acid are important to be understood. The factors that change the properties of amino acids will by default cause changes in the properties of protein. Short peptide such as insulin does a great job in our bodies. However, some amino acids if changed will dramatically change protein structure and function. The most important ones are the catalytic nucleophile. One should trust wild type protein more than artificially designed one!
- Protein from an organism is friendly to its immune system, however, foreign proteins mostly recognized as antigen, if they are different. Hence, recombinant proteins and modified ones should be well studied. Correct sequence should be well validated. In addition, the produced protein must be examined for its activity and purity.
- To escape from the patent restrictions, there is a trend to change one or more amino acids to get protein sequence other than that which was protected by patent). That should be correctly investigated, evaluated and validated. In addition, the expected side effect might not be realized until protein enter the market.

POINTS ABOUT PROTEIN PROPERTIES

- A protein is present in nature (mostly) in a folded form named native form.
- Protein unexpectedly, could do more than one function.
- More than one protein either same or different in their backbone could aggregate in a quaternary structure to do a complicated process and working as one unit.
- Protein is sensitive to any change in its surrounding area, such as pH, temperature, salt and so on.
- Protein can lose its native folding structure and become denature; however, it can regain its folding structure, if the two processes did not damage any of its backbone constituents of amino acids.
- All the amino acids in protein backbone are involved in its 3D structure, folding, refolding which by one or another way affect its function. However, certain amino acids are responsible for the catalytic activity. Any single amino acid change could dramatically change protein properties such as increasing its affinity. Such change might not be observed early.
- Proteins are responsible for functions in cells. They are catalysts that do their work faster than any known chemical catalyst.
- A protein does its work mechanically and reacts spontaneously.
- Protein resists changes in its surrounding environment but such changes could reduce or induce its activity.
- The catalytic amino acids are conserved in a single group of proteins.
- Protein can be crystallized. However, it is not fully clear if correct structure could be obtained from such crystals!
- Protein, versus, DNA and RNA could stay longer in the body before it can be degraded to its constituent of amino acids.

- Each protein is a special different case study.

Conclusion

Different proteins have different sequences of amino acids and are able to do different functions. They exist in different 3D folding structures and react on different substrates. Such differences are mainly because of their amino acids in their backbone and the environment where they exist. Single protein also could give different structures and functions based on the conditions around it. Some amino acids of active protein including enzymes are conserved and essential in catalysis. Those amino acids are named "Catalytic residues." They are responsible for doing the function. However, their function could be changed if one or more of the catalytic amino acids change. Additionally, their function can be changed if other amino acids in protein backbone change. The overall protein backbone affects protein structure and its fitness to the substrate during catalysis. Site directed and random mutagenesis enable DNA modification, hence enable protein change, which lead to a new science, protein engineering.

Lucky we are! We have two alleles enable variation but disable critical change. Where, even if serious defect in one allele happened, the other one could cover that defect. Degenerative diseases teach us that they can be simply solved by avoiding marriage from relatives. One base changes could give serious illness such as Sickle cell disease. In addition, two short peptides could save our health, such as insulin. The progress in protein science comes out through huge efforts from scientists various specialties and various backgrounds. The new instruments also added a lot; however, we should not neglect or forget basic protein science, experiments, tools and concepts, as well as the power of wild type protein. Protein's scientists should read the history of pioneers in that field of science and be thankful for them and for their efforts.

Acknowledgments

The author acknowledges the Genetic Engineering and Biotechnology Institute, City for Scientific Research and Technology Applications, for supporting this work.

References

[1] Adams, M., Joachimiak, A., Kim, R., et al. (2004). Meeting review: 2003 NIH Protein Structure Initiative Workshop in Protein Production and Crystallization for Structural and Functional Genomics. *J Struct Funct Genomics.*, *5*: 1-2.

[2] Al-Lazikani, B., Jung, J., Xiang, Z., et al. (2001). Protein structure prediction. *Curr Opin Chem Biol.*, *5*: 51-56.

[3] Amara, A. A. (2013). Pharmaceutical and industrial protein engineering: where we are? *Pak J Pharm Sci.*, *26*: 217-232.

[4] Bartlett, G, Todd, A. and Thornton, J. (2003). Inferring protein function from structure. *Methods Biochem Anal.*, *44*: 387-407.

[5] Koehl, P. (2001). Protein structure similarities. *Curr Opin Struct Biol.*, *11*: 348-353.

[6] Marti-Renom, M., Stuart, A., Fiser, A., et al. (2000). Comparative protein structure modeling of genes and genomes. *Annu Rev Biophys Biomol Struct.*, *29*: 291-325.

[7] Najmanovich, R., Torrance, J. and Thornton, J. (2005). Prediction of protein function from structure: insights from methods for the detection of local structural similarities. *Biotechniques.*, *38*: 847, 849, 851.

[8] Koch, M., Breinbauer, R. and Waldmann H. (2003). Protein structure similarity as guiding principle for combinatorial library design. *Biol Chem.*, *384*:1265-1272.

[9] Sillerud, L. and Larson, R. (2005). Design and structure of peptide and peptidomimetic antagonists of protein-protein interaction. *Curr Protein Pept Sci.*, *6*: 151-169.

[10] Borgstahl, G., Williams, D. and Getzoff, E. (1995). 1.4 A structure of photoactive yellow protein, a cytosolic photoreceptor: unusual fold, active site, and chromophore. *Biochem.*, *34*: 6278-6287.

[11] Niedzwiecka, A., Marcotrigiano, J., Stepinski, J., et al. (2002). Biophysical studies of eIF4E cap-binding protein: recognition of mRNA 5' cap structure and synthetic fragments of eIF4G and 4E-BP1 proteins. *J Mol Biol.*, *319*: 615-635.

[12] Shen, X., Mo, X., Moore, R., et al. (2006). Adhesion and structure properties of protein nanomaterials containing hydrophobic and charged amino acids. *J Nanosci Nanotechnol.*, *6*: 837-844.

[13] Bashtovyy, D., Marsh, D., Hemminga, M., et al. (2001). Constrained modeling of spin-labeled major coat protein mutants from M13 bacteriophage in a phospholipid bilayer. *Protein Sci.*, *10*: 979-987.

[14] Choinowski, T., Dyer, J., Maderegger, B., et al. (1999). Crystallization and initial X-ray analysis of rabbit mature sterol carrier protein 2. *Acta Crystallogr D Biol Crystallog.*, *55*: 1478-1480.

[15] Doxastakis, M., Sakai, V., Ohtake, S., et al. (2007). A molecular view of melting in anhydrous phospholipidic membranes. *Biophys J.*, *92*: 147-161.

[16] Oganesyan, V., Oganesyan, N., Terzyan, S., et al. (2002). The crystal structure of the endothelial protein C receptor and a bound phospholipid. *J Biol Chem.*, *277*: 24851-24854.

[17] Baldwin, R. L. (2008). The search for folding intermediates and the mechanism of protein folding. *Annu Rev Biophys.*, *37*: 1-21.

[18] Bonneau, R., Tsai, J., Ruczinski, I., et al. (2001). Functional inferences from blind ab initio protein structure predictions. *J Struct Biol.*, *134*: 186-190.

[19] Baldwin, R. L. (1975). Intermediates in protein folding reactions and the mechanism of protein folding. *Annu Rev Biochem.*, *44*: 453-475.

[20] Amara, A. A. (2010). The philosophy behind exo/endo/existing antioxidants and our built-in oxidant and antioxidant system. *Pharmazie*, *65* :711-719.

[21] Gamal, M., Hafez, E., Khaled, A., Amara, A. A. (2018). Amino acids diets as a model for investigating cancer induced by acrylamide production produced during wrong food cooking. *SOJ Biochem.*, *4*: 1-14.

[22] Germaniuk, I. L. (1984). [Structure of the gene and molecular defects of insulin in diabetes mellitus (a review of the literature)]. *Vrach Delo*: 9-14.

[23] Higgins, J. M. (2003). Structure, function and evolution of haspin and haspin-related proteins, a distinctive group of eukaryotic protein kinases. *Cell Mol Life Sci.*, *60*: 446-462.

[24] Kan, C. C. (2002). Impact of recombinant DNA technology and protein engineering on structure-based drug design: case studies of HIV-1 and HCMV proteases. *Curr Top Med Chem., 2*: 247-269.
[25] Mendel, D., Cornish, V. and Schultz, P. (1995). Site-directed mutagenesis with an expanded genetic code. *Annu Rev Biophys Biomol Struct., 24*: 435-462.
[26] Stolowich, N., Petrescu, A., Huang H., et al. (2002). Sterol carrier protein-2: structure reveals function. *Cell Mol Life Sci., 59*: 193-212.
[27] Hubbard, S. R. and Till, J. H. (2000). Protein tyrosine kinase structure and function. *Annu Rev Biochem.*, 69: 373-398.
[28] Okamoto, A. and Hirotsu, K. (1992). [Protein engineering. I: X-ray crystallographic study of aspartate aminotransferase and its mutant enzymes]. *Tanpakushitsu Kakusan Koso., 37*: 2231-2242.
[29] Peracchi, A. (2001). Enzyme catalysis: removing chemically 'essential' residues by site-directed mutagenesis. *Trends Biochem Sci., 26*: 497-503.
[30] Ramponi, G. and Stefani, M. (1997). Structural, catalytic, and functional properties of low M(r), phosphotyrosine protein phosphatases. Evidence of a long evolutionary history. *Int J Biochem Cell Biol., 29*: 279-292.
[31] Ringe, D., Wei, Y., Boino, K., et al. (2004). Protein structure to function: insights from computation. *Cell Mol Life Sci., 61*: 387-392.
[32] Amara, A. A, El-Masry, M. and Bogdady, H. (2008). Plant crude extracts could be the solution: extracts showing *in vivo* antitumorigenic activity. *Pak J Pharm Sci., 21*: 159-171.
[33] Sanger, F. (1959). Chemistry of insulin; determination of the structure of insulin opens the way to greater understanding of life processes. *Science, 129*: 1340-1344.
[34] Sanger, F. (1950). Some chemical investigations on the structure of insulin. *Cold Spring Harb Symp Quant Biol.*, 14: 153-160.
[35] Kendrew, J., Dickerson, R., Strandberg, B., et al. (1960). Structure of myoglobin: A three-dimensional Fourier synthesis at 2 A. resolution. *Nature., 185*: 422-427.
[36] Kendrew, J., Watson, H., Strandberg, B., et al. (1961). The amino-acid sequence x-ray methods, and its correlation with chemical data. *Nature, 190*: 666-670.
[37] Strandberg, B. (2009). Chapter 1: building the ground for the first two protein structures: myoglobin and haemoglobin. *J Mol Biol., 392*: 2-10.
[38] Strandberg, B. (1997). Sir John Kendrew, 1917 - 1997. *Structure., 5*: 1713-1715.
[39] Strandberg, B. (1964). People and Physicians. *Yngre Laeger, 10*: 133.
[40] Berezovskii, I. N. and Trifonov, E. N. (2001). Evolutionary aspects of protein structure and folding. *Mol Biol (Mosk)., 35*: 278-284.
[41] Berezovsky, I. N. and Trifonov, E. N. (2001). Protein structure and folding: a new start. *J Biomol Struct Dyn., 19*: 397-403.
[42] Bryan, P. N. (1995). Site-directed mutagenesis to study protein folding and stability. *Methods Mol Biol., 40*: 271-289.
[43] Daggett, V. and Fersht, A. (2003). The present view of the mechanism of protein folding. *Nat Rev Mol Cell Biol., 4*: 497-502.
[44] Englander, S., Mayne, L. and Krishna, M. M. (2007). Protein folding and misfolding: mechanism and principles. *Q Rev Biophys., 40*: 287-326.
[45] Ring, C. S. and Cohen, F. E. (1993). Modeling protein structures: construction and their applications. *FASEB J., 7*: 783-790.

[46] Shakhnovich, E. I. (1996). Modeling protein folding: the beauty and power of simplicity. *Fold Des., 1*: R50-54.

[47] Biro, J. C. (2006). Amino acid size, charge, hydropathy indices and matrices for protein structure analysis. *Theor Biol Med Model., 3*:15-26.

[48] Jones, H., Preuss, M., Wright, M., et al. (2006). The mechanism of GroEL/GroES folding/refolding of protein substrates revisited. *Org Biomol Chem., 4*: 1223-1235.

[49] Li, M., Huang, Y. and Xiao, Y. (2008). Effects of external interactions on protein sequence-structure relations of beta-trefoil fold. *Proteins, 72*: 1161-1170.

[50] Amara, A. A. (2013). The inevitability of balanced lives: genes – foods- Action-interactions. *IIOAB J., 4*: 1-27.

[51] Dorval, I., Fergelot, P., Dubourg, C., et al. (1994). Rapid detection of 1078 delT mutation by PCR-mediated site-directed mutagenesis: detection of cystic fibrosis carriers in a celtic population. *Clin Chem., 40*: 2318-2319.

[52] Huang, S., Bolser, D., Liu, H., et al. (2009). Molecular modeling of the heterodimer of human CFTR's nucleotide-binding domains using a protein-protein docking approach. *J Mol Graph Model., 27*: 822-828.

[53] Salcedo, M., Chavez, M., Ridaura, C., et al. (1993). Detection of the cystic fibrosis delta-F508 mutation at autopsy by site-directed mutagenesis. *Am J Med Genet., 46*: 268-270.

[54] Audette, G. F. and Hazes, B. (2007). Development of protein nanotubes from a multi-purpose biological structure. *J Nanosci Nanotechnol., 7*: 2222-2229.

[55] Aytuna, A., Gursoy, A. and Keskin, O. (2005). Prediction of protein-protein interactions by combining structure and sequence conservation in protein interfaces. *Bioinformatics, 21*: 2850-2855.

[56] Barua, D., Faeder, J. and Haugh, J. (2007). Structure-based kinetic models of modular signaling protein function: focus on Shp2. *Biophys J., 92*: 2290-2300.

[57] Pricl, S., Ferrone, M., Fermeglia, M., et al. (2006). Multiscale modeling of protein transport in silicon membrane nanochannels. Part 1. Derivation of molecular parameters from computer simulations. *Biomed Microdevices, 8*: 277-290.

[58] Rathore, N., Chopra, M. and de Pablo, J. (2005). Optimal allocation of replicas in parallel tempering simulations. *J Chem Phys., 122*: 024111.

[59] Vojisavljevic, V., Pirogova, E., Davidovic, D., et al. (2009). A new approach to revealing functional residues from analysis of protein primary structure. *Conf Proc IEEE Eng Med Biol Soc., 2009*: 4731-4734.

[60] Zhong, W., Altun, G., Tian, X., et al. (2004). Parallel protein secondary structure prediction based on neural networks. *Conf Proc IEEE Eng Med Biol Soc., 4*: 2968-2971.

[61] Adam, V., Carpentier, P., Violot, S., et al. (2009). Structural basis of X-ray-induced transient photobleaching in a photoactivatable green fluorescent protein. *J Am Chem Soc., 131*: 18063-18065.

[62] Archer-Lahlou, E., Tikhonova, I., Escrieut, C., et al. (2005). Modeled structure of a G-protein-coupled receptor: the cholecystokinin-1 receptor. *J Med Chem., 48*: 180-191.

[63] Balabin IA, Yang W, Beratan DN. 2009. Coarse-grained modeling of allosteric regulation in protein receptors. *Proc Natl Acad Sci.*, USA, *106*: 14253-14258.

About the Editors

Prof. Shadia M. Abdel-Aziz, PhD
Microbial Chemistry Department
Genetic Engineering and Biotechnology Research Division
National Research Centre, Dokki, Giza, Egypt
Email: abdelaziz.sm@gmail.com

Prof. Neelam Garg, PhD
Department of Microbiology
Faculty of Life Sciences
Kurukshetra University, Kurukshetra, Haryana, India
Email: grgneelam@gmail.com; nlmgarg@rediffmail.com

Dr. Abhinav Aeron, PhD
Independent Researcher
Bhartiya Colony, Muzaffarnagar, India
Email: abhinavaeron@gmail.com

Dr. Chaitanya Kumar Jha, PhD
Assistant Professor (GES-II)
Microbiology Department
Gujarat Arts and Science College, Ahmedabad
Gujarat, India
Email: jha_chaitanya@rediffmail.com

S. Chandra Nayak, PhD
Professor & Officer Incharge -ICAR (AICRP- Mysore Centre)
Department of Studies in Biotechnology
University of Mysore
Mysore, Karnataka, India
Email: moonnayak@gmail.com

Vivek Kumar Bajpai, PhD
Department of Energy & Materials Engineering,
Dongguk University-Seoul, Seoul, Republic of Korea
Email: vbajpai04@yahoo.com

INDEX

F

G

H

I

J

L

M

N

O

P

R

S

T

U

V

W

Recent Advances in Microbiology. Volume 3

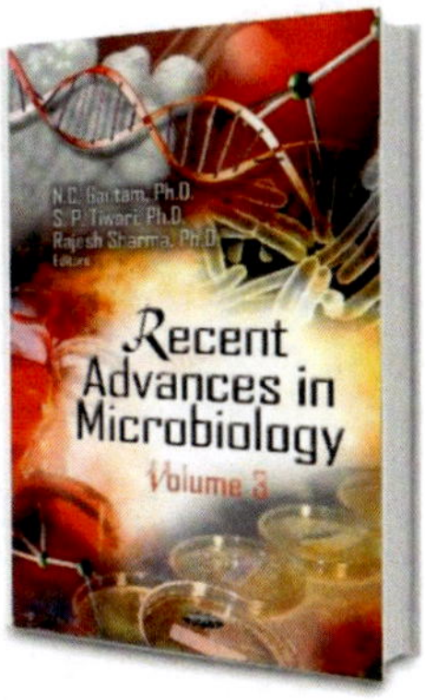

Editor: NC Gautam, Shree Prakash Tiwari, Ph.D., and Rajesh Sharma, Ph.D.

Series: Recent Advances in Microbiology

Book Description: The present volume has been complied with the help of leading microbiologists from India and other countries, with an objective of providing current developments in the field of microbiology.

Hardcover ISBN: 978-1-53614-057-6
Retail Price: $195

Bacillus subtilis: Morphology, Functions and Role in Disease Management

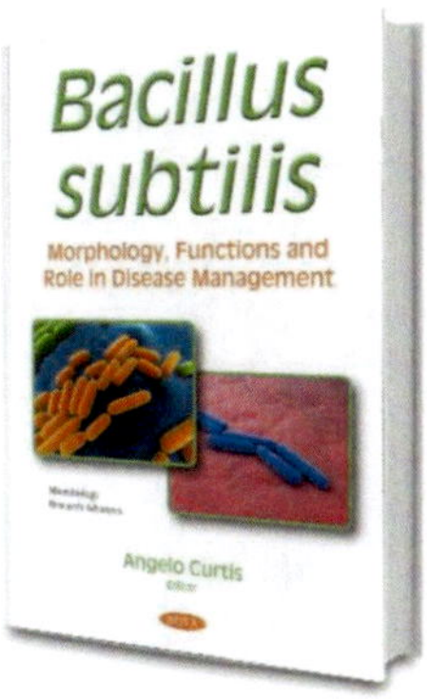

Editor: Angelo Curtis

Series: Microbiology Research Advances

Book Description: In *Bacillus subtilis: Morphology, Functions and Role in Disease Management,* the authors begin by outlining the main features of chemotaxis as a behaviour, then reviewing the molecular machinery Bacillus subtilis uses to sense and respond to chemical cues.

Hardcover ISBN: 978-1-53614-513-7
Retail Price: $95